Remote Sensing Retrievals of Optical Properties in Inland Waters and the Coastal Ocean

Remote Sensing Retrievals of Optical Properties in Inland Waters and the Coastal Ocean

Guest Editors

Chong Fang
Andrew Clive Banks
Zhidan Wen
Shaohua Lei

MDPI

Basel • Beijing • Wuhan • Barcelona • Belgrade • Novi Sad • Cluj • Manchester

Guest Editors

Chong Fang
Northeast Institute of
Geography and Agroecology
Chinese Academy of Sciences
Changchun
China

Andrew Clive Banks
Institute of Oceanography
Hellenic Centre for Marine
Research (HCMR)
Heraklion
Greece

Zhidan Wen
Northeast Institute of
Geography and Agroecology
Chinese Academy of Sciences
Changchun
China

Shaohua Lei
National Key Laboratory of
Water Disaster Prevention
Nanjing Hydraulic Research
Institute
Nanjing
China

Editorial Office
MDPI AG
Grosspeteranlage 5
4052 Basel, Switzerland

This is a reprint of the Special Issue, published open access by the journal *Remote Sensing* (ISSN 2072-4292), freely accessible at: https://www.mdpi.com/journal/remotesensing/special_issues/VD8NOTK9C4.

For citation purposes, cite each article independently as indicated on the article page online and as indicated below:

Lastname, A.A.; Lastname, B.B. Article Title. *Journal Name* **Year**, *Volume Number*, Page Range.

ISBN 978-3-7258-5975-7 (Hbk)
ISBN 978-3-7258-5976-4 (PDF)
https://doi.org/10.3390/books978-3-7258-5976-4

Contents

About the Editors

Chong Fang

Chong Fang is an Associate Researcher at the Northeast Institute of Geography and Agroecology, Chinese Academy of Sciences. She has hosted over 10 scientific research projects, including the National Natural Science Foundation Youth Fund and Joint Fund projects, receiving funding totalling over RMB 2 million. In addition to this, she has authored a total of 73 SCI-indexed academic papers, receiving over 2000 citations, including 3 ESI highly cited papers and hot papers, and an H-index of 29. She has been selected as a Class D talent in Jilin Province, as well as serving as a special editor for numerous SCI journals, such as *Remote Sensing*, and a reviewer for over 20 SCI journals in the field of environmental remote sensing, including *Global Change Biology*, *Water Research*, and *Remote Sensing*.

Andrew Clive Banks

Andrew Clive Banks is a Principal Research Scientist in the department of Satellite Oceanography and Marine Optics at the Institute of Oceanography, Hellenic Centre for Marine Research (HCMR), Crete, Greece. Dr. Banks has accrued more than twenty-five years of research experience in earth observation/remote sensing since his first publication in the field in 1997, and is now the author of over 100 papers published in peer-reviewed journals and international conference proceedings. He was a satellite oceanography researcher at HCMR for 12 years before gaining wider international experience, first at the Joint Research Centre of the European Commission as a scientific officer for satellite oceanography and then as a senior research scientist in earth observation and climate at the National Physical Laboratory of the United Kingdom. During his career, he has coordinated, been a principal investigator, or participated in more than 40 international projects. In 2018, Dr. Banks returned to Crete and HCMR as the principal researcher in satellite oceanography and marine optics and to coordinate the implementation and operations of the proposed Copernicus ocean colour system vicarious colour infrastructure in Crete. He is also a member of a number of ESA advisory committees, including the mission advisory group for the Sentinel 3 Next Generation series of Copernicus satellites. His research interests mainly comprise the validation and calibration of satellite data, the accuracy of satellite and in situ data (uncertainty and SI traceability), fiducial reference measurements (FRM), ocean colour and marine optics, Sea Surface Temperature (SST), environment and climate change monitoring, and global open ocean and coastal remote sensing, with a special interest in the Eastern Mediterranean.

Zhidan Wen

Zhidan Wen is a Research Fellow at the Northeast Institute of Geography and Agroecology, Chinese Academy of Sciences. Having hosted more than 20 scientific research projects, including the National Natural Science Foundation of China's general project and key research and development plan sub-project, she has received funding totalling over RMB 9 million. With the support of multiple projects, she has published a total of 107 SCI-indexed academic papers, with over 4000 citations, including 8 ESI highly cited papers and 2 hot papers, and currently holds an H-index of 37. In addition, she has collaborated to complete one monograph in both Chinese and English. As well as winning first prize in the Jilin Province Natural Science Award in 2020 for "Remote Sensing Model of Inland Water Quality and Eutrophication Retrieval", she has been selected as a top young talent in the "Changbai Talent Plan" of Jilin Province, is a member of the Youth Innovation Promotion Association of the Chinese Academy of Sciences, and an has been names as an "outstanding youth" in the science and technology development plan of Jilin Province. She has served as a guest editor for SCI journals such as *Remote Sensing* and *IEEE JSTAR*, as well as serving as a reviewer for over 20 SCI journals in the field of environmental remote sensing, including *Water Research*, *Journal of Hydrology*, and *Remote Sensing*. She has also served as a member of the Cartography and Geographic Information System Professional Committee of the Chinese Geographical Society, the Ecological Remote Sensing Monitoring and Evaluation Committee of the Chinese Environmental Society, and the Digital Wetland Professional Committee of the International Digital Earth Society.

Shaohua Lei

Shaohua Lei is a Senior Engineer at the Rural Water Conservancy Research Institute/Soil and Water Conservation Research Center, Nanjing Hydraulic Research Institute of the Ministry of Water Resources, Ministry of Transport, and National Energy Administration. He served as a Visiting Scholar at Indiana University-Purdue University Indianapolis (IUPUI) and Joint Supervisor for PhD Candidates at Xiamen University and Hohai University, as well as a Council Member of the Southern Society for Soil and Water Conservation. In addition to this, he is a Soil and Water Conservation Expert of Jiangsu Provincial Water Resources Department, a Registered Consulting Expert of Jiangsu Province, is undertaking the Third Level of Jiangsu High-Level Talent Training Program ("333 Project"), and has received funding from the Jiangsu Youth Science and Technology Talent Support Project. He has presided over more than 10 scientific research projects, such as national projects and those funded by the Jiangsu Provincial Natural Science Foundations, as well as having undertaken more than 20 major engineering, scientific, and technological consulting projects. He has published more than 70 academic papers, authored more than 10 national invention patents as the first inventor, published 4 monographs, participated in compiling and promoting 3 standards, served as a Young Editorial Board Member of 4 Chinese Core Journals, and acted as a reviewer for more than 30 SCI/EI indexed journals. He won numerous awards, such as the National University GIS Rising Star Award, China Telecom Feiyang Award, etc. His research achievements have won him two First Prizes (one of which for a project he led) and two Second Prizes in provincial- and ministerial-level science and technology awards, as well as four First Prizes (one of which for a project he led) in departmental-level awards.

remote sensing

MDPI

Article

Tracking Water Quality and Macrophyte Changes in Lake Trasimeno (Italy) from Spaceborne Hyperspectral Imagery

Alice Fabbretto [1,2,*], Mariano Bresciani [1], Andrea Pellegrino [1,3], Krista Alikas [2], Monica Pinardi [1], Salvatore Mangano [1], Rosalba Padula [4] and Claudia Giardino [1,5]

1 National Research Council, Institute of Electromagnetic Sensing of the Environment, 20133 Milan, Italy; bresciani.m@irea.cnr.it (M.B.); pellegrino.a@irea.cnr.it (A.P.); pinardi.m@irea.cnr.it (M.P.); mangano.s@irea.cnr.it (S.M.); giardino.c@irea.cnr.it (C.G.)
2 Tartu Observatory, Department of Remote Sensing, University of Tartu, 61602 Tõravere, Estonia; krista.alikas@ut.ee
3 University of Sapienza, Department of Engineering, 00185 Rome, Italy
4 ARPA Umbria, Regional Environmental Protection Agency, 06132 Perugia, Italy; r.padula@arpa.umbria.it
5 NBFC, National Biodiversity Future Center, 90133 Palermo, Italy
* Correspondence: fabbretto.a@irea.cnr.it

Abstract: This work aims to show the potential of imaging spectroscopy in assessing water quality and aquatic vegetation in Lake Trasimeno, Italy. Hyperspectral reflectance data from the PRISMA, DESIS and EnMAP missions (2019–2022, summer periods) were compared with in situ measurements from WISPStation and used as inputs for water quality product generation algorithms. The bio-optical model BOMBER was run to simultaneously retrieve water quality parameters (Chlorophyll-a (Chl-a) and Total Suspended Matter, (TSM)) and the coverage of submerged and emergent macrophytes (SM, EM); value-added products, such as Phycocyanin concentration maps, were generated through a machine learning approach. The results showed radiometric agreement between satellite and in situ data, with $R^2 > 0.9$, a Spectral Angle < 10° and water quality mapping errors < 30%. Both SM and EM coverage varied significantly from 2019 (135 ha, 0 ha, respectively) to 2022 (2672 ha, 343 ha), likely influenced by changes in rainfall and lake levels. The areas of greatest variability in Chl-a and TSM were identified in the littoral zones in the western side of the lake, while the highest variation in the fractional cover of SM and density of EM were observed in the south-eastern region; this information could support the water authorities' monitoring activities. To this end, further developments to improve the reference field data for the validation of water quality products are recommended.

Keywords: PRISMA; DESIS; EnMAP; shallow lake; water reflectance; phytoplankton; aquatic vegetation; bio-optical model; machine learning

1. Introduction

Lake ecosystems emerge as reliable sentinels of contemporary climate change and anthropogenic processes by providing direct and indirect climate indicators [1,2] and by reflecting alterations in climate patterns and human-induced activities. Monitoring how lakes respond to these influences is crucial because freshwater ecosystems provide a multitude of ecosystem benefits and serve as critical resources for drinking water, agricultural irrigation, biodiversity conservation, industry and recreation [3]. To this end, several regulations have been developed at national and international levels; in particular, the European Commission adopted the Water Framework Directive (WFD, Directive 2000/60/EC), which establishes a unified framework for integrated water resource management in the European Union (EU) [4,5]. Its central purpose is to establish a comprehensive framework dedicated to safeguarding and promoting the sustainable management of water resources throughout the EU member states, with the aim of collectively achieving a "good ecological status" for all water bodies. The WFD is applicable to inland, transitional and coastal surface waters, as well as groundwater, and indicates the parameters to be monitored and the frequency of monitoring, thus helping inform adaptive strategies for managing these vital natural resources [6]. Among those indicated for monitoring, the key parameters are those related to the ecological status of water bodies, in particular Chlorophyll-a (Chl-a), which is considered a proxy for phytoplankton biomass and therefore contributes to defining the trophic status of a water body [7]; Total Suspended Matter (TSM), which is a tracer for inflowing pollutants and is related to water transparency [8,9] and Phycocyanin (PC), which is a photosynthetic auxiliary pigment that is abundant in cyanobacteria [10–12].

Macrophytes are important elements of inland waters [13,14], increasing their ecosystem services (e.g., carbon sequestration, habitat provision, nutrient uptake) and fulfilling a central role in the cycling of elements in aquatic ecosystems [15,16]. The complex interactions between physical, chemical and biological processes in surface waters lead to significant challenges for in situ monitoring and often limit the ability to adequately capture the dynamics of aquatic systems and to understand their status, functioning and response to pressures. In this context, the use of remote sensing allows for wide spatial coverage and regular monitoring frequency, providing information on water conditions, bottom properties and the presence and abundance of aquatic plants, distinguishing them into different association types; this complements traditional in situ measurements [17–22]. The reviews reported in [23,24] highlighted the significant increase in remote sensing studies of inland water quality, due both to improved access to Earth Observation (EO) data and increasing computational capabilities. Particularly, imaging spectroscopy has gained wide interest [25–28], as the collection of data in narrower, contiguous bands is improving aquatic ecosystem mapping for the simultaneous retrieval of parameters describing water quality and aquatic vegetation (e.g., biomass [29] and the identification of invasive species [30]). In the last few years, hyperspectral spaceborne and airborne images have been providing significant advances in algorithm development and innovative mapping tools [31,32]. For instance, the PRecursore IperSpettrale della Missione Applicativa (PRISMA) and the DLR Earth Sensing Imaging Spectrometer (DESIS) products were exploited for the estimation of optically active parameters (Chl-a, TSM) in Italian lakes with different optical properties [33], and for one of these lakes (i.e., Trasimeno), PRISMA water quality products were evaluated with respect to Sentinel-2 in [34]. DESIS data were also used for mapping water quality parameters in lagoons and estuaries [35]. Furthermore, PRISMA spectral features enabled distinctions to be made between shallow and turbid waters [36], while in [37], PRISMA products were used to develop novel algorithms for the estimation of cyanobacteria biomass in lake Trasimeno. PRISMA products were also exploited for the detection of floating plastic materials through unsupervised and supervised machine learning approaches [38].

In the wake of these works, this study presents a relevant multitemporal dataset compiled using three flying spaceborne imaging spectroscopy missions for enabling aquatic ecosystem mapping. A set of sensor-independent algorithms for deriving water quality parameters in optically complex waters is applied to all imagery data. In particular, the study builds on 13 hyperspectral reflectance products of Lake Trasimeno (Italy) provided by PRISMA, DESIS and the Environmental Mapping and Analysis Program (EnMAP). The dataset covers the summer periods from 2019 to 2022 (from early June to mid-October) to assess changes in water quality parameters and aquatic vegetation coverage and type, upon having verified the accuracy of satellite-derived products with corresponding in situ measurements [39–45] in terms of remote sensing reflectance (Rrs), a key parameter that is widely used for aquatic applications [46]. Reference in situ measurements are provided by the WISPStation, an autonomous fixed-position radiometer that provides continuous Rrs measurements, from which an estimate of bio-geophysical parameters can also be derived [47]. The Rrs products derived from spaceborne hyperspectral observations are then exploited as inputs for the simultaneous mapping of water quality (Chl-a, TSM, PC) and aquatic vegetation. Several environmental parameters, such as water level fluctuations, are considered to discuss the spatio-temporal changes.

2. Data and Methods

2.1. Study Area

Lake Trasimeno, which is located in central Italy (43.138N, 12.113E), is the only Italian lake of tectonic origin and fourth in size, with an area of 124 km^2. The lake is nearly round in shape with a diameter of about 12 km, with three small islands. It is a closed lake, with un-stratified and very shallow waters (average bottom depth, 4.7 m; maximum depth, 6 m) characterized by banks with minimal slopes and, in the south-eastern area, an open bay colonized by aquatic vegetation: permanently submerged (e.g., *Chara globularis*) and semi-emergent macrophytes (in some seasonal phases of their life cycle, they can emerge from the water–atmosphere interface with flowers, fruits and part of the plant body, e.g., *Potamogeton pectinatus*, *P. perfoliatus*, *Myriophyllum spicatum* and *Ceratophyllum demersum*) [48]. These are among the most common species present in Italian lakes; they are also found in the large subalpine lakes of Northern Italy [49,50]. The lake is part of Tevere River Basin, which is connected to the hydrographical network of the Nestòre River by an artificial outfall built with the function of flood regulation. The gradual lowering of the lake's level, though, has made this connection non-functional for almost a decade; the waters have not reached the overflow level since 2015 [51]. Stopping—or at least drastically decreasing—water withdrawals may not be the ultimate solution to the basin's hydrological problems. Nevertheless, it would be a substantial benefit for a lake that depends on water resources that are crucial to maintaining the balance of its delicate environmental state, and to sustain several vital economic activities, such as agriculture, fishing and tourism. The lacustrine ecosystem is an area of exceptional value due to its wealth of flora and fauna and its diversity of species, and in 1995, it was declared a protected area (regional law L.R. 9, 03/03/1995). Tourism, agriculture and livestock farming are the most present activities in the Lake Trasimeno area; approximately 70% of the lake's catchment area is enveloped by cultivated land, while intensive agriculture requiring irrigation occupies 28% of this territory. As a result, these activities have contributed to the eutrophication phenomenon in the lake [52]. The lake ranges from mesotrophic to eutrophic conditions [43], where the principal critical parameters for the ecosystem status are phosphorous and nitrogen. In the past few years, the lake has hence encountered challenges in recovering equilibrium as recommended by the WFD. This situation is aggravated by the lack of natural outlets and a limited water inflow (Paganico and Pescia streams); for its water supply, the lake relies exclusively on small streams and precipitation. The water level began to decline in the 1990s, and at the end of 2022, it was 1.3 m below the hydrometric zero [53]. Therefore, the hydrographic characteristics of the lake (width, shallowness, low water exchange, etc.) determine a strong influence of temperature and light conditions, which favor an increase in Chl-a, especially starting from spring and reaching its peak at the end of summer. This phenomenon is indeed closely linked to the presence of and increase in aquatic plants and algal blooms.

2.2. Data Collection

2.2.1. In Situ Data

Reference data were obtained from the Remote Sensing for Trasimeno lake Observatory (RESTO) field infrastructure, where high-frequency in situ measurements are collected by means of a WISPStation mounted on a platform located 100 m north of Polvese island (Figure 1) [19,20].

From April 2018, the WISPStation has been collecting, every 15 min, Rrs data (in the spectral range of 350–900 nm with a spectral resolution of 4.6 nm) along with Chl-a, TSM and PC concentrations, as derived from the measured Rrs through standard water quality algorithms [54–58]. For both Rrs and water quality data, a set of three measurements closest to the acquisition time of the satellite overpasses were selected to calculate the mean and standard deviation. The Rrs data collected by the WISPStation were spectrally resampled to the band settings of PRISMA, DESIS and EnMAP to perform a radiometric comparison.

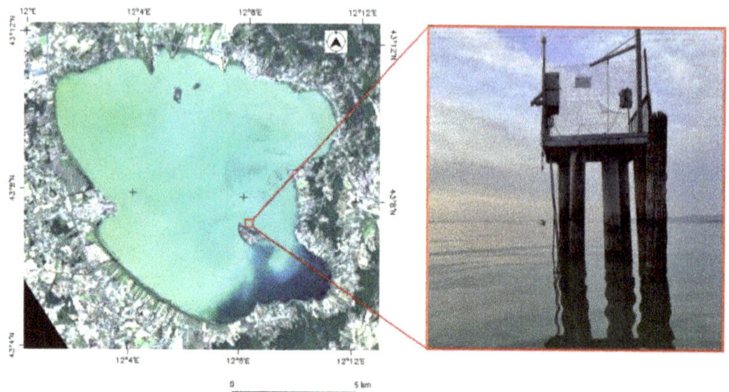

Figure 1. On the left, a true color composition image as acquired by DESIS on 4 August 2019 showing the position of the WISPStation (43.122, 12.134—red box); on the right, a picture of the platform with the WISPStation.

With ±5 days close to the dates of the images' acquisitions, ad hoc in situ campaigns were also organized to survey the optically deep and optically shallow areas, and for the latter to detect the different cover classes, with the visual GPS positioning method, as used in [59]. GPS ground truths were then used for an accuracy assessment of the satellite-derived maps.

The hydrographic service of the Umbria region [53] provided air temperature, precipitation data and lake water levels for the investigated temporal range from January 2019 to December 2022, as shown in Figure 2. The inventory of the sampling of birds feeding on macrophytes (e.g., *Anas* sp., *Aythya* sp. and *Fulica* sp.) provided by "Regione Umbria—Osservatorio Faunistico Regionale" [60] showed the following number of birds for the study period: 43′397 (in 2019), 29′377 (in 2020), 43′382 (in 2021) and 40′707 (in 2022).

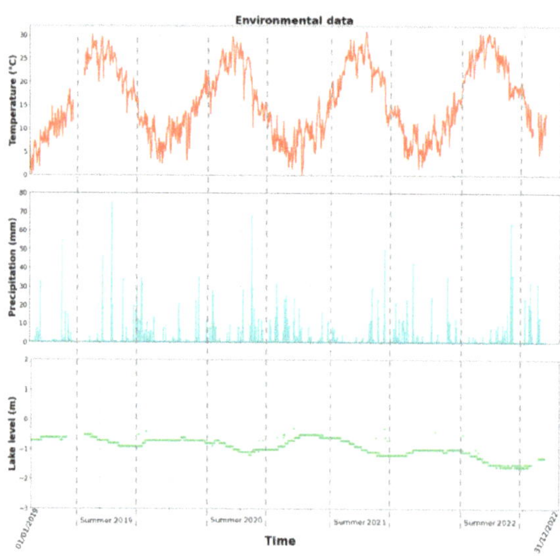

Figure 2. Time series of the environmental data: air temperature (°C) shown in orange, precipitation (mm) shown in light blue and lake level (m) shown in green. The four summer periods in which satellite images were acquired are highlighted in grey with dashed lines.

2.2.2. Spaceborne Data

The imagery data used in this study were gathered from three hyperspectral spaceborne missions: PRISMA, DESIS and EnMAP. The data can be downloaded from the respective catalogues, at different levels of processing: Level 1 (L1) are images in terms of the radiance measured at the sensor (Top of Atmosphere, TOA) and Level 2 (L2) represents the atmospherically corrected ground reflectance (Bottom of Atmosphere, BOA) data. Following a short overview of the three sensors and their main applications, the key characteristics of each are shown in Table 1.

PRISMA, operated by the Italian Space Agency (ASI), is a single satellite placed in low-Earth Sun-synchronous orbit (SSO-LTDN), with a revisit time of 29 days, a period that can be reduced to 7 days thanks to platform roll maneuvers [61]. The payload includes a medium-resolution hyperspectral camera (HYP) and a high-resolution panchromatic camera (PAN), which may support finer-scale mapping [62]. The PRISMA mission was designed to advance technology qualifications, develop applications and provide environmental observation and risk management products to both institutional and scientific users [63]. PRISMA L2 products (version 02.05) are those obtained from TOA radiance products (L1) after automatic land-based atmospheric correction based on MODTRAN v 6.0 [43,44].

DESIS is a hyperspectral instrument integrated into the Multi-User-System for Earth Sensing (MUSES) platform installed on the International Space Station (ISS). The mission is operated by the German Aerospace Center (DLR). The spectral and spatial resolution of the sensor allow new developments for a wide range of applications in very different environments, such as inland and coastal waters, the cryosphere and vegetation [64,65]. DESIS L2 products (version 2.20) are those obtained from TOA radiance products (L1) after automatic land-based atmospheric correction performed with Python-based Atmospheric Correction (PACO) [66].

EnMAP is a high-resolution imaging spectroscopy remote sensing mission, operated by the DLR, placed in SSO-LTDN. The across-track tilt-capability of 30° enables revisit times of less than four days [67]. The primary objective of the mission is to assess and examine surface variables, both qualitative and quantitative, that characterize essential Earth processes [68]. These variables are obtained on a global scale in a routine and consistent way. EnMAP L2 products (version v01.04.00) can be obtained through two individual sub-processors for the atmospheric correction of orthorectified TOA radiance both over land (PACO) and water surfaces (Modular Inversion Program, MIP [69]) [70].

Table 1. List of the main features of the PRISMA, DESIS and EnMAP payloads [71–73].

	PRISMA	DESIS	EnMAP
Launch	22 March 2019	29 June 2018	1 April 2022
Coverage	70°N to 70°S	55°N to 52°S	Global in near-nadir mode
Ground sampling distance	HYP: 30 m; PAN: 5 m	30 m	30 m
Number of bands	HYP: 240 [400–2500 nm] PAN: 1 [400–700 nm]	235 (no binning) 60 (binning) [400–1000 nm]	246 [420–2450 nm]
Radiometric resolution	12 bits	13 bits + 1 bit gain	≥14 bits
Atmospheric correction	MODTRAN v 6.0 (land based)	PACO (land)	PACO (land) MIP (water)

Overall, a total of 13 hyperspectral data with cloud coverage of less than 10% (apart from PRISMA acquired on 3 June 2020) were available for Lake Trasimeno, as presented in Table 2.

Table 2. Spaceborne data of Lake Trasimeno: sensor, date and UTC acquisition time.

Sensor	Date	UTC Time
PRISMA	4 June 2019	10:15
PRISMA	26 July 2019	10:13
DESIS	4 August 2019	13:33
DESIS	5 September 2019	06:49
PRISMA	3 June 2020	10:10
PRISMA	25 July 2020	10:07
DESIS	4 June 2021	12:20
DESIS	15 October 2021	13:47
DESIS	19 June 2022	16:10
PRISMA	20 July 2022	10:08
DESIS	7 August 2022	10:50
PRISMA	12 August 2022	10:04
EnMAP	5 October 2022	10:40

2.3. Methodology Process Flowchart

This section reports a flowchart (Figure 3) of the methodology used in this study, from the hyperspectral reflectance data gathered by PRISMA, DESIS and EnMAP to the spatio-temporal analysis of the changes in Lake Trasimeno that occurred in terms of water quality parameters and bottom coverage.

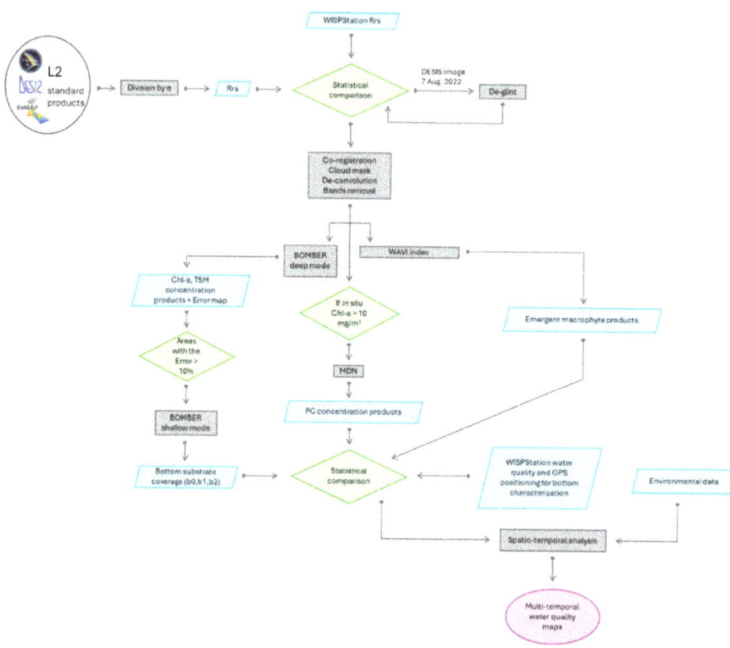

Figure 3. Flowchart of the methodology applied in the study. The oval with a black outline represents the input products where the study started from. Grey boxes indicate the methodology applied. Green diamond shapes stand for decision-making steps in the process. Blue parallelograms represent products generated, and the violet oval indicates the end point of the process.

In the following Sections (Sections 2.4–2.7) a detailed description of the methodology is presented.

2.4. Image Pre-Processing

The PRISMA, DESIS and EnMAP (MIP-based atmospheric correction) L2 products were downloaded in the visible–near-infrared (VNIR, i.e., 400–1000 nm) region of the spectrum from the respective space agencies' portals, and a division operation by π was performed to obtain the most used quantity in aquatic applications, which is Rrs. One image (DESIS, 7 August 2022) was affected by sun glint, which represents a disturbance and alteration of the Rrs. Hence, a sun glint correction method based on a negligible signal in the near-infrared (NIR) part of the spectrum was performed, using the algorithm proposed in [74]. Particularly, the relative intensity of the solar glint in the image was obtained from the NIR brightness and the light in the visible (VIS) band using a set of pixels that could be homogeneous were it not for the presence of the glint. By establishing a linear relationship between the NIR bands and each VIS band, the contribution of the sun glint could be removed. The removal was performed by applying the de-glint processor implemented in the SNAP software (version 9.0.0) "Sen2Coral toolbox", providing the DESIS L2 Rrs product as an input.

Before the Rrs data could be used as inputs in the product generation algorithms, several operations were carried out to improve the products themselves, aiming to obtain final maps that are more accurate and representative of the real state of the target [75,76]. First, co-registration of the images was carried out both to adjust the georeferencing and to identify the lake area by using the shapefile of the lake delivered by OpenStreetMap. The cloudy pixels were masked using the threshold proposed by [77]. Given the low signal of the water bodies, the machine learning algorithm could add a salt-and-pepper noise effect to the results; hence, a deconvolution—with a 5-pixel median window filter—was applied to increase the signal [78]. As recommended by [44] concerning PRISMA data specifically, the preserved spectral range (450–800 nm) was standardized for all the sensors, resulting in having 40 bands for PRISMA, 34 for DESIS and 61 for EnMAP images. The last step before product generation was to identify the portion of the lake covered by emergent macrophytes. To this end, the WAVI index [79] was calculated to detect the coverage of emergent macrophytes (positive values of WAVI) and to assess indications of their morphological traits (as a proxy of biomass index).

2.5. Algorithms for Aquatic Ecosystem Mapping

2.5.1. BOMBER

The spectral inversion procedure was performed with the BOMBER ('Bio-Optical Model-Based tool for Estimating water quality and bottom properties from Remote sensing images') code, which implements the algorithms presented in [80,81] for deriving water column and bottom properties via a non-linear optimization technique. The software employs bio-optical models for both optically deep and shallow waters; the model determines mathematical equations between the Rrs and the inherent optical properties of the optically active constituents, as well as the concentrations of water quality parameters and bottom properties (e.g., fractional cover up to three bottom types). As inputs to the code, the imagery data atmospherically corrected to Rrs, the mask of the area of interest, the definition of the model type (deep or shallow mode) and the setting of the model parameters (absorptions and backscattering coefficients, starting values, etc.) were provided. In this work, BOMBER, parameterized with the intrinsic optical properties specific for Lake Trasimeno [33], was run for all the images in deep water mode to generate Chl-a and TSM concentrations. The model was initialized with the average values of water quality parameters resulting from coincident measurements of Chl-a and TSM inferred from Rrs data gathered at the time of sensor overpasses from the WISPStation. The absorption of Colored Dissolved Organic Matter (CDOM) at 440 nm was not retrieved due to the lack of in situ data for its parametrization, but it was set constant to the average concentration of the lake and equal to 0.25 m^{-1}. The specific absorption spectra of phytoplankton for Lake Trasimeno were selected depending on the month/period, due to the change in phytoplankton concentration and composition (June: clear water without cyanobacteria,

July–October: higher phytoplankton biomass and higher presence of surface cyanobacteria). Other parameters whose change might have an impact on the retrieval of water quality parameters, such as the conversion factor of the absorption of Non-Algal Particles at 440 nm to TSM, were adjusted based on the optical closure between the modeled Rrs and measured Rrs over the deep water station of the WISPStation. In the areas where the optical closure error was greater than 10% [31], the model was then run in shallow water mode. The three bottom coverage classes—named in BOMBER as "b0", "b1" and "b2"—were thus assigned: b0 = high-albedo vegetation (semi-emergent macrophytes); b1 = low-albedo vegetation (permanently submerged macrophytes); b2 = un-colonized pixels (sand).

2.5.2. Mixture Density Network

The Mixture Density Network (MDN) algorithm was calibrated in [37] considering resampled in situ radiometric measurements for the Hyperspectral Imager for the Coastal Ocean (HICO) and PRISMA sensors. The dataset of the mentioned work consisted of a large number of samples of in situ measurements of Rrs, Chl-a and PC for water bodies worldwide (including Lake Trasimeno). The MDN uses band ratios, line heights and operational algorithms based on the correlation between Rrs and PC. Since the algorithm does not allow any setting of additional parameters, the MDN was applied in the default mode [82]. As inputs to the model, PRISMA images were provided considering the specific subset of bands from 504 to 723 nm. To exploit the entire dataset, DESIS and EnMAP images were resampled according to the spectral configuration of PRISMA between 504 and 723 nm, for which the retrieving PC algorithm was validated. PC concentration maps were generated for the imagery data showing a Chl-a concentration greater than 10 mg/m^3, as recommended by [83].

2.6. Product Validation

Common statistical metrics were used to assess the agreement of Rrs and water quality products between satellite and in situ data. In the following equations, n represents the number of concurrent observations, and x_i and y_i are the sensors-estimated Rrs data and in situ measurements, respectively. For the comparison, Regions of Interest (ROIs) with a size of 5 × 5 pixels were extracted from the satellite images and water quality maps and the mean and standard deviation were calculated.

The coefficient of determination (R^2), that is, the proportion of the variance in the dependent variable that is predictable from the independent variable, was computed as follows:

$$R^2(/) = \frac{[\Sigma (x_i - \overline{x})(y_i - \overline{y})]^2}{\Sigma (x_i - \overline{x})^2 \Sigma (y_i - \overline{y})^2}$$

The Root Mean Square Difference (RMSD), that is, a measure of the differences between values predicted by an estimator and the values observed, was calculated as follows:

$$RMSD\left(sr^{-1}\right) = \sqrt{\frac{\Sigma (x_i - \hat{y}_i)^2}{n}}$$

The Spectral Angle (SA), that is, a measure for directly comparing image spectra to in situ spectra, and determining how similar the spectra shapes are, was derived as follows:

$$SA\ (^\circ) = \cos^{-1} \frac{\Sigma y_i x_i}{\sqrt{\Sigma y_i^2}\sqrt{\Sigma x_i^2}}$$

The Mean Absolute Percentage Difference (MAPD), that is, a measure used to quantify the average percentage difference between two samples, was obtained as follows:

$$MAPD\ (\%) = \frac{100}{n} \Sigma \left| \frac{x_i - y_i}{y_i} \right|$$

Control points were used to compare in situ data and satellite-derived maps. The product accuracy was quantified using a confusion matrix, selecting four different classes.

2.7. Spatio-Temporal Analysis

Following the validation of the products, a spatio-temporal analysis was carried out to quantify the changes in primary producers' community abundance and coverage in the different summer periods investigated. To quantify the presence of aquatic vegetation, the areas of submerged macrophytes (obtained by summing the two cover classes provided by BOMBER, "b0" and "b1") and emergent macrophytes were first calculated. Additionally, a more detailed analysis of the three bottom coverage classes was carried out by calculating the fractional cover by labeling a pixel as "sand" where $b2 > 0.66$ (66%) and distinguishing three different classes of submerged macrophytes as follows: sparse ($0 < b0 + b1 < 0.4$), moderate ($0.4 < b0 + b1 < 0.7$) and dense ($b0 + b1 > 0.7$). This was carried out following similar work [50] conducted over Lake Garda, where the same macrophyte species are present as on Lake Trasimeno, and in both cases, the method for classifying the colonized substrate as "sparse", "moderate" and "dense" is generic and does not depend on the characteristics of the site, so similar thresholds can be used. Finally, to assess the variability in the water quality parameters and of the aquatic vegetation, the products obtained were integrated to perform a spatio-temporal analysis with the GRASS GIS algorithm "r.series", which allowed us to make each output pixel value a function (i.e., mean and standard deviation) of the values associated with the corresponding pixels in the input list of maps.

3. Results

3.1. Radiometric Validation

The removal of the sun glint from the image affected by this disturbance (DESIS 7 August 2022) led to the result shown in Figure 4.

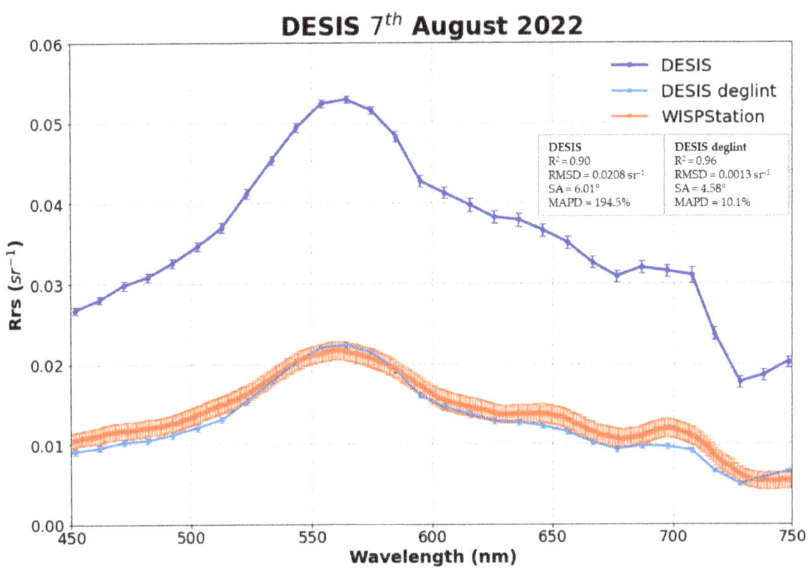

Figure 4. Comparison of DESIS and in situ Rrs data, before ("DESIS", dark blue) and after ("DESIS deglint", light blue) sun glint removal. In situ data are displayed in orange. Statistical results are displayed in the boxes.

As can be seen, the level of agreement between the satellite data and in situ measurement increased after the sun glint correction. There is an improvement in the fitting between the Rrs spectra in terms of magnitude (MAPD reduced by 95%), while the SA value, referring to the spectral similarity, remains almost constant.

Figure 5 shows the radiometric validation of all the satellite-derived Rrs products, based on the comparison with in situ data.

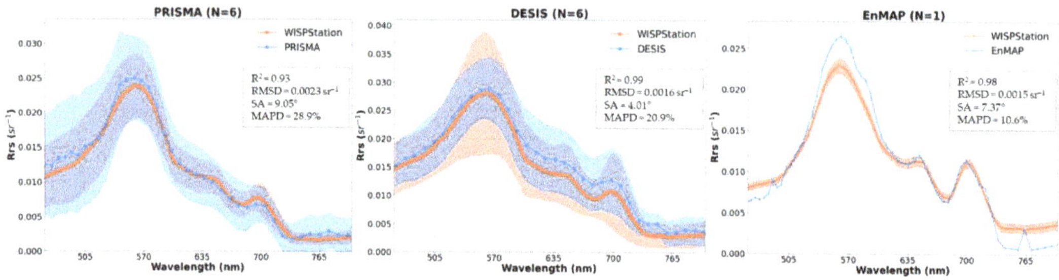

Figure 5. Comparisons of the average Rrs values gathered from the spaceborne data and corresponding in situ Rrs data. The variability in the mean spectra of PRISMA (6 images) and DESIS (6 images) is displayed as blue curves, with the shaded blue area representing the standard deviation. The mean and standard deviation of the in situ data are equivalently shown in orange. In the case of the EnMAP data, the comparison is limited to a single image, and it is shown with the same color configuration. In this case, the standard deviation refers to the variability present in the ROI and in the set of three measurements of the in situ data. The statistical results are displayed in the boxes.

Overall, the Rrs data derived from the three missions show agreement with the in situ measurements, over the entire spectrum between 450 and 800 nm, with values of SA below 10° and values of MAPD below 30%. The satellite-derived Rrs data consistently show spectral features due to the presence of PC (620, 650 nm) and Chl-a (680, 700 nm).

3.2. Water Quality Product Generation and Validation

Proceeding forward, the generated water quality maps (Chl-a, TSM, PC) are shown along with their validation (Table 3, Figure 6). The last row of the table (*) shows the validation considering samples containing all the data from the different sensors.

Table 3. Statistical validation of Chl-a, TSM and PC concentration maps, in terms of RMSD and MAPD. N is the number of the images in the sample.

	Chl-a			TSM			PC		
Product	N	RMSD (mg/m^3)	MAPD	N	RMSD (g/m^3)	MAPD	N	RMSD (mg/m^3)	MAPD
PRISMA	6	3.30	29.8%	6	3.10	19.9%	4	3.85	27.3%
DESIS	6	3.92	25.2%	6	1.85	9.6%	4	2.70	22.4%
EnMAP	1	1.42	6.5%	1	3.38	20.2%	1	2.50	25.5%
*	13	3.32	23.8%	13	2.71	15.6%	9	3.31	25.3%

Considering PRISMA and DESIS, which present a greater number of images in the dataset, for the Chl-a and TSM products generated by the BOMBER bio-optical model, the highest level of agreement is obtained in the case of the maps generated from DESIS images for the estimation of TSM (MAPD = 9.6%). Regarding the estimation of Chl-a, the level of agreement is lower, with a MAPD of over 20% both in the case of PRISMA and DESIS products. The statistical analysis of EnMAP is slightly weaker (N = 1), but it achieved RMSD and MAPD values comparable with those of PRISMA and DESIS and the highest level of agreement in the case of Chl-a estimation. In the case of the PC products generated through the MDN, the level of agreement obtained by the three sensors is similar, with slightly higher accordance in the case of the product given by EnMAP. Overall, MAPD values $\leq 30\%$ are obtained in all cases.

The validation through visual GPS positioning led to an Overall Accuracy of 89.1% for the four classes considered, as shown in Table 4.

Table 4. Confusion matrix for the validation of the four classes selected: b2 (sand), b0 + b1 (submerged aquatic vegetation), EM (emergent macrophytes), deep water.

		Spaceborne Images				
		b2	**b0 + b1**	**EM**	**Deep Water**	**Total**
In situ	b2	7	2			9
	b0 + b1	2	13			15
	EM			6		6
	Deep Water		1		15	16
	Total	9	16	6	15	
					Overall Accuracy	89.1%

Next, Figure 6 illustrates the entire dataset of the water quality products in terms of Chl-a, TSM and PC in the portion of the lake that presented optically deep waters and, in the last column, the mapping of the emergent macrophytes and the three types of bottom cover classes (sand, permanently submerged and semi-emergent macrophytes) in the portion that presented optically shallow waters. The areas of the images affected by clouds are indicated by a grey color.

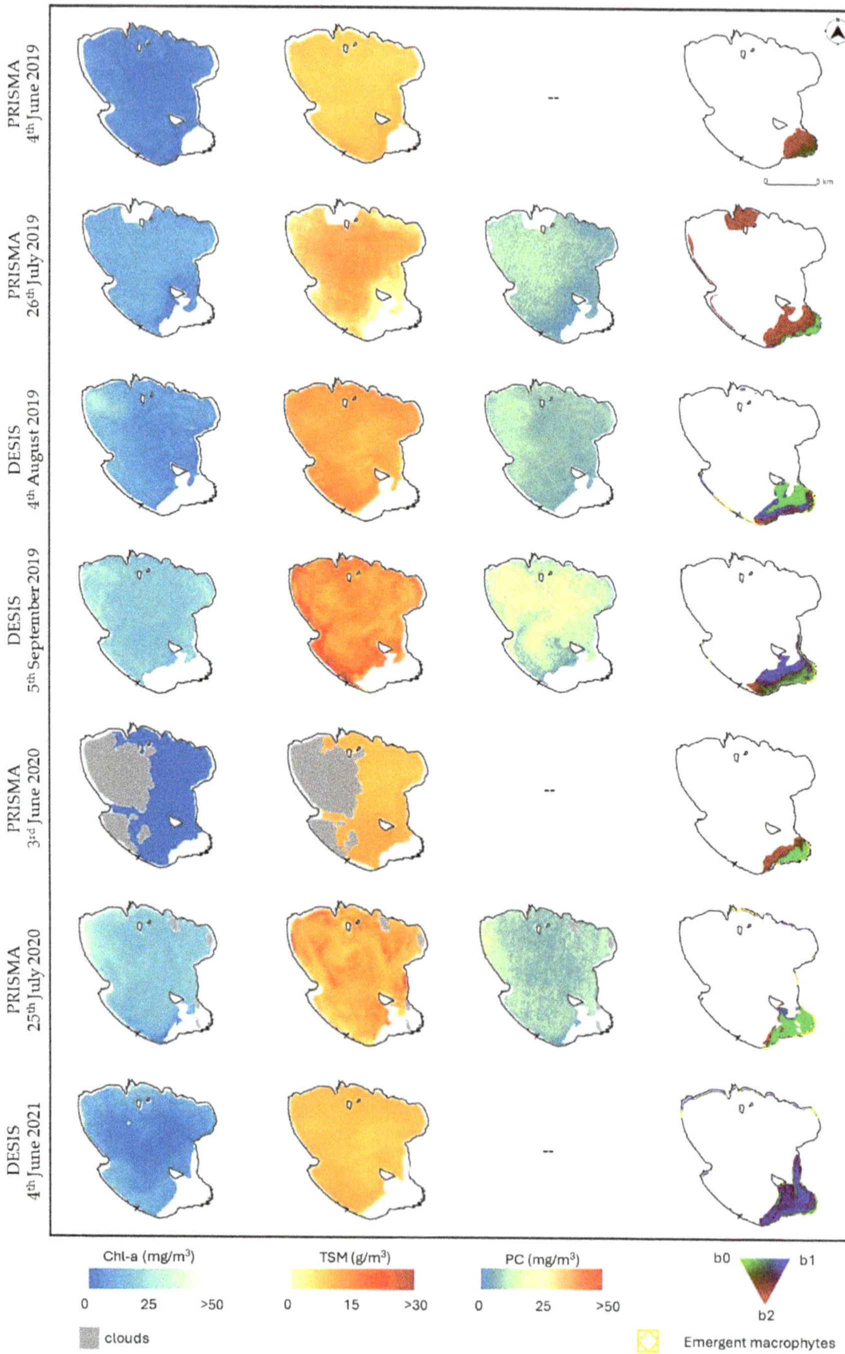

Figure 6. *Cont.*

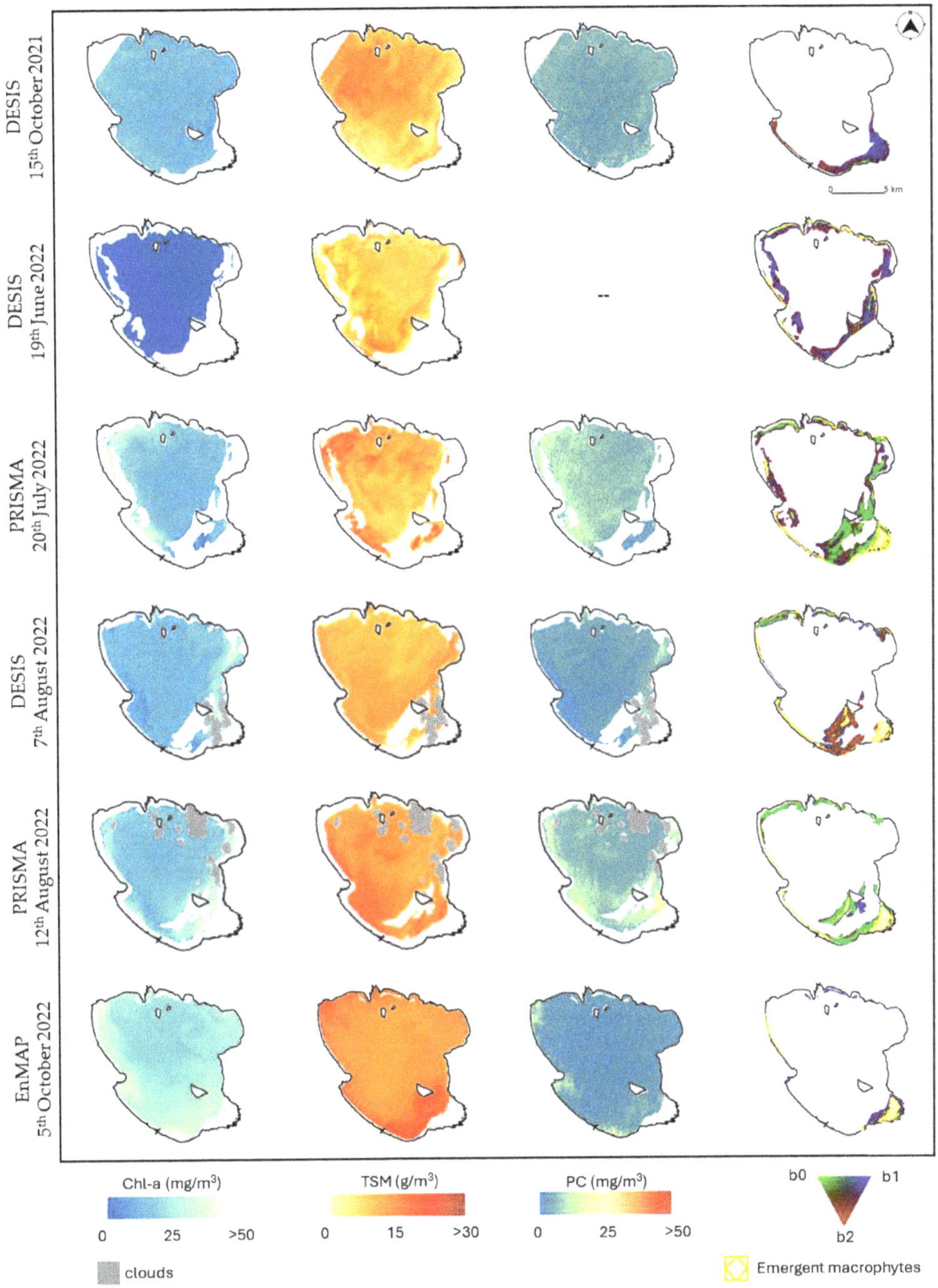

Figure 6. Water quality maps and bottom characterization for the 13 images of the available dataset. From **left** to **right**: Chl-a, TSM and PC maps; bottom characterization products, in terms of emergent macrophytes and the three cover classes: b0 (semi-emergent macrophytes), b1 (permanently submerged macrophytes), b2 (sand).

The maps show the temporal and spatial variation in the water quality parameters in the lake from June 2019 to October 2022 (summer periods). As is evident from the maps, the lowest values of Chl-a (up to 1 mg/m^3) and TSM (up to 1 g/m^3) are detected in June. Higher values, instead, occur in the following summer months for both parameters; the maximum values are generally observed in September–October. PC maps were generated for images from which a Chl-a value higher than 10 mg/m^3 was derived. Indeed, there are no PC maps for the months of June; the highest PC values are achieved in September. Regarding the optically shallow water areas, the area of submerged and emergent macrophytes increases over the years, not only in the south-easternmost part of the lake, but also along the littoral zone of the northernmost part of the lake.

3.3. Spatio-Temporal Analysis

The area of the submerged (sum of "b0" and "b1", whatever the coverage degree) and emergent macrophytes (positive WAVI) retrieved for the period investigated is shown in Table 5. The percentages were calculated with respect to the water cloud-free pixels' area.

Table 5. Area in hectares of submerged and emergent macrophytes across various dates in the dataset. In brackets, the percentages that the values represent over the total number of water pixels not covered by clouds.

Product	Submerged Macrophytes	Emergent Macrophytes
PRISMA 4 June 2019	135 ha (1.1%)	0 ha
PRISMA 26 July 2019	287 ha (2.4%)	0 ha
DESIS 4 August 2019	1140 ha (10.1%)	52 ha (0.5%)
DESIS 5 September 2019	1190 ha (10.3%)	25 ha (0.2%)
PRISMA 3 June 2020	300 ha (3.4%)	19 ha (0.2%)
PRISMA 25 July 2020	878 ha (7.9%)	37 ha (0.3%)
DESIS 4 June 2021	1523 ha (13.6%)	33 ha (0.3%)
DESIS 15 October 2021	601 ha (5.1%)	3 ha (<0.1%)
DESIS 19 June 2022	1790 ha (16.7%)	149 ha (1.4%)
PRISMA 20 July 2022	2672 ha (23.0%)	336 ha (2.9%)
DESIS 7 August 2022	1350 ha (12.3%)	324 ha (3.0%)
PRISMA 12 August 2022	1223 ha (11.7%)	343 ha (3.3%)
EnMAP 5 October 2022	344 ha (3.0%)	199 ha (1.7%)

These values reveal that, in addition to the seasonal vegetative cycle with growth starting in early summer and a senescence period in late summer for both submerged and emergent macrophytes, there is an unprecedented increase—in terms of area—in July 2022, leading to an increase of approximately 160% in submerged macrophytes and 600% in emergent macrophytes compared to the average values for the period 2019–2021. The values of the WAVI index show that in 2022, at the maximum area of emergent macrophytes (343 ha, 12 August 2022), the maximum WAVI value was recorded to be 0.38, which corresponds to an increase of about 240% compared to the date with the lowest WAVI value (0.11, on 15 October 2021).

The fractional coverage analysis during the study period is shown in Figure 7.

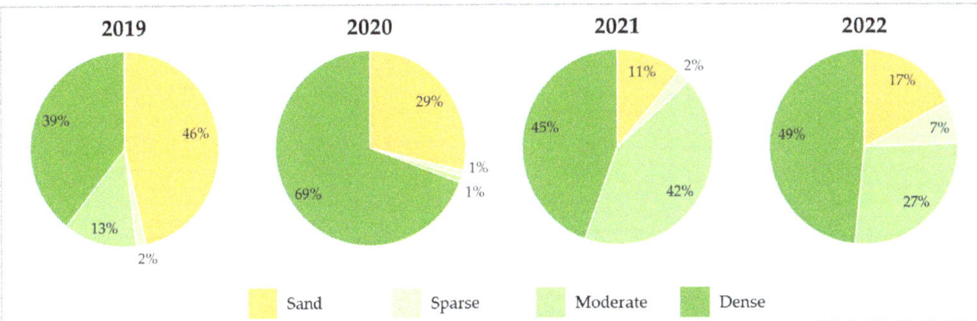

Figure 7. Distributions of sand and submerged macrophyte cover classes (sparse, moderate and dense) in the four-year study period. Un-colonized (sand) pixel percentage is represented in yellow; sparse, moderate and dense submerged macrophytes are shown with a gradient of green from lightest to darkest.

These results show how the fractional cover of the different classes changed over the study period. Un-colonized pixels representing a substrate covered by sand, for instance, decreased considerably in 2021 and 2022, compared to the initial value in 2019. At the same time, the substrate covered by submerged macrophytes increased (moderate and dense classes).

The results of the spatio-temporal analysis showing the areas of highest variability in water quality parameters (Chl-a, PC, TSM), submerged macrophytes' fractional cover and emergent macrophytes' density are presented in Figure 8.

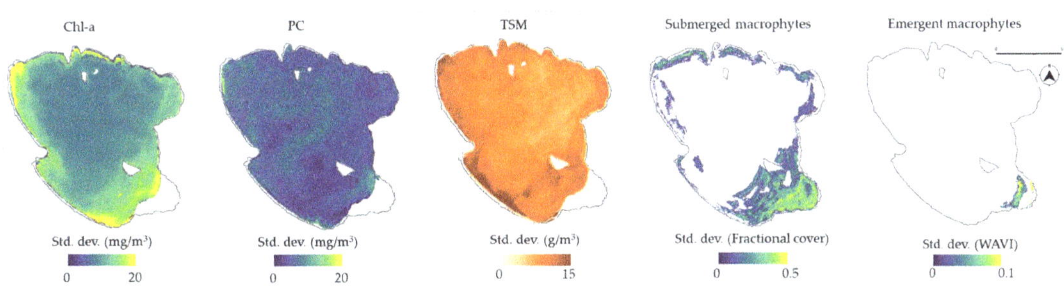

Figure 8. From **left** to **right**, standard deviation maps of Chl-a, PC, TSM, submerged macrophytes' fractional cover and emergent macrophytes' density (WAVI).

The Chl-a values show greater variability (up to 20 mg/m^3) in the portions of the lake closer to the littoral zone. The PC concentrations show higher variability in the littoral part of the lake (up to 17 mg/m^3) than for Chl-a, and in the open water part of the lake (up to 10 mg/m^3). The TSM concentrations show higher variability close to the tributary inflow in the south-western part of the lake, with values around 15 g/m^3. Macrophytes show higher variability in terms of fractional cover (up to 50%) and density in the south-eastern part of the lake where the aquatic vegetation is well established.

4. Discussion

Rrs is the core product of the optical remote sensing of water bodies and is used to retrieve water quality parameters. Given the importance of its function, it is therefore crucial to assess its accuracy [84–88]. In the present study, Rrs data from hyperspectral sensors PRISMA, DESIS and EnMAP over Lake Trasimeno were validated through comparison with in situ data, provided by the WISPStation. A total of 13 standard L2 products provided by the respective space agencies were pre-processed to ensure geographic co-registration, to mask clouds, to remove sun glint when present and to decrease the environmental noise over water. Notably, only one of the 13 images was affected by sun glint, most likely due to favorable observation/illumination geometries. Although the process of sun glint removal resulted in the loss of bands in the NIR region for the DESIS data acquired on 7 August 2022, there is a significant improvement in agreement with the in situ measurement. The MAPD value is significantly reduced, while the SA value, related to spectral similarity, remains almost constant; this result is consistent with assumption that the sun glint disturbance leads to an alteration mainly in the magnitude of the signature, rather than the shape [89]. The validation of all satellite-derived Rrs versus in situ measurements was then shown, proposing an aggregated representation for each of the three satellites. The local maximum of Rrs in the green region varies in the range of 0.020–0.030 sr^{-1} for PRISMA images, and from 0.025 to 0.035 sr^{-1} for DESIS images. The statistical results obtained in the case of PRISMA are comparable with those obtained in [34], i.e., $R^2 = 0.87$, and those obtained in [43], i.e., SA around 10° and MAPD around 30%. The in situ data synchronous with the DESIS images show greater variability than the other in situ measurements. The DESIS-derived Rrs show, on average, the best agreement in terms of spectral shape (statistically represented by SA), while the best agreement in terms of magnitude (RMSD, MAPD) is obtained for the EnMAP image. This could be attributed to the model implemented for atmospheric correction, which is designed specifically for water targets.

Once the Rrs products were validated and preliminary steps—such as the removal of noisy bands—were carried out, they were used as inputs into models for generating water quality products, which were then validated again by comparison with the in situ measurements, as performed in [90–92]. From the results of the statistical analysis, the finding that the highest level of agreement is obtained in the case of the TSM concentration maps generated by DESIS images may stem from the fact that in the radiometric statistical analysis of DESIS resulted in the lowest SA value and therefore the sensor with the highest spectral similarity compared to the in situ measurement. Generally, the maps obtained from the DESIS Rrs products seem to show greater variability in Chl-a and TSM concentrations; such an outcome may have arisen from the fact that, at the radiometric level, the standard deviation of both the satellite and in situ measurements was high, in the 700–800 nm range.

The synergetic use of the three hyperspectral sensors is a great advantage for carrying out a temporal analysis, useful, for instance, in understanding the dynamics of phytoplankton and TSM. Since the lake presents meso-eutrophic conditions with high seasonal variability, this variability can also be seen in the TSM concentrations, showing higher values near the tributary inflow, which is likely to transport suspended sediment loads from the catchment into the lake. The Chl-a values are, on average, increased in 2022, compared to previous years, at the same time as the lake level is decreasing; these two parameters are indeed negatively correlated, as stated in [93], a study conducted on Lake Trasimeno. The maximum values of PC and TSM are negatively correlated with lake levels (Pearson correlation, respectively: $r = -0.942$ and p value < 0.001; $r = -0.678$ and p value < 0.05), suggesting key roles of these parameters in lake ecology. In this work, it was possible to identify the trend of these optically active parameters and aquatic vegetation over the summer/late summer seasons of four years. The PC values range mainly from 0 to 25 mg/m^3, which represents the suggested range for the use of the MDN, as reported by [37]. Once more, the statistical results achieved by comparing the products obtained from PRISMA images with the in situ measurements show comparable values with previous comparisons with Sentinel-2 [34], e.g., RMSD = 1.23 mg/m^3 for the TSM concentration maps.

Both the coverage area and the WAVI vegetation index, a proxy for biomass density, calculated on emergent macrophytes, show a significant increase in 2022 compared to 2019, with several areas of the lake affected by this phenomenon. In a previous study [31] the aquatic vegetation of Lake Trasimeno was observed from an airborne AVIRIS image, and the results showed that the south-eastern zone was colonized by the three different submerged classes but not yet by the emergent macrophytes. In fact, this portion of the lake, in the present study, exhibits a high level of variability (i.e., high standard deviation values) both for submerged and emergent macrophytes. To improve knowledge of the complex relations between primary producers (i.e., rooted macrophytes, phytoplankton and cyanobacteria), TSM and water clarity, hyperspectral maps can be a useful tool to support lake ecosystem analysis. It is well known that the loss of aquatic vegetation threatens water transparency due to the absence of mechanical and chemical processes mediated by macrophytes (e.g., the stabilization and prevention of sediment resuspension, refuges for zooplankton that feeds on microalgae, competition for nutrients and light and the production of allelopathic substances that counteract algal proliferation and dominance) [94,95]. In the limited dataset of this work, it indeed turns out that the average TSM values are generally lower on the dates with higher area values of submerged macrophytes. The widespread expansion of submerged macrophytes and the appearance of semi-emergent macrophytes above the water surface in 2022 should be actively monitored, as this is likely to be mainly due to the sharp drop in lake levels, accompanied by a reduction in both cumulative early June rainfall (−32% compared to 2021) and total annual rainfall (−20% compared to 2021). Indeed, the lake water level is significantly negatively correlated with the area of submerged and emergent macrophytes (Pearson correlation, respectively: $r = -0.659$ and p value < 0.05; $r = -0.959$ and p value < 0.001). In addition to water levels, the component of avifauna that feeds on submerged macrophytes turns out to be an important element in the variability of submerged macrophyte cover. Aquatic plants are a fundamental element of biodiversity, i.e., the number of herbivores in Lake Trasimeno [96]. In fact, in 2020, when the submerged macrophytes were more abundant than in the previous and in the following years, the number of herbivorous birds was much lower (about 30%); this number then became more stable as the submerged macrophytes stabilized. Although the present study provides promising insights into the mapping of aquatic ecosystems with hyperspectral images, it is important to acknowledge several limitations. Firstly, to increase the robustness of the analysis, it would be necessary to assess the uncertainty related to the atmospheric correction that can propagate into errors in the retrieved satellite products. Additionally, expanding the dataset would provide wider temporal coverage, crucial for this kind of analysis. Finally, in the fundamental step of satellite products validation, it would be essential to collect more spatially distributed in situ samples to increase the robustness of the statistics, and to consider the use of water quality data as derived from laboratory analyses of water samples, which, unfortunately, were not planned to be collected at the time of satellite overpasses. According to [23], efforts have still to be made toward the integration of remote sensing with traditional monitoring programs for coordinating water authorities' strategies with the remote sensing observation requirements.

5. Conclusions

This study demonstrates the feasibility of the use of standard L2 products provided by the spaceborne hyperspectral PRISMA, DESIS and EnMAP missions to detect spatio-temporal variations in water quality parameters and aquatic vegetation in optically deep and shallow inland waters characterized by moderately high Rrs values. The validation results showed that hyperspectral satellite images can be successfully used to capture a comprehensive picture of Lake Trasimeno's water quality that can be integrated with conventional ground-based monitoring programs. The results obtained confirm the advantage of hyperspectral data, characterized by numerous narrow bands in the VNIR domain, in order to be able to simultaneously quantify concentrations of phytoplankton pigments, such as Chl-a and PC; TSM; and the extent and type of submerged and emergent macrophytes, hence contributing to the satellite-based mapping of standard (e.g., Chl-a, TSM) and advanced (e.g., PC, shallow waters) aquatic ecosystem products. The method outlined in this paper confirms the robustness of sensor-independent approaches such as the bio-optical model BOMBER, while the similar spectral setting of the three missions allowed us to adapt the MDN, originally tested for PRISMA, to DESIS and EnMAP imagery data. Nonetheless, pre-processing techniques, necessary, for instance, to remove sun glint and decrease the environmental noise of imagery data, might be necessary for applying similar approaches to other inland waters, at least for those with optical water types similar to the one analyzed in this study. These results highlight the significant degree of temporal and spatial changes in the primary producers and aquatic vegetation in Lake Trasimeno; it has been shown that the water level is a fundamental factor in determining the ecosystem conditions of Lake Trasimeno and, hence, in conditioning its biodiversity and the complex trophic and physical relations that govern the management of these areas. Further analyses would increase the robustness of this study, for instance, by testing different atmospheric correction models, increasing the dataset and collecting multiple radiometric measurements and water samples for more distributed and comprehensive validation of the satellite products. Such additional work will also boost the advantages of exploiting hyperspectral satellite data for monitoring dynamic processes and changes over time, laying the foundation for routine monitoring, in view of future operative hyperspectral missions (e.g., CHIME, SBG and PRISMA 2nd generation).

Author Contributions: Conceptualization, A.F., M.B. and C.G.; methodology, A.F., M.B., A.P. and C.G.; validation, A.F., M.B., S.M., R.P. and M.P.; formal analysis, A.F. and A.P.; investigation, A.F., M.B., A.P., M.P., K.A., S.M. and R.P.; data curation, A.F., M.B. and A.P.; writing—original draft preparation, A.F., M.B., A.P., M.P. and C.G.; writing—review and editing, A.F., M.B., A.P., M.P., K.A., R.P. and C.G.; supervision, K.A. and C.G. All authors have read and agreed to the published version of the manuscript.

Funding: This work was supported by funding from the agreement ASI-CNR, n. 20195HH0 "Attività scientifica di CAL/VAL della missione PRISMA, PRISCAV" and the project ASI-CNR, n. 2022-15-U.0 "PRISMA Products AND Applications for inland and coastal WATER, PANDA-WATER". The authors acknowledge the support of NBFC to CNR, funded by the Italian Ministry of University and Research, PNRR, Missione 4 Componente 2, "Dalla ricerca all'impresa", Investimento 1.4, Project CN00000033.

Data Availability Statement: The datasets generated and/or analyzed during the current study are available from the corresponding author on reasonable request.

Acknowledgments: We are very grateful to R. Bolpagni, E. Piaser, L. Panizza and N. Ghirardi for the useful discussion for the analysis developed in this study. We would like to thank the "Cooperativa dei Pescatori del Trasimeno" for their support in the field campaigns. We are very grateful to E. Lopinto and P. Sacco from ASI, to N. Pinnel and U. Heiden from DLR and to M. Soppa from AWI for the relevant discussions on the PRISMA, DESIS and EnMAP missions. We would like to thank the two anonymous reviewers for their help in improving the quality of the study.

Conflicts of Interest: The authors declare no conflicts of interest.

References

1. Adrian, R.; O'Reilly, C.M.; Zagarese, H.; Baines, S.B.; Hessen, D.O.; Keller, W.; Livingstone, D.M.; Sommaruga, R.; Straile, D.; Van Donk, E.; et al. Lakes as Sentinels of Climate Change. *Limnol. Oceanogr.* **2009**, *54*, 2283–2297. [CrossRef] [PubMed]
2. Rose, K.C.; Bierwagen, B.; Bridgham, S.D.; Carlisle, D.M.; Hawkins, C.P.; Poff, N.L.; Read, J.S.; Rohr, J.R.; Saros, J.E.; Williamson, C.E. Indicators of the Effects of Climate Change on Freshwater Ecosystems. *Clim. Chang.* **2023**, *176*, 23. [CrossRef]
3. Chapman, D.V.; Sullivan, T. The Role of Water Quality Monitoring in the Sustainable Use of Ambient Waters. *One Earth* **2022**, *5*, 132–137. [CrossRef]
4. Carvalho, L.; Mackay, E.B.; Cardoso, A.C.; Baattrup-Pedersen, A.; Birk, S.; Blackstock, K.L.; Borics, G.; Borja, A.; Feld, C.K.; Ferreira, M.T.; et al. Protecting and Restoring Europe's Waters: An Analysis of the Future Development Needs of the Water Framework Directive. *Sci. Total Environ.* **2019**, *658*, 1228–1238. [CrossRef] [PubMed]
5. Alikas, K.; Kangro, K.; Randoja, R.; Philipson, P.; Asuküll, E.; Pisek, J.; Reinart, A. Satellite-Based Products for Monitoring Optically Complex Inland Waters in Support of EU Water Framework Directive. *Int. J. Remote Sens.* **2015**, *36*, 4446–4468. [CrossRef]
6. European Commission Water Framework Directive. Available online: https://environment.ec.europa.eu/topics/water/water-framework-directive_en (accessed on 5 March 2024).
7. Kratzer, S.; Harvey, E.T.; Canuti, E. International Intercomparison of In Situ Chlorophyll-a Measurements for Data Quality Assurance of the Swedish Monitoring Program. *Front. Remote Sens.* **2022**, *3*, 866712. [CrossRef]
8. Zhai, M.; Zhou, X.; Tao, Z.; Lv, T.; Zhang, H.; Li, R.; Huang, Y. Retrieve of Total Suspended Matter in Typical Lakes in China Based on Broad Bandwidth Satellite Data: Random Forest Model with Forel-Ule Index. *Front. Environ. Sci.* **2023**, *11*, 1132346. [CrossRef]

9. Wen, Z.; Wang, Q.; Ma, Y.; Jacinthe, P.A.; Liu, G.; Li, S.; Shang, Y.; Tao, H.; Fang, C.; Lyu, L.; et al. Remote Estimates of Suspended Particulate Matter in Global Lakes Using Machine Learning Models. *Int. Soil Water Conserv. Res.* **2024**, *12*, 200–216. [CrossRef]
10. Lyu, L.; Song, K.; Wen, Z.; Liu, G.; Fang, C.; Shang, Y.; Li, S.; Tao, H.; Wang, X.; Li, Y.; et al. Remote Estimation of Phycocyanin Concentration in Inland Waters Based on Optical Classification. *Sci. Total Environ.* **2023**, *899*, 166363. [CrossRef] [PubMed]
11. Cazzaniga, I.; Zibordi, G.; Mélin, F. Spectral Features of Ocean Colour Radiometric Products in the Presence of Cyanobacteria Blooms in the Baltic Sea. *Remote Sens. Environ.* **2023**, *287*, 113464. [CrossRef]
12. Roelfsema, C.; Dennison, B.; Phinn, S.; Dekker, A.; Brando, V. Remote Sensing of a Cyanobacterial Bloom (*Lyngbya majuscula*) in Moreton Bay, Australia. In Proceedings of the IGARSS 2001. Scanning the Present and Resolving the Future. Proceedings. IEEE 2001 International Geoscience and Remote Sensing Symposium (Cat. No.01CH37217), Sydney, Ausralia, 9–13 July 2001; pp. 613–615.
13. Jeppesen, E.; Peder Jensen, J.; Søndergaard, M.; Lauridsen, T.; Junge Pedersen, L.; Jensen, L. Top-down Control in Freshwater Lakes: The Role of Nutrient State, Submerged Macrophytes and Water Depth. In *Shallow Lakes '95*; Springer: Dordrecht, The Netherlands, 1997; pp. 151–164.
14. Piaser, E.; Berton, A.; Bolpagni, R.; Caccia, M.; Castellani, M.B.; Coppi, A.; Vecchia, A.D.; Gallivanone, F.; Sona, G.; Villa, P. Impact of Radiometric Variability on Ultra-High Resolution Hyperspectral Imagery Over Aquatic Vegetation: Preliminary Results. *IEEE J. Sel. Top. Appl. Earth Obs. Remote Sens.* **2023**, *16*, 5935–5950. [CrossRef]
15. Liang, S.; Gong, Z.; Wang, Y.; Zhao, J.; Zhao, W. Accurate Monitoring of Submerged Aquatic Vegetation in a Macrophytic Lake Using Time-Series Sentinel-2 Images. *Remote Sens.* **2022**, *14*, 640. [CrossRef]
16. Carr, J.; D'Odorico, P.; McGlathery, K.; Wiberg, P. Stability and Bistability of Seagrass Ecosystems in Shallow Coastal Lagoons: Role of Feedbacks with Sediment Resuspension and Light Attenuation. *J. Geophys. Res. Biogeosci.* **2010**, *115*, G3. [CrossRef]
17. Adjovu, G.E.; Stephen, H.; James, D.; Ahmad, S. Overview of the Application of Remote Sensing in Effective Monitoring of Water Quality Parameters. *Remote Sens.* **2023**, *15*, 1938. [CrossRef]
18. Cao, Q.; Yu, G.; Qiao, Z. Application and Recent Progress of Inland Water Monitoring Using Remote Sensing Techniques. *Environ. Monit. Assess.* **2023**, *195*, 125. [CrossRef] [PubMed]
19. Odermatt, D.; Gitelson, A.; Brando, V.E.; Schaepman, M. Review of Constituent Retrieval in Optically Deep and Complex Waters from Satellite Imagery. *Remote Sens. Environ.* **2012**, *118*, 116–126. [CrossRef]
20. Tyler, A.N.; Hunter, P.D.; Spyrakos, E.; Groom, S.; Constantinescu, A.M.; Kitchen, J. Developments in Earth Observation for the Assessment and Monitoring of Inland, Transitional, Coastal and Shelf-Sea Waters. *Sci. Total Environ.* **2016**, *572*, 1307–1321. [CrossRef] [PubMed]
21. Birk, S.; Ecke, F. The Potential of Remote Sensing in Ecological Status Assessment of Coloured Lakes Using Aquatic Plants. *Ecol. Indic.* **2014**, *46*, 398–406. [CrossRef]
22. Phinn, S.; Roelfsema, C.; Dekker, A.; Brando, V.; Anstee, J. Mapping Seagrass Species, Cover and Biomass in Shallow Waters: An Assessment of Satellite Multi-Spectral and Airborne Hyper-Spectral Imaging Systems in Moreton Bay (Australia). *Remote Sens. Environ.* **2008**, *112*, 3413–3425. [CrossRef]
23. Topp, S.N.; Pavelsky, T.M.; Jensen, D.; Simard, M.; Ross, M.R.V. Research Trends in the Use of Remote Sensing for Inland Water Quality Science: Moving Towards Multidisciplinary Applications. *Water* **2020**, *12*, 169. [CrossRef]
24. Samarinas, N.; Spiliotopoulos, M.; Tziolas, N.; Loukas, A. Synergistic Use of Earth Observation Driven Techniques to Support the Implementation of Water Framework Directive in Europe: A Review. *Remote Sens.* **2023**, *15*, 1983. [CrossRef]
25. Rast, M.; Painter, T.H. Earth Observation Imaging Spectroscopy for Terrestrial Systems: An Overview of Its History, Techniques, and Applications of Its Missions. *Surv. Geophys.* **2019**, *40*, 303–331. [CrossRef]
26. Dey, J.; Vijay, R. A Critical and Intensive Review on Assessment of Water Quality Parameters through Geospatial Techniques. *Environ. Sci. Pollut. Res.* **2021**, *28*, 41612–41626. [CrossRef] [PubMed]
27. Leiper, I.; Phinn, S.; Roelfsema, C.; Joyce, K.; Dekker, A. Mapping Coral Reef Benthos, Substrates, and Bathymetry, Using Compact Airborne Spectrographic Imager (CASI) Data. *Remote Sens.* **2014**, *6*, 6423–6445. [CrossRef]
28. Vahtmäe, E.; Paavel, B.; Kutser, T. How Much Benthic Information Can Be Retrieved with Hyperspectral Sensor from the Optically Complex Coastal Waters? *J. Appl. Remote Sens.* **2020**, *14*, 1. [CrossRef]
29. Turpie, K.R.; Klemas, V.V.; Byrd, K.; Kelly, M.; Jo, Y.-H. Prospective HyspIRI Global Observations of Tidal Wetlands. *Remote Sens. Environ.* **2015**, *167*, 206–217. [CrossRef]
30. Chander, S.; Pompapathi, V.; Gujrati, A.; Singh, R.P.; Chaplot, N.; Patel, U.D. GROWTH OF INVASIVE AQUATIC MACROPHYTES OVER TAPI RIVER. *Int. Arch. Photogramm. Remote Sens. Spat. Inf. Sci.* **2018**, *XLII-5*, 829–833. [CrossRef]
31. Giardino, C.; Bresciani, M.; Valentini, E.; Gasperini, L.; Bolpagni, R.; Brando, V.E. Airborne Hyperspectral Data to Assess Suspended Particulate Matter and Aquatic Vegetation in a Shallow and Turbid Lake. *Remote Sens. Environ.* **2015**, *157*, 48–57. [CrossRef]
32. Adam, E.; Mutanga, O.; Rugege, D. Multispectral and Hyperspectral Remote Sensing for Identification and Mapping of Wetland Vegetation: A Review. *Wetl. Ecol. Manag.* **2010**, *18*, 281–296. [CrossRef]
33. Bresciani, M.; Giardino, C.; Fabbretto, A.; Pellegrino, A.; Mangano, S.; Free, G.; Pinardi, M. Application of New Hyperspectral Sensors in the Remote Sensing of Aquatic Ecosystem Health: Exploiting PRISMA and DESIS for Four Italian Lakes. *Resources* **2022**, *11*, 8. [CrossRef]

34. Niroumand-Jadidi, M.; Bovolo, F.; Bruzzone, L. Water Quality Retrieval from PRISMA Hyperspectral Images: First Experience in a Turbid Lake and Comparison with Sentinel-2. *Remote Sens.* **2020**, *12*, 3984. [CrossRef]
35. Soppa, M.A.; Dinh, D.A.; Silva, B.; Steinmetz, F.; Alvarado, L.; Bracher, A. INTERCOMPARISON OF DESIS, SENTINEL-2 (MSI) AND SENTINEL-3 (OLCI) DATA FOR WATER COLOUR APPLICATIONS. *Int. Arch. Photogramm. Remote Sens. Spat. Inf. Sci.* **2022**, *XLVI-1/W1-2021*, 69–72. [CrossRef]
36. Katlane, R.; Doxaran, D.; ElKilani, B.; Trabelsi, C. Remote Sensing of Turbidity in Optically Shallow Waters Using Sentinel-2 MSI and PRISMA Satellite Data. *PFG—J. Photogramm. Remote Sens. Geoinf. Sci.* **2023**, 1–17. [CrossRef]
37. O'Shea, R.E.; Pahlevan, N.; Smith, B.; Bresciani, M.; Egerton, T.; Giardino, C.; Li, L.; Moore, T.; Ruiz-Verdu, A.; Ruberg, S.; et al. Advancing Cyanobacteria Biomass Estimation from Hyperspectral Observations: Demonstrations with HICO and PRISMA Imagery. *Remote Sens. Environ.* **2021**, *266*, 112693. [CrossRef]
38. Taggio, N.; Aiello, A.; Ceriola, G.; Kremezi, M.; Kristollari, V.; Kolokoussis, P.; Karathanassi, V.; Barbone, E. A Combination of Machine Learning Algorithms for Marine Plastic Litter Detection Exploiting Hyperspectral PRISMA Data. *Remote Sens.* **2022**, *14*, 3606. [CrossRef]
39. Justice, C.; Belward, A.; Morisette, J.; Lewis, P.; Privette, J.; Baret, F. Developments in the "validation" of Satellite Sensor Products for the Study of the Land Surface. *Int. J. Remote Sens.* **2000**, *21*, 3383–3390. [CrossRef]
40. Concha, J.A.; Bracaglia, M.; Brando, V.E. Assessing the Influence of Different Validation Protocols on Ocean Colour Match-up Analyses. *Remote Sens. Environ.* **2021**, *259*, 112415. [CrossRef]
41. Sterckx, S.; Brown, I.; Kääb, A.; Krol, M.; Morrow, R.; Veefkind, P.; Boersma, K.F.; De Mazière, M.; Fox, N.; Thorne, P. Towards a European Cal/Val Service for Earth Observation. *Int. J. Remote Sens.* **2020**, *41*, 4496–4511. [CrossRef]
42. Giardino, C.; Bresciani, M.; Braga, F.; Fabbretto, A.; Ghirardi, N.; Pepe, M.; Gianinetto, M.; Colombo, R.; Cogliati, S.; Ghebrehiwot, S.; et al. First Evaluation of PRISMA Level 1 Data for Water Applications. *Sensors* **2020**, *20*, 4553. [CrossRef] [PubMed]
43. Braga, F.; Fabbretto, A.; Vanhellemont, Q.; Bresciani, M.; Giardino, C.; Scarpa, G.M.; Manfè, G.; Concha, J.A.; Brando, V.E. Assessment of PRISMA Water Reflectance Using Autonomous Hyperspectral Radiometry. *ISPRS J. Photogramm. Remote Sens.* **2022**, *192*, 99–114. [CrossRef]
44. Pellegrino, A.; Fabbretto, A.; Bresciani, M.; de Lima, T.M.A.; Braga, F.; Pahlevan, N.; Brando, V.E.; Kratzer, S.; Gianinetto, M.; Giardino, C. Assessing the Accuracy of PRISMA Standard Reflectance Products in Globally Distributed Aquatic Sites. *Remote Sens.* **2023**, *15*, 2163. [CrossRef]
45. Mélin, F. Validation of Ocean Color Remote Sensing Reflectance Data: Analysis of Results at European Coastal Sites. *Remote Sens. Environ.* **2022**, *280*, 113153. [CrossRef]
46. Zibordi, G.; Ruddick, K.; Ansko, I.; Moore, G.; Kratzer, S.; Icely, J.; Reinart, A. In Situ Determination of the Remote Sensing Reflectance: An Inter-Comparison. *Ocean Sci.* **2012**, *8*, 567–586. [CrossRef]
47. Bresciani, M.; Pinardi, M.; Free, G.; Luciani, G.; Ghebrehiwot, S.; Laanen, M.; Peters, S.; Della Bella, V.; Padula, R.; Giardino, C. The Use of Multisource Optical Sensors to Study Phytoplankton Spatio-Temporal Variation in a Shallow Turbid Lake. *Water* **2020**, *12*, 284. [CrossRef]
48. Landucci, F.; Gigante, D.; Venanzoni, R. An Application of the Cocktail Method for the Classiication of the Hydrophytic Vegetation at Lake Trasimeno (Central Italy). *Fitosociologia* **2011**, *48*, 3–22.
49. Bolpagni, R. Macrophyte Richness and Aquatic Vegetation Complexity of the Lake Idro (Northern Italy). *Ann. Bot.* **2013**, 35–43.
50. Bresciani, M.; Bolpagni, R.; Braga, F.; Oggioni, A.; Giardino, C. Retrospective Assessment of Macrophytic Communities in Southern Lake Garda (Italy) from in Situ and MIVIS (Multispectral Infrared and Visible Imaging Spectrometer) Data. *J. Limnol.* **2012**, *71*, 180–190. [CrossRef]
51. Melelli, A.; Fatichenti, F. Bacini idrografici e sfruttamento delle acque in Umbria. Tra passato e presente. Percorsi di ricerca, problemi, proposte. In *L'acqua in Umbria. Disponibilità, Consumo e Salute. Le Rappresentazioni e gli Atteggiamenti dei Cittadini*; ARPA Umbria: Terni, Italy, 2013; pp. 109–132.
52. Álvarez, X.; Valero, E.; Santos, R.M.B.; Varandas, S.G.P.; Sanches Fernandes, L.F.; Pacheco, F.A.L. Anthropogenic Nutrients and Eutrophication in Multiple Land Use Watersheds: Best Management Practices and Policies for the Protection of Water Resources. *Land. Use Policy* **2017**, *69*, 1–11. [CrossRef]
53. Regione Umbria Servizio Idrografico. Available online: https://annali.regione.umbria.it/# (accessed on 13 March 2024).
54. Peters, S.; Laanen, M.; Groetsch, P.; Ghezehegn, S.; Poser, K.; Hommersom, A.; De Reus, E.; Spaias, L. WISPstation: A New Autonomous above Water Radiometer System. In Proceedings of the Ocean Optics XXIV Conference, Dubrovnik, Croatia, 7 October 2018; pp. 7–12.
55. Riddick, C.; Tyler, A.; Hommersom, A.; Alikas, K.; Kangro, K.; Ligi, M.; Bresciani, M.; Antilla, S.; Vaiciute, D.; Bucas, M.; et al. EOMORES D5.3: Final Validation Report. 2019. Available online: https://zenodo.org/records/4057057 (accessed on 9 May 2024).
56. Gons, H.J.; Rijkeboer, M.; Ruddick, K.G. Effect of a Waveband Shift on Chlorophyll Retrieval from MERIS Imagery of Inland and Coastal Waters. *J. Plankton Res.* **2004**, *27*, 125–127. [CrossRef]
57. Rijkeboer, M. *Algoritmen Voor Het Bepalen van de Concentratie Chlorofyl-a En Zwevend Stof Met de Optische Teledetectie Methode in Verschillende Optische Watertypen*; IVM Report; No. O-00/08; Dept. of Spatial Analysis and Decision Support: Amsterdam, The Netherlands, 2000.
58. Simis, S.G.H. Blue-Green Catastrophe: Remote Sensing of Mass Viral Lysis of Cyanobacteria. Ph.D. Thesis, Vrije Universiteit Amsterdam, Amsterdam, The Netherlands, 2006.

59. Giardino, C.; Bresciani, M.; Fava, F.; Matta, E.; Brando, V.; Colombo, R. Mapping Submerged Habitats and Mangroves of Lampi Island Marine National Park (Myanmar) from in Situ and Satellite Observations. *Remote Sens.* **2015**, *8*, 2. [CrossRef]
60. Regione Umbria Osservatorio Faunistico Regionale. Available online: https://www.regione.umbria.it/turismo-attivita-sportive/osservatorio-faunistico (accessed on 4 March 2024).
61. Cogliati, S.; Sarti, F.; Chiarantini, L.; Cosi, M.; Lorusso, R.; Lopinto, E.; Miglietta, F.; Genesio, L.; Guanter, L.; Damm, A.; et al. The PRISMA Imaging Spectroscopy Mission: Overview and First Performance Analysis. *Remote Sens. Environ.* **2021**, *262*, 112499. [CrossRef]
62. Kremezi, M.; Kristollari, V.; Karathanassi, V.; Topouzelis, K.; Kolokoussis, P.; Taggio, N.; Aiello, A.; Ceriola, G.; Barbone, E.; Corradi, P. Pansharpening PRISMA Data for Marine Plastic Litter Detection Using Plastic Indexes. *IEEE Access* **2021**, *9*, 61955–61971. [CrossRef]
63. Candela, L.; Formaro, R.; Guarini, R.; Loizzo, R.; Longo, F.; Varacalli, G. The PRISMA Mission. In Proceedings of the 2016 IEEE International Geoscience and Remote Sensing Symposium (IGARSS), Beijing, China, 10–15 July 2016; pp. 253–256.
64. Alonso, K.; Bachmann, M.; Burch, K.; Carmona, E.; Cerra, D.; de los Reyes, R.; Dietrich, D.; Heiden, U.; Hölderlin, A.; Ickes, J.; et al. Data Products, Quality and Validation of the DLR Earth Sensing Imaging Spectrometer (DESIS). *Sensors* **2019**, *19*, 4471. [CrossRef] [PubMed]
65. Krutz, D.; Müller, R.; Knodt, U.; Günther, B.; Walter, I.; Sebastian, I.; Säuberlich, T.; Reulke, R.; Carmona, E.; Eckardt, A.; et al. The Instrument Design of the DLR Earth Sensing Imaging Spectrometer (DESIS). *Sensors* **2019**, *19*, 1622. [CrossRef] [PubMed]
66. de los Reyes, R.; Langheinrich, M.; Schwind, P.; Richter, R.; Pflug, B.; Bachmann, M.; Müller, R.; Carmona, E.; Zekoll, V.; Reinartz, P. PACO: Python-Based Atmospheric Correction. *Sensors* **2020**, *20*, 1428. [CrossRef] [PubMed]
67. Habermeyer, M.; Pinnel, N.; Storch, T.; Honold, H.P.; Tucker, P.; Guanter, L.; Segl, K.; Fischer, S. The EnMAP Mission: From Observation Request to Data Delivery. In Proceedings of the IGARSS 2019—2019 IEEE International Geoscience and Remote Sensing Symposium, Yokohama, Japan, 28 July–2 August 2019; pp. 4507–4510.
68. Guanter, L.; Kaufmann, H.; Segl, K.; Foerster, S.; Rogass, C.; Chabrillat, S.; Kuester, T.; Hollstein, A.; Rossner, G.; Chlebek, C.; et al. The EnMAP Spaceborne Imaging Spectroscopy Mission for Earth Observation. *Remote Sens.* **2015**, *7*, 8830–8857. [CrossRef]
69. Kiselev, V.; Bulgarelli, B.; Heege, T. Sensor Independent Adjacency Correction Algorithm for Coastal and Inland Water Systems. *Remote Sens. Environ.* **2015**, *157*, 85–95. [CrossRef]
70. Storch, T.; Honold, H.-P.; Chabrillat, S.; Habermeyer, M.; Tucker, P.; Brell, M.; Ohndorf, A.; Wirth, K.; Betz, M.; Kuchler, M.; et al. The EnMAP Imaging Spectroscopy Mission towards Operations. *Remote Sens. Environ.* **2023**, *294*, 113632. [CrossRef]
71. ASI—Italian Space Agency PRISMA Algorithm Theoretical Basis Document (ATBD). Available online: http://prisma.asi.it/missionselect/docs.php (accessed on 18 January 2024).
72. DLR—German Space Agency DESIS Instrument. Available online: https://www.dlr.de/eoc/desktopdefault.aspx/tabid-13622/23667_read-54280/ (accessed on 4 March 2024).
73. DLR—German Space Agency EnMAP Specification. Available online: https://www.enmap.org/mission/ (accessed on 4 March 2024).
74. Hedley, J.D.; Harborne, A.R.; Mumby, P.J. Technical Note: Simple and Robust Removal of Sun Glint for Mapping Shallow-water Benthos. *Int. J. Remote Sens.* **2005**, *26*, 2107–2112. [CrossRef]
75. Campbell, G.; Stuart, R. Phinn An Assessment of the Accuracy and Precision of Water Quality Parameters Retrieved with the Matrix Inversion Method. *Limnol. Ocean. Methods* **2010**, *8*, 16–29.
76. Soomets, T.; Uudeberg, K.; Jakovels, D.; Brauns, A.; Zagars, M.; Kutser, T. Validation and Comparison of Water Quality Products in Baltic Lakes Using Sentinel-2 MSI and Sentinel-3 OLCI Data. *Sensors* **2020**, *20*, 742. [CrossRef] [PubMed]
77. Vanhellemont, Q.; Ruddick, K. Acolite for Sentinel-2: Aquatic Applications of MSI Imagery. In Proceedings of the 2016 ESA Living Planet Symposium, Prague, Czech Republic, 9 May 2016.
78. Sagayam, K.M.; Bruntha, P.M.; Sridevi, M.; Renith Sam, M.; Kose, U.; Deperlioglu, O. A Cognitive Perception on Content-Based Image Retrieval Using an Advanced Soft Computing Paradigm. In *Advanced Machine Vision Paradigms for Medical Image Analysis*; Elsevier: Amsterdam, The Netherlands, 2021; pp. 189–211.
79. Villa, P.; Pinardi, M.; Tóth, V.R.; Hunter, P.D.; Bolpagni, R.; Bresciani, M. Remote Sensing of Macrophyte Morphological Traits: Implications for the Management of Shallow Lakes. *J. Limnol.* **2017**, *76*, 109–126. [CrossRef]
80. Lee, Z.; Carder, K.L.; Mobley, C.D.; Steward, R.G.; Patch, J.S. Hyperspectral Remote Sensing for Shallow Waters I A Semianalytical Model. *Appl. Opt.* **1998**, *37*, 6329. [CrossRef] [PubMed]
81. Lee, Z.; Carder, K.L.; Mobley, C.D.; Steward, R.G.; Patch, J.S. Hyperspectral Remote Sensing for Shallow Waters: 2 Deriving Bottom Depths and Water Properties by Optimization. *Appl. Opt.* **1999**, *38*, 3831. [CrossRef] [PubMed]
82. Github (MDN). Available online: https://github.com/STREAM-RS/STREAM-RS (accessed on 4 March 2024).
83. Kutser, T.; Metsamaa, L.; Strömbeck, N.; Vahtmäe, E. Monitoring Cyanobacterial Blooms by Satellite Remote Sensing. *Estuar. Coast. Shelf Sci.* **2006**, *67*, 303–312. [CrossRef]
84. Zibordi, G.; Mélin, F.; Berthon, J.-F.; Holben, B.; Slutsker, I.; Giles, D.; D'Alimonte, D.; Vandemark, D.; Feng, H.; Schuster, G.; et al. AERONET-OC: A Network for the Validation of Ocean Color Primary Products. *J. Atmos. Ocean. Technol.* **2009**, *26*, 1634–1651. [CrossRef]
85. Bailey, S.W.; Franz, B.A.; Werdell, P.J. Estimation of Near-Infrared Water-Leaving Reflectance for Satellite Ocean Color Data Processing. *Opt. Express* **2010**, *18*, 7521. [CrossRef] [PubMed]

86. McCarthy, S.; Gould, R.; Richman, J.; Kearney, C.; Lawson, A. Impact of Aerosol Model Selection on Water-Leaving Radiance Retrievals from Satellite Ocean Color Imagery. *Remote Sens.* **2012**, *4*, 3638–3665. [CrossRef]
87. Schaepman, M.E.; Jehle, M.; Hueni, A.; D'Odorico, P.; Damm, A.; Weyermann, J.; Schneider, F.D.; Laurent, V.; Popp, C.; Seidel, F.C.; et al. Advanced Radiometry Measurements and Earth Science Applications with the Airborne Prism Experiment (APEX). *Remote Sens. Environ.* **2015**, *158*, 207–219. [CrossRef]
88. González Vilas, L.; Brando, V.E.; Concha, J.A.; Goyens, C.; Dogliotti, A.I.; Doxaran, D.; Dille, A.; Van der Zande, D. Validation of Satellite Water Products Based on HYPERNETS in Situ Data Using a Match-up Database (MDB) File Structure. *Front. Remote Sens.* **2024**, *5*, 1330317. [CrossRef]
89. Zhang, H.; Wang, M. Evaluation of Sun Glint Models Using MODIS Measurements. *J. Quant. Spectrosc. Radiat. Transf.* **2010**, *111*, 492–506. [CrossRef]
90. Di Francesco, S.; Venturi, S.; Casadei, S. An Integrated Water Resource Management Approach for Lake Trasimeno, Italy. *Hydrol. Sci. J.* **2023**, *68*, 630–644. [CrossRef]
91. Chebud, Y.; Naja, G.M.; Rivero, R.G.; Melesse, A.M. Water Quality Monitoring Using Remote Sensing and an Artificial Neural Network. *Water Air Soil Pollut.* **2012**, *223*, 4875–4887. [CrossRef]
92. Su, Y.-F.; Liou, J.-J.; Hou, J.-C.; Hung, W.-C.; Hsu, S.-M.; Lien, Y.-T.; Su, M.-D.; Cheng, K.-S.; Wang, Y.-F. A Multivariate Model for Coastal Water Quality Mapping Using Satellite Remote Sensing Images. *Sensors* **2008**, *8*, 6321–6339. [CrossRef] [PubMed]
93. Giardino, C.; Bresciani, M.; Villa, P.; Martinelli, A. Application of Remote Sensing in Water Resource Management: The Case Study of Lake Trasimeno, Italy. *Water Resour. Manag.* **2010**, *24*, 3885–3899. [CrossRef]
94. Marzocchi, U.; Benelli, S.; Larsen, M.; Bartoli, M.; Glud, R.N. Spatial Heterogeneity and Short-term Oxygen Dynamics in the Rhizosphere of *Vallisneria Spiralis*: Implications for Nutrient Cycling. *Freshw. Biol.* **2019**, *64*, 532–543. [CrossRef]
95. Licci, S.; Nepf, H.; Delolme, C.; Marmonier, P.; Bouma, T.J.; Puijalon, S. The Role of Patch Size in Ecosystem Engineering Capacity: A Case Study of Aquatic Vegetation. *Aquat. Sci.* **2019**, *81*, 41. [CrossRef]
96. Janssen, A.B.G.; Hilt, S.; Kosten, S.; de Klein, J.J.M.; Paerl, H.W.; Van de Waal, D.B. Shifting States, Shifting Services: Linking Regime Shifts to Changes in Ecosystem Services of Shallow Lakes. *Freshw. Biol.* **2021**, *66*, 1–12. [CrossRef]

remote sensing

MDPI

Article

A Chlorophyll-a Concentration Inversion Model Based on Backpropagation Neural Network Optimized by an Improved Metaheuristic Algorithm

Xichen Wang, Jianyong Cui * and Mingming Xu

College of Oceanography and Space Informatics, China University of Petroleum (East China), Qingdao 266580, China; 2116020109@s.upc.edu.cn (X.W.); 20180043@upc.edu.cn (M.X.)
* Correspondence: cui_jianyong@upc.eud.cn; Tel.: +86-158-631-17130

Abstract: Chlorophyll-a (Chl-a) concentration monitoring is very important for managing water resources and ensuring the stability of marine ecosystems. Due to their high operating efficiency and high prediction accuracy, backpropagation (BP) neural networks are widely used in Chl-a concentration inversion. However, BP neural networks tend to become stuck in local optima, and their prediction accuracy fluctuates significantly, thus posing restrictions to their accuracy and stability in the inversion process. Studies have found that metaheuristic optimization algorithms can significantly improve these shortcomings by optimizing the initial parameters (weights and biases) of BP neural networks. In this paper, the adaptive nonlinear weight coefficient, the path search strategy "Levy flight" and the dynamic crossover mechanism are introduced to optimize the three main steps of the Artificial Ecosystem Optimization (AEO) algorithm to overcome the algorithm's limitation in solving complex problems, improve its global search capability, and thereby improve its performance in optimizing BP neural networks. Relying on Google Earth Engine and Google Colaboratory (Colab), a model for the inversion of Chl-a concentration in the coastal waters of Hong Kong was built to verify the performance of the improved AEO algorithm in optimizing BP neural networks, and the improved AEO algorithm proposed herein was compared with 17 different metaheuristic optimization algorithms. The results show that the Chl-a concentration inversion model based on a BP neural network optimized using the improved AEO algorithm is significantly superior to other models in terms of prediction accuracy and stability, and the results obtained via the model through inversion with respect to Chl-a concentration in the coastal waters of Hong Kong during heavy precipitation events and red tides are highly consistent with the measured values of Chl-a concentration in both time and space domains. These conclusions can provide a new method for Chl-a concentration monitoring and water quality management for coastal waters.

Keywords: water quality; Artificial Ecosystem Optimization; chlorophyll-a; backpropagation neural network; Hong Kong; Google Earth Engine

1. Introduction

In the past few decades, the challenges faced by coastal ecosystems have been intensifying continuously [1]. Nearshore waters are rich in nutrient salts that can easily lead to environmental harms such as bottom water hypoxia and red tides, threatening the balance of marine ecosystems [2]. Continuous water quality monitoring is essential for ensuring reasonable utilization of water resources and maintaining the health and stability of marine ecosystems. As an important parameter reflecting phytoplankton biomass, algal activity and the degree of eutrophication, algal activity is the core indicator of water quality [3].

Traditional shipboard sampling methods have limitations in terms of spatial coverage and measuring frequency, making it difficult for them to fully capture the dynamic spatiotemporal changes in Chl-a concentration in estuarine waters and rendering them unable to meet the needs of high-resolution spatiotemporal data and real-time information analysis in current scientific research [1,4]. In comparison, remote sensing plays a revolutionary role in water quality monitoring by using the spectral reflectance characteristics of water [5]. It can achieve comprehensive and continuous water quality monitoring by analyzing surface spectral data [6] and effectively compensate for the shortcomings of traditional monitoring methods, providing a large-scale, real-time and cost-effective monitoring method [7].

Existing Chl-a concentration inversion models can be classified into three main categories, namely, empirical, semi-empirical, and artificial intelligence-based models [8,9]. Examples of empirical and semi-empirical methods include ocean color (OC) algorithms [10], the three-band reflectance difference algorithm [11], and the Normalized Difference Chlorophyll Index (NDCI) [12]. These methods are practical, but their prediction accuracy may be limited due to complex and volatile environmental conditions. In recent years, Support Vector Regression (SVR) and Random Forest (RF) have been applied to the retrieval of Chl-a concentration. However, due to the complexity and large volume of data required during model training, overfitting can easily occur, leading to a decrease in model performance [13]. Compared to the aforementioned methods, neural networks have powerful parallel processing capabilities, superior learning performance, and are suitable for nonlinear problems, making them more suitable for Chl-a concentration retrieval scenarios [13,14]. Among them, BP neural networks, due to their training mechanism of backpropagation, can flexibly adjust network weights and thresholds, which is advantageous when modeling complex nonlinear relationships [15,16]. However, this method still suffers from issues such as susceptibility to local optima and large fluctuations in model accuracy [17,18].

The performance of BP neural networks has strong potential for improvement [19], and existing research has confirmed that metaheuristic algorithms can effectively optimize BP neural networks. For example, Deng et al. used particle swarm optimization (PSO) to optimize the initial parameters of BP neural networks and improve the accuracy and robustness of BP neural network models [20]; Zhu et al. used genetic algorithms to optimize the structure of BP neural networks [21]; Huang et al. use an improved PSO algorithm to optimize BP neural networks and built a model with better prediction performance in different application scenarios [22]. Their model has improved performance, but its global search capability and stability still need to be improved.

To address the problems mentioned above, a BP neural network optimized using the Synthesized Artificial Ecosystem Optimization (SAEO) algorithm is proposed herein, which is called SAEO-BP neural network. The SAEO algorithm, which is based on the AEO algorithm [23], integrates the adaptive nonlinear weight coefficient, the "Levy flight" search strategy, and the dynamic crossover mechanism to improve its performance in solving complex problems and its global search capability.

In order to verify the effectiveness of the SAEO-BP neural network, the spectral data collected from Sentinel-2 remote sensing images were selected and used as input features, and experiments were conducted on the estimation of Chl-a concentration in the coastal waters of Hong Kong. The performance of the SAEO-BP neural network was evaluated, and the SAEO-BP neural network was compared with 17 BP neural networks optimized using different algorithms including the standard AEO algorithm, three variants of the AEO algorithm, and 13 well-known metaheuristic optimization algorithms. In addition, the prediction accuracy of the SAEO-BP neural network was investigated under conditions where Chl-a concentration fluctuates greatly during heavy precipitation events and red rides. The results show that the SAEO-BP neural network performs excellently in terms of prediction accuracy and stability. This finding provides an important scientific basis for water resources management and protection in coastal areas and valuable support for decision making.

2. Materials and Methods

This section mainly elaborates on the preprocessing steps of experimental data and the construction steps of the SAEO-BP neural network. A total of 5754 valid observation records of water quality parameters and 756 high-quality cloud-free remote sensing images were collected in the experiment. There were 612 records with water quality data collected within 2 h of satellite overpasses. Based on the sampling time and coordinates of these water quality data, 12 bands of reflectance of corresponding image pixels were extracted, forming 612 sets of data for model training. The preprocessing operations of the data were conducted on the Google Earth Engine (GEE) cloud platform, and the training task of the model was completed on the cloud-based Google Colab computing platform (https://colab.research.google.com/, accessed on 15 January 2024).

As an advanced cloud-based geospatial analysis platform, GEE can gain instant access to massive volumes of historical and real-time remote sensing data, thus significantly reducing the time and storage space consumed during data preprocessing. In addition, it has a strong parallel processing capability and can efficiently complete the cropping, stitching and complex analysis of large-scale data sets [24]. As a cloud-based development platform, Colab is preconfigured with a series of libraries widely used in data science and machine learning, thus simplifying the time-consuming and complex environment configuration process and improving the overall experimental efficiency [14]. In addition, Colab is tightly integrated with Google Drive, providing an efficient and convenient way to store and recall data sets.

2.1. Study Area

Hong Kong is a highly urbanized subtropical region between latitude 22°08′N and 22°35′N, longitude 113°49′E and 114°31′E. In terms of natural conditions, Hong Kong's climate is characterized by significant seasonal variations. Affected by monsoons, the precipitation in Hong Kong during the rainy season (from April to September) can reach 80% of the total annual precipitation. Heavy precipitation can cause the Chl-a concentration in coastal waters to decrease abruptly. In addition, factors such as river runoff, wind and tide continuously act on the coastal waters of Hong Kong and affect the nutrient content in water. In terms of human activities, urban wastewater discharge in Hong Kong (a highly developed and densely populated city) can further affect coastal water quality. The nutrient salts in urban wastewater can easily cause eutrophication, increasing the probability of occurrence of red tides.

In order to comparatively analyze the prediction performance of different models, four coastal areas of Hong Kong with typical characteristics and abundant monitoring data were selected as experimental areas. Experimental Area A covers the Deep Bay area. This area is enriched with microalgae due to the inflow of large amounts of nutrients from the Pearl River [25], and due to its semi-enclosed form, eutrophication is common [26]. Experimental Areas B and C are located in the southern waters of Hong Kong. These areas are affected by river runoff and monsoons, where large-scale algae blooms often occur in summer and red tides occur frequently [25]. Experimental Area D covers the Tolo Harbor area, which is almost completely surrounded by land. In this area, the tidal force is weak, resulting in limited water exchange, and nutrient salts tend to accumulate in this area, leading to eutrophication problems [27]. The specific geographical locations of these four experimental areas (A–D) are shown in Figure 1.

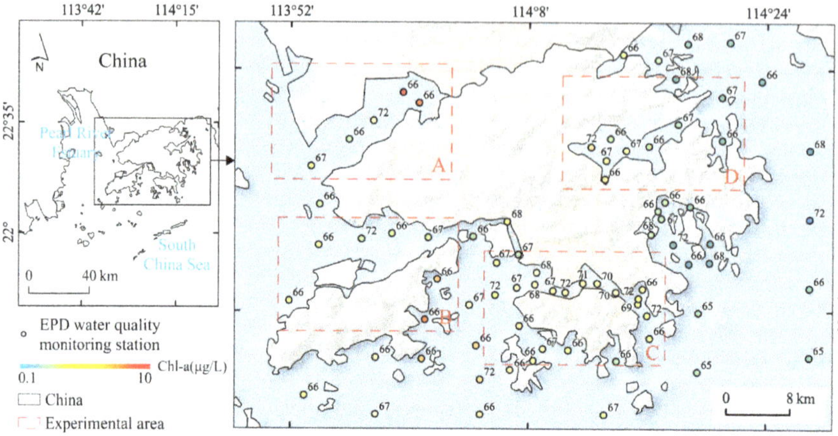

Figure 1. Study area and selected experimental areas.

2.2. Data Sources and Processing

This section details the sources and processing of model training data (Table 1). The measured water quality data used for experimental purposes were acquired and downloaded from the open data platform of Hong Kong Government (https://data.gov.hk/, accessed on 1 August 2023). The remote sensing data were acquired from the Sentinel-2 L2A collection and preprocessed on GEE. The other data processing steps and subsequent model building and training tasks were efficiently executed on Colab.

Table 1. The data used and their sources in the study.

Data	Data Source
Water Quality Data	Hong Kong Environmental Protection Department (https://data.gov.hk/, accessed on 1 August 2023)
Remote-Sensing Data	Google Earth Engine (https://earthengine.google.com/, accessed on 1 August 2023)
Climate Data	Hong Kong Observatory (https://www.weather.gov.hk/, accessed on 15 January 2024)
Red Tide Data	Hong Kong Red Tide Database (https://redtide.afcd.gov.hk/urtin/#/maps, accessed on 15 January 2024)

2.2.1. Water Quality Data

The complexity and variability of Hong Kong's coastal environment make it necessary to estimate the water quality parameters of the coastal waters of Hong Kong in a highly accurate and adaptive manner. Such estimation requires an accurate prediction model to be built based on a large amount of real-time monitoring data. Since 1986, the Environmental Protection Department (EPD) of Hong Kong has established 76 widely distributed water quality monitoring stations and implemented a comprehensive marine water quality monitoring plan to maintain the health of Hong Kong's marine ecosystem. A total of 5754 valid observation records of surface water were selected from the set of measured Chl-a concentration data for the period from 2017 to 2022 provided by Hong Kong EPD. The measured data set is characterized by wide coverage and even distribution in space and high continuity in time, providing high-quality data for building a highly accurate and adaptable Chl-a concentration inversion model. Figure 1 illustrates the specific locations of 76 water quality monitoring stations. The color of each monitoring station intuitively reflects its average Chl-a concentration level, while the annotated numbers represent the amount of valid data obtained at each site.

2.2.2. Remote Sensing Data

The remote sensing data used for experimental purposes are sourced from the Sentinel-2 L2A collection of the European Space Agency, preprocessed using radiation calibration and atmospheric correction. The accuracy of its spatial coordinates is suitable for small- and medium-scale research. A total of 756 high-quality images were selected from the remote sensing images for the period from 28 March 2017 to 31 December 2022 depending on cloud cover and used for analysis. Considering that some images are affected by sun glints, reference was made to the research idea of Kay et al. [28], and a simple correction strategy was employed. The idea of the correction strategy is to subtract 50% of the reflectance of the shortwave infrared (SWIR) band (i.e., Band 11) from the reflectance of each band. This method is based on the fact that the scattered radiation in water has a very slight effect on the SWIR band; therefore, the high signal intensity of this band in the image is believed to result mainly from the influence of sun glints. The correction values were selected based on the previous analysis of surface reflectance performed by Kuhn et al. [29] and manual inspections of the images of adjacent strips from different perspectives to ensure that the interference of sun glints can be effectively eliminated.

Finally, based on the corrected data, the Modified Normalized Difference Water Index (MNDWI) [30] of each pixel was calculated using the reflection values of the SWIR band and the green band, and only the pixels with MNDWI values greater than 0 were retained, thereby accurately extracting water quality information from remote sensing images.

2.2.3. Data Set for Model Building

Based on the condition of a two-hour difference between satellite overpass time and observation time, pairing and screening were conducted between remote sensing images and field-acquired water quality data. Through spatial overlay operations, the positions of 76 water monitoring points were precisely matched with the images, and the spectral values of 12 bands corresponding to each point were extracted. The experiment eventually identified 612 valid data records, forming the training set foundation required to build the Chl-a concentration inversion model. Subsequently, to standardize the data scale and improve model training efficiency, normalization preprocessing was applied to the acquired band data. Finally, the normalized dataset was randomly divided into two parts, a training set and a validation set, in a 7:3 ratio.

2.3. Methodology

This subsection details how to train a neural network using the backpropagation algorithm, briefly describes the basic principle of the AEO algorithm, deeply analyzes the scope and mechanism of improvement of the SAEO algorithm proposed in this paper, and describes in detail how to use the SAEO algorithm to optimize a BP neural network and how to train the Chl-a concentration inversion model.

2.3.1. BP Neural Networks

BP neural networks are multilayer feedforward networks trained using the backpropagation algorithm, which consists of input, hidden and output layers [31]. The workflow of a BP neural network can be divided into two major processes, namely, forward propagation and backward propagation of errors (backpropagation). In the forward propagation process, the input layer receives external input signals and passes them to the hidden layer; the hidden layer performs nonlinear signal transformation; the processed information is transmitted to the output layer, and the output layer outputs the result. The activation function of a single neuron can be mathematically expressed by Equation (1).

$$x_j^{l+1} = f\left(\sum_i w_{ij}^l x_i^l + w_{bj}^{l+1}\right) \tag{1}$$

$$E = \frac{1}{2}\sum_j \left(x_j^{l+1} - f(x_i)\right)^2 \tag{2}$$

where x_i^l is the input signal of the ith neuron in the lth layer, w_{ij}^l is the weight of the connection from the ith neuron in the lth layer to the jth neuron in the (l + 1)st layer, w_{bj}^{l+1} is the bias term of the jth neuron in the (l + 1)st layer, and $f(*)$ is a nonlinear activation function. When there is an error E between the actual output $f(x_i)$ from the output layer and the expected output, the system enters the backpropagation stage. In the backpropagation stage, an error is propagated backward from the output layer through the hidden layer(s) toward the input layer, and the weights of connections between neurons in each layer are updated according to the principle of error gradient descent. This iterative process is repeated until the weights of all connections in the network are optimized so that the network model's output gradually approaches the actual target value.

The above principle indicates that the initialization of model weights is crucial for training neural networks. Choosing an appropriate initialization method is advantageous for improving the convergence speed and performance of the model.

2.3.2. AEO Algorithm

The Artificial Ecosystem Optimization (AEO) algorithm created by Zhao et al. [23] is a metaheuristic optimization algorithm based on the principles of ecosystem energy dynamics. This algorithm represents a model based on three stages, namely, production, consumption, and decomposition. The populations in the model include producers representing the worst solution, decomposers representing the optimal solution, and multilevel consumers ranked by fitness, among which consumers simulate the herbivores, carnivores, and omnivores in the food chain depending on their fitness.

Production stage: In the production stage, the worst individual (producer) is updated based on the upper and lower bounds information in the search space and the characteristics of the best individual (decomposer), the step control parameter α is adjusted through linear decrease to achieve search balance, and a new individual with higher exploration potential is generated. This process can be mathematically expressed as:

$$x_1(t+1) = (1-\alpha)x_n(t) + \alpha x_{\text{rand}}(t) \quad (3)$$

$$\alpha = (1 - t/T)r_1 \quad (4)$$

$$x_{\text{rand}} = r(U-L) + L \quad (5)$$

where T is the maximum number of iterations, r_1 is a random number within the range of [0, 1], a is a coefficient used for linear weighting, n is the population size, r is a random vector within the range of [0, 1], U is the upper bound, and L is the lower bound.

Consumption stage: In the consumption stage, consumers obtain "nutrition" from other organisms based on their position in the food chain to improve the solution. In order to avoid complexity in parameter adjustment, a simple, parameter-free random walk strategy with the characteristics of Levy flight is proposed, which is called the consumption factor C. In this stage, different types of consumers are updated according to different rules. Specifically, herbivores are updated according to Equation (7), carnivores are updated according to Equation (8), and omnivores are updated according to Equation (9).

$$C = \frac{1}{2}\frac{v_1}{|v_2|}, v_1 \sim N(0,1), v_2 \sim N(0,1) \quad (6)$$

$$x_i(t+1) = x_i(t) + C.(x_i(t) - x_1(t)), i \in [2, \ldots, n] \quad (7)$$

$$x_i(t+1) = x_i(t) + C.(x_i(t) - x_j(t)), i \in [3, \ldots, n], j = randi([2, i-1]) \quad (8)$$

$$x_i(t+1) = x_i(t) + C \cdot (r_2 \cdot (x_i(t) - x_1(t)) \\ +(1-r_2) \cdot (x_i(t) - x_j(t))), i = 3, \ldots, n, j = rand\ i(\ [2, i-1]) \tag{9}$$

where r_2 denotes uniformly distributed random numbers between 0 and 1.

Decomposition stage: This stage is similar to the process of decomposition of biological remains by decomposers in an ecosystem, and its purpose is to enhance the exploration and exploitation of the neighborhood of the optimal solution. This process can be mathematically expressed as:

$$x_i(t+1) = x_n(t) + D \cdot (e \cdot x_n(t) - h \cdot x_i(t)) \tag{10}$$

$$D = 3u_1, u_1 \sim N(0,1) \tag{11}$$

$$e = r_3 \cdot randi([1 \quad 2]) - 1 \tag{12}$$

$$h = 2 \cdot r_3 - 1 \tag{13}$$

2.3.3. SAEO Algorithm

In this paper, an improved metaheuristic algorithm based on the AEO algorithm, which is called the SAEO algorithm, is proposed to improve the performance of the original AEO algorithm.

Production stage: The original AEO algorithm adjusts the step control parameter α through linear decrement during the production phase to achieve search balance. However, this method has limitations when simulating complex nonlinear ecological evolution [32]. The SAEO algorithm addresses this limitation by introducing a nonlinear function to dynamically update the α value during the development phase. This optimization strategy aims to enhance the algorithm's ability to solve nonlinear problems.

$$\alpha = r_1 \left(1 - \frac{1}{\cos^{-1}(0)} (\cos^{-1}(1 - t/T))\right) \tag{14}$$

Consumption stage: The original AEO algorithm adopts a simple, parameter-less random walk strategy, which lacks flexibility when dealing with complex problems. The SAEO algorithm addresses this limitation by optimizing it. During each iteration, each individual has a 50% chance of evolving according to the basic AEO algorithm. Simultaneously, there is a 50% chance of incorporating global exploration using Levy flight trajectories based on Equation (15) [33]. This optimization strategy aims to enhance the algorithm's global exploration capability.

$$x_i(t+1) = x_i(t) + step(t+1) \cdot (x_i(t) - x_n(t)) \tag{15}$$

where *step* is generated based on the Levy distribution, and the parameter β of this distribution is randomly selected in the range of [0, 2] [34,35]. The related mathematical formulas are given below.

$$s = \frac{u}{|v|^{1/\beta}} \tag{16}$$

$$u \sim N(0, \sigma_u^2), v \sim N(0, \sigma_v^2) \tag{17}$$

$$\sigma_u = \frac{\Gamma(1+\beta) sin \frac{\pi\beta}{2}}{\Gamma\left(\frac{1+\beta}{2}\right) \beta 2^{\frac{\beta-1}{2}}}, \sigma_v = 1 \tag{18}$$

$$\Gamma(1+\beta) = \int_0^\infty \frac{k^\beta}{e} dk \quad (19)$$

$$step(t) = 0.01 \cdot s(t) \cdot uniform(0,1) \quad (20)$$

Decomposition stage: The original AEO algorithm employs a single strategy for evolutionary operations, which may lead to overfitting risks. The SAEO algorithm optimizes this by introducing a dynamic crossover strategy. During each iteration, each individual has a 50% chance of evolving according to the basic AEO algorithm, and there is a 50% chance of adopting the dynamic crossover strategy. This strategy involves randomly selecting an individual, x_R, from the entire population and updating it according to Equation (21). This optimization strategy aims to ensure that the population can continuously optimize and move towards the global optimal state throughout the iterations.

$$CL = \begin{cases} m \cdot x_R + (1-m) \cdot x_I & \text{rand} < 0.5 \\ m \cdot x_I + (1-m) \cdot x_R & \text{rand} \geq 0.5 \end{cases} \quad (21)$$

$$m = \alpha^{max} - \left(\alpha^{max} - \alpha^{min}\right) t/T \quad (22)$$

The operation process of the SAEO algorithm is shown in Figure 2.

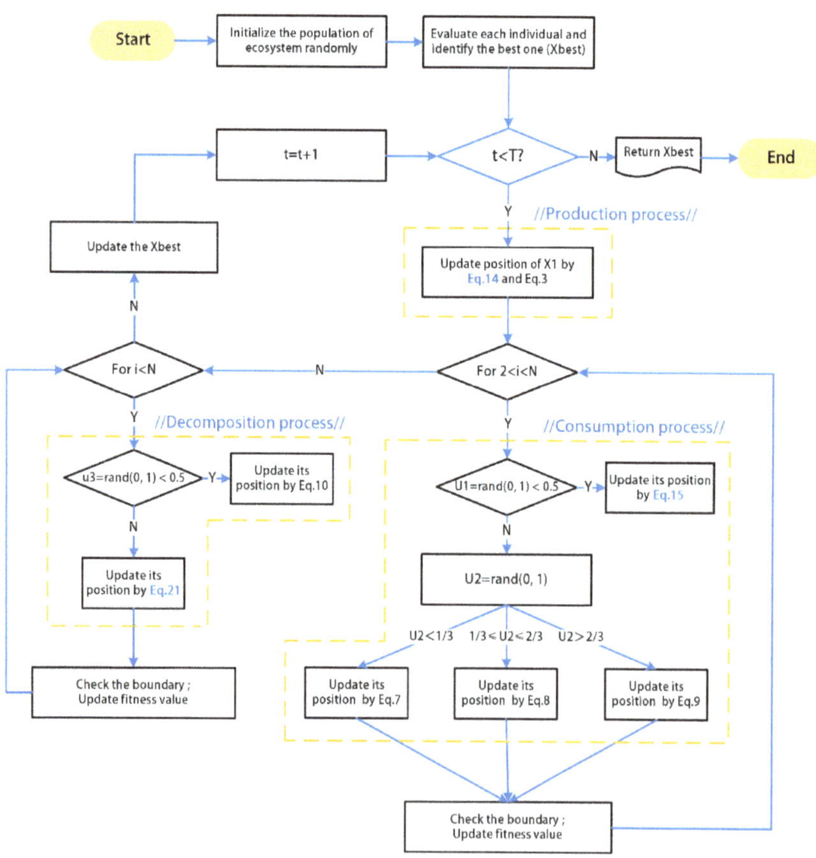

Figure 2. Flowchart of the SAEO algorithm.

2.3.4. Building of the SAEO-BP Neural Network Model

In this study, the Chl-a concentration prediction model was trained using the SAEO algorithm in combination with a BP neural network. The complete process of this model consists of three major links, namely, data preprocessing, model building, and model performance evaluation (Figure 3).

In the data preprocessing stage, a total of 612 valid data records were obtained and divided into a training set and a test set. The training set was used for model building and training, and the test set was used for model evaluation.

In the model building stage, the spectral values of 12 bands in the training set were used as input features, and the SAEO-BP neural network was used to train the Chl-a concentration prediction model. Specifically, the weights and biases in the network were set as decision variables of the optimization problem, and the mean squared error (MSE) for the objective function was set. Subsequently, the SAEO algorithm was used to iteratively optimize these parameters until the preset maximum number of iterations was reached. When the optimal solution was produced, the optimal parameter vector was extracted and used to train the BP neural network. After training was completed, it was applied on the test set to evaluate the performance of the SAEO-BP neural network.

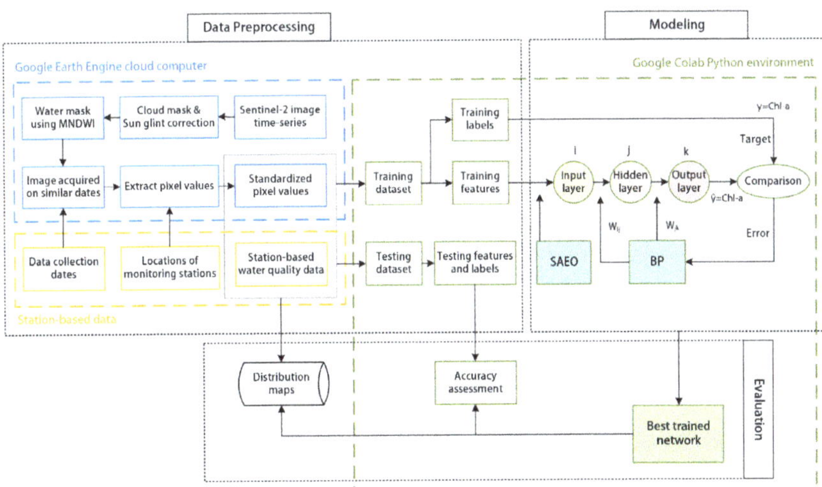

Figure 3. Flowchart for predicting water quality parameters using the SAEO-BP neural network model.

In the model evaluation stage, the trained SAEO-BP neural network was quantitatively evaluated using a series of performance indicators widely used in the field of remote sensing, including the coefficient of determination (R^2), mean squared error (MSE), mean absolute error (MAE), and root mean squared error (RMSE). The mathematical expressions of these indicators are given below.

$$R^2 = 1 - \frac{\sum_{i=1}^{N}(y_i - \hat{y}_i)^2}{\sum_{i=1}^{N}(y_i - \overline{y})^2} \quad (23)$$

$$MSE = \frac{1}{N}\sum_{i=1}^{N}(y_i - \hat{y}_i)^2 \quad (24)$$

$$MAE = \frac{1}{N}\sum_{i=1}^{N}|y_i - \hat{y}_i| \quad (25)$$

$$RMSE = \sqrt{\frac{1}{N}\sum_{i=1}^{N}(y_i - \hat{y}_i)^2} \quad (26)$$

where $\hat{y}_i$ is the predicted value of Chl-a concentration,y_i is the actual (measured) value of Chl-a concentration, $\overline{y}$ is the average of all measured values of Chl-a concentration, and N is the total sample size.

Finally, the Chl-a concentration prediction model was applied on the Sentinel-2 satellite imagery data selected for a particular period, including the data of heavy precipitation events and red tides, to produce spatial distribution maps for Chl-a concentration in the experimental areas. The spatial and temporal variations in Chl-a concentration during particular events were visually analyzed to demonstrate the model's actual performance in application scenarios and its capability to monitor environmental changes.

3. Results

First, the performance of the SAEO-BP neural network model was evaluated. Subsequently, the model was applied in scenarios involving large fluctuations in Chl-a concentration resulting from heavy precipitation events and red tides, the patterns of spatiotemporal variation in Chl-a concentration in the coastal waters of Hong Kong were identified through inversion, and the model's prediction accuracy in scenarios involving dramatic changes was verified.

3.1. *Model Performance Evaluation*

This section focuses on investigating the performance of the SAEO-BP neural network in predicting Chl-a concentration. First, an ablation experiment was conducted to determine the performance of the optimization scheme. Then, the performance of the SAEO-BP neural network was compared with that of the BP neural network optimized using the AEO algorithm (AEO-BP neural network). Finally, 16 representative metaheuristic optimization algorithms were selected for comparison, and the performance of different neural network models in Chl-a concentration prediction was comparatively analyzed in an in-depth manner.

During the experiment, all metaheuristic optimization algorithms to be compared were programmed and implemented using the open source Python library MEALPY 2.4.1 [36] to ensure the consistent execution of different optimization strategies and effective comparison of the results produced.

3.1.1. Ablation Experiment

An ablation experiment was conducted to investigate the feasibility of the AEO algorithm-based optimization scheme and demonstrate the specific meaning of each module. Figure 4 shows the training losses of the BP neural networks optimized using different metaheuristic optimization algorithms, including the original AEO algorithm and AEO algorithms incorporating one or more of the nonlinear weight coefficient, the "Levy flight path" search strategy, and the dynamic crossover mechanism. The experiment showed that when these mechanisms were, respectively, integrated with three algorithms, the balance between global exploration and local exploitation was affected, resulting in training loss fluctuations, slower convergence, and poor convergence performance. When two of these mechanisms were incorporated into one algorithm, the optimization performance was not stable; when these three mechanisms were integrated with the same algorithm, the model's convergence performance tended to become stable, and its training loss was significantly reduced. These results show that each of the three mechanisms plays an important role in optimizing the AEO algorithm.

3.1.2. Comparative Analysis of the Performance of the SAEO-BP and AEO-BP Neural Networks

In this study, the performance of the SAEO-BP and AEO-BP neural networks was evaluated using the following four indicators: R^2, MSE, MAE, and RMSE. All models were trained for 10 times with the same settings (epoch = 500, pop_size = 50), and the average values of all indicators was recorded (Table 2). The experimental results show that the SAEO-BP neural network performs better than the AEO-BP neural network in the training process, and the superior performance of the former is reflected in higher R^2 (0.9214) and lower MSE (1.1201), MAE (0.2774) and RMSE (1.0581). This indicates that the SAEO-BP neural network can better describe the data set and produces smaller prediction errors in the learning process. For the testing process, the R^2 value of the SAEO-BP neural network is 0.7307, which is greater than that of the AEO-BP neural network (0.6828), and the SAEO-BP neural network has the lowest values of MSE, MAE and RMSE, which clearly demonstrates its high prediction accuracy. Figure 5 depicts a line graph of the validation dataset, which includes both the measured values of Chl-a concentration and the predicted values obtained using two models. It is observed that compared to the AEO-BP neural network model, the predictions of the SAEO-BP neural network model exhibit higher overall consistency with the measured values. The SAEO-BP model provides more accurate descriptions of both extreme values and minor fluctuations, enabling it to predict short-term rapid changes in Chl-a concentration more effectively.

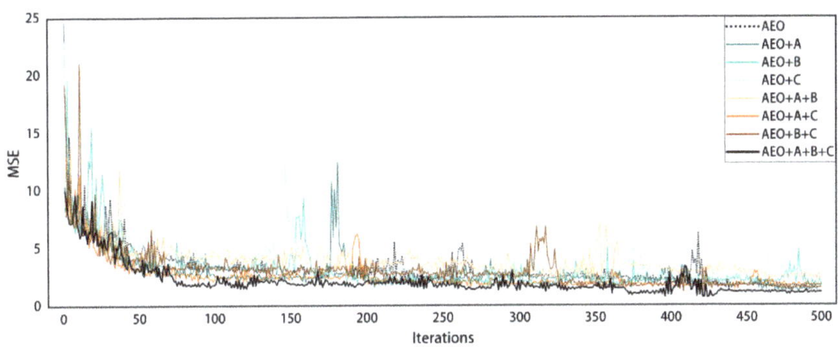

Figure 4. Training losses of BP neural networks optimized using different metaheuristic optimization algorithms. A denotes the adaptive nonlinear weight coefficient, B denotes the "Levy flight" path search strategy, and C denotes the dynamic crossover mechanism. In the legend, "AEO + A" refers to the AEO algorithm incorporating the adaptive nonlinear weight coefficient, and so forth.

Table 2. Average values of performance indicators for the AEO-BP and SAEO-BP neural network models. The unit of Chl-a concentration is µg/L.

Model	Training Set				Test Set			
	R^2	MSE	MAE	RMSE	R^2	MSE	MAE	RMSE
AEO-BP	0.906	1.3384	0.3258	1.1543	0.6828	3.5948	0.9851	1.892
SAEO-BP	0.9214	1.1201	0.2774	1.0581	0.7307	3.0521	0.9185	1.746

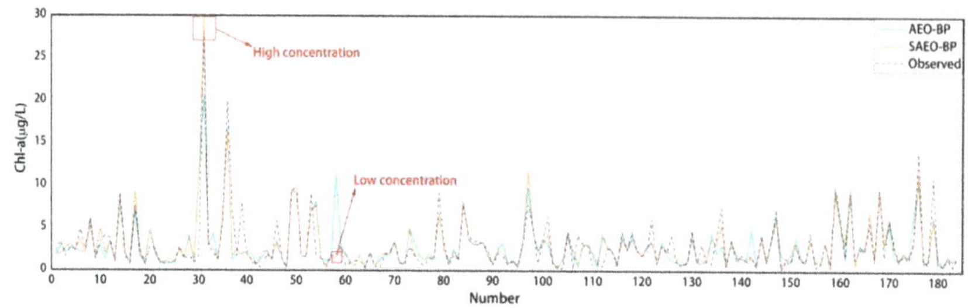

Figure 5. Comparison between the measured and predicted values of Chl-a concentration. AEO-BP neural network and SAEO-BP neural network.

In addition, the results from 10 occasions of training were visually presented in box plots and compared (Figure 6). The comparison results show that the SAEO-BP neural network performs more stably than the AEO-BP neural network.

3.1.3. Comparative Analysis of the Performance of the SAEO-BP Neural Network and Other BP Neural Networks Based on Metaheuristic Optimization Algorithms

Other BP neural network models optimized using 16 metaheuristic optimization algorithms were used to predict Chl-a concentration. All models were trained for 10 times with the same settings (epoch = 500, pop_size = 50), and the average values of all indicators was recorded. The results in Table 3 show that the SAEO-BP neural network has advantages in the training set, and these advantages are further enhanced in the validation set. The SAEO-BP neural network has achieved the best results in terms of R^2 (0.7307), MSE (3.0521), MAE (0.9185), and RMSE (1.746). Figure 7 illustrates the distribution of validation set data after training with the SAEO-BP neural network, along with the fitting lines of validation set data after training with 17 other algorithms, providing visual evidence of the superiority of the SAEO-BP neural network in predicting Chl-a concentration.

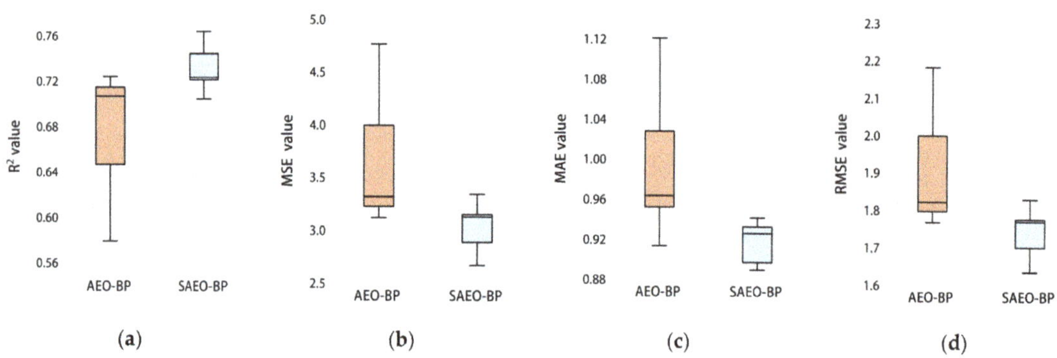

Figure 6. Comparison of the stability of the AEO-BP and SAEO-BP neural networks. Four performance indicators, including R^2 (**a**), MSE (**b**), MAE (**c**) and RMSE (**d**), are considered. The unit of MSE is $(\mu g/L)^2$, while the units of RMSE and MAE are μg/L.

Table 3. Average values of performance indicators for 17 models. Note: SAEO-BP neural network and BP neural networks optimized using other metaheuristic optimization algorithms. The unit of Chl-a concentration is μg/L.

Model	Training Set				Test Set			
	R^2	MSE	MAE	RMSE	R^2	MSE	MAE	RMSE
PSO-BP [37]	0.9068	1.3281	0.3470	1.1455	0.6295	4.1992	1.0367	2.039
GWO-BP [38]	0.9132	1.2361	0.3460	1.1096	0.5313	5.3121	1.0844	2.286
DE-BP [39]	0.9021	1.3947	0.3508	1.1758	0.6756	3.6771	1.0197	1.915
CRO-BP [40]	0.9115	1.2605	0.357	1.1208	0.407	6.7210	1.1778	2.586
TWO-BP [41]	0.9276	1.0319	0.2589	1.0131	0.6624	3.8261	1.0254	1.955
MVO-BP [36]	0.9195	1.1472	0.3242	1.0655	0.6599	3.8550	1.0553	1.954
TLO-BP [42]	0.9249	1.0696	0.3033	1.0257	0.6619	3.8325	1.0343	1.947
LCO-BP [36]	0.9169	1.1835	0.2905	1.0844	0.6579	3.8771	1.0072	1.953
SMA-BP [36]	0.9181	1.1668	0.3213	1.0766	0.5135	5.5148	1.1069	2.336
VCS-BP [36]	0.9203	1.1347	0.3147	1.0625	0.6803	3.6234	1.0159	1.899
SCA-BP [36]	0.9126	1.2451	0.2857	1.1148	0.6484	3.9848	1.0446	1.990
AOA-BP [43]	0.9167	1.1862	0.2851	1.0863	0.6961	3.4441	0.9882	1.851
WCA-BP [44]	0.9127	1.2436	0.3502	1.1145	0.6656	3.7904	1.0468	1.944
EAEO-BP [45]	0.8995	1.4314	0.3161	1.1935	0.6066	4.4589	1.0577	2.099
IAEO-BP [32]	0.8976	1.4586	0.3517	1.2045	0.6909	3.5034	0.9999	1.871
AAEO-BP [46]	0.9096	1.2873	0.3324	1.1313	0.594	4.6015	1.0899	2.136
SAEO-BP	0.9214	1.1201	0.2774	1.0581	0.7307	3.0521	0.9185	1.746

The stability of different models was comparatively analyzed in an all-round way using box plots showing the results of 17 models trained on the test set for 10 times (Figure 8). The results show that, in terms of MAE, the SAEO-BP neural network has the highest stability, followed by the IAEO-BP and VCS-BP neural networks; in terms of the other three performance indicators, the SAEO-BP has a high stability, second only to the IAEO-BP neural network.

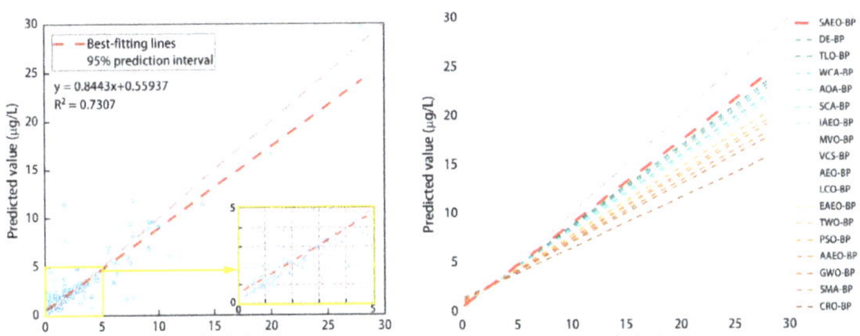

(a) (b)

Figure 7. The relationship between the measured and predicted values of Chl-a concentration in the validation set: (**a**) scatter plot and fitting line under the SAEO-BP neural network algorithm, (**b**) comparison of fitting lines under the SAEO-BP neural network and BP neural network algorithm optimized using other metaheuristic algorithms.

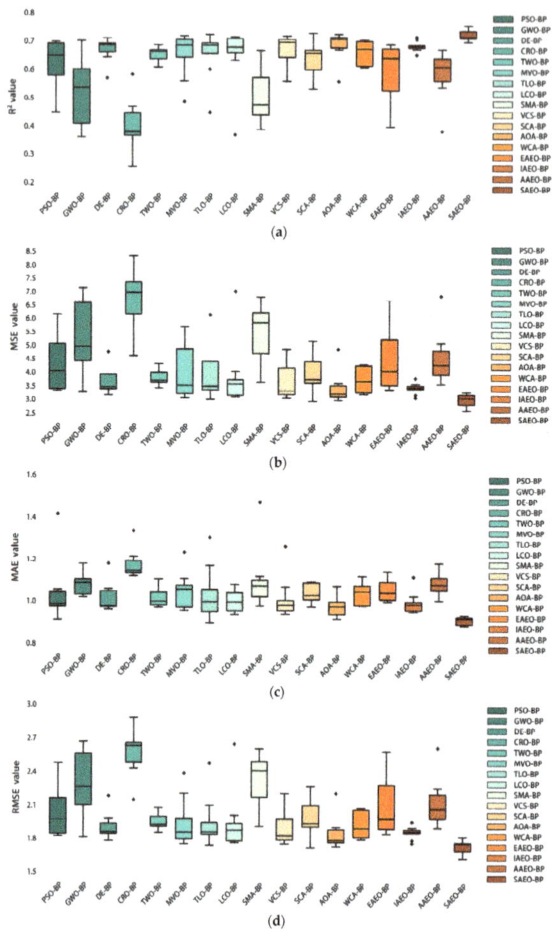

Figure 8. Comparison of the stability of 17 models. Four performance indicators, including R^2 (**a**), MSE (**b**), MAE (**c**) and RMSE (**d**), are considered. The unit of MSE is $(\mu g/L)^2$, while the units of RMSE and MAE are $\mu g/L$.

In summary, the SAEO-BP neural network proposed in this paper performs better than the BP neural networks optimized using other metaheuristic optimization algorithms in terms of prediction accuracy and stability, providing a good solution to predicting Chl-a concentration under complex environmental conditions.

3.2. Investigation of the Inversion Performance of the SAEO-BP Neural Network under Conditions of Significant Changes in Chl-a Concentration

As mentioned earlier, the SAEO-BP neural network model established based on historical data performed well during the evaluation phase. To assess whether the model can accurately predict short-term drastic changes in chlorophyll-a concentration caused by extreme events, this study evaluates the model's prediction results for extreme low values during heavy rainfall periods and extremely high values during red tide outbreaks in the study area. Specifically, predictions were made for the Chl-a concentration during heavy rainfall periods in regions A, B, and C from 6 October 6 to 10 November 2021, as well as during the red tide outbreak period in region D from 8 March to 23 March 2021. By comparing the model's predicted values with the actual scenarios, and assessing the consistency between the maximum and minimum values at the same observation location on different days within the experimental period, the paper aims to determine the model's predictive accuracy.

3.2.1. Analysis of the Accuracy of the SAEO-BP Neural Network and Its Sensitivity to Spatial and Temporal Changes in Chl-a Concentration during Extreme Precipitation Events

According to the research findings of Zhou et al. [25], Chl-a concentration will decrease during typhoons, but in a few days after a typhoon, new conditions favorable for algae growth will be established and Chl-a concentration will increase. In order to verify the sensitivity of the SAEO-BP neural network to the spatial and temporal (dynamic) changes in Chl-a concentration, the remote sensing images for the period in which Hong Kong was affected by tropical cyclones Lionrock and Kompasu (October 2021) were selected for inversion and analysis, with the focus on the spatial and temporal changes in Chl-a concentration in experimental Areas A, B, and C.

In October 2021, affected by the two typhoons mentioned above, the cumulative precipitation in Hong Kong during this period increased significantly and reached 631.1 mm, which is more than five times the average precipitation for the same period in history. According to the climate data provided by the Hong Kong Observatory, after Lionrock made landfall on 7 October 2021, Hong Kong experienced continued heavy rainfall; as the typhoon moved away on 11 October 2021, the precipitation decreased, and the weather cleared up. Kompasu approached Hong Kong on 12 October 2021, resulting in widespread precipitation, and the weather gradually cleared up on 16 October 2021.

Based on these changes in precipitation, two groups of key points in time were selected. The time points in the first group are 6, 11 and 16 October 2021, which were selected to reveal the changes in Chl-a concentration within the study area on the day before the onset of heavy precipitation, during heavy precipitation, and on the day when heavy precipitation ended. The time point in the second group is 10 November 2021, which was selected to observe the changes in Chl-a concentration within a certain period of time after heavy precipitation ended. Since the cumulative precipitation in the period from 16 October to 10 November 2021 is less than 10 mm, its impact on Chl-a concentration in the study area can be ignored.

The study employed the trained SAEO-BP neural network model to retrieve the Chl-a concentration from images taken on different dates in each experimental area (Figure 9). Table 4 lists the mean Chl-a concentration obtained through inversion from each image and the records of the cumulative precipitation in Hong Kong for several five-day periods are given in Table 5. The tabulated data show that the cumulative precipitation in the first five-day period before the onset of heavy precipitation (from 2 October to 6 October 2021) is only 1.9 mm and the average values of Chl-a concentration in the three experimental areas are at levels higher than 7 μg/L. The cumulative precipitation in the five-day period of the first heavy rainfall event (from 7 October to 11 October 2021) reached 549 mm, resulting in abrupt decreases in the average Chl-a concentration in the three experimental areas. The inversion results for 11 October 2021 show that the average Chl-a concentration is 3.90 μg/L in experimental Area A, 3.12 μg/L in experimental Area B, and 1.38 μg/L in experimental Area C. Subsequently, the second heavy rainfall event occurred from 12 October to 16 October 2021, resulting in cumulative precipitation of 75.8 mm in five days and the average values of Chl-a concentration in the three experimental areas decreased again. The average Chl-a concentration as of 16 October 2021 is 3.08 μg/L in experimental Area A, 1.93 μg/L in experimental Area B, and 1.22 μg/L in experimental Area C. As the second heavy rainfall event ended, the Chl-a concentration in the coastal waters began to increase. The average Chl-a concentration as of 10 November 2021 is 4.81 μg/L in experimental Area A, 4.73 μg/L in experimental Area B, and 3.16 μg/L in experimental Area C. These results are consistent with the monitoring results (https://khoyinivan.users.earthengine.app/view/marine-water-quality-hk, accessed on 15 January 2024) [22] and demonstrate that the SAEO-BP neural network model can effectively capture large fluctuations/changes in Chl-a concentration under heavy precipitation conditions.

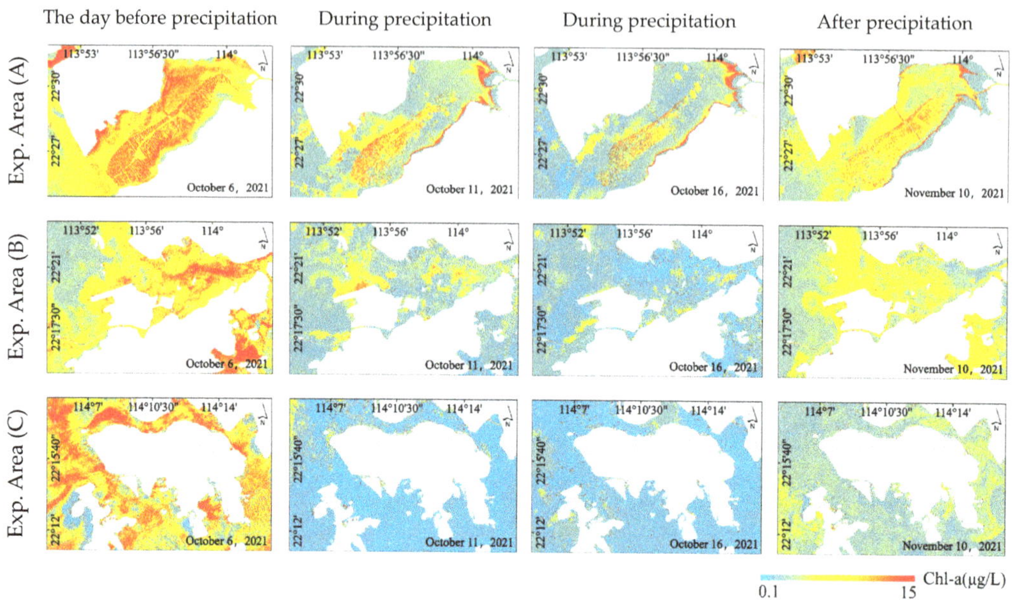

Figure 9. Images of Chl-a concentration in experimental Areas A, B, and C. Three experimental areas (A, B, and C) and four observation dates (6 October 2021, 11 October 2021, 16 October 2021, and 10 November 2021).

Table 4. Average predicted values of Chl-a concentration in various experimental areas.

Study Area	Average Predicted Values of Chl-a Concentration (μg/L)			
	6 October 2021	11 October 2021	16 October 2021	10 November 2021
(A)	**7.02**	3.90	3.08	4.81
(B)	**7.12**	3.12	1.93	4.73
(C)	**7.92**	1.38	1.22	3.16

Table 5. Cumulative precipitation in Kong Hong in five days.

Study Area	Cumulative Precipitation (mm) in Five-Day Periods			
	2 October to 6 October 2021	7 October to 11 October 2021	12 October to 16 October 2021	2 October to 10 November 2021
Hong Kong	1.9	**549**	75.8	2

3.2.2. Simulation of Changes in Chl-a Concentration during Red Tides and Its Spatial Distribution Characteristics Using the SAEO-BP Neural Network Model

In addition, the effectiveness and stability of the SAEO-BP neural network model under complex environmental conditions were further verified by comparing the data acquired during red tides; this study further confirmed the effectiveness and stability of the SAEO-BP neural network model under complex environmental conditions. According to information from the Hong Kong Red Tide Database (https://redtide.afcd.gov.hk/urtin/#/maps, accessed on 15 January 2024), a continuous red tide caused by Gonyaulax polygramma occurred in the Tolo Channel northeast of Hong Kong during the period from 8 March to 23 March 2021.

Chl-a concentration is usually used to represent phytoplankton biomass. In this study, the SAEO-BP neural network model was used to perform the inversion of Chl-a concentration in experimental Area D, and the Chl-a concentration distribution in this area on 17 March and 25 March 2021 was visually displayed (Figure 10) to demonstrate the spatial variations in Chl-a concentration during and after the red tide.

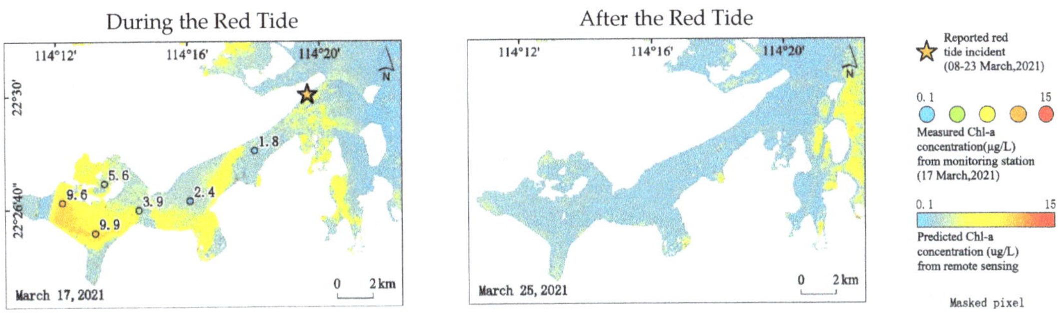

Figure 10. Images of Chl-a concentration in experimental Area D obtained by inversion.

It can be seen that the Chl-a concentration map for 17 March 2021 has accurately captured the distribution characteristics (high levels) of Chl-a concentration in the Tolo Channel on that day, and the predicted values of Chl-a concentration in the surrounding areas are basically consistent with the measured data from the monitoring station on that day. The Chl-a concentration map for 25 March 2021 shows the distribution of Chl-a concentration in the Tolo Channel and surrounding areas after the red tide dissipated. It can be seen that the Chl-a concentration in the Tolo Channel decreased significantly on 25 March 2021. These results show that the SAEO-BP neural network model can effectively reflect water conditions during red tides.

4. Discussion

In this section, we first introduce the results of optimizing the BP neural network using the SAEO algorithm. Secondly, we describe the effectiveness of the Chl-a inversion model constructed using the SAEO-BP neural network model, confirming the advantages of the SAEO-BP neural network model in prediction accuracy and stability. Finally, by comparing it with traditional regression algorithms, simplified physical model algorithms (QAA), and commonly used machine learning algorithms (SVR), we verify the effectiveness of the SAEO algorithm in optimizing the BP neural network model.

4.1. Performance of the SAEO Algorithm in Optimizing BP Neural Networks

In the AEO algorithm, producers are mainly responsible for exploring new possible solutions, consumers further explore the solution space through three mechanisms, and decomposers improve the exploration efficiency by referring to existing solutions. Such configuration enables the AEO algorithm to further expand the scope of search for solutions while maintaining efficient exploration. On this basis, the nonlinear weight coefficient, the "Levy flight" path search strategy, and the dynamic crossover mechanism were incorporated into the AEO algorithm to further improve its performance in optimizing BP neural networks, and an ablation experiment was conducted. The results show that the incorporation of these three mechanisms can effectively balance global exploration and local exploitation, stabilize the loss in the training process (training loss), and thereby improve the convergence rate and performance.

The SAEO algorithm was compared with other 17 metaheuristic optimization algorithms to verify its performance in optimizing BP neural networks. The results show that the BP neural network optimized using the improved AEO algorithm has advantages in terms of prediction accuracy and stability. In addition, these results demonstrate the powerful performance of the original AEO algorithm and prove that the SAEO algorithm can effectively overcome the limitations of BP neural networks, such as their tendency to become stuck in local optima and large fluctuations in accuracy.

4.2. Performance of the Chl-a Concentration Inversion Model Built Based on the SAEO-BP Neural Network

In water environment research, water quality monitoring plays a vital role, and Chl-a concentration is widely recognized as a key measure of water quality. In this study, an innovative optimization algorithm based on the AEO algorithm, namely the SAEO algorithm, was proposed and used to optimize the BP neural network to improve the accuracy of Chl-a concentration prediction. The SAEO-BP neural network model for predicting Chl-a concentration in the coastal waters of Hong Kong was built. The results of comparison and verification show that the SAEO-BP neural network model is more accurate and stable than the neural networks optimized using the AEO algorithm and other metaheuristic optimization algorithms.

By inverting and analyzing the remote sensing data of Hong Kong waters for the period in which Hong Kong was affected by Lionrock and Kompasu (October 2021), previous theories on the impact of typhoons on marine ecosystems were verified. Specifically, heavy precipitation and wind caused drastic changes in the water environment of the study area, significantly enhanced vertical mixing, accelerated the dilution process, and increased the water exchange rate, resulting in rapid short-term (for the period from 6 October to 11 October 2021 and the period from 12 October to 16 October 2021) decreases in Chl-a concentration from high levels (above 7 μg/L) to around 3 μg/L or even lower levels. After typhoons ended, with the decrease in precipitation and the weakening of vertical mixing, a large amount of Pearl River water (runoff) rich in nitrogen-containing nutrient salts flew into the ocean, providing favorable conditions for algae reproduction. Therefore, for a certain period of time (such as 10 November 2021), the average levels of Chl-a concentration in various monitored coastal areas of Hong Kong rose to 4.81 μg/L, 4.73 μg/L, and 3.16 μg/L, respectively.

In addition, a red tide that occurred in the Tolo Channel northeast of Hong Kong in the period from 8 March to 23 March 2021 was investigated. The results of the inversion of remote sensing images show that the Chl-a concentration in the Tolo Channel increased abruptly on 17 March 2021, which is consistent with the occurrence of the red tide in this area, and on 25 March 2021, the Chl-a concentration near the Tolo Channel decreased to a normal level, which is consistent with the reported dissipation of the red tide. These results further verify the accuracy of the SAEO-BP neural network model in predicting Chl-a concentration during the occurrence and dissipation of red tides.

4.3. Validation of the Effectiveness of SAEO Algorithm Optimized BP Neural Network Model

In research related to the inversion of water quality parameters in small samples, traditional algorithmic models are simple to establish and computationally fast, but their accuracy is relatively low. Physical methods have clear mechanisms, but the models are overly complex, requiring numerous parameters, and exhibit unstable prediction accuracy across different water bodies. Machine learning algorithms are suitable for solving nonlinear problems, but finding the optimal parameter settings for the model is challenging, although their accuracy is relatively higher compared to the first two types of algorithms. This paper compares the proposed SAEO-BP neural network model with empirical algorithms, commonly used simplified physical model algorithms (QAA [47]), and the Support Vector Regression (SVR) algorithm (Figure 11), which is widely used in machine learning algorithms, to verify the effectiveness of the SAEO algorithm in optimizing the BP neural network model for data prediction.

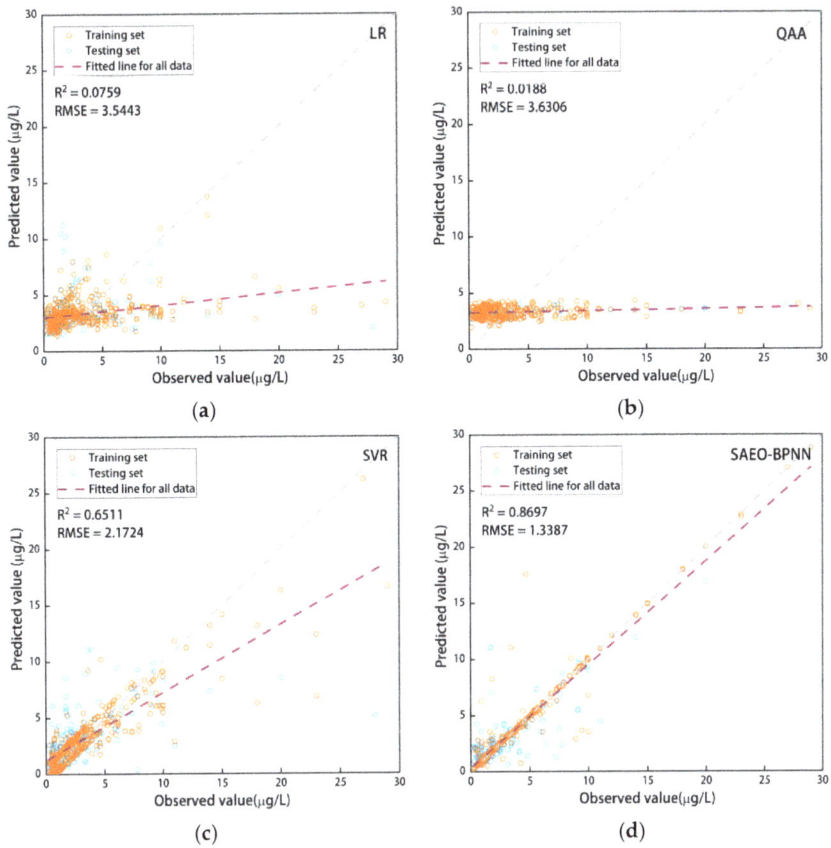

Figure 11. Scatter plots of predicted and measured values for Chl-a concentration from the complete sample dataset are plotted for four models: (**a**) LR model, (**b**) QAA model, (**c**) SVR model, and (**d**) SAEO-BPNN model.

In the experiment, the model trained with the training dataset will be applied to the dataset covering all samples. By comparing the distribution of predicted values with actual measurements and plotting fitting curves, the study found significant differences between the predicted values calculated using the linear regression algorithm and QAA algorithm and the actual measurements, indicating a lower accuracy of the algorithms. Machine learning algorithms demonstrated superiority in predicting accuracy, especially the model based on SAEO-BP neural network, which showed the best predictive performance and significantly outperformed the other three methods, validating the effectiveness of the SAEO-BP neural network algorithm in data prediction.

5. Conclusions

In order to improve the accuracy and stability of the Chl-a concentration prediction model, 5754 valid observation records of Chl-a concentration in the coastal waters of Hong Kong for the period from 2017 to 2022 and 500 high-quality Sentinel-2 remote sensing images for the same period were collected, and 612 pieces of valid data were selected and used for model training and validation. Based on this data set, the levels of Chl-a concentration in the coastal waters of Hong Kong were predicted and analyzed through inversion. The main achievements of this study are summarized below.

1. **Algorithm innovation and model building**: An improved metaheuristic optimization algorithm, namely, the Synthesized Artificial Ecosystem Optimization (SAEO) algorithm, was proposed. First, the algorithm was integrated with the adaptive nonlinear weight coefficient, the "Levy flight" path search strategy, and the dynamic crossover mechanism to overcome its limitations in solving complex problems and to enhance its global search capability. Then, a novel model named the SAEO-BP neural network model was built by combining the SAEO algorithm with a BP neural network and was used for Chl-a concentration prediction.
2. **Performance evaluation and comparative analysis**: The experimental results show that, compared with the AEO-BP neural network model, the SAEO-BP neural network model produced smaller prediction errors in the training and testing stages and performed more stably in the testing stage. In addition, the SAEO-BP neural network model has obvious advantages over 16 BP neural networks optimized using different metaheuristic optimization algorithms in terms of all key performance indicators. The comparative analysis of the stability of these models shows that the SAEO-BP neural network model has the highest stability in terms of MAE and is highly stable in terms of the other three indicators.
3. **Prediction and evaluation of abrupt changes in Chl-a concentration**: The SAEO-BP neural network model was used to identify the patterns of spatiotemporal variation in Chl-a concentration in the coastal waters of Hong Kong during heavy precipitation events and red tides. The results show that the SAEO-BP neural network model can predict the spatiotemporal changes in Chl-a concentration under the influence of heavy precipitation caused by typhoons, identify the spatial variations in Chl-a concentration during and after red tides, and predict abrupt Chl-a concentration changes and extrema in an accurate and stable manner.

There are still some shortcomings in this study. The Chl-a concentration inversion model built using Sentinel-2 remote sensing data still has some inherent limitations. The discontinuity of satellite revisit cycles and cloud cover pose obstacles to obtaining continuous and high-quality remote sensing data. Moreover, although Sentinel-2 satellite remote sensing imagery data provide a rich information basis for Chl-a inversion, the generalizability of algorithms to other marine areas under different geographical conditions, lighting conditions, water characteristics, and other environmental variables still needs to be considered.

Future research should focus on integrating multi-source remote sensing data resources, optimizing atmospheric correction algorithms, and improving the quantity and quality of remote sensing data applicable to coastal water quality monitoring. Additionally, combining field sampling data, considering key water quality parameters such as suspended sediment and nutrient concentration that affect the optical properties of water bodies, thoroughly analyzing the complex relationships between water quality parameters, constructing coupled models, and more accurately describing the water quality status and variation characteristics of coastal waters will provide a more reliable data source for water environment monitoring and management.

Author Contributions: Conceptualization, X.W. and J.C.; methodology, X.W.; software, X.W.; validation, X.W.; formal analysis, X.W. and J.C.; investigation, X.W.; resources, J.C. and M.X.; data curation, X.W.; writing—original draft preparation, X.W.; writing—review and editing, X.W.; visualization, X.W.; supervision, J.C. and M.X.; project administration, X.W. All authors have read and agreed to the published version of the manuscript.

Funding: This research was funded by the National Natural Science Foundation of China under Grant U1906217, 62071491 and the Fundamental Research Funds for the Central Universities 22CX01004A-5, 22CX01004A-4.

Data Availability Statement: The measured water quality data are available from HKEPD (https://www.epd.gov.hk/epd/sc_chi/environmentinhk/water/hkwqrc/waterquality/marine.html) (accessed on 1 August 2023). The surface reflectance datasets of Sentinel-2 are available on GEE (https://earthengine.google.com) (accessed on 1 August 2023). The precipitation data are available from the Hong Kong Observatory. (https://www.weather.gov.hk/tc/wxinfo/pastwx/mws/mws.htm) (accessed on 15 January 2024). Hong Kong red tide data can be obtained through the following link (https://redtide.afcd.gov.hk/urtin/#/maps) (accessed on 15 January 2024).

Acknowledgments: We thank the developers for their tools and the respective agencies that provided the data. We further thank all editors and anonymous reviewers for spending their time working on the manuscript.

Conflicts of Interest: The authors declare no conflicts of interest.

References

1. Gholizadeh, M.H.; Melesse, A.M.; Reddi, L. A Comprehensive Review on Water Quality Parameters Estimation Using Remote Sensing Techniques. *Sensors* **2016**, *16*, 1298. [CrossRef]
2. Poddar, S.; Chacko, N.; Swain, D. Estimation of Chlorophyll-a in Northern Coastal Bay of Bengal Using Landsat-8 OLI and Sentinel-2 MSI Sensors. *Front. Mar. Sci.* **2019**, *6*, 598. [CrossRef]
3. Flores-Anderson, A.I.; Griffin, R.; Dix, M.; Romero-Oliva, C.S.; Ochaeta, G.; Skinner-Alvarado, J.; Ramirez Moran, M.V.; Hernandez, B.; Cherrington, E.; Page, B.; et al. Hyperspectral Satellite Remote Sensing of Water Quality in Lake Atitlán, Guatemala. *Front. Environ. Sci.* **2020**, *8*, 7. [CrossRef]
4. Pizani, F.M.C.; Maillard, P.; Ferreira, A.F.F.; de Amorim, C.C. Estimation of water quality in a reservoir from Sentinel-2 MSI and Landsat-8 OLI sensors. *ISPRS Ann. Photogramm. Remote Sens. Spat. Inf. Sci.* **2020**, *3*, 401–408. [CrossRef]
5. Chen, J.; Chen, S.; Fu, R.; Li, D.; Jiang, H.; Wang, C.; Peng, Y.; Jia, K.; Hicks, B.J. Remote Sensing Big Data for Water Environment Monitoring: Current Status, Challenges, and Future Prospects. *Earth's Future* **2022**, *10*, e2021EF002289. [CrossRef]
6. Wang, X.; Yang, W. Water Quality Monitoring and Evaluation Using Remote Sensing Techniques in China: A Systematic Review. *Ecosyst. Health Sustain.* **2019**, *5*, 47–56. [CrossRef]
7. Sobel, R.S.; Kiaghadi, A.; Rifai, H.S. Modeling Water Quality Impacts from Hurricanes and Extreme Weather Events in Urban Coastal Systems Using Sentinel-2 Spectral Data. *Environ. Monit. Assess.* **2020**, *192*, 307. [CrossRef]
8. Gitelson, A.A.; Dall'Olmo, G.; Moses, W.; Rundquist, D.C.; Barrow, T.; Fisher, T.R.; Gurlin, D.; Holz, J. A Simple Semi-Analytical Model for Remote Estimation of Chlorophyll-a in Turbid Waters: Validation. *Remote Sens. Environ.* **2008**, *112*, 3582–3593. [CrossRef]
9. Yang, Y.; Zhang, X.; Gao, W.; Zhang, Y.; Hou, X. Improving Lake Chlorophyll-a Interpreting Accuracy by Combining Spectral and Texture Features of Remote Sensing. *Environ. Sci. Pollut. Res.* **2023**, *30*, 83628–83642. [CrossRef] [PubMed]
10. O'Reilly, J.E.; Werdell, P.J. Chlorophyll Algorithms for Ocean Color Sensors—OC4, OC5 & OC6. *Remote Sens. Environ.* **2019**, *229*, 32–47. [CrossRef]
11. Hu, C.; Feng, L.; Lee, Z.; Franz, B.A.; Bailey, S.W.; Werdell, P.J.; Proctor, C.W. Improving Satellite Global Chlorophyll a Data Products Through Algorithm Refinement and Data Recovery. *J. Geophys. Res. Ocean.* **2019**, *124*, 1524–1543. [CrossRef]
12. Rodríguez-López, L.; Duran-Llacer, I.; González-Rodríguez, L.; Abarca-del-Rio, R.; Cárdenas, R.; Parra, O.; Martínez-Retureta, R.; Urrutia, R. Spectral Analysis Using LANDSAT Images to Monitor the Chlorophyll-a Concentration in Lake Laja in Chile. *Ecol. Inform.* **2020**, *60*, 101183. [CrossRef]
13. Hafeez, S.; Wong, M.S.; Ho, H.C.; Nazeer, M.; Nichol, J.; Abbas, S.; Tang, D.; Lee, K.H.; Pun, L. Comparison of Machine Learning Algorithms for Retrieval of Water Quality Indicators in Case-II Waters: A Case Study of Hong Kong. *Remote Sens.* **2019**, *11*, 617. [CrossRef]
14. Kwong, I.H.Y.; Wong, F.K.K.; Fung, T. Automatic Mapping and Monitoring of Marine Water Quality Parameters in Hong Kong Using Sentinel-2 Image Time-Series and Google Earth Engine Cloud Computing. *Front. Mar. Sci.* **2022**, *9*, 871470. [CrossRef]
15. Zhu, W.-D.; Qian, C.-Y.; He, N.-Y.; Kong, Y.-X.; Zou, Z.-Y.; Li, Y.-W. Research on Chlorophyll-a Concentration Retrieval Based on BP Neural Network Model-Case Study of Dianshan Lake, China. *Sustainability* **2022**, *14*, 8894. [CrossRef]
16. Jiang, B.; Liu, H.; Xing, Q.; Cai, J.; Zheng, X.; Li, L.; Liu, S.; Zheng, Z.; Xu, H.; Meng, L. Evaluating Traditional Empirical Models and BPNN Models in Monitoring the Concentrations of Chlorophyll-A and Total Suspended Particulate of Eutrophic and Turbid Waters. *Water* **2021**, *13*, 650. [CrossRef]
17. Li, Z.; Zhao, X. BP Artificial Neural Network Based Wave Front Correction for Sensor-Less Free Space Optics Communication. *Opt. Commun.* **2017**, *385*, 219–228. [CrossRef]
18. Zheng, B.-H. Material Procedure Quality Forecast Based on Genetic BP Neural Network. *Mod. Phys. Lett. B* **2017**, *31*, 1740080. [CrossRef]

19. Pyo, J.; Duan, H.; Ligaray, M.; Kim, M.; Baek, S.; Kwon, Y.S.; Lee, H.; Kang, T.; Kim, K.; Cha, Y.; et al. An Integrative Remote Sensing Application of Stacked Autoencoder for Atmospheric Correction and Cyanobacteria Estimation Using Hyperspectral Imagery. *Remote Sens.* **2020**, *12*, 1073. [CrossRef]
20. Deng, Y.; Xiao, H.; Xu, J.; Wang, H. Prediction Model of PSO-BP Neural Network on Coliform Amount in Special Food. *Saudi J. Biol. Sci.* **2019**, *26*, 1154–1160. [CrossRef]
21. Zhu, C.; Zhang, J.; Liu, Y.; Ma, D.; Li, M.; Xiang, B. Comparison of GA-BP and PSO-BP Neural Network Models with Initial BP Model for Rainfall-Induced Landslides Risk Assessment in Regional Scale: A Case Study in Sichuan, China. *Nat. Hazards* **2020**, *100*, 173–204. [CrossRef]
22. Huang, Y.; Xiang, Y.; Zhao, R.; Cheng, Z. Air Quality Prediction Using Improved PSO-BP Neural Network. *IEEE Access* **2020**, *8*, 99346–99353. [CrossRef]
23. Zhao, W.; Wang, L.; Zhang, Z. Artificial Ecosystem-Based Optimization: A Novel Nature-Inspired Meta-Heuristic Algorithm. *Neural Comput. Appl.* **2020**, *32*, 9383–9425. [CrossRef]
24. Li, J.; Peng, B.; Wei, Y.; Ye, H. Accurate Extraction of Surface Water in Complex Environment Based on Google Earth Engine and Sentinel-2. *PLoS ONE* **2021**, *16*, e0253209. [CrossRef]
25. Zhou, W.; Yin, K.; Harrison, P.J.; Lee, J.H.W. The Influence of Late Summer Typhoons and High River Discharge on Water Quality in Hong Kong Waters. *Estuar. Coast. Shelf Sci.* **2012**, *111*, 35–47. [CrossRef]
26. Xu, J.; Yin, K.; Lee, J.H.W.; Liu, H.; Ho, A.Y.T.; Yuan, X.; Harrison, P.J. Long-Term and Seasonal Changes in Nutrients, Phytoplankton Biomass, and Dissolved Oxygen in Deep Bay, Hong Kong. *Estuaries Coasts* **2010**, *33*, 399–416. [CrossRef]
27. Deng, T.; Duan, H.-F.; Keramat, A. Spatiotemporal Characterization and Forecasting of Coastal Water Quality in the Semi-Enclosed Tolo Harbour Based on Machine Learning and EKC Analysis. *Eng. Appl. Comput. Fluid Mech.* **2022**, *16*, 694–712. [CrossRef]
28. Kay, S.; Hedley, J.D.; Lavender, S. Sun Glint Correction of High and Low Spatial Resolution Images of Aquatic Scenes: A Review of Methods for Visible and Near-Infrared Wavelengths. *Remote Sens.* **2009**, *1*, 697–730. [CrossRef]
29. Kuhn, C.; De Matos Valerio, A.; Ward, N.; Loken, L.; Sawakuchi, H.O.; Kampel, M.; Richey, J.; Stadler, P.; Crawford, J.; Striegl, R.; et al. Performance of Landsat-8 and Sentinel-2 Surface Reflectance Products for River Remote Sensing Retrievals of Chlorophyll-a and Turbidity. *Remote Sens. Environ.* **2019**, *224*, 104–118. [CrossRef]
30. Xu, H. Modification of Normalised Difference Water Index (NDWI) to Enhance Open Water Features in Remotely Sensed Imagery. *Int. J. Remote Sens.* **2006**, *27*, 3025–3033. [CrossRef]
31. Azwar; Rashid, M.M.; Hussain, M.A. Design of AI Neural Network Based Controller for Controlling Dissolved Oxygen Concentration in a Sequencing Batch Reactor. *Int. J. Knowl.-Based and Intell. Eng. Syst.* **2008**, *12*, 121–136. [CrossRef]
32. Rizk-Allah, R.M.; El-Fergany, A.A. Artificial Ecosystem Optimizer for Parameters Identification of Proton Exchange Membrane Fuel Cells Model. *Int. J. Hydrog. Energy* **2021**, *46*, 37612–37627. [CrossRef]
33. Chegini, S.N.; Bagheri, A.; Najafi, F. PSOSCALF: A New Hybrid PSO Based on Sine Cosine Algorithm and Levy Flight for Solving Optimization Problems. *Appl. Soft Comput.* **2018**, *73*, 697–726. [CrossRef]
34. Amirsadri, S.; Mousavirad, S.J.; Ebrahimpour-Komleh, H. A Levy Flight-Based Grey Wolf Optimizer Combined with Back-Propagation Algorithm for Neural Network Training. *Neural Comput. Appl.* **2018**, *30*, 3707–3720. [CrossRef]
35. Jensi, R.; Jiji, G.W. An Enhanced Particle Swarm Optimization with Levy Flight for Global Optimization. *Appl. Soft Comput.* **2016**, *43*, 248–261. [CrossRef]
36. Van Thieu, N.; Mirjalili, S. MEALPY: An Open-Source Library for Latest Meta-Heuristic Algorithms in Python. *J. Syst. Archit.* **2023**, *139*, 102871. [CrossRef]
37. Ghasemi, M.; Akbari, E.; Rahimnejad, A.; Razavi, S.E.; Ghavidel, S.; Li, L. Phasor Particle Swarm Optimization: A Simple and Efficient Variant of PSO. *Soft Comput.* **2019**, *23*, 9701–9718. [CrossRef]
38. Obadina, O.O.; Thaha, M.A.; Althoefer, K.; Shaheed, M.H. Dynamic Characterization of a Master–Slave Robotic Manipulator Using a Hybrid Grey Wolf–Whale Optimization Algorithm. *J. Vib. Control* **2022**, *28*, 1992–2003. [CrossRef]
39. Zhang, J.; Sanderson, A.C. JADE: Adaptive Differential Evolution with Optional External Archive. *IEEE Trans. Evol. Comput.* **2009**, *13*, 945–958. [CrossRef]
40. Nguyen, T.; Nguyen, T.; Nguyen, B.M.; Nguyen, G. Efficient Time-Series Forecasting Using Neural Network and Opposition-Based Coral Reefs Optimization. *Int. J. Comput. Intell. Syst.* **2019**, *12*, 1144–1161. [CrossRef]
41. Kaveh, A.; Almasi, P.; Khodagholi, A. Optimum Design of Castellated Beams Using Four Recently Developed Meta-Heuristic Algorithms. *Iran. J. Sci. Technol. Trans. Civ. Eng.* **2023**, *47*, 713–725. [CrossRef]
42. Rao, R.; Patel, V. An Elitist Teaching-Learning-Based Optimization Algorithm for Solving Complex Constrained Optimization Problems. *Int. J. Ind. Eng. Comput.* **2012**, *3*, 535–560. [CrossRef]
43. Abualigah, L.; Diabat, A.; Mirjalili, S.; Abd Elaziz, M.; Gandomi, A.H. The Arithmetic Optimization Algorithm. *Comput. Methods Appl. Mech. Eng.* **2021**, *376*, 113609. [CrossRef]
44. Eskandar, H.; Sadollah, A.; Bahreininejad, A.; Hamdi, M. Water Cycle Algorithm—A Novel Metaheuristic Optimization Method for Solving Constrained Engineering Optimization Problems. *Comput. Struct.* **2012**, *110–111*, 151–166. [CrossRef]
45. Eid, A.; Kamel, S.; Korashy, A.; Khurshaid, T. An Enhanced Artificial Ecosystem-Based Optimization for Optimal Allocation of Multiple Distributed Generations. *IEEE Access* **2020**, *8*, 178493–178513. [CrossRef]

46. Van Thieu, N.; Deb Barma, S.; Van Lam, T.; Kisi, O.; Mahesha, A. Groundwater Level Modeling Using Augmented Artificial Ecosystem Optimization. *J. Hydrol.* **2023**, *617*, 129034. [CrossRef]
47. Lee, Z.; Shang, S.; Hu, C.; Du, K.; Weidemann, A.; Hou, W.; Lin, J.; Lin, G. Secchi Disk Depth: A New Theory and Mechanistic Model for Underwater Visibility. *Remote Sens. Environ.* **2015**, *169*, 139–149. [CrossRef]

remote sensing

MDPI

Article

Deep-Learning-Based Automatic Extraction of Aquatic Vegetation from Sentinel-2 Images—A Case Study of Lake Honghu

Hangyu Gao [1,2], Ruren Li [1], Qian Shen [2,3,4,*], Yue Yao [2,3], Yifan Shao [2,3], Yuting Zhou [2,3], Wenxin Li [2], Jinzhi Li [2], Yuting Zhang [2,3,4] and Mingxia Liu [5]

1 School of Transportation and Geomatics Engineering, Shenyang Jianzhu University, Shenyang 110168, China; ghy1415@stu.sjzu.edu.cn (H.G.); cerrli@sjzu.edu.cn (R.L.)
2 Key Laboratory of Digital Earth Science, Aerospace Information Research Institute, Chinese Academy of Sciences, Beijing 100094, China; yaoyue@aircas.ac.cn (Y.Y.); shaoyifan23@mails.ucas.ac.cn (Y.S.); 2020210404@jou.edu.cn (Y.Z.); li_599@stu.xust.edu.cn (W.L.); 2020211444@jsnu.edu.cn (J.L.); zhangyuting211@mails.ucas.ac.cn (Y.Z.)
3 International Research Center of Big Data for Sustainable Development Goals, Beijing 100094, China
4 University of Chinese Academy of Sciences, Beijing 100049, China
5 Department of Rail Transportation Affairs, Shenyang Center for Urban and Rural Construction, Shenyang 110168, China; gdb-scxjszx@mail.shenyang.gov.com
* Correspondence: shenqian@radi.ac.cn

Abstract: Aquatic vegetation is an important component of aquatic ecosystems; therefore, the classification and mapping of aquatic vegetation is an important aspect of lake management. Currently, the decision tree (DT) classification method based on spectral indices has been widely used in the extraction of aquatic vegetation data, but the disadvantage of this method is that it is difficult to fix the threshold value, which, in turn, affects the automatic classification effect. In this study, Sentinel-2 MSI data were used to produce a sample set (about 930 samples) of aquatic vegetation in four inland lakes (Lake Taihu, Lake Caohai, Lake Honghu, and Lake Dongtinghu) using the visual interpretation method, including emergent, floating-leaved, and submerged vegetation. Based on this sample set, a DL model (Res-U-Net) was used to train an automatic aquatic vegetation extraction model. The DL model achieved a higher overall accuracy, relevant error, and kappa coefficient (90%, 8.18%, and 0.86, respectively) compared to the DT method (79%, 23.07%, and 0.77) and random forest (78%,10.62% and 0.77) when utilizing visual interpretation results as the ground truth. When utilizing measured point data as the ground truth, the DL model exhibited accuracies of 59%, 78%, and 91% for submerged, floating-leaved, and emergent vegetation, respectively. In addition, the model still maintains good recognition in the presence of clouds with the influence of water bloom. When applying the model to Lake Honghu from January 2017 to October 2023, the obtained temporal variation patterns in the aquatic vegetation were consistent with other studies. The study in this paper shows that the proposed DL model has good application potential for extracting aquatic vegetation data.

Keywords: aquatic vegetation; deep learning; automatic extraction; Lake Honghu; Sentinel-2 MSI

1. Introduction

Aquatic vegetation is an important part of the ecosystem and plays an important role in climate regulation, carbon sequestration and emission reduction, the purification of water bodies, the protection of biodiversity, and the maintenance of ecosystem balance [1,2]. According to their morphological characteristics and ecological habits, species of aquatic vegetation can be generally categorized into three types—emergent vegetation, floating-leaved vegetation, and submerged vegetation—and are widely distributed in coastal and inland lakes [3].

Compared with traditional methods, remote sensing technology has the advantages of timeliness, low cost, fast monitoring speeds, and a wide range and has become an important means of monitoring aquatic vegetation [4–6]. The apparent spectral reflectance of vegetation is influenced by a combination of the physiological characteristics of the plant itself and the environment in which it grows [7]. Aquatic vegetation in remotely sensed images can be classified and extracted based on its reflectance and combined with texture features. Currently, the remote sensing techniques used for monitoring aquatic vegetation are mainly based on methods such as unsupervised classification [8,9], supervised classification [10,11], decision trees [12,13], and bio-optical models [14]. There are also intelligent methods that utilize machine learning [15,16], such as support vector machines [11,17], random forests [18,19], and deep learning [20,21]; among them, the decision tree classification method is the most mainstream approach for classifying aquatic vegetation. Meanwhile, deep learning is increasingly being used for the classification of aquatic vegetation because of its high accuracy, ease of use, and lower susceptibility to noise. With the continuous improvements in computer hardware and software, the increase in the availability of remote sensing image data provides a solid foundation for deep learning in the field of aquatic vegetation extraction.

The decision tree classification method has been the mainstream method for aquatic vegetation classification for a long time due to its simplicity, efficiency, and ease of expression [6]. This method analyzes the spectral characteristics of different types of vegetation, establishes spectral indices to construct a decision tree, and adjusts the thresholds in order to achieve aquatic vegetation classification [22–24]. However, determining a suitable threshold for the automatic extraction of aquatic vegetation has been a challenging problem due to a variety of factors such as the shooting time, shooting angle, weather conditions, water turbidity, and water surface conditions [25]. Typically, false-color composite images are used to determine the optimal segmentation thresholds [26], but this method is subject to large errors due to subjective judgments and is time consuming when performing long-time-series monitoring. In order to reduce humans' subjective influence and achieve fast extraction, researchers usually use fixed-threshold methods. For example, Oyama [27] and Luo et al. [28] classified and extracted the aquatic vegetation from their selected regions using fixed-threshold approaches; however, significant errors are still produced in the practical application of these methods. Therefore, researchers have attempted to automatically obtain thresholds for different images using various methods, such as gradient descent [29], bimodal thresholding [19], and the Otsu method [30]. In addition, researchers have made use of bio-optical models to extract aquatic vegetation data. This type of model simulates the transmission process of light in aquatic plants and their surrounding media, and analyzes the influence of environmental factors, such as water depth, water quality, and plant biomass, on the spectral characteristics of aquatic plants, thereby improving the accuracy of remote sensing monitoring. For example, Rotta et al. [31] revealed the relationship between submerged plants and light radiation through field investigation, remote sensing technology, and a bio-optical model and provided valuable information for water management. However, these methods have limitations and are unable to address all of the challenges of image thresholding when confronted with images such as clouds, fog, and shadows, resulting in low accuracy in the results.

Researchers are much less interested in aquatic vegetation in lakes compared to marine plants [32] and wetland vegetation [33,34]. In addition, marine vegetation is located in a body of water that is clearer than inland lakes, making the extraction of marine vegetation less difficult than that of aquatic vegetation in inland lakes. The remote sensing characteristics of wetland vegetation are very similar to those of water-supported vegetation, so the existing remote sensing extraction methods for wetland vegetation can generally be directly migrated to water-supported vegetation extraction. However, floating-leaved vegetation floats on the surface of the water and is subject to the neighboring effect of the water surface, while submerged vegetation is completely immersed within the water body. The reflectance of these two types of aquatic vegetation is quite different from the performance characteristics of wetland vegetation/terrestrial vegetation on remote sensing images, and the extraction of these two types of aquatic vegetation via remote sensing is difficult. For this reason, the existing remote sensing extraction methods used for wetland vegetation cannot be applied. Zhang et al. [35], Wang et al. [36] classified and extracted emergent vegetation, floating-leaved foliage vegetation, and submerged vegetation in the study area.

Obviously, the extraction of aquatic vegetation data from inland lakes is more complex. Some scholars try to use artificial intelligence methods, such as machine learning algorithms and deep learning algorithms, to extract aquatic vegetation. Machine learning is a broad approach to artificial intelligence that covers various techniques for learning patterns and regularities from data to make predictions or decisions. Some researchers try to use machine learning to classify and extract aquatic vegetation. For example, Huber et al. [32] conducted large-scale monitoring of aquatic vegetation in Sweden through machine learning algorithms. However, due to the weak generalization ability and feature learning ability of the machine learning model, the existing research has not applied the model to multiple regions. At the same time, machine learning models performed relatively poorly on complex tasks, and the researchers did not assess the robustness of their respective models under disturbed conditions (e.g., clouds, water blooms). Deep learning, as a subfield of machine learning, focuses on using deep neural networks to learn complex features and representations, and although it requires more computational resources, it has more powerful feature learning capabilities than traditional machine learning methods. For example, Zac Yung-Chun Liu et al. [37] used deep learning technology to conduct the semantic segmentation of satellite images and identify freshwater vegetation in Senegal. Chen et al. [38] extracted aquatic vegetation in Heilongjiang Basin in China through the DeepPlabv 3+ network. However, the above studies did not perform a more detailed division of the types of aquatic vegetation. In summary, due to the difficulty in distinguishing aquatic vegetation, the low attention, and the high computational resource requirements of deep learning methods, there are few studies on the classification extraction of aquatic vegetation in inland lakes based on deep learning.

In this study, with the goal of developing a set of widely applicable and robust automatic classification and extraction models for aquatic vegetation, we propose the use of Sentinel-2 satellite Multispectral Imager(MSI) launched by the European Space Agency (ESA) as the remote sensing data source and construct a set of sample datasets containing three types of aquatic vegetation, covering the image data from four inland lakes (Lake Taihu, Lake Caohai, Lake Honghu, and Poyanghu Lake) in different growing seasons. We use these datasets to train a model based on a deep learning semantic partitioning network (Res-U-Net), apply the model to the extraction of three types of aquatic vegetation data, and share the results. In addition, we use the model to test images that contain the effects of clouds and water blooms to explore the applicability of the model under complex conditions.

2. Study Area and Data

2.1. Study Area

The lakes concerned in this study are shown in Figure 1: Lake Honghu (39°29′~30° 12′N, 113°7′–114°5′E), Lake Taihu (30°55′~31°32′N, 119°52′~120°36′E), Lake Caohai (26°49′N~53′, 104°12′~104°18′E), and Lake Dongtinghu (27°39′~29°51′N, 111°19′~113°34′E). These four lakes belong to the Yangtze River system, are rich in ecological resources, play important ecological, economic, and social roles, and need continuous attention and protection.

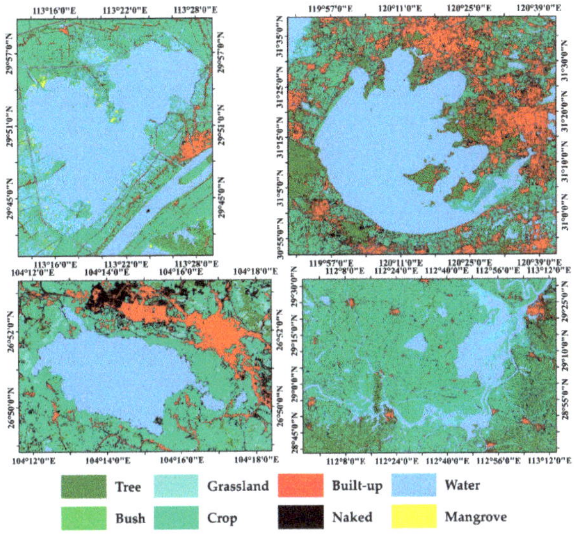

Figure 1. Land use of the study area. Downloaded from ESA.

2.2. Data

2.2.1. Remote Sensing Image Data

The data source for this study is the Sentinel-2 imagery data from the European Space Agency (ESA) (https://scihub.copernicus.eu/dhus/#/home, accessed on 15 October 2023), which provides multiband high-resolution Earth observation data in 13 bands. In addition to the visible, near-infrared, and short-wave infrared bands common to optical remote sensing satellites, the Sentinel-2 image data contain three "red-edge" bands used in vegetation monitoring and an enhanced narrow near-infrared band. Sentinel-2's satellites make observations of a specific area every five days. The spatial resolution of each band varies from 10 to 60 m, but can be uniformly upgraded to 10 m after resampling. The higher spatial resolution can clearly show the distribution and texture characteristics of different types of aquatic vegetation and is particularly suitable for monitoring large lakes. It has sufficient potential for aquatic vegetation data extraction.

Some of the raw data used in this study are geometrically corrected L1C-level top-of-atmosphere reflectance data only, which need to be atmospherically corrected using the Sen2corV02.08.00 model provided by ESA to obtain L2A-level surface reflectance data. The nearest neighbor interpolation method was used to resample the L2A level data to 10 m in SNAP 5.0 software, and the resampled L2A level data were spliced, band-synthesized, and cropped with the support of SNAP5.0 software and ENVI5.3 software.

2.2.2. Measured Data

The remote sensing classification and validation of aquatic vegetation requires reference to ground-truthed spectral information and sample point location information. The measured spectral data for aquatic vegetation and water bodies used in this study, as well as the location information of the sample sites, were obtained from four field observation experiments: the April 2023 Lake Dongtinghu and Poyang Lake experiment, the August 2023 Lake Honghu experiment, the September 2023 Lake Taihu experiment, and the October 2023 Caohai experiment. Data were obtained for 178 locations, including 29 submerged locations, 23 floating-leaved locations, 97 emergent locations, and 29 water locations. A total of 80 valid spectral data were collected. Each point was delineated in an area of 10 m * 10 m, and an experienced staff member was used to make a comprehensive on-site judgment of the coverage of aquatic vegetation, including the presence of four types of aquatic vegetation, in ranges of (0–25%), (25–50%), (50–75%), and (75–100%).

The spectra of the water bodies and aquatic vegetation were measured using an ASD portable spectrometer (wavelength range: 350 nm to 2500 nm) using the abovementioned surface spectroscopic measurements (ASD) [39]. The upward water column radiation (L_u), and gray-slab radiation (L_{blank}) are measured uniformly to facilitate reader understanding and be consistent with the atmospheric correction evaluation and model data input. The measured results are calculated via Formula (1) as the surface reflectance of the target. Affected by the measurement conditions, after removing the invalid data, the surface reflectance curves of the four types of target objects are shown in Figure 2.

$$SR = \frac{L_u - P_{blank}}{L_{blank}} \tag{1}$$

where p_f is the Fresnel reflection of the water surface, which is usually taken to be 0.022 for calm water surfaces, and P_{blank} in Equation (1) is the standard reflectance of the whiteboard, which is provided by its manufacturer.

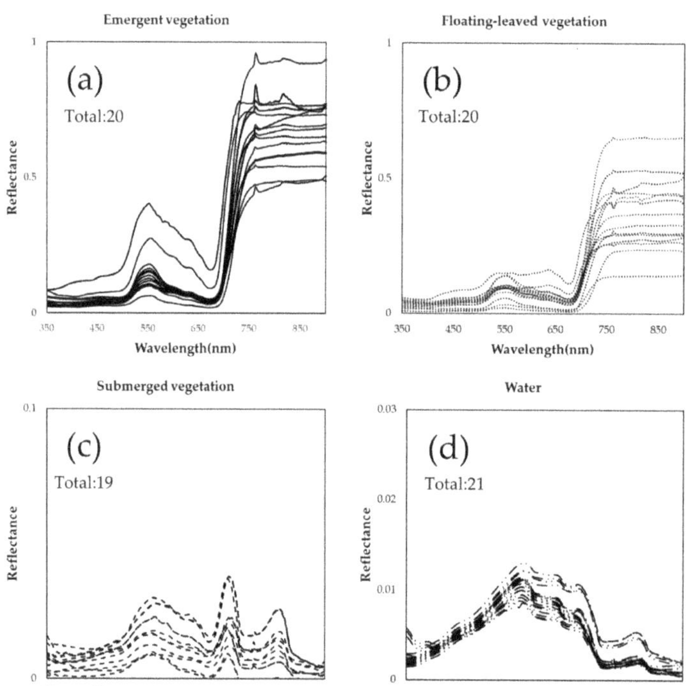

Figure 2. Different types of aquatic vegetation with their spectral characteristics: (**a**) emergent vegetation; (**b**) floating-leaved vegetation; (**c**) submerged vegetation; (**d**) water.

3. Method

3.1. Overall Technical Process

The workflow of this study is shown in Figure 3. First, the Sentinel-2 MSI raw data were preprocessed to obtain surface reflectance data with a 10 m resolution. Then, cloud detection was carried out using the visual interpretation method to exclude the influence of the cloud level and mask the water body. By combining the spectral index and visual interpretation, the aquatic vegetation data in each image were accurately extracted to construct the aquatic vegetation dataset. Then, the images were cut into 512 × 512 sub-images, and the dataset was divided into a training set and a validation set according to the ratio of 8:2; these sets were used for the training of the deep learning network model. Finally, based on the validation set, the precision of the prediction results of the deep learning network model was evaluated to verify its performance and effectiveness.

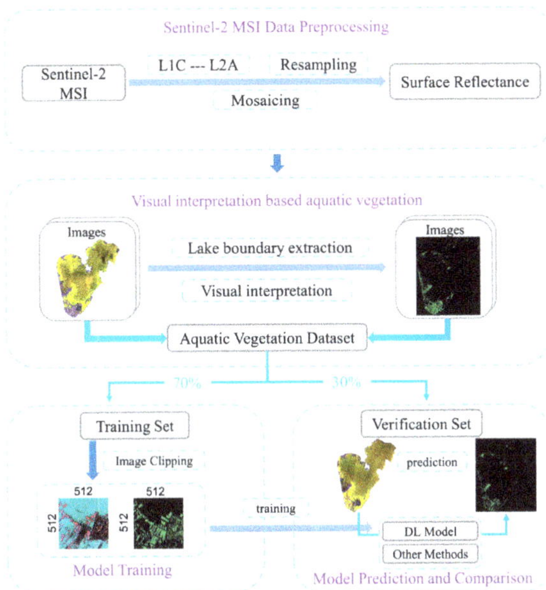

Figure 3. Overall workflow of the aquatic vegetation data extraction process.

3.2. Sentinel-2 MSI Data Preprocessing

In order to obtain band surface spectral reflectance data at a 10 m spatial resolution, we preprocessed the raw Sentinel-2 MSI data: first, atmospheric corrections were performed on the L1C data; second, band resampling and image stitching were conducted on the L2A data.

3.2.1. Atmospheric Correction of Sentinel-2 MSI Data

Since the European Space Agency has only provided L1C-class products for data up to 2018, we performed an atmospheric correction process on these early L1C-class data. This correction process was implemented using the official Sen2Cor plugin, which relies on the LlibRadtran Laboratory's theoretical model of atmospheric radiative transfer, as well as the parameter information associated with the MSI sensor, based on the L2A_Process command line tool. The reflectance data for the atmosphere and ground objects collected in the field were compared with the remote sensing images. Since not all of the band data from the Sentinel-2 remote sensing data were required in this study, only the accuracy of the atmospheric corrections from Band 1 to Band 9 was evaluated.

3.2.2. Lake Boundary Data Extraction

Since terrestrial vegetation and emergent vegetation show a high similarity in terms of spectral features, this similarity is very close in remote sensing images, which can easily lead to the misclassification of the two types of vegetation. In order to facilitate the subsequent extraction of aquatic vegetation data, we delineated the boundary of the study area. Based on the lake shoreline boundary range, the lake water range, and aquatic vegetation growth area in the current high-spatial-resolution remote sensing images, the islands, buildings, farmland, fish ponds, and other elements in the lake area in the images were maximally eliminated in order to obtain a more complete representation of the lake watershed covering the area of aquatic vegetation growth, as shown in Figure 4 [40].

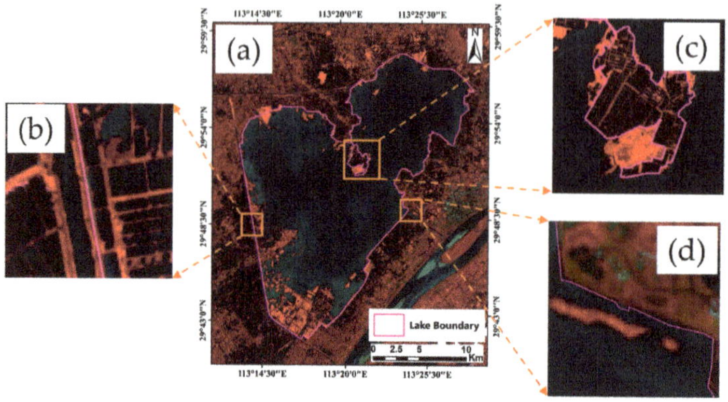

Figure 4. The boundary of Lake Honghu. (**a**) The finalized extent of the lake; (**b**) using the inner side of the surrounding canals as the boundary in the absence of a natural shoreline; (**c**) excluding the islands and paddy fields in the lake; (**d**) the natural shoreline.

3.2.3. Visual Interpretation of Aquatic Vegetation Characteristics

Obtaining information regarding the actual distribution of aquatic vegetation is a prerequisite for deep learning model training and evaluation. Aquatic vegetation can show different spectral and textural characteristics from water in images, so the visual interpretation of remote sensing images is usually used to determine a more accurate distribution of aquatic vegetation [41–43]. As shown in Figure 5a, the emergent vegetation appeared as red and pink on the false-color (N-G-R) images, and dark green and grass green on the real-color (R-G-B) images, as shown in Figure 5b, mainly in the lake shore area in a facultative distribution. As shown in Figure 5c, the floating-leaved vegetation appeared as pink and bright pink on the false-color image, and bright green on the real-color image (as shown in Figure 5d), mainly in the center of the lake and the area around the lake shore in a patchy distribution. As shown in Figure 5e, the submerged vegetation was shown as dark red and light pink on the false-color image. As shown in Figure 5f, the water was shown as dark green and blue–green on the real-color image in a flocculent distribution.

3.2.4. Sample Production

Based on the final generated aquatic vegetation results, the submerged vegetation, floating-leaved vegetation, and emergent vegetation pixel values were assigned as 1, 2, and 3, respectively, and the non-aquatic vegetation pixels were designated as 0. The results extracted through visual interpretation were then used as the real dataset for the training and validation of the extraction method. The image data from this dataset were obtained from the Sentinel-2 false-color dataset (Green(Band3)–Red(Band4)–NIR(Band8)). We randomly divided the dataset into training and validation sets in a ratio of 8:2. The training set was used to train the aquatic vegetation extraction deep learning model and the validation set was used for precision evaluation and method comparison.

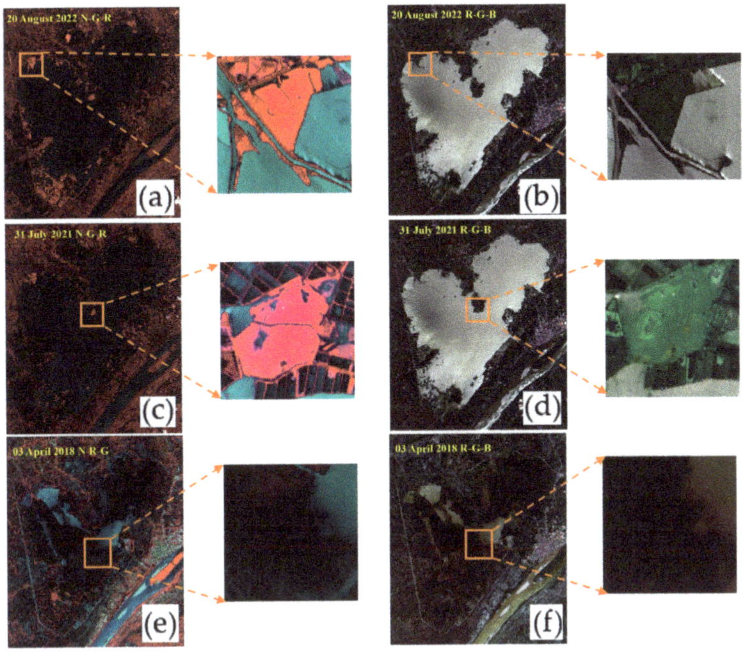

Figure 5. Distinction between different types of aquatic vegetation and background water bodies under different waveband synthesis methods. Panels (**a**,**c**,**e**) show the effect of the false-color image display, and panels (**b**,**d**,**f**) show the effect of the true-color image display. Comparing the magnified images of local areas reveals that the spectral and textural features of aquatic vegetation are more obvious under false-color compositing (N-R-G). R: red; G: green; B: blue; N: near-infrared.

The original images and their corresponding labels from the training dataset were converted to obtain images with 24-bit and 8-bit depths, respectively. The converted images were cropped into multiple images and corresponding labels with a size of 512 × 512 pixels. These images and labels were used as the input data for model training. Given the relatively small amount of submerged vegetation in the Lake Honghu area, it was difficult to meet the requirements for establishing a sample library of this type of vegetation, which may have led to a significant reduction in the model training effect. Thus, we added the data from Lake Taihu and Lake Caohai to satisfy the requirements for establishing a sample library.

After removing invalid images and labels, a total of 930 sets of image–label pairs were selected, of which 7 typical examples are shown below. In deep learning, sample balancing is crucial for improving the stability and generalization ability of the model. A balanced sample set ensures that each category is adequately learned and prevents the model from overly focusing on a large number of categories, resulting in the misclassification of a small number of categories. The proportion of pixels of the different samples in the sample set selected in this study is shown in Figure 6. In order to meet the requirements for sample balance and model follow-up training [44], in this experiment, the sample set was divided into a training set and test set and the division ratio was 8:2. After division, the training set contains 20,597,613 pixels of emergent vegetation, 16,489,963 pixels of floating-leaved vegetation, and 15,901,538 pixels of submerged vegetation. The number of pixels for all categories in the sample set is indicated below the picture sample scale.

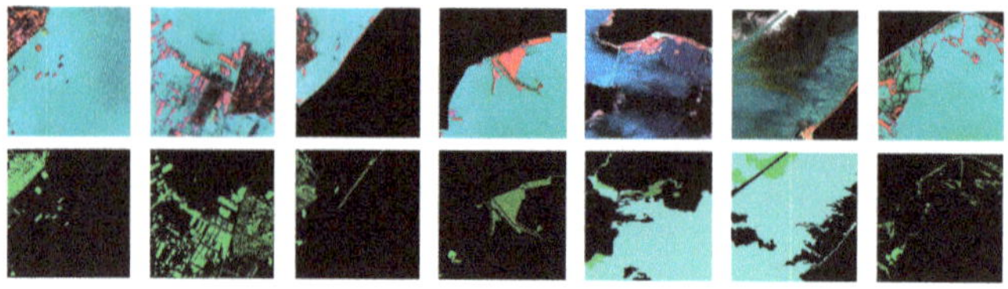

(a)

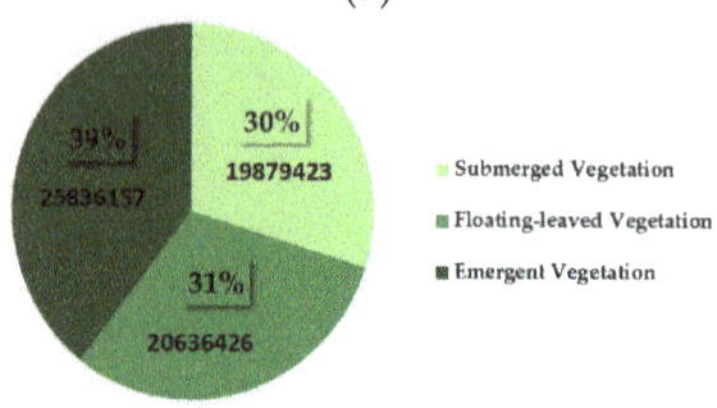

(b)

Figure 6. Schematic diagram of the sample display and sample proportion. (**a**) Example images from the aquatic vegetation training set. (**b**) The proportion of the number of pixels in the sample pool for each category; the number of pixels in this category is marked below.

3.3. Deep-Learning-Based Training of Aquatic Vegetation Models

Res-U-Net (Residual U-Net, as shown in Figure 7) is a deep learning neural network model used for image segmentation tasks that is capable of both capturing the local features of an image and overcoming the problem of gradient vanishing in deep network training [45]. This network can improve the performance of image segmentation tasks, which is especially effective when facing large-scale datasets and complex tasks [46,47]. We also used an adaptive moment estimation (Adam) optimizer [48] for model optimization. Adam is a popular stochastic gradient descent optimization algorithm that combines the advantages of momentum and the adaptive learning rate and has shown a good performance and convergence speed in many deep learning tasks.

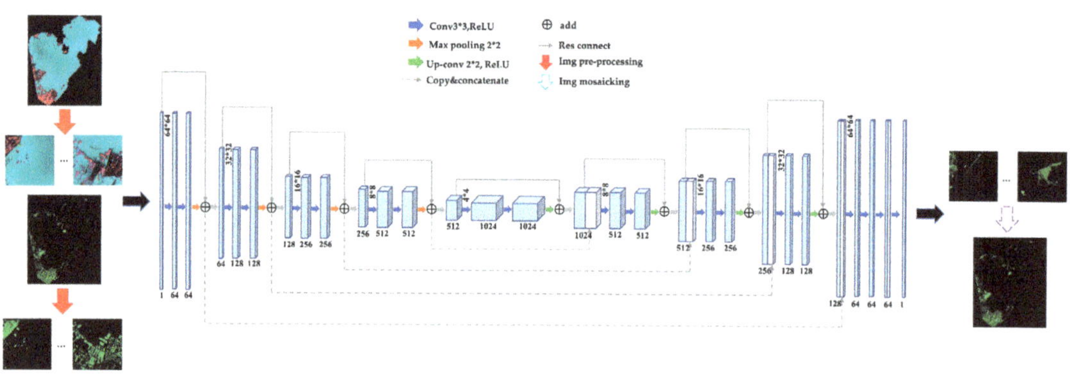

Figure 7. Schematic diagram of the Res-U-Net network model.

Res-U-Net usually uses a pixel-level loss function to measure the segmentation performance of the model, and in this study, dice loss was used [49]. The mathematical expression for dice loss is as follows:

$$\text{Dice Loss} = \mathbf{1} - \frac{\mathbf{2}|X \cap Y|}{|X| + |Y|} \quad (2)$$

where $|X \cap Y|$ denotes the number of elements in the intersection between X and Y, and $|X|$ and $|Y|$ denote the number of elements in X and Y, respectively.

In this study, we set the initial learning rate to 0.0001 and gradually decreased the learning rate via cosine annealing [50]. The training and prediction in this study were performed on a computer equipped with an Intel Core i5-13400F CPU and an Nvidia GeForce RTX 3060 GPU.

3.4. Precision Evaluation

3.4.1. Atmospheric Correction Precision

The mean relative error (MRE) is an indicator commonly used to evaluate the accuracy of a prediction model, measuring the average relative deviation between the predicted value and the actual observed value. It is usually calculated by summing the relative error between each predicted value and the corresponding actual observed value, and then averaging it.

$$\mathrm{MRE} = \frac{1}{n}\sum_{i=1}^{n} - \frac{|y_i - \hat{y}_i|}{y_i} \tag{3}$$

where n is the number of samples; y_i is the true value; and $\hat{y}_i$ is the predicted value.

The root mean square error (RMSE) is the square root of the mean square error with the formula:

$$\mathrm{RMSE} = \sqrt{\frac{1}{n}\sum_{i=1}^{n} + (y_i - \hat{y}_i)^2} \tag{4}$$

where n is the number of samples; y_i is the true value; and $\hat{y}_i$ is the predicted value.

R-squared (R^2) is a statistical measure used to evaluate the goodness of fit of a regression model. The formula for calculating R^2 is as follows:

$$R^2 = 1 - \frac{\Sigma(y_i - \hat{y}_i)^2}{\Sigma(y_i - \overline{y_i})^2} \tag{5}$$

where y_i is the true value; $\hat{y}_i$ is the predicted value; and $\overline{y_i}$ is the average of the true observed values.

3.4.2. Model Training Precision

In image segmentation tasks, it is common to use the mean intersection over union (MIoU) measure to evaluate the precision of the model's prediction results for each category [51]. The MIoU is the average of the intersection over union (IoU) of all of the categories and can be used to evaluate the model's prediction performance for the entire dataset. The MIoU is calculated using the following formula:

$$\mathrm{MIoU} = \frac{\mathrm{IoU_1} + \mathrm{IoU_2} + \cdots + \mathrm{IoU_n}}{\mathrm{N}} \tag{6}$$

where $\mathrm{IoU_1}$, $\mathrm{IoU_2}$,..., $\mathrm{IoU_n}$ denote the IoUs of each category, respectively, and N denotes the total number of categories.

The IoU is calculated by calculating the ratio of overlapping and concatenated regions between the predicted results and the true labels for each category. For each category, the IoU is calculated as follows:

$$\mathrm{IoU} = \frac{\text{Intersection Area}}{\text{Union Area}} \tag{7}$$

where the intersection area is the number of pixels in the intersection area of the prediction result and the real label, and the union area is the number of pixels in the concatenation area of the prediction result and the real label.

The kappa coefficient is a metric used for consistency testing. In classification problems, it tests whether the predicted results of the model are consistent with the actual classification results. The closer it is to 1, the more accurate the model is. For multivariate classification tasks, it can be expressed as

$$\text{Kappa} = \frac{p_0 + p_e}{1 - p_e} \tag{8}$$

where p_0 is the sum of the number of correctly classified samples in each category divided by the total number of samples, which is the overall classification precision. Suppose the true number of samples in each category is $a_1, a_2, \ldots, a_n$, and the predicted number of samples in each category is $b_1, b_2, \ldots, b_n$ and the total number of samples is n; then, p_e can be expressed as

$$p_e = \frac{a_1 \times b_1 + a_2 \times b_2 + \cdots + a_n \times b_n}{n \times n} \tag{9}$$

3.4.3. Model Prediction Precision

The optimization model was used to predict the aquatic vegetation in the set of validation images, and the precision of the extraction results was assessed based on the pixel-level-based precision assessment metrics, including precision (P), recall, F1 coefficient (F1), and relative error (RE).

$$\text{P} = \frac{\text{TP}}{\text{TP} + \text{FP}} \tag{10}$$

$$\text{Recall} = \frac{\text{TP}}{\text{TP} + \text{FN}} \tag{11}$$

$$\text{F1} = 2 \times \frac{\text{R} \times \text{P}}{\text{P} + \text{R}} \tag{12}$$

$$\text{RE} = 100\% \times \frac{|\text{T} - \text{P}|}{\text{T}} \tag{13}$$

where TP (true positive) is the number of correctly predicted aquatic vegetation pixels; FP (false positive) is the number of incorrectly predicted aquatic vegetation pixels; FN (false negative) is the number of incorrectly predicted non-aquatic vegetation pixels; T is the number of aquatic vegetation pixels in the real dataset; and P is the number of aquatic vegetation pixels in the prediction results. In F1, P denotes the precision rate, and R denotes the recall rate. It is important to note that we can calculate a set of precision assessment indices based on each image, but for images with few or no pixels of aquatic vegetation, the assessment indices will show large errors. Therefore, to avoid this problem, we used the number of pixels in all validation images within the validation set to calculate the total precision assessment index.

4. Results

4.1. *Evaluation of Atmospheric Correction Precision*

Problems can be encountered during the atmospheric correction of Sentinel-2 MSI data for aquatic environments, and the relationship between errors in aquatic classification and poor atmospheric correction quality needs to be eliminated. Thus, the atmospheric correction accuracy was evaluated by using each category index listed in Section 3.4.1, and the results are shown in Table 1. The accuracy of the atmospheric correction varied from high to low for emergent vegetation, floating-leaved vegetation, and submerged vegetation. The R^2 of all vegetation types was above 0.60, which met the requirements for follow-up treatment.

Table 1. Results of atmospheric correction evaluation for various aquatic vegetation types.

	RMSE	**MRE**	**R^2**
Emergent vegetation	0.13	20.03	0.84
Floating-leaved vegetation	0.15	25.61	0.76
Submerged vegetation	0.19	39.00	0.69

4.2. Model Training and Prediction Precision

We trained 930 pairs of images and labels according to the method presented in Section 3.4 and performed 300 iterations. In order to ensure that the model did not overfit during the training and testing, the fivefold k-fold cross-validation method was used for verification. The average IoU of the final model was 0.70 for submerged vegetation, 0.67 for floating-leaved vegetation, 0.77 for emergent vegetation, and 0.99 for background categories. The MIoU of the model training was 0.79, and the kappa was 0.87. In general, the model exhibited high training accuracy and is therefore suitable for the prediction of aquatic vegetation.

Since the growth of different types of aquatic vegetation in Lake Honghu is strongly influenced by seasons, with floating-leaved and emergent vegetation dominating in summer and submerged and emergent vegetation dominating in spring, we used images of two views of Lake Honghu in summer and spring for the prediction accuracy assessment. Figure 8 and Table 2 show the extraction results and prediction accuracies for each type of aquatic vegetation in Lake Honghu based on the DL model, respectively. In terms of the total area extraction, for 31 July 2021, the difference between the two is large at approximately 2.9 km^2; for 28 April 2017, the difference between the two is relatively small at approximately 1.01 km^2. The kappa coefficients for the two view images are 0.83 and 0.85, respectively. In terms of visual interpretation data validation, by comparing the prediction accuracy of different types of aquatic vegetation, it can be found that the prediction accuracy of submerged vegetation is slightly lower.

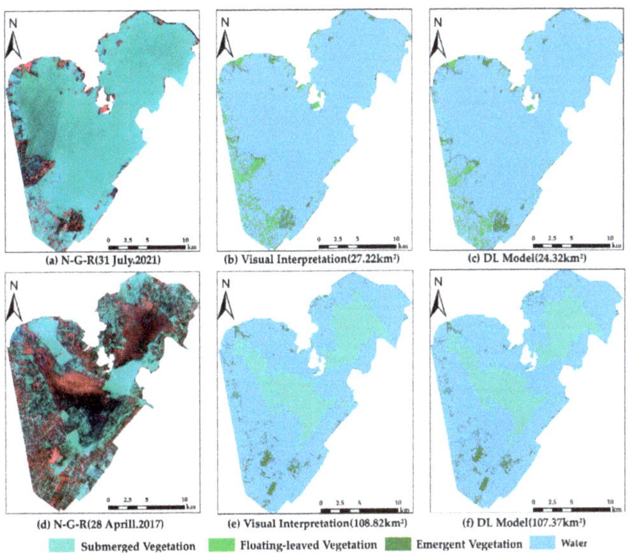

Figure 8. Aquatic vegetation visual interpretation results and DL prediction results. (**a**) The (N-R-G) band synthetic image; (**b**) the visual interpretation results; (**c**) the DL model results; (**d**) the (N-R-G) band synthetic image; (**e**) the visual interpretation results; (**f**) the DL model results (parentheses represent the total area extracted by aquatic vegetation; N: near-infrared; R: red; G: green).

4.3. Evaluation of Model Precision Based on Measured Data

To verify the classification accuracy of the model, we conducted five field investigations in April 2023 and between August and October 2023, covering Lake Caohai, Poyanghu Lake, Lake Dongtinghu, Lake Taihu, and Lake Honghu, respectively. The experimental sites were laid out prior to the expedition using the DL model identification results and historical documentation, as well as point accessibility. Visual, photographic, and salvage methods were used to estimate the aquatic vegetation cover upon arrival at the point. A total of 29 waterbody sites, 29 submerged vegetation sites, 23 floating-leaved vegetation sites, and 97 emergent vegetation sites were assessed, as shown in Figure 9a–e. The classification results were objectively evaluated using the confusion matrix, as shown in Figure 9f, with a kappa coefficient of 0.78, and the accuracy of the extracted results in each category was 100% (background/water body), 59% (submerged vegetation), 78% (floating-leaved vegetation), and 91% (emergent vegetation), respectively. The submerged vegetation had the lowest accuracy for each category compared to the emergent and floating-leaved vegetation.

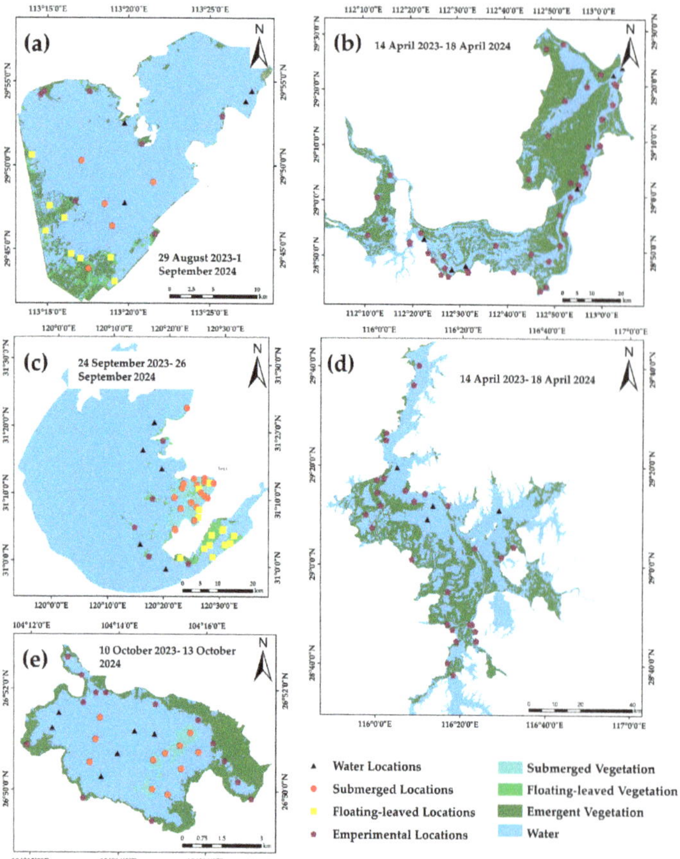

Confusion Matrix	Water Locations	Submerged Locations	Floating-leaved Locations	Emergent Locations
Water Locations	29	0	0	0
Submerged Locations	8	17	4	0
Floating-leaved Locations	0	1	18	4
Emergent Locations	8	0	1	88

(f)

Figure 9. Results for the verification of field sites in each lake. (**a**–**e**) Distribution of verified sites in each lake, with the date of the experiment and the name of the experimental lake in the margins; (**f**) confusion matrix of the verified sites, with each column denoting the predicted category, and each row denoting the true number of that category.

Table 2. Evaluation of the certainty of the prediction results of Lake Honghu.

Dates	Assessment Indicators	Submerged Vegetation	Floating-Leaved Vegetation	Emergent Vegetation
31 July 2021	Recall	/	0.83	0.91
	Precision	/	0.79	0.86
	F1 score	/	0.76	0.81
	RE	/	20.64%	14.45%
28 April 2017	Recall	0.76	/	0.90
	Precision	0.80	/	0.83
	F1 score	0.81	/	0.90
	RE	19.70%	/	16.15%

In the field experiments conducted on Lake Caohai and Lake Taihu, the amount of submerged vegetation was found to be the lowest compared to the emergent vegetation and floating-leaved vegetation. Here, 17 of the 29 submerged vegetation sites were correctly categorized; 8 of the 12 sites were incorrectly categorized as background (water body), and the remaining 4 were incorrectly categorized as other types of aquatic vegetation. In this regard, we investigated the causes of misclassification through field experiments in the more typical Lake Caohai and Lake Taihu.

During our field experiments at Lake Caohai, we found that, as shown in Figure 10, there were six points where the submerged vegetation was misclassified as a water body, and the vegetation coverage of these points was 0–25%. The more sparse submerged vegetation did not occupy a full image element in the remote sensing image, making recognition difficult [52]. The five correctly identified submerged vegetation points had a vegetation coverage of 75–100%; the submerged vegetation at these locations grows closely together and has more obvious spatial continuity in the image, making recognition easier.

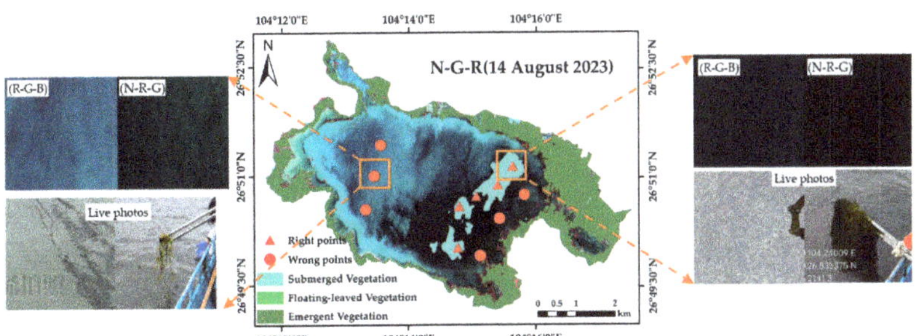

Figure 10. Submerged vegetation in Lake Caohai that cannot be effectively identified using remote sensing (**left**) and that can be effectively identified (**right**).

In Lake Taihu, as shown in Figure 11, we conducted a field experiment on 18 submerged vegetation sites. A total of four sites were misclassified: one was misclassified as background/water body due to low submerged vegetation cover, and the other three were misclassified as floating-leaved vegetation. The misclassification of the water body was due to the coverage of submerged vegetation, similar to the case of Lake Caohai. The other three were misclassified as floating-leaved vegetation mainly because Lake Taihu regularly salvages and clears submerged vegetation to ensure the quality of the water body; some of the cleared vegetation floats on the surface of the water and resembles floating-leaved vegetation in the remote sensing images, thus causing the misclassification.

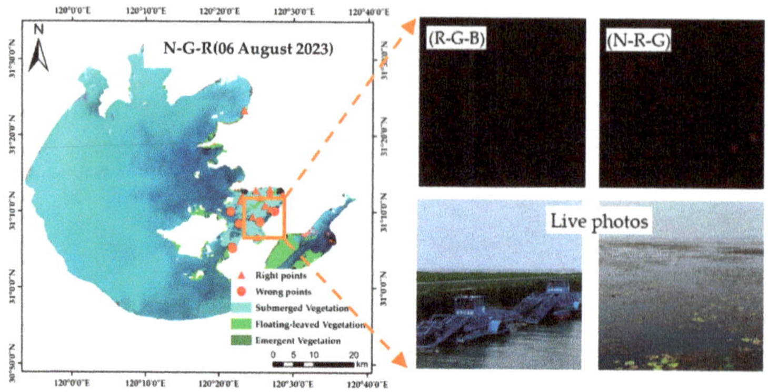

Figure 11. Remote sensing images of Lake Taihu, from which vegetation cannot be effectively identified.

4.4. Spatial and Temporal Changes in Aquatic Vegetation in Lake Honghu, 2017–2023

In order to analyze the spatial and temporal distribution patterns in aquatic vegetation in Lake Honghu, we extracted aquatic vegetation classification information from all available Sentinel-2 images in the Lake Honghu area from 2017 to 2023, as shown in Figure 12.

The temporal distribution pattern in the aquatic vegetation in Lake Honghu is quite obviously affected by the season, especially the submerged vegetation. As shown in Figure 12b, the submerged vegetation can be monitored over a large area from November to April each year, with the maximum value during our monitoring period being 96.49 km^2 on 28 April 2017; however, outside of this timeframe, i.e., from May to October, it is difficult to monitor the distribution of submerged vegetation over a large area due to the influence of the rise in the water level of Lake Honghu and the increase in the turbidity of the water body [53]. In contrast to the submerged vegetation, the most vigorous growth period for the aquatic and floating-leaved vegetation was from May to October each year, with maximum values of 38.05 km^2 (9 September 2023; aquatic vegetation) and 51.24 km^2 (22 July 2019; floating-leaved vegetation), respectively. Overall, the amount of aquatic vegetation in Lake Honghu showed a clear downward trend from 2017 to 2021, decreasing from the mean value of 45.68 km^2 in 2017 to 21.40 km^2 in 2021. After 2022, a slow growth trend was observed for aquatic vegetation; however, the recovery trend for the submerged vegetation was still not optimistic, with its area decreasing from the mean value in 2017 of 29.71km^2 to 0.8km^2 in 2022 and 0.31km^2 in 2023. Thus, it can be seen that more time and energy need to be invested in the recovery of submerged vegetation in Lake Honghu.

The vegetation examined in this study first decreased from 2017 to 2021, and later showed a slowly increasing trend in the period 2022–2023. This result is more consistent with other research findings: for example, Song [54] conducted a multi-year study of submerged vegetation in Lake Honghu from 2006 to 2016 and concluded that this vegetation type would decline. Since 2022, the Chinese government has initiated a pilot water ecology assessment program in the Yangtze River Basin [55]. This also indirectly explains the positive trend in water-supporting vegetation and floating-leaved vegetation in Lake Honghu since 2022, as it could be due to this policy.

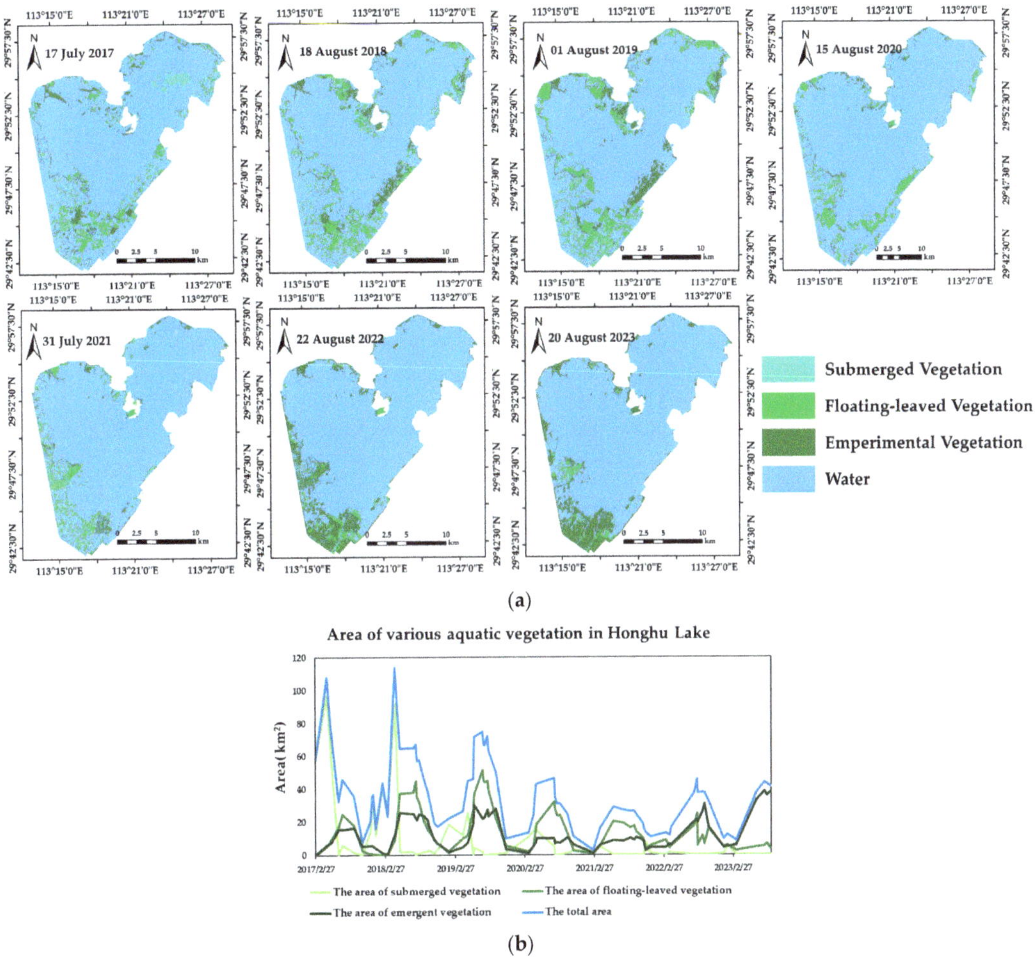

Figure 12. Spatial and temporal changes in aquatic vegetation in Lake Honghu. (**a**) Spatial distribution of aquatic vegetation in Lake Honghu from 2017 to 2023; (**b**) trend in aquatic vegetation area change in Lake Honghu from 2017 to 2023.

5. Discussion

5.1. Applicability of the Model

5.1.1. Applicability of Single-Aquatic-Vegetation-Type Lakes

Lake Dongtinghu mainly contains a single vegetation type and is not affected by cyanobacterial blooms, so we used this lake to test the extraction of a single type of aquatic vegetation. The kappa coefficient of the recognition result was 0.91, and the precision was 96%, indicating that the model shows good recognition performance for Lake Dongtinghu; the results are shown in Table 3. This model can be used to assess the generalization ability when migrating from lakes with complex vegetation types to lakes with a single vegetation type.

Table 3. Evaluation of the certainty of the prediction results for Lake Dongtinghu.

	Recall	Precision	F1 Score	RE
Background	0.99	0.99	0.99	0.29%
Emergent vegetation	0.98	0.95	0.97	3.61%

5.1.2. Applicability in Complex Situations

The situation for Lake Taihu is more complicated than for the other lakes. The aquatic vegetation types found in Lake Taihu are rich; moreover, as a lake with high eutrophication, Lake Taihu has often experienced cyanobacterial blooms over recent years, and as the spectral features of cyanobacterial blooms and aquatic vegetation are similar, misjudgments can easily occur. In order to verify the model's effectiveness in recognizing aquatic vegetation in more complex situations, we selected a remote sensing image of Lake Taihu for testing. The kappa coefficient of the Lake Taihu recognition result was 0.86, and the extraction precision of each category was above 80%, indicating that the model is suitable for dealing with the more complex Lake Taihu. The specific recognition results are shown in Table 4.

Table 4. Evaluation of the results for the Lake Taihu predictions.

	Recall	Precision	F1 Score	RE
Background	0.99	0.99	0.99	0.57%
Submerged vegetation	0.95	0.90	0.92	9.49%
Floating-leaved vegetation	0.80	0.94	0.87	5.73%
Emergent vegetation	0.74	0.83	0.65	7.47%

Unlike other lakes, Lake Taihu, as a eutrophic lake, has been seriously affected by water bloom [56], resulting in changes in the optical properties of the water body including an increase in turbidity, a change in water color, a decrease in transparency, and a change in spectral reflectance, ultimately leading to color confusion and spatial information distortion in remote sensing data [57,58]. This situation makes it difficult to effectively segment the water color and aquatic vegetation of Lake Taihu using traditional decision tree methods. On the other hand, cloud occlusion also increases the difficulty of aquatic vegetation data extraction. Cloud cover causes the loss or distortion of information in remote sensing images, which limits the feasibility of aquatic vegetation data extraction. To overcome this difficulty, cloud masking algorithms are usually employed to remove cloudy pixels [59], multi-temporal data are utilized to expand the effective observation time window, and data fusion or synthetic image techniques are used to mitigate the effects of clouds. Although these strategies help to improve the precision and reliability of aquatic vegetation information, they usually require large amounts of time and computational resources.

We selected the remote sensing image of Lake Taihu taken on 30 September 2021 to verify the performance of the model in terms of aquatic vegetation recognition under complex environmental conditions. The model did not misclassify the bloom region and performed well in the cloud-obscured region. Although the characteristics of aquatic vegetation are similar to those of water bloom, the model did not make any misclassifications in this region; when facing the influence of clouds, the model was able to effectively classify along the edges of the cloud obscuration.

5.2. Comparison with Other Methods

We select a Sentinel-2 remote sensing image of Lake Honghu that was not affected by clouds or water blooms for the evaluation of the aquatic vegetation data extraction results. The results showed that the deep learning method demonstrated higher precision and reliability in the aquatic vegetation data extraction task. As shown in Figure 13, the recognition accuracies of the DL model for the three vegetation types were 90% (submerged vegetation), 83% (floating-leaved vegetation), and 71% (emergent vegetation), which were better than those of the decision tree, which were 84% (submerged vegetation), 63% (floating-leaved vegetation), and 67% (emergent vegetation). In addition, the kappa coefficients of the DL results (0.84) were better than those of the decision tree (0.70). Compared with the random forest model, it was found that the accuracies for the submerged vegetation (91%) and floating-leaved vegetation (96%) were slightly higher than that of the DL model, while the accuracy of recall was lower than that of the DL model. These results indicate that random forest performs well in the classification of negative samples when classifying aquatic vegetation, but may be less effective in the identification of true positive samples. This indicates that the DL model has a significant advantage in terms of recognizing aquatic vegetation in complex environments.

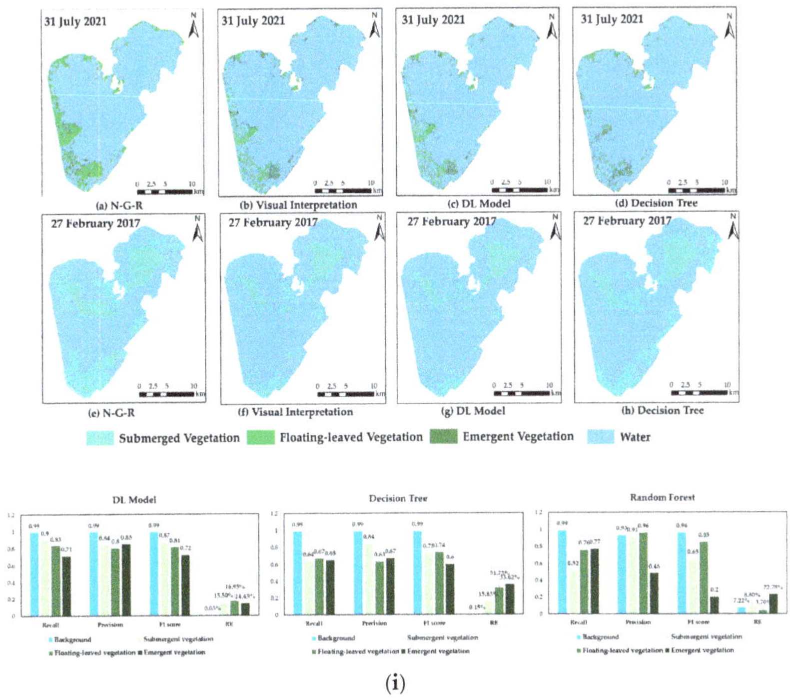

Figure 13. Comparison of DL model and decision tree and random forest. (**a–h**) The result of different interpretation methods; (**i**) precision comparison of the results of different interpretation methods.

6. Conclusions

In this study, a sample set containing the distribution of various types of aquatic vegetation was established for Lake Honghu, Lake Taihu, Lake Caohai, and Poyanghu Lake by utilizing Sentinel-2 MSI data. The sample set was generated through visual interpretation using cloud-free and cloudless reflectance images of the inland lakes. The sample sets were divided into a training set (70%) and a validation set (30%); the aquatic vegetation extraction model constructed based on Res-U-Net was trained using the training set in order to automatically extract and classify the aquatic vegetation data, and the following research results were obtained.

We evaluated the accuracy of the DL model in terms of training accuracy, prediction accuracy, and measured accuracy using visual interpretation and measured data as true values. In terms of the training accuracy, the average IoU of the submerged vegetation was 0.70, the average IoU of the floating-leaved vegetation was 0.67, the average IoU of the emergent vegetation was 0.77, and the IoU of the background category was 0.99. The MIoU of the model training was 0.79, with a kappa of 0.87. In terms of the prediction accuracy, the average accuracy of the DL model's extracted results was 0.80 with a relative error (RE) of 0.87. The average kappa coefficient was 0.86. In terms of the measured accuracy, the accuracies of the verified points for each category were 59% (submerged vegetation), 78% (floating-leaved vegetation), and 91% (emergent vegetation).

We applied the DL model to Lake Honghu and obtained and analyzed the trend of aquatic vegetation from 2017 to 2023. The results showed that the type of aquatic vegetation in Lake Honghu was greatly influenced by the season: the most dominant vegetation type from November to April each year was submerged vegetation, and from May to October each year, the dominant types were emergent and floating-leaved vegetation. In terms of the area covered by aquatic vegetation in Lake Honghu, there was a decreasing trend from 2017 to 2021, and the overall area has rebounded since 2022. This shows the great potential of the DL model in classifying and extracting aquatic vegetation data from lakes.

Regarding the applicability of the model, we used images of Lake Dongtinghu, which has a single vegetation type, and Lake Taihu, which is rich in vegetation types and has interference from submerged vegetation, for the classification and extraction analysis. We found that the DL model could avoid the interference of submerged vegetation and part of the clouds to a large extent, which is difficult to achieve using traditional methods. However, the model also has some limitations, and some submerged vegetation has no obvious features in the remote sensing image due to its low density, so the ideal extraction effect for this type of aquatic vegetation is not achieved.

Author Contributions: Conceptualization, Q.S.; methodology, H.G.; software, Y.S.; validation, H.G.; formal analysis, Q.S. and Y.Y.; investigation, H.G.; resources, Y.Z. (Yuting Zhou); data curation, W.L.; writing—original draft preparation, H.G.; writing—review and editing, R.L., Q.S. and Y.Y.; visualization, Y.Z. (Yuting Zhang); supervision, J.L. and Q.S.; project administration, M.L. and Q.S.; funding acquisition, Q.S. All authors have read and agreed to the published version of the manuscript.

Funding: This research was funded by the National Key Research and Development Program of China, grant number 2021YFB3901101.

Data Availability Statement: The data presented in this study are available on request from the corresponding author.

Acknowledgments: The authors would like to thank Qian Shen from the Key Laboratory of Digital Earth Science, Aerospace Information Research Institute, Chinese Academy of Sciences, for the help with writing. We are also thankful to all of the anonymous reviewers for their constructive comments on the study. All authors have read and agreed to the published version of the manuscript.

Conflicts of Interest: The authors declare no conflicts of interest.

References

1. Zhang, Y.; Jeppesen, E.; Liu, X.; Qin, B.; Shi, K.; Zhou, Y.; Thomaz, S.M.; Deng, J. Global Loss of Aquatic Vegetation in Lakes. *Earth-Sci. Rev.* **2017**, *173*, 259–265. [CrossRef]
2. Pu, J.; Song, K.; Lv, Y.; Liu, G.; Fang, C.; Hou, J.; Wen, Z. Distinguishing Algal Blooms from Aquatic Vegetation in Chinese Lakes Using Sentinel 2 Image. *Remote Sens.* **2022**, *14*, 1988. [CrossRef]
3. Zhao, J.; Liu, C.; Li, H.; Liu, J.; Jiang, T.; Yan, D.; Tong, J.; Dong, L. Review on Ecological Response of Aquatic Plants to Balanced Harvesting. *Sustainability* **2022**, *14*, 12451. [CrossRef]
4. Wang, Z.H.; Xin, C.L.; Sun, Z.; Luo, J.H.; Ma, R.H. Automatic Extraction Method of Aquatic Vegetation Types in Small Shallow Lakes Based on Sentinel-2 Data: A Case Study of Cuiping Lake. *Remote Sens. Inf.* **2019**, *34*, 132–141. [CrossRef]
5. Zou, W.; Yuan, L.; Zhang, L. Analyzing the Spectral Response of Submerged Aquatic Vegetation in a Eutrophic Lake, Shanghai, China. *Ecol. Eng.* **2013**, *57*, 65–71. [CrossRef]
6. Luo, J.; Yang, J.; Duan, H.; Sun, Z.; Xin, Y. Research progress of aquatic vegetation remote sensing in shallow lakes. *Natl. Remote Sens. Bull.* **2022**, *26*, 68–76. [CrossRef]
7. Silva, T.S.F.; Costa, M.P.F.; Melack, J.M.; Novo, E.M.L.M. Remote Sensing of Aquatic Vegetation: Theory and Applications. *Environ. Monit. Assess.* **2007**, *140*, 131–145. [CrossRef] [PubMed]
8. Albright, T.P.; Ode, D.J. Monitoring the Dynamics of an Invasive Emergent Macrophyte Community Using Operational Remote Sensing Data. *Hydrobiologia* **2011**, *661*, 469–474. [CrossRef]
9. Ackleson, S.; Klemas, V. Remote Sensing of Submerged Aquatic Vegetation in Lower Chesapeake Bay: A Comparison of Landsat MSS to TM Imagery. *Remote Sens. Environ.* **1987**, *22*, 235–248. [CrossRef]

10. Cheruiyot, E.K.; Mito, C.; Menenti, M.; Gorte, B.; Koenders, R.; Akdim, N. Evaluating MERIS-Based Aquatic Vegetation Mapping in Lake Victoria. *Remote Sens.* **2014**, *6*, 7762–7782. [CrossRef]
11. Hunter, P.D.; Gilvear, D.J.; Tyler, A.N.; Willby, N.J.; Kelly, A. Mapping Macrophytic Vegetation in Shallow Lakes Using the Compact Airborne Spectrographic Imager (CASI). *Aquat. Conserv. Mar. Freshw. Ecosyst.* **2010**, *20*, 717–727. [CrossRef]
12. Feng, L.; Hou, X.; Liu, J.; Zheng, C. Unrealistic Phytoplankton Bloom Trends in Global Lakes Derived from Landsat Measurements. *EarthArXiv* **2020**, preprint. [CrossRef]
13. Dai, Y.; Feng, L.; Hou, X.; Tang, J. An automatic Classification Algorithm for Submerged Aquatic Vegetation in Shallow Lakes Using Landsat Imagery. *Remote Sens. Environ.* **2021**, *260*, 112459. [CrossRef]
14. Malthus, T. Chapter 9. Bio-Optical Modeling and Remote Sensing of Aquatic Macrophytes. In *Bio-optical Modeling and Remote Sensing of Inland Waters*; Elsevier: Amsterdam, The Netherlands, 2017. [CrossRef]
15. Piaser, E.; Villa, P. Evaluating Capabilities of Machine Learning Algorithms for Aquatic Vegetation Classification in Temperate Wetlands Using Multi-Temporal Sentinel-2 Data. *Int. J. Appl. Earth Obs. Geoinform.* **2023**, *117*, 103202. [CrossRef]
16. Pearson, R.M.; Collier, C.J.; Brown, C.J.; Rasheed, M.A.; Bourner, J.; Turschwell, M.P.; Sievers, M.; Connolly, R.M. Remote estimation of aquatic light environments using machine learning: A new management tool for submerged aquatic vegetation. *Sci. Total Environ.* **2021**, *782*, 146886. [CrossRef]
17. Pande-Chhetri, R.; Abd-Elrahman, A.; Jacoby, C. Classification of Submerged Aquatic Vegetation in Black River Using Hyperspectral Image Analysis. *GEOMATICA* **2014**, *68*, 169–182. [CrossRef]
18. Martínez-Santos, P.; Aristizábal, H.; Díaz-Alcaide, S.; Gómez-Escalonilla, V. Predictive mapping of aquatic ecosystems by means of support vector machines and random forests. *J. Hydrol.* **2021**, *595*, 126026. [CrossRef]
19. Singh, G.; Reynolds, C.; Byrne, M.; Rosman, B. A Remote Sensing Method to Monitor Water, Aquatic Vegetation, and Invasive Water Hyacinth at National Extents. *Remote Sens.* **2020**, *12*, 4021. [CrossRef]
20. Patel, M.; Jernigan, S.; Richardson, R.; Ferguson, S.; Buckner, G. Autonomous Robotics for Identification and Management of Invasive Aquatic Plant Species. *Appl. Sci.* **2019**, *9*, 2410. [CrossRef]
21. Piaser, E.; Villa, P. Comparing Machine Learning Techniques for Aquatic Vegetation Classification Using Sentinel-2 Data. In Proceedings of the 2022 IEEE 21st Mediterranean Electrotechnical Conference (MELECON), Palermo, Italy, 14–16 June 2022.
22. Yao, F.; Wang, J.; Wang, C.; Crétaux, J.-F. Constructing Long-Term High-Frequency Time Series of Global Lake and Reservoir Areas Using Landsat Imagery. *Remote Sens. Environ.* **2019**, *232*, 111210. [CrossRef]
23. Rowan, G.S.L.; Kalacska, M. A Review of Remote Sensing of Submerged Aquatic Vegetation for Non-Specialists. *Remote Sens.* **2021**, *13*, 623. [CrossRef]
24. Rouse, R.; Haas, J.; Deering, D.W. 'aper A 20 Monitoring Vegetation Systems in The Great Plains with ERTS. 1974. Available online: https://api.semanticscholar.org/CorpusID:133358670 (accessed on 23 February 2024).
25. Xu, L.; Zhang, S.; He, Z.; Guo, Y. The comparative study of three methods of remote sensing image change detection. In Proceedings of the 2009 17th International Conference on Geoinformatics, Fairfax, VA, USA, 12–14 August 2009; pp. 1–4.
26. Valta-Hulkkonen, K.; Partanen, S.; Kanninen, A. Remote Sensing as a Tool in the Aquatic Macrophyte Mapping of a Eutrophic Lake: A Comparison Between Visual and Digital Classification. In Proceedings of the ScanGIS'2003—The 9th Scandinavian Research Conference on Geographical Information Science, Espoo, Finland, 4–6 June 2003. Proceedings DBLP, 2003.
27. Oyama, Y.; Matsushita, B.; Fukushima, T. Distinguishing surface cyanobacterial blooms and aquatic macrophytes using Landsat/TM and ETM + shortwave infrared bands. *Remote Sens. Environ.* **2015**, *157*, 35–47. [CrossRef]
28. Luo, J.; Ni, G.; Zhang, Y.; Wang, K.; Shen, M.; Cao, Z.; Qi, T.; Xiao, Q.; Qiu, Y.; Cai, Y.; et al. A New Technique for Quantifying Algal Bloom, Floating/Emergent and Submerged Vegetation in Eutrophic Shallow Lakes Using Landsat Imagery. *Remote Sens. Environ.* **2023**, *287*, 113480. [CrossRef]
29. Luo, J.; Li, X.; Ma, R.; Li, F.; Duan, H.; Hu, W.; Qin, B.; Huang, W. Applying Remote Sensing Techniques to Monitoring Seasonal and Interannual Changes of Aquatic Vegetation in Taihu Lake, China. *Ecol. Indic.* **2016**, *60*, 503–513. [CrossRef]
30. Fan, Y.; Chen, Y.; Chen, X.; Zhang, H.; Liu, C.; Duan, Q. Estimating the Aquatic-Plant Area on a Pond Surface Using a Hue-saturation-Component Combination and an Improved Otsu Method. *Comput. Electron. Agric.* **2021**, *188*, 106372. [CrossRef]
31. Rotta, L.H.D.S. Estimation of Submerged Aquatic Vegetation Height and Distribution in Nova Avanhandava Reservoir (São Paulo State, Brazil) Using Bio-Optical Modeling. 2015. Available online: http://hdl.handle.net/11449/123843 (accessed on 23 February 2024).
32. Huber, S.; Hansen, L.B.; Nielsen, L.T.; Rasmussen, M.L.; Sølvsteen, J.; Berglund, J.; von Friesen, C.P.; Danbolt, M.; Envall, M.; Infantes, E.; et al. Novel Approach to Large-Scale Monitoring of Submerged Aquatic Vegetation: A Nationwide Example from Sweden. *Integr. Environ. Assess. Manag.* **2022**, *18*, 909–920. [CrossRef] [PubMed]
33. Faza, S.; Nababan, E.B.; Efendi, S.; Basyuni, M.; Rahmat, R.F. An Initial Study of Deep Learning for Mangrove Classification. *IOP Conf. Ser. Mater. Sci. Eng.* **2018**, *420*, 012093. [CrossRef]
34. Guo, M.; Yu, Z.; Xu, Y.; Huang, Y.; Li, C. ME-Net: A Deep Convolutional Neural Network for Extracting Mangrove Using Sentinel-2A Data. *Remote Sens.* **2021**, *13*, 1292. [CrossRef]
35. Zhang, P.; Zhang, F.; Li, J.; Xle, Y.; Zhang, B. Aquatic vegetation extraction of Yugiao Reservoir Based on Sentinel-2 image feature optimizati. *Ecol. Sci.* **2023**, *42*, 40. [CrossRef]
36. Wang, S.; Zhang, Y.; Yu, Z. Classification of Honghe Wetland Remote Sensing Image Based on Random Forests. *Geomat. Spat. Inf. Technol.* **2014**, *37*, 4.

37. Liu, Z.Y.-C.; Chamberlin, A.J.; Tallam, K.; Jones, I.J.; Lamore, L.L.; Bauer, J.; Bresciani, M.; Wolfe, C.M.; Casagrandi, R.; Mari, L.; et al. Deep Learning Segmentation of Satellite Imagery Identifies Aquatic Vegetation Associated with Snail Intermediate Hosts of Schistosomiasis in Senegal, Africa. *Remote Sens.* **2022**, *14*, 1345. [CrossRef]
38. Chen, M.; Zhang, R.; Jia, M.; Cheng, L.; Zhao, C.; Li, H.; Wang, Z. Accurate and Rapid Extraction of Aquatic Vegetation in the China Side of the Amur River Basin Based on Landsat Imagery. *Remote Sens.* **2024**, *16*, 654. [CrossRef]
39. Wang, X.Y.; Tang, J.W.; Li, T.J.; Zhou, H. Key Technologies of water spectra measurements with above-water method. *Ocean. Technol* **2012**, *31*, 72–76. [CrossRef]
40. Merwade, V.M. An Automated GIS Procedure for Delineating River and Lake Boundaries. *Trans. GIS* **2007**, *11*, 213–231. [CrossRef]
41. Cao, M.; Qing, S.; Jin, E.; Hao, Y.; Zhao, W. A Spectral Index for the Detection of Algal Blooms Using Sentinel-2 Multispectral Instrument (MSI) Imagery: A Case Study of Hulun Lake, China. *Int. J. Remote Sens.* **2021**, *42*, 4514–4535. [CrossRef]
42. Xie, Y.; Sha, Z.; Yu, M. Remote Sensing Imagery in Vegetation Mapping: A Review. *J. Plant Ecol.* **2008**, *1*, 9–23. [CrossRef]
43. Zhang, X. On the Estimation of Biomass of Submerged Vegetation Using Landsat Thematic Mapper (TM) Imagery: A Case Study of the Honghu Lake, PR China. *Int. J. Remote Sens.* **1998**, *19*, 11–20. [CrossRef]
44. Chen, X.; Weng, J.; Luo, W.; Lu, W.; Wu, H.; Xu, J.; Tian, Q. Sample Balancing for Deep Learning-Based Visual Recognition. *IEEE Trans. Neural Networks Learn. Syst.* **2020**, *31*, 3962–3976. [CrossRef] [PubMed]
45. He, K.; Zhang, X.; Ren, S.; Sun, J. Deep residual learning for image recognition. In Proceedings of the 2016 IEEE Conference on Computer Vision and Pattern Recognition (CVPR), Las Vegas, NV, USA, 27–30 June 2016; pp. 770–778. [CrossRef]
46. Sharifi, A.; Mahdipour, H.; Moradi, E.; Tariq, A. Agricultural Field Extraction with Deep Learning Algorithm and Satellite Imagery. *J. Indian Soc. Remote Sens.* **2022**, *50*, 417–423. [CrossRef]
47. Liu, Z.; Feng, R.; Wang, L.; Zhong, Y.; Cao, L. D-Resunet: Resunet and Dilated Convolution for High Resolution Satellite Imagery Road Extraction. In Proceedings of the IGARSS 2019—2019 IEEE International Geoscience and Remote Sensing Symposium, Yokohama, Japan, 28 July–2 August 2019.
48. Kingma, D.P.; Ba, J. Adam: A Method for Stochastic Optimization. *arXiv* **2014**. [CrossRef]
49. Li, X.; Sun, X.; Meng, Y.; Liang, J.; Wu, F.; Li, J. Dice Loss for Data-Imbalanced NLP Tasks. In Proceedings of the 58th Annual Meeting of the Association for Computational Linguistics, Online, 5–10 July 2020; pp. 465–476. [CrossRef]
50. Loshchilov, I.; Hutter, F. SGDR: Stochastic Gradient Descent with Warm Restarts. In Proceedings of the International Conference on Learning Representations, San Juan, Puerto Rico, 2–4 May 2016.
51. Loshchilov, I.; Hutter, F. Decoupled Weight Decay Regularization. *arXiv* **2017**. [CrossRef]
52. Luo, J.; Ma, R.; Duan, H.; Hu, W.; Zhu, J.; Huang, W.; Lin, C. A New Method for Modifying Thresholds in the Classification of Tree Models for Mapping Aquatic Vegetation in Taihu Lake with Satellite Images. *Remote Sens.* **2014**, *6*, 7442–7462. [CrossRef]
53. Wu, H.; Han, H.; Zhi, Y.; Deng, Z.; Zhou, J.; Li, W.; Yuan, L.; Cao, Y. Enclosure study on the limiting factors of submerged macrophyte growth in restored sites of Lake Honghu. *Plant Sci. J.* **2023**, *41*, 583–593. [CrossRef]
54. Xinxin, S.; Xiaobin, C.; Zhi, W.; Enhua, L.; Xuelei, W. Community change of dominant submerged macrophyte in Lake Honghu since 1950s. *J. Lake Sci.* **2016**, *28*, 859–867. [CrossRef]
55. Four departments jointly issued assessment index scoring rules for the Yangtze River Basin water ecology "score". *Environ. Monit. Manag. Technol.* **2023**, *35*, 6.
56. Wang, S.; Wang, W.; Chen, J.; Zhang, B.; Zhao, L.; Jiang, X. Characteristics of Dissolved Organic Matter and Its Role in Lake Eutrophication at the Early Stage of Algal Blooms—A Case Study of Lake Taihu, China. *Water* **2020**, *12*, 2278. [CrossRef]
57. Shen, L.; Xu, H.; Guo, X. Satellite Remote Sensing of Harmful Algal Blooms (HABs) and a Potential Synthesized Framework. *Sensors* **2012**, *12*, 7778–7803. [CrossRef] [PubMed]
58. Warner, R.A.; Fan, C. Optical Spectra of Phytoplankton Cultures for Remote Sensing Applications: Focus on Harmful Algal Blooms. *Int. J. Environ. Sci. Dev.* **2013**, *4*, 94–98. [CrossRef]
59. Baetens, L.; Desjardins, C.; Hagolle, O. Validation of Copernicus Sentinel-2 Cloud Masks Obtained from MAJA, Sen2Cor, and FMask Processors Using Reference Cloud Masks Generated with a Supervised Active Learning Procedure. *Remote Sens.* **2019**, *11*, 433. [CrossRef]

remote sensing

MDPI

Article

Remote Sensing Estimation of CDOM and DOC with the Environmental Implications for Lake Khanka

Sining Qiang [1,2], Kaishan Song [2], Yingxin Shang [2], Fengfa Lai [2], Zhidan Wen [2], Ge Liu [2], Hui Tao [2] and Yunfeng Lyu [1,*]

1 School of Geographic Science, Changchun Normal University, Changchun 130102, China; qiangsining@iga.ac.cn
2 Northeast Institute of Geography and Agroecology, Chinese Academy of Sciences, Changchun 130102, China; songkaishan@iga.ac.cn (K.S.); shangyingxin@iga.ac.cn (Y.S.); laifengfa@iga.ac.cn (F.L.); wenzhidan@iga.ac.cn (Z.W.); liuge@iga.ac.cn (G.L.); taohui@iga.ac.cn (H.T.)
* Correspondence: lvyunfeng@ccsfu.edu.cn

Abstract: Chromophoric dissolved organic matter (CDOM) is a significant contributor to the biogeochemical cycle and energy dynamics within aquatic ecosystems. Hence, the implementation of a systematic and comprehensive monitoring and governance framework for the CDOM in inland waters holds significant importance. This study conducted the retrieval of CDOM in Lake Khanka. Specifically, we use the GBDT (R^2 = 0.84) algorithm which performed best in retrieving CDOM levels and an empirical relationship based on the situ data between CDOM and dissolved organic carbon (DOC) to indicate the distribution of DOC indirectly. The performance of the CDOM-DOC retrieval scheme was reasonably good, achieving an R^2 value of 0.69. The empirical algorithms were utilized for the analysis of Sentinel-3 datasets from the period 2016 to 2020 in Lake Khanka. The potential factors that contributed to the sources of DOM were also analyzed with the humification index (HIX). The significant relationship between CDOM and DOC (HIX and chemical oxygen demand (COD)) indicated the potential remote sensing application of water quality monitoring for water management. An analysis of our findings suggests that the water quality of the Great Khanka is superior to that of the Small Khanka. Moreover, the distribution of diverse organic matter exhibits a pattern where concentrations are generally higher along the shoreline compared to the center of the lake. Efficient measures should be promptly implemented to safeguard the water resources in international boundary lakes such as Lake Khanka and comprehensive monitoring systems including DOM distribution, DOM sources, and water quality management would be essential for water resource protection and government management.

Keywords: chromophoric dissolved organic matter; dissolved organic carbon; chemical oxygen demand; water quality; remote sensing

1. Introduction

Lakes play a significant role within the terrestrial hydrosphere. In addition to serving as the foundation for sustaining human life and supporting production activities, they also play a crucial role in hydrological regulation and the preservation of ecosystem functions. The lake ecosystem serves as a dependable medium for the storage, transformation, and transportation of carbon within inland areas [1]. Dissolved organic carbon (DOC) is an important component of the global organic carbon pool as well as a major component of dissolved organic matter (DOM) [2], and DOC in inland waters plays an important role in the global carbon cycle. Chromophoric dissolved organic matter (CDOM) refers to the light-sensitive component of DOM. CDOM is a prominent optically active compound found in water. It exhibits notable properties such as a high absorption of ultraviolet and visible blue light. The strong absorption of this substance has a notable impact on the optical properties of the water's surface. However, CDOM serves to reduce the transmission and absorption of ultraviolet and blue light in aquatic environments [3]. The primary function of the spread is to mitigate the detrimental effects of UV-B radiation on aquatic organisms [4]. Additionally, it contributes to safeguarding aquatic ecosystems to a certain degree within the water column. As the colored part of DOM, CDOM is prone to photochemical degradation and microbial decomposition in the natural environment, releasing greenhouse gases such as carbon dioxide and methane to exacerbate global warming. The source, composition, degradation and mineralization of CDOM are related to the global carbon cycle [5].

The investigation of the source and composition of CDOM can provide insights into the movement and alteration of organic pollutants, as CDOM possesses the capacity to adsorb metal ions and organic pollutants [6]. The absorption coefficient of CDOM is considered as a potential measure of dissolved organic carbon, which can help estimate the release of carbon dioxide and assess carbon pools in carbon cycle studies at the same time [7]. CDOM assumes a significant optical role within DOC. By examining the optical properties of CDOM, one can effectively assess the distribution and variations of CDOM within inland waters. This, in turn, serves as an indicator for the distribution and movement of DOC [8]. Additionally, CDOM contributes to the global carbon cycle. As the components of CDOM are complex and easy to degrade, the absorption coefficient at a specific wavelength is usually used as an indicator of its concentration [9,10], and a(350) is usually used to represent the concentration of CDOM and the biogeochemical cycle processes [11,12]. The 3D-EEMs (3Dimensions Excitation-Emission Matrices) were considered as a sensitive way to evaluate the sources and composition of DOM rapidly [13]. The fluorescence index can reflect the source and component characteristics of DOM. HIX semi-quantitatively represents the degree of humification of DOM and the possible duration of organic matter in the environment. Some scholars have found an obvious positive correlation between DOC concentration and CDOM in inland waters [8,10,14,15]. DOC is a reliable regional proxy indicator in these studies [16–19]. The water quality status can be inferred indirectly by examining the relationship between CDOM and its composition parameters and water environment indicators [20–22]. Organic pollutants are also known as oxygen-consuming organic matter, such as chemical oxygen demand (COD) which indicates the amount of oxygen required for the oxidation of organic matter in lakes, which is related to DOM to some extent.

Remote sensing technology is a highly efficient tool for generating maps that depict the distribution of CDOM with both high temporal and spatial resolution [10,14,15]. Studies have shown [23,24] that adding 600 nm to the CDOM inversion model can significantly improve the accuracy of CDOM estimation in turbid waters, and using the band search at 800 nm may indirectly prompt CDOM. Increasing the number of bands used as variables is expected to enhance the performance of the model, resulting in improved fitting results and increased robustness. Zhang [20] found that in the main organic pollutant index of Lake Taihu, there was a significant correlation between COD and CDOM uptake. In their study, Huang et al. [25] were able to effectively obtain the spatial distribution of CDOM and COD in the Liaodong Bay region. This was achieved by utilizing the H J-1/CCD environmental satellite. The establishment of this empirical model has promoted the application of water color remote sensing in the monitoring of water environmental pollution components. Wu et al. [26] evaluated seasonal water quality in Lake Taihu by measuring water quality parameters, such as water temperature, dissolved oxygen (DO), turbidity (Tur), COD, total phosphorus (TP), total nitrogen (TN), and assigning weights to them to evaluate the seasonal water quality of Lake Taihu. Sentinel-3 ocean and land color imager (OLCI) is equipped with a full resolution capability of 300 m for earth observation. Additionally, the revisit period for this instrument is approximately 1.5 days. This dataset contains atmospheric radiation in 21 spectral bands, so Sentinel-3 has the advantage of a short revisit period, and the spatial resolution is suitable for the breadth of the 4380 km^2 water area of Lake Khanka. As is known, atmospheric correction methods that are designed for either ocean color or land applications often result in low quality or even no surface reflectance data for coastal and inland waters. Machine learning algorithms have experienced significant advancements and widespread adoption in recent years, and GBDT (Gradient Boosting Decision Tree) is a kind of machine learning algorithm. It is an integrated learning method based on the CART decision tree. The algorithm integrates and iterates multiple tree models obtained through learning and generalization [27,28]; therefore, machine learning algorithms have been widely used in water color remote sensing research in recent years [29–31]. We developed and validated a novel approach to determine long term CDOM records for Lake Khanka using OLCI unsaturated spectral bands following Rayleigh scattering correction in the present study.

Lake Khanka serves as the official boundary between the nations of China and Russia. Water plays a crucial and indispensable role in various domains [32], including the provision of drinking water, support for industrial and agricultural activities, facilitation of waterway transportation, regulation of climate patterns, and preservation of biodiversity. In recent years, the water environment of Lake Khanka has been affected by global climate change and human activities. Hence, the exploration of the CDOM-DOC inversion model holds immense importance in the context of monitoring and safeguarding the water environment in Lake Khanka. The presence of a significant quantity of organic matter and an excessive amount of nutrients in inland water are poised to result in the discoloration and unpleasant odor of the water body. This phenomenon is known as water pollution, which not only disrupts the regional carbon cycle mechanism essential for maintaining water ecological balance but also poses a threat to human survival and development. The primary aim of this study is to create a remote sensing model for the absorption coefficient of CDOM to establish the linkage for understanding the sources and distribution of DOM. This model will be used to assess the potential for monitoring the water quality of Lake Khanka and will even provide further scientific support and a theoretical basis for protection and governance through remote sensing techniques. Additionally, the study seeks to enhance our understanding of the dynamics between CDOM and DOC to offer a comprehensive and interdisciplinary perspective on CDOM in inland waters.

2. Materials and Methods

2.1. Study Area

Lake Khanka serves as an international boundary lake, spanning across both China and Russia. The lake under consideration is the largest freshwater lake located within the Ussuri River Basin. Furthermore, Lake Khanka has an open water surface and is divided into two parts by a sand embankment. Small Lake Khanka covers an area of about 176 km^2, all of which are located in China, and Great Khanka covers an area of about 4380 km^2, located in China and Russia, as shown in Figure 1f,g. The western region of Lake Khanka is characterized by the presence of several rivers, including the Muling River, Lev River, Bailing River, Xintu River, Takhrezh River, and Beyqi River. Additionally, there are over 20 small- and medium-sized rivers, such as the Men River. On the eastern side, the Songacha River serves as the sole outlet for excess water from the lake. As the largest boundary lake in China, Lake Khanka, with its huge volume, has multi-directional functions such as water supply in the two basins, the regulation of river runoff, irrigation for agricultural production, and the maintenance of biodiversity protection ecosystem balance.

2.2. Field Sampling Data

The in situ data samples were collected during field expeditions conducted in September 2017, July 2018, October 2020, and September and October 2021, respectively. In the field, a total of 170 surface water samples were collected at depths ranging from 0 to 50 cm. The spatial distribution of the sampling points can be observed in Figure 1a–e. This distribution includes 99 water samples from Small Khanka and 71 water samples from Great Khanka. We used a polyethylene container with a capacity of 2 L to seal and store water away from light and used a mixed fiber filter membrane (pore diameter 0.45 μm) to complete the preliminary filtration of the water sample within 12 h. The water samples were stored in a refrigerator for further measurement in the laboratory of CDOM absorption, DOC concentration, and other water quality parameters. Additionally, we recorded the latitude and longitude coordinates, altitude, wind speed, atmospheric pressure, and weather conditions at the sampling point.

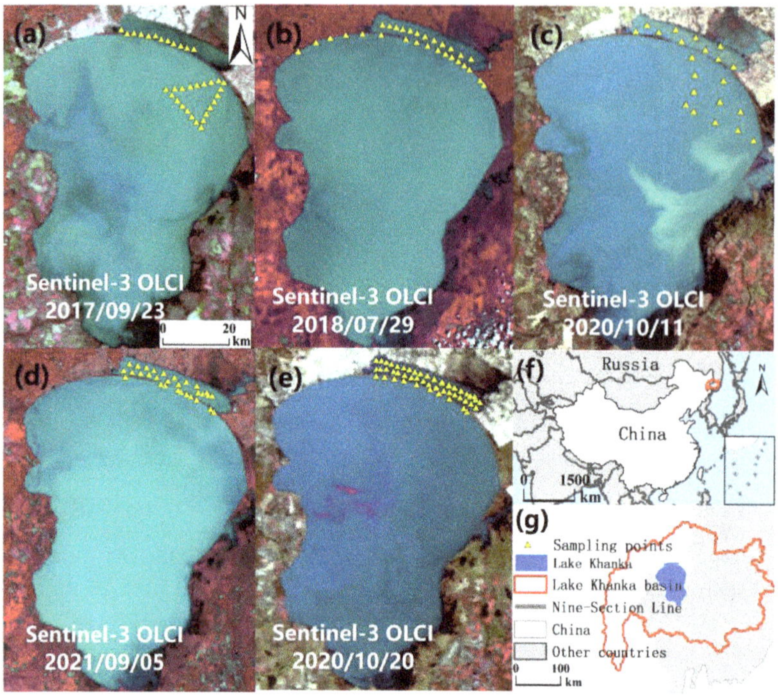

Figure 1. (**a–e**) Pseudo-color displays Sentinel-3 OLCI images showing the landscape and sample point locations of the study area, (**f**,**g**) Lake Khanka, and surrounding drainage basin in the world.

2.3. Measurement for CDOM Properties and Water Quality Parameters

The water samples collected in the field were initially filtered using a polycarbonate filter membrane with a pore size of 0.22 μm. This filtration process was performed to obtain the samples specifically for CDOM testing. Subsequently, a Shimadzu UV-spectrophotometer (UV-6200) was utilized to measure the absorbance spectrum in the range of 200–800 nm. Ultrapure water was used as the reference for this measurement. The DOC concentration is obtained by testing the water sample filtered by the Whatman GF/F glass fiber membrane with a pore size of 0.45 μm through the Shimadzu Total Organic Carbon Analyzer (TOC-VCPN). The calculation formula for the CDOM absorption coefficient is as follows:

$$a_{CDOM}(\lambda) = \frac{2.303A(\lambda)}{L} \tag{1}$$

In Equation (1), A(λ) is the absorbance at λ nm, and the optical path L is 1 cm. The absorption spectrum of CDOM in the ultraviolet-visible band exhibits an exponential decay as the wavelength increases; a(350) is usually used to indicate the CDOM concentration and biogeochemical cycle process in lakes and rivers [11,12], so we select a(350) [33] as the parameter referring to the CDOM concentration.

Parameters associated with the absorption properties of chromophoric dissolved organic matter (CDOM) in water are frequently employed to indirectly assess the distribution, origin, composition, fate, and attributes of CDOM involved in geochemical cycles. The parameter $SUVA_{254}$, which represents the ratio between the absorption coefficient of ultraviolet light at a wavelength of 254 nm and the concentration of dissolved organic carbon

(DOC), primarily indicates the presence of aromatic compounds [34]. The 3D-EEMs was measured with a Hitachi F-7000 fluorescence spectrometer and the HIX was calculated using MATLAB [35]. Other water quality parameters such as chlorophyll (Chla) [36], total phosphorus (TP), total nitrogen (TN) [37], chemical oxygen demand (COD_{Mn}) [38], turbidity (Turb) [26], dissolved organic carbon (DOC) [39], total suspended solids (TSM) concentration [40], etc., were obtained by laboratory analysis and testing.

2.4. Remote Sensing Data and Preprocessing

The remote sensing data of Lake Khanka from 2016 to 2022 was obtained from the Sentinel3-OLCI sensor. These data were sourced from the Copernicus Data Center, which can be accessed at https://scihub.copernicus.eu/ (accessed on 1 January 2020). The optimized design of the OLCI incorporates 21 spectral channels that cover the visible to near-infrared spectral range. This design enables a more comprehensive and accurate representation of the spectral characteristics of complex water bodies. Partial atmospheric correction is performed using SeaDAS8.2.0 (https://seadas.gsfc.nasa.gov/) (accessed on 1 January 2020). There is currently no reliable atmospheric correction that can produce accurate OLCI Rrs data for Lake Khanka. Therefore, a partial atmospheric correction to correct for the gaseous absorption and Rayleigh (molecular) scattering effects was applied to the Level-1B data using routines and look up tables (LUTs) embedded in the software (SeaDAS, version 8.2.0), where Equation (2) was used to calculate the Rayleigh-corrected reflectance $R_{rc}(\lambda)$:

$$R_{rc}(\lambda) = \pi / F_0(\lambda) / cos\theta \times (L_t(\lambda) / t_{g_{sol}}(\lambda) / t_{g_{sen}}(\lambda) - L_r(\lambda) - TL_g(\lambda)) \quad (2)$$

where $L_t(\lambda)$ represents the irradiance of the top of the atmosphere (TOA), $L_r(\lambda)$ represents the ray radiance, $TL_g(\lambda)$ represents the flash radiance of TOA, θ represents the solar zenith angle, $F_0(\lambda)$ represents the average solar flux, $t_{g_{sol}}(\lambda)$ represents the gaseous transmittance from the sun to the surface, and $t_{g_{sen}}(\lambda)$ represents gaseous transmittance to the sun in Equation (2).

A total of 70 non-cloud images should be acquired within a one-week time frame or on the same day, aligning with the five designated sampling times between May and October. The software ArcGIS 10.7 is utilized to extract the Rayleigh-corrected reflectance of the image at the specified sampling point, denoted as $R_{rc}(\lambda)$.

2.5. Model Construction and Evaluation

The final inversion model utilized a selection of 17 bands, specifically Oa1 to Oa17, covering a wavelength range of 400 to 865 nm as the input variables. After the model was trained and verified successfully, error analysis was carried out, and the result prediction and batch inversion were largely completed. The correlation analysis of Sentinel-3 remote sensing reflectance for each band and $a(350)$ absorption coefficient was conducted using matlab2016 software programming. This analysis includes various approaches such as single band analysis, band ratio analysis, and other band combinations. The objective is to develop a linear model that is appropriate for the Lake Khanka CDOM absorption inversion regression model. MATLAB (2016) software trained the correlation between single band, band ratio, and other band combinations of 17 OLCI $R_{rc}(\lambda)$ randomly and the absorption coefficient of CDOM. The highest correlation coefficient R is recorded in the addition of each band combination.

In order to determine the most suitable model, we conducted a comparative analysis of five machine-learning techniques. These techniques were applied to Sentinel-3 simulated reflectance data to retrieve CDOM. The following are examples of popular machine learning algorithms: 1. Gradient Boosting Decision Tree (GBDT); 2. eXtreme Gradient Boosting (XGBoost); 3. Support Vector Regression (SVR); 4. Back Propagation Neural Network (BPNN); 5. Random Forest (RF). The Random Forest Regressor package in Python works to build a Gradient Boosting Decision Tree model. The number of decision trees affects the complexity and fitting performance of the model to a certain extent; thus, the hyperparameters are used to obtain the optimal parameter combination of the model. The independent variable utilized in the OLCI data reflectance is the input from bands B1-B17, while the dependent variable is the CDOM absorption *a*(350) at a wavelength of 350 nm. Two thirds of the data are used as the modeling dataset, and the remaining one third is used as the verification dataset. The established linear regression model and machine learning model were compared and evaluated using R, RMSE, MAE, and other indicators.

$$RMSE = \sqrt{\frac{\sum_{i=1}^{n}(X_{estimated} - X_{observed})^2}{n}} \tag{3}$$

$$MAE = \frac{1}{n}\sum_{m=1}^{n}|(X_{estimated} - X_{observed})| \tag{4}$$

3. Results

3.1. Characteristics of CDOM and Water Quality Parameters

The concentration of CDOM is characterized by the absorption coefficient at 350 nm. The average value of the overall absorption coefficient of Lake Khanka is 4.69 ± 2.28 m^{-1}. The value is 4.77 ± 1.26 m^{-1} in the Small Khanka, while it is 2.99 ± 1.14 m^{-1} in the Great Khanka. The $SUVA_{254}$ in Lake Khanka is 4.99 ± 1.38, which is 4.85 ± 0.77 in the Small Khanka and 5.51 ± 1.82 in the Great Khanka. The overall range of water sample HIX obtained from five field experiments from 2017 to 2021 is 2.16 ± 1.41, including 2.39 ± 1.59 for the Small Khanka and 1.67 ± 0.79 for the Great Khanka. The concentration of Chla in Lake Khanka water was measured five times between 2016 and 2021. The average concentration was found to be 10.36 ± 9.29µm/L. Similarly, the total nitrogen (TN) concentration was measured to be 0.57 ± 0.21 mg/L, total phosphorus (TP) concentration was 0.14 ± 0.73 mg/L, chemical oxygen demand (COD_{Mn}) was 3.30 ± 0.19 mg/L, total suspended matter (TSM) was 92.69 ± 48.09 mg/L, turbidity was 68.16 ± 25.16 NTU, and DOC concentration was 5.08 ± 1.41 mg/L.

3.2. Correlation between Lake Environmental Hydro-Chemical Characteristics and DOM

The relationship between parameters associated with CDOM in inland water and DOC is widely acknowledged and supported by substantial evidence, as depicted in Figure 2 [14,17,41–44]. The indirect inversion of the DOC concentration using the absorption properties of the optically active substance CDOM in water mainly relies on the conservative mixing relationship between CDOM and DOC. The data collected during five instances of field sampling in Lake Khanka between 2017 and 2021 revealed a noteworthy linear correlation between the absorption coefficient of CDOM at a wavelength of 350 nm and the concentration of DOC. The correlation coefficient value was determined to be 0.823 ($p < 0.01$), indicating a strong relationship. This finding suggests that the absorption coefficient *a*(350) of CDOM in Lake Khanka can serve as a reliable indicator of DOC. Consequently, we have opted to utilize the absorption coefficient of *a*(350) in conjunction with the measured DOC concentration to develop an empirical model for estimating the distribution of DOC concentration in Lake Khanka. In addition, the results of HIX show a strong correlation between HIX and CDOM (R = 0.78, $p < 0.01$), and CDOM absorption can be used as an indirect inversion of the spatial distribution of HIX.

The correlation analysis of water quality parameters involved the calculation of the Spearman correlation coefficient and the corresponding significance test p value for each pair of water quality parameters. In addition to DOC, positive correlation factors with *a*(350) include COD, SDD (Secchi-Disk Depth), Chla, pH, and salinity, and the linear correlation coefficient with COD is as high as 0.625 ($p < 0.01$). The CDOM inversion results reflect the temporal and spatial distribution and change trend of COD to a certain extent.

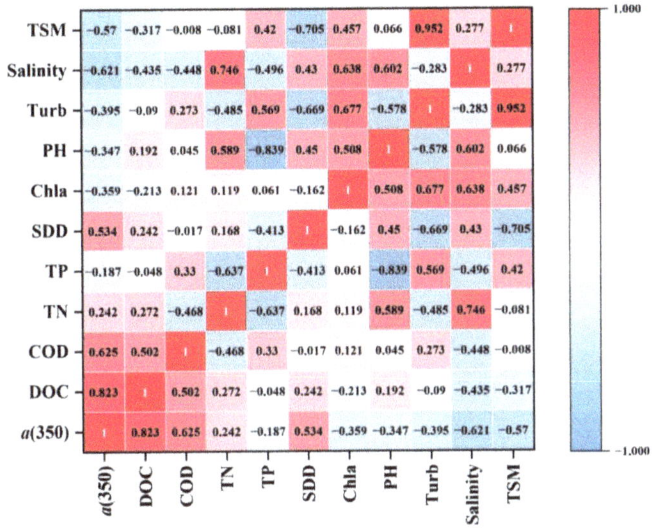

Figure 2. Correlation heat map of lake environmental water chemical parameters.

3.3. Modeling and Verification of CDOM and DOC

This study utilized two-thirds of the measured CDOM and the corresponding remote sensing reflectance data at sampling points to establish a concentration retrieving model for a wavelength of 350 nm. The remaining one third of the data were allocated to verify the accuracy of the model. We tried to use Matlab2016 and several machine learning algorithms to fit the relationship between the RrcB1-B17 band and CDOM absorption coefficient $a(350)$. The initial step involved determining the appropriate model that correlates CDOM and single-band reflectance $R_{rc}(\lambda)$ within the spectral range of 400–865 nm, or considering the various combinations of each band. After a huge amount of training, including, but not limited to, what is shown in Table 1, we found that $[Rrc(412.5) + Rrc(490)]/[Rrc(400)/Rrc(490)]$ had the closest correlation with CDOM absorption (modeling R^2 = 0.38).

Table 1. Performance of different algorithms in Lake Khanka.

Linear Model	Modeling R^2	Verify R^2
$a(350) = -87.54b5 + 12.28$	0.28	0.13
$a(350) = -82.74b4 + 11.38$	0.27	0.13
$a(350) = -90.71b6 + 14.06$	0.27	0.11
$a(350) = -45.71(b4 + b6) + 13.14$	0.29	0.13
$a(350) = -64.01\ (b6 - b13) + 10.71$	0.18	0.08
$a(350) = -462.9(b4 \times b6) + 8.60$	0.29	0.14
$a(350) = -29.57(b15/b16) + 28.89$	0.14	0.04
$a(350) = 52.34\ [(b2 + b4)/(b1/b4)] + 14.56$	0.38	0.20

Upon comparing the linear model (Table 1) with the machine learning model (Table 2), it has been observed that the machine learning model, constructed through nonlinear modeling using training data, exhibits superior performance and stability compared to the linear regression model. Consequently, machine learning has been chosen as the preferred approach for establishing the final model and implementing the inversion method.

Table 2. The accuracy evaluation of machine learning algorithm modeling and verification.

ML Algorithms	Modeling R^2	Verify R^2	RMSE (m^{-1})	MAE (m^{-1})
SVR	0.47	0.43	1.05	0.73
BP	0.68	0.59	0.81	0.55
XGBoost	0.74	0.56	0.73	0.53
RF	0.82	0.72	0.60	0.42
GBDT	0.95	0.84	0.31	0.23

The best fitting model is created by GBDT (modeling R^2 = 0.95) which is better than that of the band combination linear model produced by Matlab2016 and any other machine learning algorithms we have ever tried. Consequently, in this study, the machine learning approach was employed to model and retrieve the CDOM concentration of Lake Khanka (as shown in Figures 3–7).

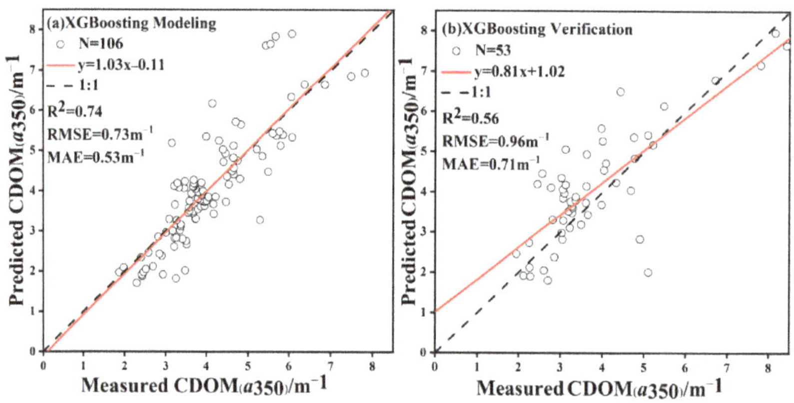

Figure 3. Modeling and validating scatterplots of eXtreme Gradient Boosting.

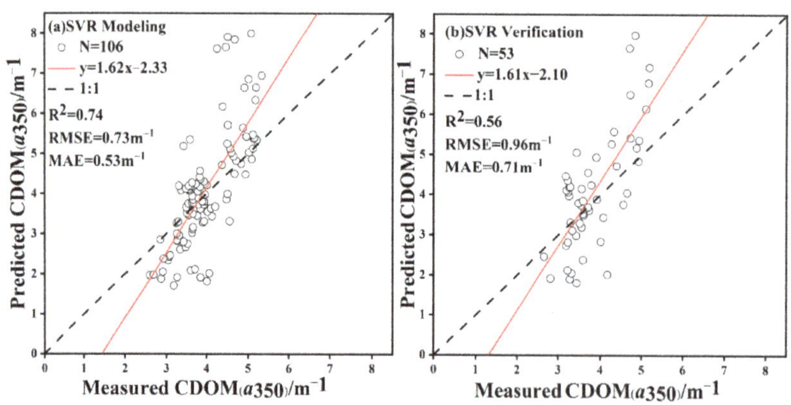

Figure 4. Modeling and validating scatterplots of Support Vector Regression.

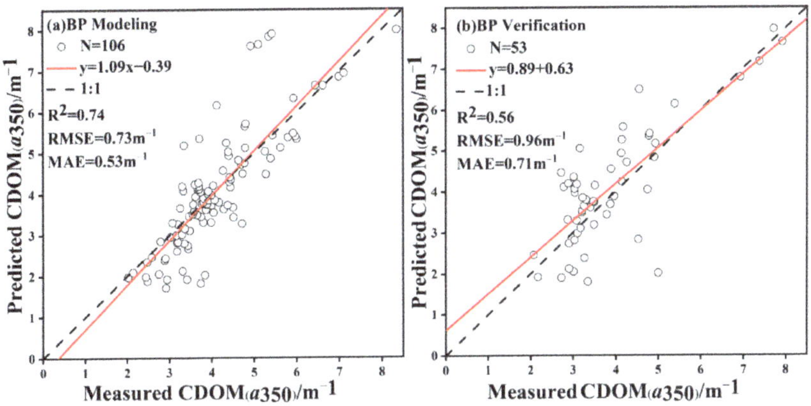

Figure 5. Modeling and validating scatterplots of Back Propagation.

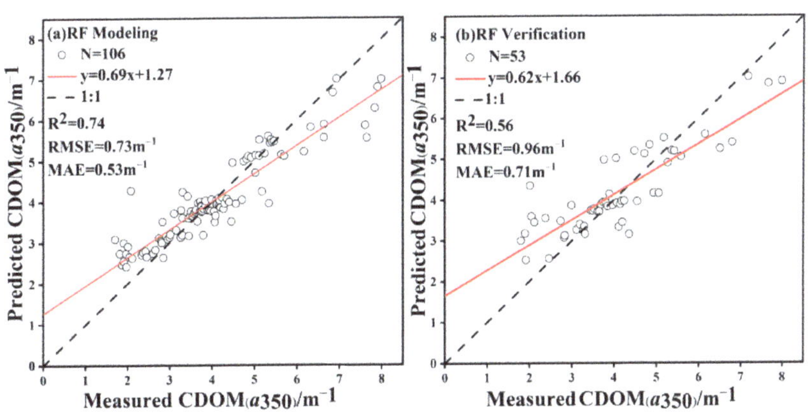

Figure 6. Modeling and validating scatterplots of Random Forest.

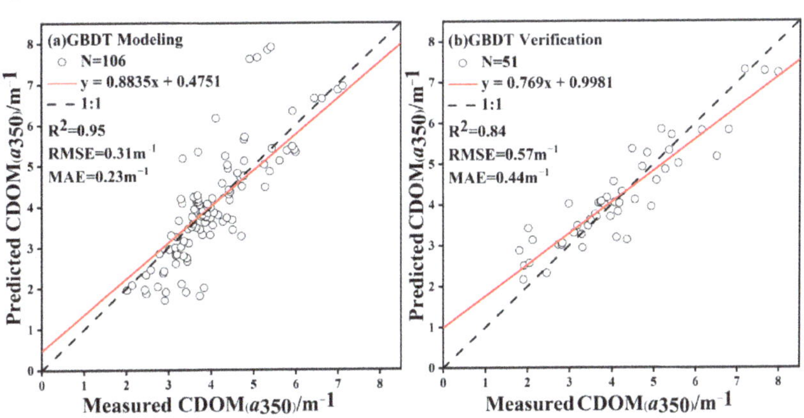

Figure 7. Modeling and validating scatterplots of Gradient Boosting Decision Tree.

Based on the analysis mentioned in Section 3.2, a prediction model for DOC concentration was established using two thirds of the measured CDOM absorption and DOC concentration data selected randomly. The accuracy of the model was then verified using the remaining one third of the data. The independent variable selected for establishing the linear regression equation between CDOM and DOC was the index with the highest correlation within the absorption parameters of CDOM. This selection was made using SPSS (version 26.0). The model accuracy is shown in Figure 8. $R^2 = 0.69$, indicated that the CDOM variable is the standard measure of DOC.

$$DOC = 0.7779aCDOM(350) + 1.8968 \quad (5)$$

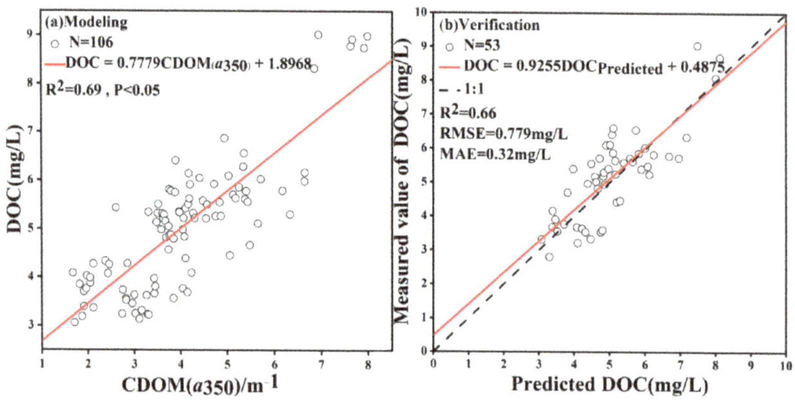

Figure 8. Construction and verification of CDOM absorption coefficient and DOC concentration model.

3.4. The Distribution of DOC for Lake Khanka

Our efforts were focused on the selection of cloudless or minimally cloudy images of Sentinel-3 from the period of 2016 to 2022. Subsequently, we applied the GBDT model established earlier to invert the CDOM value of a single image. The inversion results are integrated into the spatial distribution map representing the annual mean value. Then, according to the good linear correlation between CDOM-DOC, the grid calculator was used to enter the model formula to obtain the spatial distribution of the annual average DOC concentration in 2016–2022. The results are presented in Figure 9. The annual mean value of CDOM absorption coefficient for Lake Khanka falls within the range of 3.26–4.09 m^{-1}. Additionally, the annual mean value of DOC ranges from 4.56 to 5.19 mg/L.

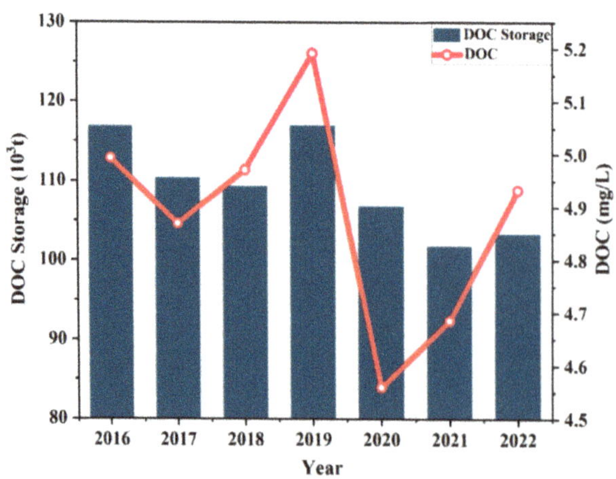

Figure 9. Annual mean value fluctuation and reserves of DOC in lake Khanka during 2016–2022.

The interannual variation in DOC concentration in Lake Khanka is minimal, and the overall level remains consistent over time. The provided illustration demonstrates the fluctuation of DOC value over a specific period. From 2016 to 2019, the value of DOC experienced an initial increase. However, in the subsequent period of 2019 to 2021, the value declined. Eventually, in 2022, the value stabilized at an average of 4.89 mg/L. Simultaneously, we estimated the DOC reserves of the lake to be 109.31 $\pm$ 5.56 (10^3 t). The normal DOC reserves [45] of Lake Khanka shown in Figure 10 are consistent with the annual DOC concentration changes. We also established a regression model (R^2 = 0.61, RMSE = 0.88) based on good correlation between CDOM and HIX to inverse the spatial distribution and mean of HIX. The linear model can be found in the Supplementary Materials.

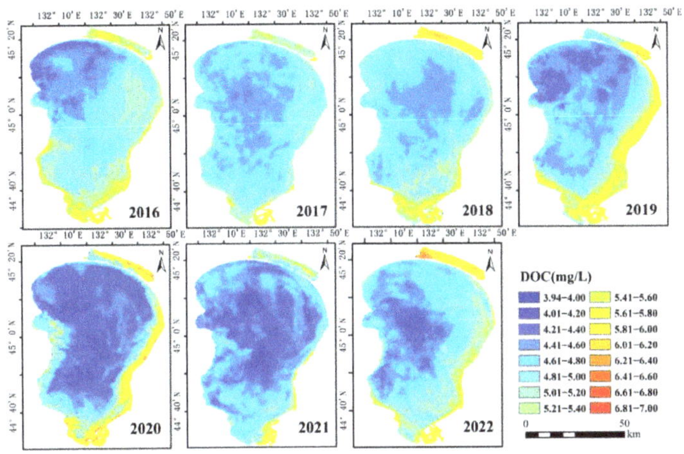

Figure 10. Inversion results of DOC concentration distribution in lake Khanka during 2016–2022.

From the spatial dimension (Figure 10), according to the DOC concentration inversion results, the overall trend is that the value of DOC in Small Khanka is bigger. The occurrence of low values was observed exclusively in Great Khanka, indicating that the overall water quality of Great Khanka is comparatively superior to that of Small Khanka. The concentration of DOC in the northwest of Small Khanka was tiny, while that in the southeast was great.

4. Discussion

4.1. Advantages and Disadvantages of the Model

The GBDT algorithm in machine learning is a classification prediction technique that relies on decision trees. It has the capability to combine multiple tree models and minimize errors through the gradient boosting process [46]. CDOM represents a fraction of the overall organic matter present in water, while DOC serves as a measure of the concentration of organic matter by quantifying its carbon content. CDOM is considered to be subordinate to DOC in its nature. Therefore, the indirect inversion of the temporal and spatial distribution of DOC through the relationship between CDOM and DOC has a theoretical basis.

When considering error estimation, it is important to note that various atmospheric correction algorithms have distinct impacts on OLCI remote sensing images [47]. The Rayleigh scattering correction we use may not be able to eliminate the possible adjacency effect absolutely on the coastal area of Lake Khanka. Due to the signal-to-noise ratio in the NIR bands, eventual residual NIR contributions from the bottom resuspension, and anisotropic reflectance of the land surface may influence adjacency effects in the NIR [48]. Unfortunately, an accurate estimate and likely correction of adjacency effects relies on the determination of the actual land albedo at the same time, and, at the same bands, our atmospheric correction cannot overcome the low SNR of the OLCI sensor. Future missions targeted at the observation of inland waters should take this issue into account. In the long run, improved atmospheric correction is still required to obtain reliable *Rrs* data under both optimal and non-optimal observing conditions, as most algorithms do rely on *Rrs* data. In the GBDT modeling process, the input bands involved in the calculation have a limited ability to eliminate the influence of the atmosphere on the image quality; thus, the original input variables may lead to some inevitable deviations to the real values of the images and bands. Additionally, the optical complexity of inland water is commonly influenced by elevated levels of phytoplankton biomass, Chla, non-algae particulate matter, and other factors. This overlapping absorption poses challenges in discerning the signals of multiple optically active substances during remote sensing [49–51]. One additional aspect to consider is the impact of CDOM and total suspended matter (TSM) on the attenuation of light penetration in water, which subsequently affects the phenomenon known as the water adjacency effect [50,52,53]. In that, high concentrations of TSM can cause an overestimation of CDOM, which should be considered to improve atmospheric correction in Case 2 water and reduce the signals of other optical components in NIR by employing a novel band ratio combination. Last but not least, in recent years, most algorithms for estimating CDOM through remote sensing technology have been empirical or semi-empirical [17,22,54], lacking the theoretical basis of radiative transfer, and the extension of the model is contingent upon the limitations imposed by specific regional conditions, as determined through measurements and experiments.

4.2. *Factors Affecting the Change in DOM*

CDOM-DOC exhibits significant heterogeneity on both temporal and spatial scales. Its dynamic fluctuations are primarily influenced by the combined impacts of diverse human activities, including agricultural runoff, industrial pollution, and natural factors within the watershed. These natural factors encompass solar radiation, temperature (TEMP), precipitation (PRCP), Wind Speed (WDSP), Station Pressure (STP), as well as disturbances, e.g., wastewater discharge, material transport, and energy conversion in water and sediments caused by human activities.

4.2.1. Natural Environmental Attributions

We collected the daily and monthly natural factors at the meteorological station in JiXi with DOC concentration for multiple regression analysis (Figure 11). Multiple regression analysis was performed in SPSS on the natural factors (TEMP, STP, WDSP, and PRCP) corresponding to every single date. In the multiple regression model, Beta represents the standardized regression coefficient, representing the degree to which the dependent variable can be explained. Thus, Beta represents the degree of impact of such factors on DOC. From the perspective of distinct time scales, TEMP has the greatest impact within a day, followed by PRCP, WDSP, and STP; but, within a month, the influencing factors in descending order are STP, TEMP, and WDSP. The long-term influence of various natural factors results in a consistent effect on the concentration of DOC. Specifically, the summer season is typically characterized by high temperatures and abundant rainfall in the Xingkai–Muleng basin. This climatic pattern results in a significant influx of organic matter from the upstream river, leading to an elevated concentration of organic pollutants. Over one day, temperature and precipitation have the greatest impact. The phenomenon of high-intensity precipitation induces the soil-leaching effect, resulting in an influx of water supply to the lake and a rapid increase in runoff into the lake, which dilute the concentration of organic pollutants in the current lake, accelerate the metabolism of water and also the reproduction and death of microorganisms, consume a large amount of organic matter, and, as a result, reduce the concentration of CDOM, DOC, and COD. But, within 30 days, the combination of temperature, pressure, and wind speed can explain a greater impact. The temperature can affect the decomposition rate of organic matter. Since Lake Khanka is relatively shallow (1.70–3.99 m), the action of strong winds and atmospheric pressure on the underlying surface will cause sediment resuspension, which may release a large amount of organic matter.

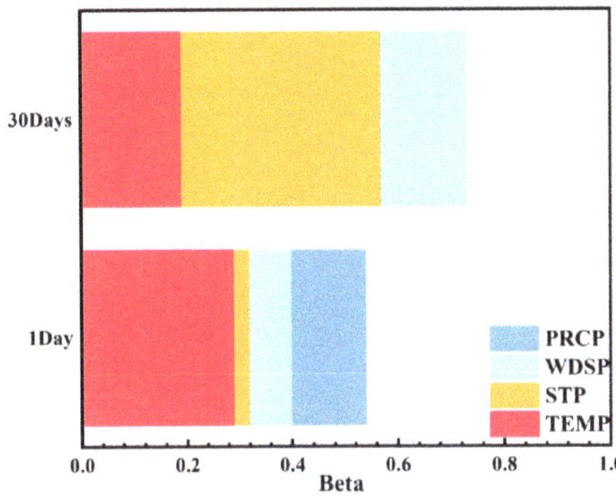

Figure 11. Contribution rate of natural factors (PRCP, WDSP, STP, and TEMP) to DOC changes in 1 day and 30 days in multiple regression.

Considering the hydrological conditions, the Muleng River, one of the main tributaries of the Ussuri River, flows through Jixi City (a coal and industrial city with a population of 1.6 million) with tremendous DOC which is severely polluted by industrial wastewater and sewage. Songacha River, the only discharge river to the east of Lake Khanka, has a minimum water depth of only 1 m, and a natural overflow weir has been formed at the mouth of the river, which seriously hinders the discharge of the lake area, causing the lake in the east of Lake Khanka to inherit more organic matter. From the perspective of topography, the topography DEM of the Muleng–Xingkai watershed where Lake Khanka is located is generally higher in the northwest (250–500 m) and lower in the southeast (50–72.5 m), which results in the flow of the Lake Khanka system from west to east. However, these natural factors cannot explain 100% of the changes that might happen in DOC, and other natural and human factors require profound exploration.

4.2.2. Human Activity Attributions

Due to the deficiency of accurate data on interannual changes in human factors from 2016 to 2022, they are insufficient to support data analysis. However, we can still deconstruct the impact of human activities on CDOM, DOC, and the water environment from human traces in land use patterns. The land use on the west bank of Lake Khanka is mainly cultivated land and artificial surface. The excessive application of agricultural chemical fertilizers, as well as the presence of aquaculture and livestock breeding in the vicinity, result in agricultural runoff. This runoff carries a significant quantity of organic matter into Lake Khanka, leading to the enrichment of various nutrients and organic pollutants in the lake. The east bank of Lake Khanka is dominated by large wetland grasslands, and the surrounding wetland ecosystem has become the main source of DOC on the east bank of Lake Khanka because of its rich organic matter and huge carbon storage capacity. This is consistent with the results of the research of Liu [55]. Notably, 2020 was the first year of the new coronavirus outbreak. The epidemic has led to a reduction in tourism and agricultural activities, and the average annual DOC concentration has dropped sharply compared to 2019. In 2021, this new epidemic appeared to be easing, and the DOC concentration gradually increased as production and life resumed. HIX is almost consistent with CDOM and DOC changes (Figure 12). The weak HIX (HIX < 4) indicated that Lake Khanka was abundant with more biological or recent autochthonous components [35] and that the large abnormal fluctuations (HIX > 4) might be caused by interference from human activities. Consequently, this process directly or indirectly contributes to the conversion of these substances into CDOM or DOC, among other compounds. Additionally, SUVA provides insights into the level of humification of organic matter in water; it is shown to be a useful parameter for estimating the dissolved aromatic carbon content in aquatic systems [56]. $SUVA_{254}$ of the Great Khanka is greater than that of the Small Khanka, and the large standard deviation indicates that the Great Khanka is more affected by terrestrial aromatic organic carbon input.

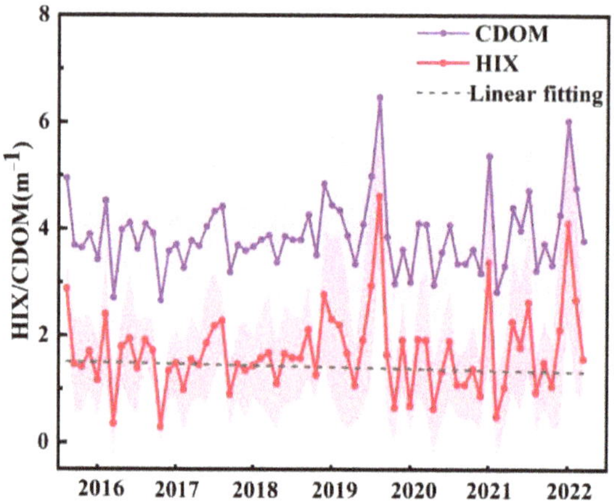

Figure 12. CDOM and HIX average values in the inversion results of Lake Khanka images in chronological order.

4.3. Environmental Significance and Prospect of Lake Potential Water Quality Parameter Evaluation

4.3.1. Implications for Potential Water Quality Parameters and Carbon Cycling in Lakes

The water quality of Lake Khanka significantly influences the economic development and ecological equilibrium between China and Russia. The measurement of total nitrogen, total phosphorus, and permanganate index COD are crucial water quality parameters for assessing the level of eutrophication of lakes. Because these indicators do not have optical properties, the indirect inversion of these substances by remote sensing is used to infer their concentrations. A notable correlation exists between the primary indicators of organic pollution, namely COD and CDOM absorption coefficient, and the concentration of DOC in the Great Khanka and Small Khanka water bodies. The estimation of CDOM-DOC by means of remote sensing can further estimate the concentration and distribution of COD indirectly, and then conduct more comprehensive testing and analysis of water quality conditions. The correlation between COD and DOC exhibits a stronger relationship ($R = 0.78$, $p < 0.05$). Consequently, it is crucial to prioritize the management of DOC in efforts aimed at mitigating the migration and transformation of pollutants in the water environment.

The application of remote sensing technology for the large-scale inversion of DOC concentration continues to encounter significant challenges. Many studies [57–62] have shown that the current relationship between DOC concentration and carbon flux is unprecedented, and relevant hypotheses indicate that the continued rise in DOC levels will have unpredictable consequences for the global carbon cycle [63,64]. Long-term DOC increases may have wide-ranging impacts on freshwater biota, drinking water quality, and inland water ecosystems. More specifically, the agricultural and fishery benefits of the farms around Lake Khanka will be severely affected, and humans will face more risks regarding drinking water and food safety along with unprecedented threats to ecological balance.

4.3.2. Shortcomings and Prospects of the Study

CDOM exhibits a specific vertical divergence pattern within the aquatic environment. The concentration and composition of CDOM in the surface layer of water frequently exhibit significant disparities compared to the deep layer. This discrepancy arises primarily from the process of photochemical degradation. The optical signals detected by remote sensing sensors in deep water are strongly correlated with water quality parameters, specifically water transparency and turbidity. Excluding the sensor's sensitivity to water optical signals, Lake Khanka has low transparency (21.50 ± 5.89 cm) and high turbidity (68.16 ± 25.16 NTU). The main optical signal of the water body comes from optically active substances such as surface CDOM; however, the concentration and distribution of substances in the lower layer have little effect on the optical response of the surface. In this particular scenario, it is recommended that further investigation be directed towards the longitudinal distribution of CDOM in aquatic environments. When employing CDOM as an indirect means to assess the lateral spatial distribution of DOC, this variability introduces inconsistencies in CDOM measurements.

Since the latitude of Lake Khanka varies from 43°55′ to 46°30′, the lake is located in the north temperate monsoon climate zone where the winter is cold and lasts long, and thus the freezing period is about 168 days [65]. The presence of a thick ice layer hinders the remote sensing inversion of CDOM. Consequently, the temporal and spatial distribution of CDOM in Lake Khanka may be overestimated if the annual average is not taken into account and glacial image data are excluded. More attention should be paid to considering these conditions in the future monitoring of CDOM and DOC.

The OLCI data possess a significant time resolution that enables efficient utilization of both macro and micro time scales for the purpose of monitoring lake water quality. In the future, the integration of daily, monthly, and annual variations in water parameters will be employed to assess alterations in lake water quality and biogeochemical cycles. The limited duration of Sentinel-3′s launch, spanning from 2016 to present, has resulted in an insufficient time frame for the available annual remote sensing data. Additionally, there is an inadequate quantity of matched driving factor data to support comprehensive data analysis. As a result, the further exploration of natural factors and human activities that have influenced the spatial and temporal variation in CDOM-DOC should be implemented properly.

5. Conclusions

Results from the research indicate that the modeling approach utilizing machine learning GBDT outperforms the linear regression model based on band combination in accurately predicting the concentration of CDOM in Lake Khanka from 2016 to 2022. A comprehensive analysis was conducted on a monthly basis from 2016 to 2021, specifically focusing on the nonglacial period. The parameters examined encompassed both optical and non-optical characteristics of inland water, providing an accurate representation of the water quality conditions. The analysis of the spatial distribution of CDOM and DOC reveals that the Small Khanka Lake exhibits a larger size compared to the Great Khanka Lake. Additionally, the perimeter of the lake is generally greater than its central region. The concentration of DOC in the Great Khanka is reported to be 4.74 ± 0.12 mg/L, while in the Small Khanka it is 5.50 ± 0.30 mg/L. These measurements were recorded between the years 2016 and 2022.

The establishment of the CDOM-DOC relationship enables the potential prediction of DOM distribution and sources. Additionally, the correlation coefficient with COD and HIX also further indicates the occurrence of organic matter in water. These variations in size and distribution may be attributed to the impact of natural factors, human industrial and agricultural activities along the coast, as well as the succession of the aquatic ecological environment. Meteorological factors including temperature, wind speed, precipitation, and pressure, play a role in the distribution of CDOM in the short term. In order to ensure comprehensive planning and sustainable management, it is imperative to prioritize the examination of long-term land use alterations and potential impacts on the water environment resulting from human activities. Given the significant geographical location, ecological functions, and economic value attributed to Lake Khanka, ensuring the effective monitoring and treatment of water quality within its watershed is crucial for promoting sustainable development for both human and natural systems. This study provides valuable insights into the analysis of organic matter in lakes and the distribution of carbon sources. Additionally, it establishes a strong foundation for future investigations into the source and composition of carbon in lakes.

Supplementary Materials: The following supporting information can be downloaded at https://www.mdpi.com/article/10.3390/rs15245707/s1, Figure S1. CDOM absorption coefficient curve of samples from 200 to 800 nm. Figure S2. Sentinel 3 OLCI reflectance *Rrc* after atmospheric correction between Oa1–Oa16. Figure S3. Accuracy verification of HIX and CDOM concentration model. Table S1. Sentinel-3 Ocean and Land Color Instrument (S3-OLCI) spectral and spatial specifications. Table S2. Descriptive statistics of water quality parameters.

Author Contributions: S.Q.: Conceptualization, data curation, formal analysis, methodology, visualization, roles/writing—original draft. K.S.: funding acquisition, project administration, resources. Y.S.: funding acquisition, project administration, supervision, writing—review and editing. F.L.: software, investigation, resources. Z.W.: investigation, resources. G.L.: software, resources. H.T.: investigation, resources. Y.L.: funding acquisition, project administration, investigation. All authors have read and agreed to the published version of the manuscript.

Funding: This research was funded by the Strategic Priority Research Program of the Chinese Academy of Sciences (XDA28100100), Natural Science Foundation of China (No. 41231390 and NO.41971322), the Science and Technology development plan project of Jilin Province (No. 20220508017RC), the Special Postdoctoral Fellowship of Jilin Province of China granted for Dr. Yingxin Shang, the Youth Innovation Promotion Association of Chinese Academy of Sciences (2020234), Young Scientist Group Project of Northeast Institute of Geography and Agroecology, Chinese Academy of Sciences (2022QNXZ03), Municipal Academy of Science and Technology Innovation Cooperation Project, Changchun Surface Water Quality Sky-Ground Integrated Remote Sensing Monitoring Technology Research and Development (21SH10), and Jilin Province and Chinese Academy of Sciences Science and Technology Cooperation High-Tech Industrialization Special Fund Project (2021SYHZ0002).

Data Availability Statement: Data are contained within the article and Supplementary Materials.

Conflicts of Interest: The authors declare no conflict of interest.

References

1. Casas-Ruiz, J.P.; Jakobsson, J.; del Giorgio, P.A. The role of lake morphometry in modulating surface water carbon concentrations in boreal lakes. *Environ. Res. Lett.* **2021**, *16*, 11. [CrossRef]
2. Wang, H.J.; Wang, Y.; Han, X.; Zhang, J.; Liu, J.L.; Xiang, Y.F.; Zhou, M.X.; Guo, P.L.; Tang, F.; Liu, A.L. Monitoring DOM in drinking water supply systems using DOC, CODMn, UV and fluorescence measurements. *Environ. Sci.-Water Res. Technol.* **2021**, *7*, 2307–2320. [CrossRef]
3. Lyu, L.L.; Wen, Z.D.; Jacinthe, P.A.; Shang, Y.X.; Zhang, N.; Liu, G.; Fang, C.; Hou, J.B.; Song, K.S. Absorption characteristics of CDOM in treated and non-treated urban lakes in Changchun, China. *Environ. Res.* **2020**, *182*, 109084. [CrossRef] [PubMed]
4. Williamson, C.E.; Neale, P.J.; Hylander, S.; Rose, K.C.; Figueroa, F.L.; Robinson, S.A.; Häder, D.P.; Wängberg, S.; Worrest, R.C. The interactive effects of stratospheric ozone depletion, UV radiation, and climate change on aquatic ecosystems. *Photochem. Photobiol. Sci.* **2019**, *18*, 717–746. [CrossRef] [PubMed]
5. Minor, E.C.; Oyler, A.R. Dissolved organic matter in large lakes: A key but understudied component of the carbon cycle. *Biogeochemistry* **2023**, *164*, 295–318. [CrossRef]
6. Wan, D.; Sharma, V.K.; Liu, L.; Zuo, Y.G.; Chen, Y. Mechanistic Insight into the Effect of Metal Ions on Photogeneration of Reactive Species from Dissolved Organic Matter. *Environ. Sci. Technol.* **2019**, *53*, 5778–5786. [CrossRef] [PubMed]
7. Kallio, K.; Koponen, S.; Ylostalo, P.; Kervinen, M.; Pyhalahti, T.; Attila, J. Validation of MERIS spectral inversion processors using reflectance, IOP and water quality measurements in boreal lakes. *Remote Sens. Environ.* **2015**, *157*, 147–157. [CrossRef]
8. Brezonik, P.L.; Olmanson, L.G.; Finlay, J.C.; Bauer, M.E. Factors affecting the measurement of CDOM by remote sensing of optically complex inland waters. *Remote Sens. Environ.* **2015**, *157*, 199–215. [CrossRef]
9. Nelson, N.B.; Siegel, D.A. The Global Distribution and Dynamics of Chromophoric Dissolved Organic Matter. In *Annual Review of Marine Science*; Carlson, C.A., Giovannoni, S.J., Eds.; Annual Reviews: Palo Alto, CA, USA, 2013; Volume 5, pp. 447–476.

10. Massicotte, P.; Asmala, E.; Stedmon, C.; Markager, S. Global distribution of dissolved organic matter along the aquatic continuum: Across rivers, lakes and oceans. *Sci. Total Environ.* **2017**, *609*, 180–191. [CrossRef]
11. Lambert, T.; Teodoru, C.R.; Nyoni, F.C.; Bouillon, S.; Darchambeau, F.; Massicotte, P.; Borges, A.V. Along-stream transport and transformation of dissolved organic matter in a large tropical river. *Biogeosciences* **2016**, *13*, 2727–2741. [CrossRef]
12. Osburn, C.L.; Wigdahl, C.R.; Fritz, S.C.; Saros, J.E. Dissolved organic matter composition and photoreactivity in prairie lakes of the U.S. Great Plains. *Limnol. Oceanogr.* **2011**, *56*, 2371–2390. [CrossRef]
13. Sgroi, M.; Roccaro, P.; Korshin, G.V.; Greco, V.; Sciuto, S.; Anumol, T.; Snyder, S.A.; Vagliasindi, F.G.A. Use of fluorescence EEM to monitor the removal of emerging contaminants in full scale wastewater treatment plants. *J. Hazard. Mater.* **2017**, *323*, 367–376. [CrossRef] [PubMed]
14. Griffin, C.G.; Finlay, J.C.; Brezonik, P.L.; Olmanson, L.; Hozaiski, R.M. Limitations on using CDOM as a proxy for DOC in temperate lakes. *Water Res.* **2018**, *144*, 719–727. [CrossRef] [PubMed]
15. Song, K.S.; Zhao, Y.; Wen, Z.D.; Fang, C.; Shang, Y.X. A systematic examination of the relationships between CDOM and DOC in inland waters in China. *Hydrol. Earth Syst. Sci.* **2017**, *21*, 5127–5141. [CrossRef]
16. Shao, T.T.; Song, K.S.; Du, J.; Zhao, Y.; Liu, Z.M.; Zhang, B. Retrieval of CDOM and DOC Using In Situ Hyperspectral Data: A Case Study for Potable Waters in Northeast China. *J. Indian Soc. Remote Sens.* **2016**, *44*, 77–89. [CrossRef]
17. Cao, F.; Tzortziou, M.; Hu, C.M.; Mannino, A.; Fichot, C.G.; Del Vecchio, R.; Najjar, R.G.; Novak, M. Remote sensing retrievals of colored dissolved organic matter and dissolved organic carbon dynamics in North American estuaries and their margins. *Remote Sens. Environ.* **2018**, *205*, 151–165. [CrossRef]
18. Herrault, P.A.; Gandois, L.; Gascoin, S.; Tananaev, N.; Le Dantec, T.; Teisserenc, R. Using High Spatio-Temporal Optical Remote Sensing to Monitor Dissolved Organic Carbon in the Arctic River Yenisei. *Remote Sens.* **2016**, *8*, 803. [CrossRef]
19. Vantrepotte, V.; Danhiez, F.P.; Loisel, H.; Ouillon, S.; Meriaux, X.; Cauvin, A.; Dessailly, D. CDOM-DOC relationship in contrasted coastal waters: Implication for DOC retrieval from ocean color remote sensing observation. *Opt. Express* **2015**, *23*, 33–54. [CrossRef]
20. Zhang, Y.L.; Yang, L.Y.; Qin, B.Q.; Gao, G.; Liu, M.L. Spatial distribution of COD and the correlations with other parameters in the northern region of Lake Taihu. *Huan Jing Ke Xue* **2008**, *29*, 1457–1462.
21. Coelho, C.; Heim, B.; Foerster, S.; Brosinsky, A.; de Araujo, J.C. In Situ and Satellite Observation of CDOM and Chlorophyll-a Dynamics in Small Water Surface Reservoirs in the Brazilian Semiarid Region. *Water* **2017**, *9*, 913. [CrossRef]
22. Olmanson, L.G.; Brezonik, P.L.; Finlay, J.C.; Bauer, M.E. Comparison of Landsat 8 and Landsat 7 for regional measurements of CDOM and water clarity in lakes. *Remote Sens. Environ.* **2016**, *185*, 119–128. [CrossRef]
23. Zhu, W.N.; Yu, Q.; Tian, Y.Q.; Becker, B.L.; Zheng, T.; Carrick, H.J. An assessment of remote sensing algorithms for colored dissolved organic matter in complex freshwater environments. *Remote Sens. Environ.* **2014**, *140*, 766–778. [CrossRef]
24. Attila, J.; Koponen, S.; Kallio, K.; Lindfors, A.; Kaitala, S.; Ylostalo, P. MERIS Case II water processor comparison on coastal sites of the northern Baltic Sea. *Remote Sens. Environ.* **2013**, *128*, 138–149. [CrossRef]
25. Huang, M.F.; Xing, X.F.; Zhao, Z.L.; Li, Z.Q.; Wang, X.X. Inversion of CDOM and COD in water using HJ-1/CCD data. In Proceedings of the 35th International Symposium on Remote Sensing of Environment (ISRSE35), Inst Remote Sensing & Digital Earth, Beijing, China, 22–26 April 2013.
26. Wu, Z.S.; Wang, X.L.; Chen, Y.W.; Cai, Y.J.; Deng, J.C. Assessing river water quality using water quality index in Lake Taihu Basin, China. *Sci. Total Environ.* **2018**, *612*, 914–922. [CrossRef] [PubMed]
27. Yan, X.G.; Li, J.; Smith, A.R.; Yang, D.; Ma, T.Y.; Su, Y.T.; Shao, J.H. Evaluation of machine learning methods and multi-source remote sensing data combinations to construct forest above-ground biomass models. *Int. J. Digit. Earth* **2023**, *16*, 4471–4491. [CrossRef]
28. Devos, L.; Meert, W.; Davis, J. Fast Gradient Boosting Decision Trees with Bit-Level Data Structures. In Proceedings of the European Conference on Machine Learning and Principles and Practice of Knowledge Discovery in Databases (ECML PKDD), Wurzburg, Germany, 16–20 September 2019; pp. 590–606.
29. Ruescas, A.B.; Hieronymi, M.; Mateo-Garcia, G.; Koponen, S.; Kallio, K.; Camps-Valls, G. Machine Learning Regression Approaches for Colored Dissolved Organic Matter (CDOM) Retrieval with S2-MSI and S3-OLCI Simulated Data. *Remote Sens.* **2018**, *10*, 786. [CrossRef]
30. Aurin, D.; Mannino, A.; Lary, D.J. Remote Sensing of CDOM, CDOM Spectral Slope, and Dissolved Organic Carbon in the Global Ocean. *Appl. Sci.* **2018**, *8*, 2687. [CrossRef]
31. Ruescas, A.B.; Hieronymi, M.; Koponen, S.; Kallio, K.; Camps-Vallsi, G. Retrieval of coloured dissolved organic matter with machine learning methods. In Proceedings of the IEEE International Geoscience & Remote Sensing Symposium, Fort Worth, TX, USA, 23–28 July 2017; pp. 2187–2190.
32. Yu, S.L.; Li, X.Y.; Wen, B.L.; Chen, G.S.; Hartleyc, A.; Jiang, M.; Li, X.J. Characterization of Water Quality in Xiao Xingkai Lake: Implications for Trophic Status and Management. *Chin. Geogr. Sci.* **2021**, *31*, 558–570. [CrossRef]
33. Zhang, Y.L.; Zhou, L.; Zhou, Y.Q.; Zhang, L.Q.; Yao, X.L.; Shi, K.; Jeppesen, E.; Yu, Q.; Zhu, W.N. Chromophoric dissolved organic matter in inland waters: Present knowledge and future challenges. *Sci. Total Environ.* **2021**, *759*, 13. [CrossRef]
34. Spencer, R.G.M.; Butler, K.D.; Aiken, G.R. Dissolved organic carbon and chromophoric dissolved organic matter properties of rivers in the USA. *J. Geophys. Res.-Biogeosci.* **2012**, *117*. [CrossRef]

35. Shang, Y.X.; Song, K.S.; Lai, F.F.; Lyu, L.; Liu, G.; Fang, C.; Hou, J.B.; Qiang, S.N.; Yu, X.F.; Wen, Z.D. Remote sensing of fluorescent humification levels and its potential environmental linkages in lakes across China. *Water Res.* **2023**, *230*, 13. [CrossRef] [PubMed]
36. Lyu, L.; Song, K.S.; Wen, Z.D.; Liu, G.; Fang, C.; Shang, Y.X.; Li, S.J.; Tao, H.; Wang, X.; Li, Y.; et al. Remote estimation of phycocyanin concentration in inland waters based on optical classification. *Sci. Total Environ.* **2023**, *899*, 14. [CrossRef]
37. Li, X.; Zhang, J.; Yu, W.; Liu, L.; Wang, W.; Cui, Z.; Wang, W.; Wang, R.; Li, Y. How Landscape Patterns Affect River Water Quality Spatially and Temporally: A Multiscale Geographically Weighted Regression Approach. *J. Environ. Inform.* **2023**, *42*, 158–172. [CrossRef]
38. Wei, Q.; Wei, Q.; Li, S.Y.; Xu, J.Z.; Yang, Z.H.; Liu, Z.Y.; Chen, P.; Liu, Y.Z.; Ding, Y.M.; Tan, J.Y.; et al. Evaluation of surface water quality in Heilongjiang Province, China: Based on different quantities of water quality indicators. *Ecol. Indic.* **2023**, *154*. [CrossRef]
39. Wen, Z.D.; Song, K.S.; Liu, G.; Shang, Y.X.; Hou, J.B.; Lyu, L.L.; Fang, C. Impact factors of dissolved organic carbon and the transport in a river-lake continuum in the Tibet Plateau of China. *J. Hydrol.* **2019**, *579*, 10. [CrossRef]
40. Li, S.J.; Song, K.S.; Zhao, Y.; Mu, G.Y.; Ma, J.H. Absorption Characteristics of Particulates and CDOM in Waters of Chagan Lake and Xinlicheng Reservoir in Autumn. *Huan Jing Ke Xue* **2016**, *37*, 112–122.
41. Danhiez, F.P.; Vantrepotte, V.; Cauvin, A.; Lebourg, E.; Loisel, H. Optical properties of chromophoric dissolved organic matter during a phytoplankton bloom. Implication for DOC estimates from CDOM absorption. *Limnol. Oceanogr.* **2017**, *62*, 1409–1425. [CrossRef]
42. Fichot, C.G.; Benner, R. A novel method to estimate DOC concentrations from CDOM absorption coefficients in coastal waters. *Geophys. Res. Lett.* **2011**, *38*. [CrossRef]
43. Hestir, E.L.; Brando, V.; Campbell, G.; Dekker, A.; Malthus, T. The relationship between dissolved organic matter absorption and dissolved organic carbon in reservoirs along a temperate to tropical gradient. *Remote Sens. Environ.* **2015**, *156*, 395–402. [CrossRef]
44. Joshi, I.D.; D'Sa, E.J.; Osburn, C.L.; Bianchi, T.S.; Ko, D.S.; Oviedo-Vargas, D.; Arellano, A.R.; Ward, N.D. Assessing chromophoric dissolved organic matter (CDOM) distribution, stocks, and fluxes in Apalachicola Bay using combined field, VIIRS ocean color, and model observations. *Remote Sens. Environ.* **2017**, *191*, 359–372. [CrossRef]
45. Song, K.S.; Wen, Z.D.; Shang, Y.X.; Yang, H.; Lyu, L.L.; Liu, G.; Fang, C.; Du, J.; Zhao, Y. Quantification of dissolved organic carbon (DOC) storage in lakes and reservoirs of mainland China. *J. Environ. Manag.* **2018**, *217*, 391–402. [CrossRef] [PubMed]
46. Zhang, C.S.; Liu, C.C.; Zhang, X.L.; Almpanidis, G. An up-to-date comparison of state-of-the-art classification algorithms. *Expert Syst. Appl.* **2017**, *82*, 128–150. [CrossRef]
47. Toming, K.; Kutser, T.; Laas, A.; Sepp, M.; Paavel, B.; Noges, T. First Experiences in Mapping Lake Water Quality Parameters with Sentinel-2 MSI Imagery. *Remote Sens.* **2016**, *8*, 640. [CrossRef]
48. Bulgarelli, B.; Zibordi, G. On the detectability of adjacency effects in ocean color remote sensing of mid-latitude coastal environments by SeaWiFS, MODIS-A, MERIS, OLCI, OLI and MSI. *Remote Sens. Environ.* **2018**, *209*, 423–438. [CrossRef] [PubMed]
49. Adhikari, A.; Menon, H.B.; Lotliker, A. Coupling of hydrography and bio-optical constituents in a shallow optically complex region using ten years of in-situ data. *ISPRS J. Photogramm. Remote Sens.* **2023**, *202*, 499–511. [CrossRef]
50. Zhang, H.; Yao, B.; Wang, S.R.; Huang, Y.Q. Understanding the changes of optically active substances (OACs) in Hulun Lake in the past 35 years and its indication to the degradation of aquatic ecology. *J. Clean. Prod.* **2022**, *377*, 18. [CrossRef]
51. Li, H.; He, X.Q.; Shanmugam, P.; Bai, Y.; Wang, D.F.; Huang, H.Q.; Zhu, Q.K.; Gong, F. Radiometric Sensitivity and Signal Detectability of Ocean Color Satellite Sensor Under High Solar Zenith Angles. *IEEE Trans. Geosci. Remote Sens.* **2019**, *57*, 8492–8505. [CrossRef]
52. Soppa, M.A.; Pefanis, V.; Hellmann, S.; Losa, S.N.; Hölemann, J.; Martynov, F.; Heim, B.; Janout, M.A.; Dinter, T.; Rozanov, V.; et al. Assessing the Influence of Water Constituents on the Radiative Heating of Laptev Sea Shelf Waters. *Front. Mar. Sci.* **2019**, *6*, 13. [CrossRef]
53. Tilstone, G.H.; Angel-Benavides, I.M.; Pradhan, Y.; Shutler, J.D.; Groom, S.; Sathyendranath, S. An assessment of chlorophyll-a algorithms available for SeaWiFS in coastal and open areas of the Bay of Bengal and Arabian Sea. *Remote Sens. Environ.* **2011**, *115*, 2277–2291. [CrossRef]
54. Kutser, T.; Pierson, D.C.; Kallio, K.Y.; Reinart, A.; Sobek, S. Mapping lake CDOM by satellite remote sensing. *Remote Sens. Environ.* **2005**, *94*, 535–540. [CrossRef]
55. Li-Zhen, L.; Qi, H.; Yong-Ming, W.U.; Dai-She, W.U.; Hai-Lin, Y.; Mi, D. Optical Absorption Properties of Chromophoric Dissolvable Organic Matter (CDOM) and Their Quantitative Relationships With Dissolved Organic Carbon in the Poyang Lake in Dry Season. *J. Ecol. Rural Environ.* **2017**, *33*, 762–768.
56. Philibert, M.; Luo, S.M.; Moussanas, L.; Yuan, Q.Q.; Filloux, E.; Zraick, F.; Murphy, K.R. Drinking water aromaticity and treatability is predicted by dissolved organic matter fluorescence. *Water Res.* **2022**, *220*, 12. [CrossRef]
57. Freeman, C.; Fenner, N.; Ostle, N.J.; Kang, H.; Dowrick, D.J.; Reynolds, B.; Lock, M.A.; Sleep, D.; Hughes, S.; Hudson, J. Export of dissolved organic carbon from peatlands under elevated carbon dioxide levels. *Nature* **2004**, *430*, 195–198. [CrossRef] [PubMed]
58. Findlay, S.E.G. Increased carbon transport in the Hudson River: Unexpected consequence of nitrogen deposition? *Front. Ecol. Environ.* **2005**, *3*, 133–137. [CrossRef]
59. Cheng, J.B.; Chen, Y.C.; He, T.B.; Liao, R.J.; Liu, R.L.; Yi, M.; Huang, L.; Yang, Z.M.; Fu, T.L.; Li, X.Y. Soil nitrogen leaching decreases as biogas slurry DOC/N ratio increases. *Appl. Soil Ecol.* **2017**, *111*, 105–113. [CrossRef]

60. Clark, J.M.; Ashley, D.; Wagner, M.; Chapman, P.J.; Lane, S.N.; Evans, C.D.; Heathwaite, A.L. Increased temperature sensitivity of net DOC production from ombrotrophic peat due to water table draw-down. *Glob. Chang. Biol.* **2009**, *15*, 794–807. [CrossRef]
61. Shi, S.W.; Yang, M.X.; Hou, Y.; Peng, C.H.; Wu, H.B.; Zhu, Q.A.; Liang, Q.; Xie, J.F.; Wang, M. Simulation of dissolved organic carbon concentrations and fluxes in Chinese monsoon forest ecosystems using a modified TRIPLEX-DOC model. *Sci. Total Environ.* **2019**, *697*, 15. [CrossRef]
62. Fichot, C.G.; Tzortziou, M.; Mannino, A. Remote sensing of dissolved organic carbon (DOC) stocks, fluxes and transformations along the land-ocean aquatic continuum: Advances, challenges, and opportunities. *Earth-Sci. Rev.* **2023**, *242*, 27. [CrossRef]
63. Liu, F.T.; Wang, D. Dissolved organic carbon concentration and biodegradability across the global rivers: A meta-analysis. *Sci. Total Environ.* **2022**, *818*, 8. [CrossRef]
64. Hansell, D.A. Recalcitrant Dissolved Organic Carbon Fractions. In *Annual Review of Marine Science*; Carlson, C.A., Giovannoni, S.J., Eds.; Annual Reviews: Palo Alto, CA, USA, 2013; Volume 5, pp. 421–445.
65. Ke, C.Q.; Cai, Y.; Xiao, Y. Passive microwave remote sensing monitoring of ice phenological changes in Xingkai Lake from 1979 to 2019. *J. Remote Sens.* **2022**, *26*, 201–210.

remote sensing

MDPI

Article

Satellite Estimation of pCO_2 and Quantification of CO_2 Fluxes in China's Chagan Lake in the Context of Climate Change

Ruixue Zhao [1,2], Qian Yang [1,2], Zhidan Wen [1,2,*], Chong Fang [2], Sijia Li [2], Yingxin Shang [2], Ge Liu [2], Hui Tao [2], Lili Lyu [2] and Kaishan Song [2,3]

1 School of Geomatics and Prospecting Engineering, Jilin Jianzhu University, Changchun 130118, China; zhaoruixue@student.jlju.edu.cn (R.Z.); jluyangqian10@hotmail.com (Q.Y.)
2 Northeast Institute of Geography and Agroecology, Chinese Academy of Sciences, Changchun 130102, China; fangchong1991@gmail.com (C.F.); lisj983@nenu.edu.cn (S.L.); goodlucksyx27@163.com (Y.S.); liugers@163.com (G.L.); taohui_gis@163.com (H.T.); lvlili0814@sina.com (L.L.); songkaishan@iga.ac.cn (K.S.)
3 School of Environment and Planning, Liaocheng University, Liaocheng 252000, China
* Correspondence: wenzhidan@iga.ac.cn

Abstract: The massive increase in the amount of greenhouse gases in the atmosphere, especially carbon dioxide (CO_2), has had a significant impact on the global climate. Research has revealed that lakes play an important role in the global carbon cycle and that they can shift between the roles of carbon sources and sinks. This study used Landsat satellite images to analyze the spatiotemporal characteristics and factors influencing the CO_2 changes in Chagan Lake in China. We conducted six water sampling campaigns at Chagan Lake in 2020–2021 and determined the partial pressure of carbon dioxide (pCO_2) from 110 water samples. Landsat surface reflectance was matched with water sampling events within ±7 days of satellite overpasses, yielding 75 matched pairs. A regression analysis indicated strong associations between pCO_2 and both the band difference model of the near-infrared band and green band (Band 5-Band 3, R^2 = 0.83, RMSE = 27.55 μatm) and the exponential model $[((\exp(b3) - \exp(b5))^2/(\exp(b3) + \exp(b5))^2$, R^2 = 0.82, RMSE = 27.99 μatm]. A comparison between the performances of a linear regression model and a machine learning model found that the XGBoost model had the highest fitting accuracy (R^2 = 0.94, RMSE = 16.86 μatm). We used Landsat/OLI images acquired mainly in 2021 to map pCO_2 in Chagan Lake during the ice-free period. The pCO_2 in the surface water of Chagan Lake showed considerable spatiotemporal variability within a range of 0–200 μatm. pCO_2 also showed significant seasonal variations, with the lowest and highest mean values in autumn (30–50 μatm) and summer (120–150 μatm), respectively. Spatially, the pCO_2 values in the southeast of Chagan Lake were higher than those in the northwest. The CO_2 fluxes were calculated based on the pCO_2 and ranged from −3.69 to −2.42 mmol/m^2/d, indicating that Chagan Lake was absorbing CO_2 (i.e., it was a weak carbon sink). Temperature, chlorophyll a, total suspended matter, and turbidity were found to have reinforcing effects on the overall trend of pCO_2, while the Secchi disk depth was negatively correlated with pCO_2. The results of this study provide valuable insights for assessing the role of lakes in the carbon cycle in the context of climate change.

Keywords: CO_2 flux; remote sensing; CO_2 budget; carbon cycle; Chagan Lake

1. Introduction

For decades, the release of greenhouse gases (GHGs), especially carbon dioxide (CO_2), into the atmosphere by human activities and environmental changes has increased dramatically [1,2]. There is a consensus that the greenhouse effect caused by the continuous rise in atmospheric CO_2 concentration is the principal factor causing global climatic change, which is seriously threatening sustainable societal and economic development [3]. The reduction in GHGs has become a critical topic of intense research focus [4–6].

A large number of studies have shown that lakes, as important contributors to inland waters, play a nonnegligible role in the global carbon cycle [7]. Although the total surface area of global lakes accounts for only 3.7% of the global non-glacial land area [8], lake waters are often a significant CO_2 emission source due to their relative saturated dissolved atmospheric CO_2 concentrations (cCO_2) [9–11], whose emission potential may offset about 22% of the net CO_2 absorption of terrestrial ecosystems [12]. This plays a key role in the global carbon budget. At the same time, many lakes in the world are rich trophic lakes. Theoretically, phytoplankton growing vigorously in these lakes consume large amounts of water CO_2 through photosynthesis and absorb atmospheric CO_2 through respiration, thus increasing the carbon exchange at the water–air interface and acting as a carbon sink [13]. It is important to study the carbon emissions at the lake water–air interface for carbon emission reduction and regulation of the role of lakes in global warming. Currently, most lakes around the world are oversaturated, but the amount of CO_2 that lakes contribute to the atmosphere varies from 0.11 to 0.57 $Pg{\cdot}Cy^{-1}$ according to different studies [10,14,15]. Such a large variation in the estimates is mainly caused by the lack of field monitoring data, especially for small- and medium-sized lakes [16,17]. Therefore, comprehensive and accurate monitoring of lake GHG emissions is an important prerequisite for regulating lake CO_2 emissions and is required for further evaluating the influence of lakes on the pattern of atmospheric GHG sources and sinks.

Studies of lake CO_2 flux based on measured data provided a clearer understanding of the source–sink patterns of global inland water carbon [18,19]. The flux of lake CO_2 can be directly measured by the flux chamber, but the more common approach is to indirectly calculate it from in situ monitoring data of the partial pressure of CO_2 (pCO_2) and Henry's law [20–22]. However, in situ monitoring over long time periods and vast geographical regions is challenging, and water pCO_2 has strong spatial heterogeneity and time variability [23]. When conducting spatial-scale extrapolation, time integration, and water–air interface flux estimation using monitoring data, the routine statistical analysis methods adopted in environment sciences and ecology have obvious shortcomings. Remote sensing data have multiple spatial and temporal scales and are multi-spectral, which enables continuous spatiotemporal observation of water parameters. In recent years, there has been a technological breakthrough in the space satellite remote sensing observation of watercolor elements of inland waters, and a series of inversion algorithms and models suitable for the study of spatiotemporal variations of lake water quality parameters have been developed, including the concentration of chlorophyll a and suspended matter in water, the depth of the true light layer, the lake surface water temperature, and the colored dissolved organic matter (CDOM) [24–29]. However, since there is no optical signal from the water body cCO_2 and it is affected by atmospheric absorption and scattering due to signal receiving by sensors and other surface processes, remote sensing technology cannot directly capture the flux characteristics of a water body and water–air interface, and can only achieve pCO_2 remote sensing inversion by building an indirect relationship between pCO_2 and the physical or biological parameters that can be retrieved from remote sensing. For the remote estimation of inland water body CO_2, exploratory work has been carried out overseas based on large measured datasets, but research on using the remote sensing technology to estimate inland water body CO_2 flux is still in the exploration stage. Although the method itself has certain limitations, because of its ability to cover considerable spatiotemporal scales, it is gradually becoming an internationally growing research focus [30].

Chagan Lake is the seventh-largest freshwater lake in China and the largest natural lake in Jilin Province, with a surface area of 420 km^2 and a mean water depth of 2.5 m [31,32]. The main input source of Chagan Lake is the Songhua River, and the lake is one of the eutrophication lakes in China [33]. Chagan Lake is endowed with rich natural resources and is a well-known fishery and reed industry base. Although the main economic value of the lake lies in its fish resources, it is also important for agricultural and recreational use.

The specific objectives of this study are: (1) to determine the factors impacting the pCO_2 variability and establish a remote sensing model for pCO_2; (2) to evaluate the variations in the lake pCO_2 based on image data and multi-site measurements; and (3) to quantify the

Chagan Lake's contribution to the atmospheric CO_2 budget. Chagan Lake was selected as a typical lake in Northeast China to perform remote sensing CO_2 flux estimations for further understanding of the lake carbon cycle. The long-term monitoring of the lake is important for agricultural and recreational uses.

2. Materials and Methods

2.1. Study Area and Field Data Acquisition

Chagan Lake is the seventh-largest freshwater lake in China. It is located in the northwest of Jilin Province; in the Golden Triangle area of the Inner Mongolia Autonomous Region, Heilongjiang Province, and Jilin Province; at the end of the Huolin River; and at the confluence of the three rivers of Songhua River, Songhua River south source, and Nenjiang river (Figure 1). The lake has a water surface area of 420 km^2 and 2.5 m of mean water depth [32]. The ice cover period of Chagan Lake is about 4–5 months, freezing in the middle of November every winter and melting in April of the following year. The major features of the geomorphology of Chagan Lake are high in the southeast, slightly higher in the southwest, and low in the central part and the northeast. The Chagan Lake area is rich in natural resources; it is the largest natural lake and important fishery base in Jilin Province. However, large amounts of receding water from the newly built saline–alkali irrigation area are discharged into Chagan Lake, resulting in the deterioration of lake water quality and aggravation of eutrophication, which poses a threat to the ecological security of lake water and the sustainability of its ecological products [33,34].

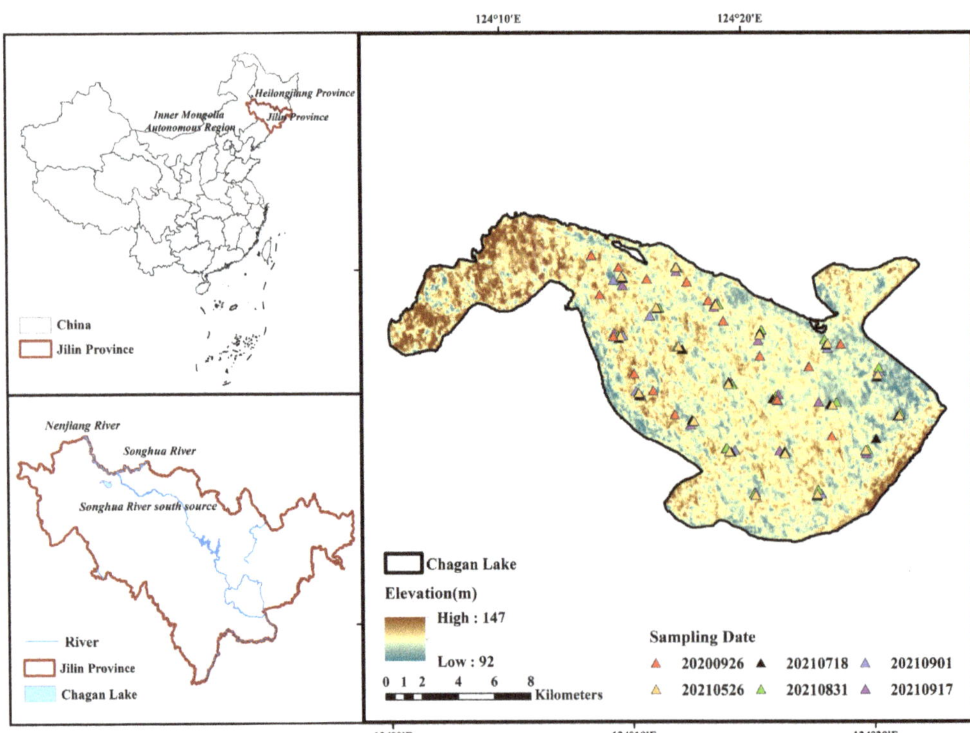

Figure 1. Locations of sampling sites in Chagan Lake.

Lake water sample collection was conducted during the ice-free periods (May to October) between 2020 and 2021. Surface water samples from depths of 0.1–0.5 m were acquired. The locations of sampling points are shown in Figure 1. The specific sampling times were 26 September 2020, 26 May 2021, 18 July 2021, 31 August 2021, 1 September 2021, and 17 September 2021. A total of 110 sampling points were selected. After a water sample was collected, it was immediately put in a car refrigerator at 4 °C and delivered to the laboratory in this state. The water samples were kept in the laboratory at 4 °C in a refrigerator until the analytical tests were complete, which took 2–3 days. Generally, 3–4 L of water was collected at each point to measure various water quality indicators in the laboratory. In this study, the collection of CO_2 in lakes was carried out using the headspace equilibrium method [35,36], and 250 mL atmospheric air was equilibrated with 250 mL water by shaking vigorously in a gas bottle for 1 min. The headspace gas was transferred by a 30 mL plastic syringe to a gas collection bag (Delin, China) and carried to the laboratory within 24–48 h. The measurements of CO_2 concentrations were made by a gas chromatography analyzer (Agilent, 7890B). The measurement for each sampling site was repeated three times using a similar operation, and the value used in the calculation was the mean value [37]. In the calculation, we needed to convert the volume fraction into the molarity through the basic unit conversion, subtract the air CO_2 concentration from the measured concentration of CO_2 in water, multiply the volume ratio of air above the bottle and water to obtain the final CO_2 dissolution in the water, and then convert it into partial pressure. For measuring the in situ water temperature (T), PH, dissolved oxygen (DO) and water transparency, the HQ40D Portable Multi-Parameter Meter (HQ40d, Hach) was used. Wind speed data were also measured simultaneously [38,39].

2.2. Data Acquisition and Processing

Landsat 8 images were downloaded to coincide with the in situ measurement dates. Detailed information on the selection of the Landsat 8 satellite images is provided (Table 1). Landsat 8 images (Landsat 8 OLI/TIRS Collection 2 Tier 1 atmospherically corrected surface reflectance) with no clouds or less cloud cover (which have been processed by atmospheric correction) were selected on the Google Earth Engine (GEE) platform and used in subsequent model inversions. The Landsat5/7 Surface Reflectance Collection 2 Tier 1 products were created with the Landsat Ecosystem Disturbance Adaptive Processing System (LEDAPS) algorithm (version 3.4.0) according to the 6S radiative transfer model to obtain the surface reflectance products. The Landsat 8 Surface Reflectance Collection 2 Tier 1 products were created with the Land Surface Reflectance Code (LaSRC). Image reflectance was equal to the image DN values multiplied by 0.0000275 and with 0.2 subtracted, and the model was constructed based on the image surface reflectance and the measured partial pressure value of carbon dioxide. Landsat surface reflectance was matched with water sampling events within ±7 days of satellite overpasses, yielding 75 matched pairs. Because the number of suitable images with less cloud cover in 2021 was insufficient, we downloaded supplementary images from 2020 to 2022 (Table 1). For images that were only used to match measured data, the cloud cover was ignored. Band surface reflectance conversion between Landsat TM/ETM+ and Landsat OLI was available [40].

Weather research and forecasting (WRF) is a widely used numerical weather prediction and atmospheric research tool jointly developed by the National Center for Atmospheric Research, the National Oceanic and Atmospheric Administration, and their partners. It is used for short-term weather forecasting, atmospheric processes, and long-term climate simulations. The WRF is more accurate than traditional numerical weather prediction models and has a higher spatiotemporal resolution, making it a popular tool in the meteorological and atmospheric research community. FNL (Final) Operational Global Analysis data is a global reanalysis database with a spatial accuracy of $1° \times 1°$ provided by the National Center for Environmental Prediction (NCEP)/National Center for Atmospheric Research (NCAR). In this paper, the FNL data were used as input data to the WRF; then, temperature, wind speed, and precipitation data were extracted from the WRF simulations. The highly accurate WRF simulation results were compared to the experimental site data. Furthermore, we downloaded the ground meteorological data for China, available from the National Climatic Data Center (NCDC), which is a part of the National Oceanic and

Atmospheric Administration (NOAA). Using this dataset, the 1990 average temperature and U10 wind speed ware calculated for flux estimations. The annual global atmospheric carbon dioxide data were also downloaded from the Global Monitoring Laboratory (GML).

Table 1. The date and quantity of images downloaded.

Image	Date	Resolution	Cloud Cover
LC08_119029_20200722	22 July 2020	30 m	<5%
LC08_119029_20201010	10 October 2020	30 m	<5%
LC08_119029_20210420	20 April 2021	30 m	<5%
LC08_119029_20210522	22 May 2021	30 m	<5%
LC08_119029_20210607	7 June 2021	30 m	<5%
LC08_119029_20210623	23 June 2021	30 m	<20%
LC08_119029_20210709	9 July 2021	30 m	-
LC08_120028_20210902	2 September 2021	30 m	-
LC08_119029_20210911	11 September 2021	30 m	-
LC08_119029_20211013	13 October 2021	30 m	<5%
LC08_119029_20221016	16 October 2022	30 m	<5%
LT05_119029_19890802	2 August 1989	30 m	<20%
LT05_119029_19891021	21 October 1989	30 m	<5%
LT05_119029_19900704	4 July 1990	30 m	<20%
LT05_119029_19900906	6 September 1990	30 m	<5%
LT05_119029_19901024	24 October 1990	30 m	<5%
LT05_119029_19910418	18 April 1991	30 m	<5%
LT05_119029_19910504	4 May 1991	30 m	<5%
LT05_119029_19910605	5 June 1991	30 m	<5%
LT05_119029_19910824	24 August 1991	30 m	<20%
LT05_119029_19910909	9 September 1991	30 m	<5%
LT05_119029_19911027	27 October 1991	30 m	<5%

2.3. pCO_2 Modeling, Model Calibration, and Validation

2.3.1. Band Ratio Modeling

The measured data were randomly divided at a 2:1 ratio into 50 training samples and 25 testing samples. To select the appropriate model parameters, the correlation between the 50 pCO_2 values and five bands (B1–B5) or multiple band combinations of the Landsat 8 images were analyzed using linear regression in the Statistical Package for the Social Sciences (SPSS). The factors with strong correlations were selected to form a band ratio model.

2.3.2. Machine-Learning Modeling

The random forest regression model establishes multiple unrelated decision trees using randomly selected samples and features, and makes parallel predictions. Extreme gradient boosting (XGBoost) uses multiple Classification and Regression Trees (CART) for prediction and has good generalization capabilities, which is very useful for solving complex nonlinear regression problems. K-fold cross-validation can be used to check the stability of a model and effectively avoid the influence of information leakage on the model's hyperparameters. Cross-validation divides the available data into K parts, uses each of them in turn as the testing data and the other K-1 parts as the training sets, and assesses the accuracy of the model multiple times to obtain the average accuracy of the model. XGBoost traverses the search with a 5-fold cross-validation and grid method.

2.3.3. Model Calibration and Validation

The 25 test sampling points data were used to model calibration and validation. The errors between the inverted values and the measured values of the linear regression model were compared and analyzed. The coefficient of determination (R^2), root mean square error (*RMSE*), and mean absolute percentage error (*MAPE*) were used as accuracy evaluation indices.

$$RMSE = \sqrt{\frac{\sum_{i=1}^{N} (y_i' - y_i)^2}{N}} \tag{1}$$

$$MAPE = \frac{100\%}{N} \sum_{i=1}^{N} \left| \frac{y_i' - y_i}{y_i} \right| \tag{2}$$

where N is the number of samples in the dataset and y_i and y_i' are in situ-measured pCO_2 and model-predicted pCO_2 values. The determination coefficient (R^2) is often used to represent model prediction performance. *MAPE* is also used to examine agreement between the measured and predicted values.

2.4. Data Analysis

pCO_2 in the surface water of lakes was calculated using Henry's law [20].

$$pCO_2 = C_B \times k_C \tag{3}$$

where p is the gas partial pressure in the surface water of lake (atm), which has C_B as the molarity ($mol{\cdot}m^{-3}$); and k_C is Henry's constant [$atm{\cdot}(mol{\cdot}m^{-3})^{-1}$].

The areal (per unit surface area) CO_2 flux (FCO_2, in $mmol{\cdot}m^{-2}{\cdot}d^{-1}$) across the water–air interface was estimated based on a theoretical diffusion model:

$$FCO_2 = k \times k_H \times (pCO_2 - pCO_{2air}) \tag{4}$$

where pCO_2 is the gas partial pressure of CO_2 in surface water (μatm); pCO_2air is the concentration of CO_2 in air (μatm); k_H ($mol{\cdot}L^{-1}{\cdot}atm^{-1}$) is the solubility for CO_2 corrected for temperature (T, °C) [41] and pressure (μatm) (Equation (4)); and k is the gas transfer velocity ($cm{\cdot}h^{-1}$) [42]. The estimation of k ($cm{\cdot}h^{-1}$) was related to k_{600} (gas transfer velocity normalized to a Schmidt number of 600 in fresh water) and U_{10} (wind speed measured at 10 m height) (Equations (5)–(8)) [43–45]. Then, 0.24 times k ($cm{\cdot}h^{-1}$) obtained k ($m{\cdot}d^{-1}$), which is used in Equation (4). A $-2/3$ power dependence of the Schmidt number (*Sc*) was assumed below a wind speed of 3 m/s (measured at 1 m height) with the smooth lake surface, and a $-1/2$ power dependence of *Sc* was assumed above 3 m/s for a surface with waves [44,46]. S_C was calculated from the surface water temperature T (°C) using Equation (9) [45]:

$$k_H = 10^{-(1.11+0.016T-0.00007T^2)} \tag{5}$$

$$k = k_{600} \times (S_c/600)^{-1/2or-2/3} \tag{6}$$

$$k_{600} = 0.251 \times U_{10}^2 \ (U_{10} > 3.7\ m/s) \tag{7}$$

$$k_{600} = 2.07 + 0.215 \times U_{10}^{1.7} \ (U_{10} < 3.7\ m/s) \tag{8}$$

$$S_C = 1911.1 - 118.11T + 3.4527T^2 - 0.04132T^3 \tag{9}$$

WRF was used to simulate the Chagan Lake selected from the FNL data. The annual average temperature and U_{10} wind speed data of the station were extracted from the simulation results as known quantities into the carbon dioxide flux calculation formula, and the final carbon dioxide flux results were obtained. In 2021, the U_{10} wind speed was 0.718923 m/s and the temperature was 4.1810225 °C. In 1990, the temperature was

3.149914 °C and the U_{10} wind speed was 0.045115 m/s. And the CO_2 in the atmosphere was 354.45 ppm in 1990 and 416.45 ppm in 2021.

3. Results

3.1. pCO_2 Model Calibration and Validation

The correlations between the pCO_2 values and the five bands and multiple band combinations of the Landsat 8 images are listed in Table 2. The potential model factors were selected from among those with strong correlations, indicated in Table 2, to form a one-way linear regression model and an exponential regression model (Table 3).

Table 2. Pearson correlation coefficient (p) between pCO_2 and bands of Landsat 8 images.

Factors	B1	B2	B3	B4	B5	B6	B7
p	−0.252	−0.168	0.182	−0.048	−0.382	0.046	0.228
Factors	**B3 − B2**	**B5 − B3**	**B5 − B4**	**B4 − B3**	**exp(B3) − exp(B5))/ (exp(B3) + exp(B5)**		**(B5 − B2)/ (B5 + B2)**
p	0.703	−0.911	−0.636	−0.600	−0.910		−0.621

Table 3. Regression models.

Regression Model	Formula	R^2
Band difference model	y = −4639.749 × (B5 − B3) − 353.845	0.83
	y = 3824.755 × (B3 − B2) − 109.721	0.48
	y = −3976.502 × (B5 − B4) + 157.229	0.39
Band ratio model	y = −93.365 × ((B5 − B2)/(B5 + B2)) + 22.669	0.37
	y = −156.216 × ((B5-B2)/(B5 + B2)) + 1928.883 × ((B3 + B4)/2) − 214.638	0.67
Exponential model	y = 93273.487 × (exp(B3) − exp(B5))^2/(exp(B3) + exp(B5))^2 − 127.066	0.82

Pure band calculations for regression models inevitably produce negative inversion values in the imagery, which skew the overall results. Nonlinear data are not well fitted. The model was optimized and parameterized using multiple experiments, and the final parameter values are shown in Table 4. Other parameters were assigned default values. Using the same training samples as in the linear model, the random forest and XGBoost pCO_2 models were constructed. The pCO_2 values predicted by the models vs. those measured in situ are plotted in Figure 2, which reflects the differences between the two types of results for each model. The closer the verification sample points are to the diagonal line, the smaller the model error.

Table 4. Important model parameter values.

Random Forest Parameters	Values	XGBoost Parameters	Values
n_estimators	80	n_estimators	50
random_state	110	learning_rate	0.11
max_features	sqrt	booster	gbtree
max_depth	5	max_depth	5
min_samples_leaf	3	gamma	2
n_jobs	−1	lambda	7

All the tested models from pCO_2 collected at 110 sites across Chagan Lake achieved accurate calibration, with R^2 ranging from 0.82 to 0.97. Validation results also exhibited stable performance (R^2 = 0.87–0.94) (Figure 2). The linear model and band difference model (Figure 2a), i.e., unary one-time models, had R^2, RMSE, and MAPE values of 0.83, 27.54 µatm, and 32.25%, respectively. The quadratic model (Figure 2b) showed validation accuracy, with R^2, RMSE, and MAPE values of 0.90, 22.13 µatm, and 40.63%, respectively.

In general, the accuracy of the linear models was similar and relatively good. The accuracy of the random forest model was high, with R^2, RMSE, and MAPE of 0.93, 17.50 μatm, and 21.09% (Figure 2c), respectively, but the validation accuracy was not high, with R^2 of 0.87, RMSE of 25.42 μatm, and MAPE of 49.74%. The XGBoost model (Figure 2d) had significantly smaller errors, and its R^2, RMSE, and MAPE of 0.97, 11.40 μatm, and 11.76%, respectively, were better than those of other models. The validation accuracy (R^2 = 0.94 and RMSE = 16.87 μatm) was also better than those of the remaining three models. The XGBoost model showed the best performance, indicating that it effectively utilized various feature information supporting its nonlinear regression prediction ability.

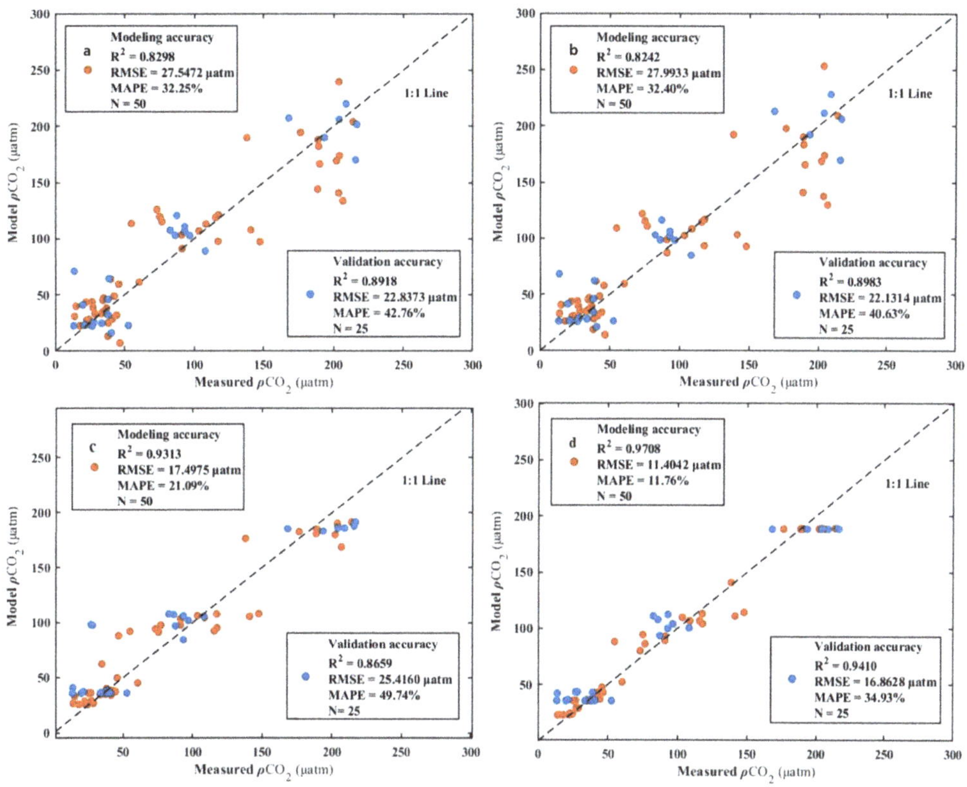

Figure 2. Calibration and validation of pCO_2 model for Chagan Lake based on band surface reflectance. (**a**) Band difference regression model; (**b**) exponential regression model; (**c**) random forest model; (**d**) XGBoost model.

3.2. Spatiotemporal Analysis of pCO_2 Based on the Inversion Results

We applied the XGBoost model to map the pCO_2 of Chagan Lake in 2021 and obtained satisfactory results (Figure 3). The corresponding inversion result accuracy of the sampling points was also better (R^2 = 0.80) (Figure 4). A full-scale assessment of pCO_2 in Chagan Lake in 2021 was conducted using the Landsat OLI surface reflectance product in the ice-free season. The pCO_2 values of Chagan Lake ranged from 0 to 200 μatm, with 25–75% of them concentrated between 30 and 150 μatm. The minimum value changes were small, concentrated in the range of 15–25 μatm. The spatial distribution of the pCO_2 values was higher in the southeast of Chagan Lake than in the northwest. Because of the occurrence of large individual extrema in the models, the maximum and minimum values were not included in the following trend analysis.

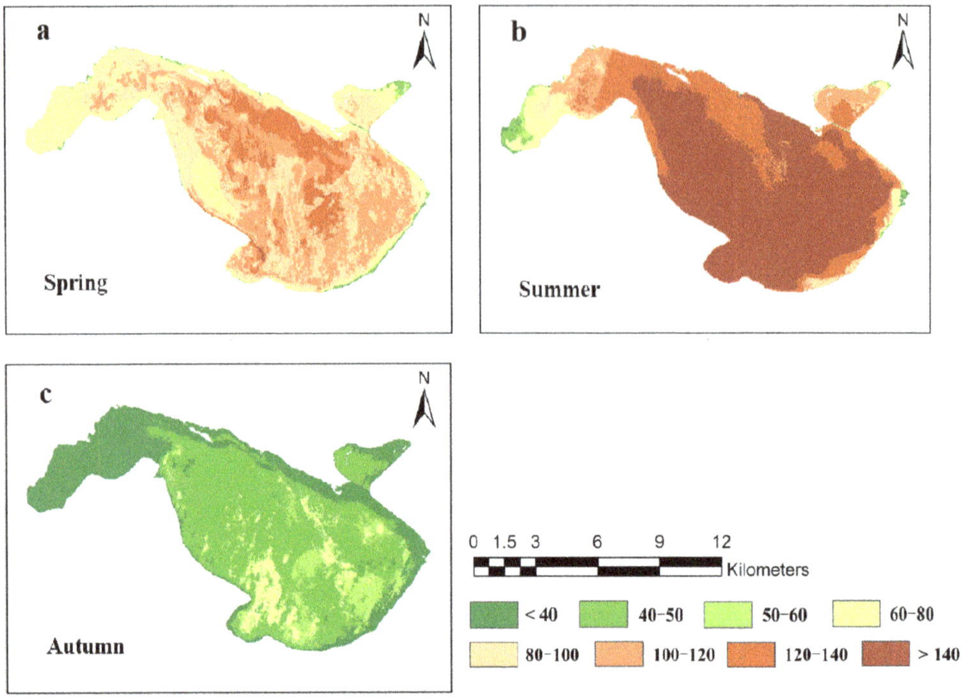

Figure 3. pCO_2 variation in different seasons in 2021 based on inversion values. (**a**) Spring; (**b**) summer; (**c**) autumn. Unit: µatm.

A seasonal analysis of the model results was conducted (Figure 3, spring: March–May; summer: June–August, autumn: September–November). The highest seasonal mean pCO_2 values in Chagan Lake were in the summer (mean = 132.45 ± 24.44 µatm), followed by spring (mean = 98.11 ± 18.24 µatm), and the lowest was in autumn (mean = 44.03 ± 8.46 µatm). The characteristic values were analyzed (Figures 5 and 6) and it was revealed that the lowest pCO_2 was found in autumn, followed by spring, and highest value was found in summer. The trends of the eigenvalues were essentially consistent (Figure 6a). The mean, lower quartile, median, and upper quartile changes were the highest in summer (120–150 µatm), followed by spring (80–110 µatm) and autumn (30–50 µatm) (Figure 6b). It was observed that the autumn pCO_2 values were the lowest; the spring pCO_2 values were about twice those of autumn; and the summer pCO_2 values were thrice those of autumn and about 1.5 times those of spring. Then, the XGBoost model was used to invert the images and the ENVI Quick Stats function was adopted to obtain the eigenvalues of the inversion results, including the maximum and minimum values; mean; and the lower, median, and upper quartiles, as shown in the Figure 5.

The annual mean pCO_2 in Chagan Lake was calculated based on the XGBoost model results. The values significantly increased between 1990 (mean = 42.45 ± 8.50 µatm) and 2021 (mean = 91.52 ± 14.31 µatm) (Figure 7). This trend of annual mean pCO_2 was consistent with the increase in nutrient levels in Chagan Lake over the same years. We can see that the values were higher at the boundary of the lake than on the surface in 1990, while in 2021, this was the opposite, as seen in Figure 7. The reason may be that in 2021, the lake biomass of Chagan Lake was large, most of it was concentrated on the lake water surface, and the low biomass at the lake boundary led to low pCO_2 values. In the 1980s and 1990s, on the other hand, Chagan Lake was in the process of "introducing loose water into the lake"; the water area was in a period of rapid growth [47]; the development of fishery was slow; there was overfishing; and the overall biomass was small, resulting in high pCO_2 values only in the southeast region and low ones overall. The abnormally high value of red in the eastern part of the lake in 1990 was due to the fact that the water area of Chagan Lake was fluctuant between 1990 and 2021. Moreover, we used the same shapefile of Chagan lake, and the boundary of lake may include the transition zone from the lake to the land. In 1990 and 2021, the overall pattern was the same, with the southeast being higher than the northwest.

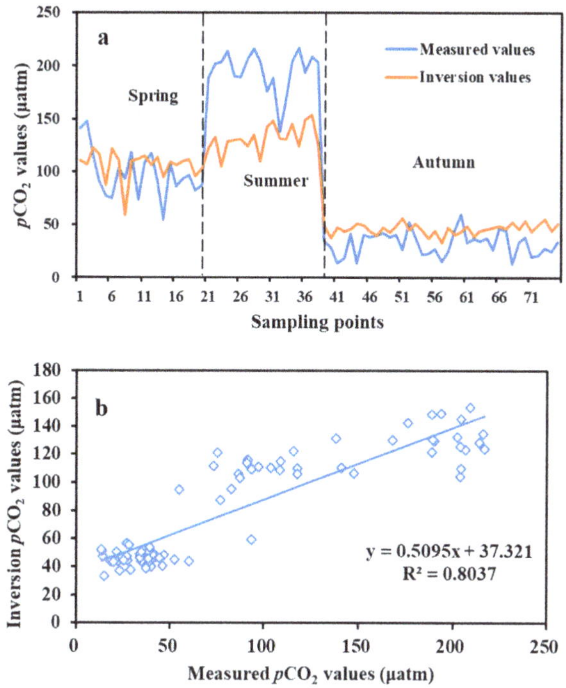

Figure 4. Quarterly comparison and precision of measured pCO_2 and inversion values at sampling points. (**a**) Quarterly comparison of measured pCO_2 and inversion values at sampling points; (**b**) Precision of measured pCO_2 and inversion values at sampling points.

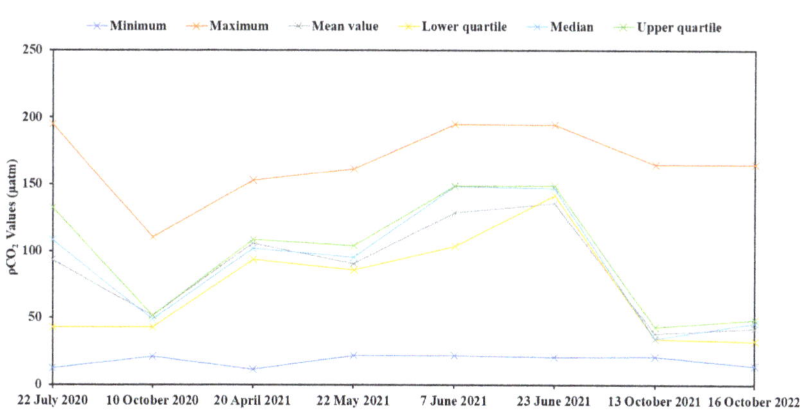

Figure 5. Monthly series variation analysis of pCO_2 eigenvalues.

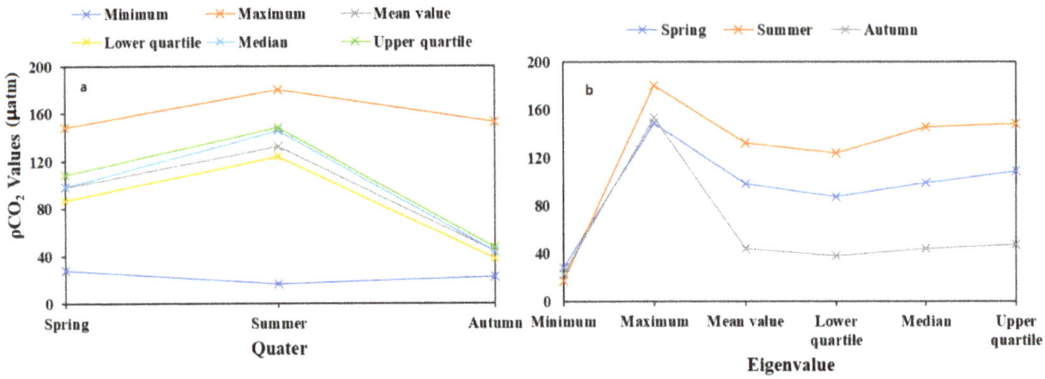

Figure 6. Quarterly series variation analysis of pCO_2 eigenvalues. (**a**) The variation of pCO_2 eigenvalues with quarter; (**b**) Comparison of the variation trend of pCO_2 eigenvalues in each quater.

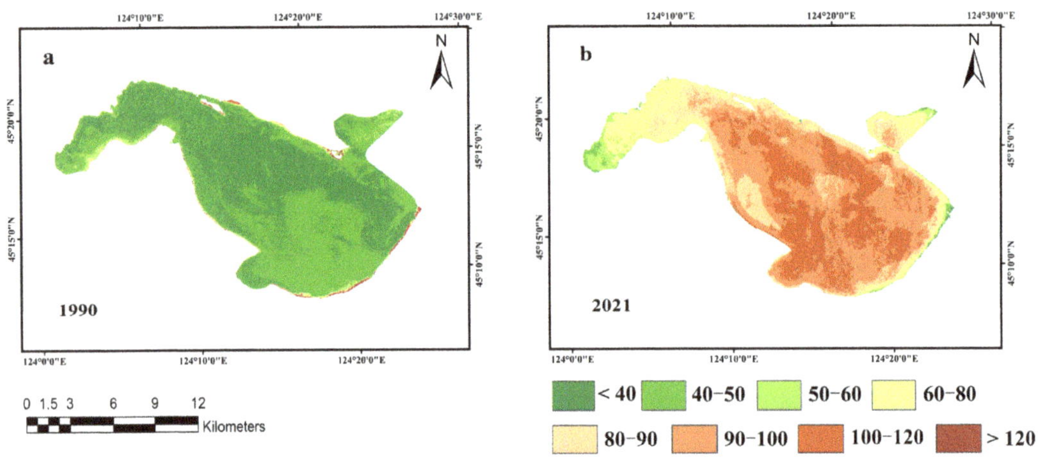

Figure 7. Variations in annual mean pCO_2 values in 1990 and 2021. (**a**) 1990; (**b**) 2021. Unit: μatm.

3.3. CO_2 Flux through Water–Atmosphere Interface of Chagan Lake

CO_2 flux through the water–atmosphere interface of Chagan Lake was calculated based on the modeled pCO_2 values. This negative value of CO_2 flux values indicates that Chagan Lake was absorbing CO_2 from atmosphere (i.e., the lake functioned as a weak carbon sink). The CO_2 flux was lowest in summer, ranging between 2.4 and 2.64 mmol/m^2/d, followed by spring (2.64–3.12 mmol/m^2/d), and it was highest in autumn, with values above 3.36 mmol/m^2/d (Figure 8). The study also compared these CO_2 fluxes to those in 1990 (Figure 9), which showed that Chagan Lake was a carbon sink in 1990 and 2021 and the CO_2 flux increased between 1990 (mean = -2.53 ± 0.07 mmol/m^2/d) and 2021 (mean = -2.96 ± 0.13 mmol/m^2/d).

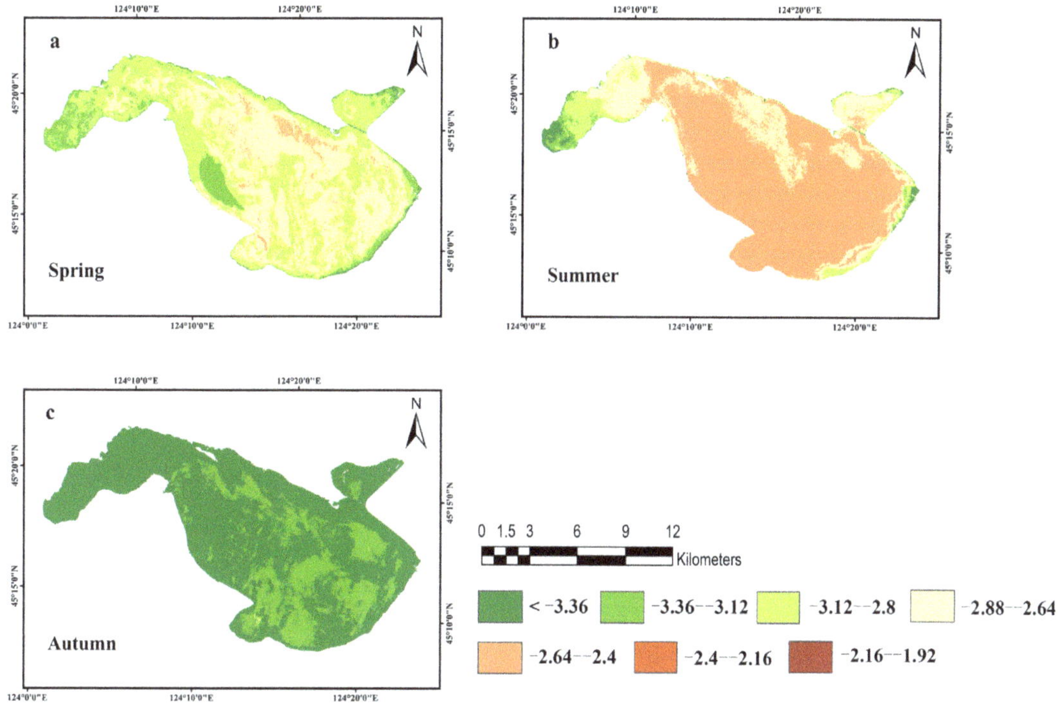

Figure 8. CO_2 flux variation in different seasons, calculated from the inversion pCO_2 value. (**a**) Spring; (**b**) summer; (**c**) autumn. Unit: mmol/m^2/d.

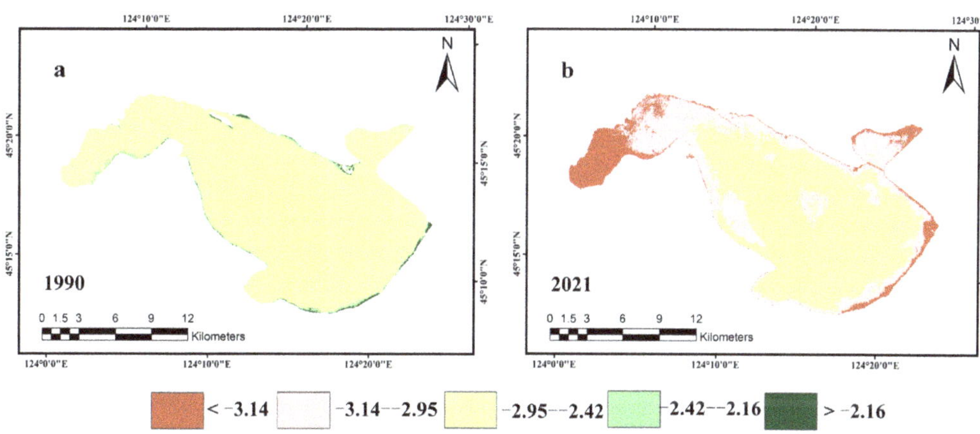

Figure 9. Variations in annual mean CO_2 fluxes in 1990 and 2021. (**a**) 1990; (**b**) 2021. Unit: mmol/m^2/d.

3.4. Correlation between pCO$_2$ and Its Influencing Factors

The climatic factors, including temperature, wind speed, and precipitation, were extracted using WRF. The climatic factors were positively correlated with the overall trend of monthly pCO_2 with the water temperature showing the highest correlation. However, the relationships between wind speed, precipitation, and pCO_2 were not very conclusive (Figure 10). In addition, water quality parameters, chlorophyll a, total suspended solids, and turbidity showed high positive correlations ($R^2 > 0.8$) with the pCO_2 values, while the

Secchi disk depth (SDD), total nitrogen (TN) and total phosphorus (TP) were all negatively correlated with them. TN varied from 0.74–1.35 mg/L and TP varied from 0.03–0.22 mg/L (Figure 11); this indicates a high degree of eutrophication in the lake.

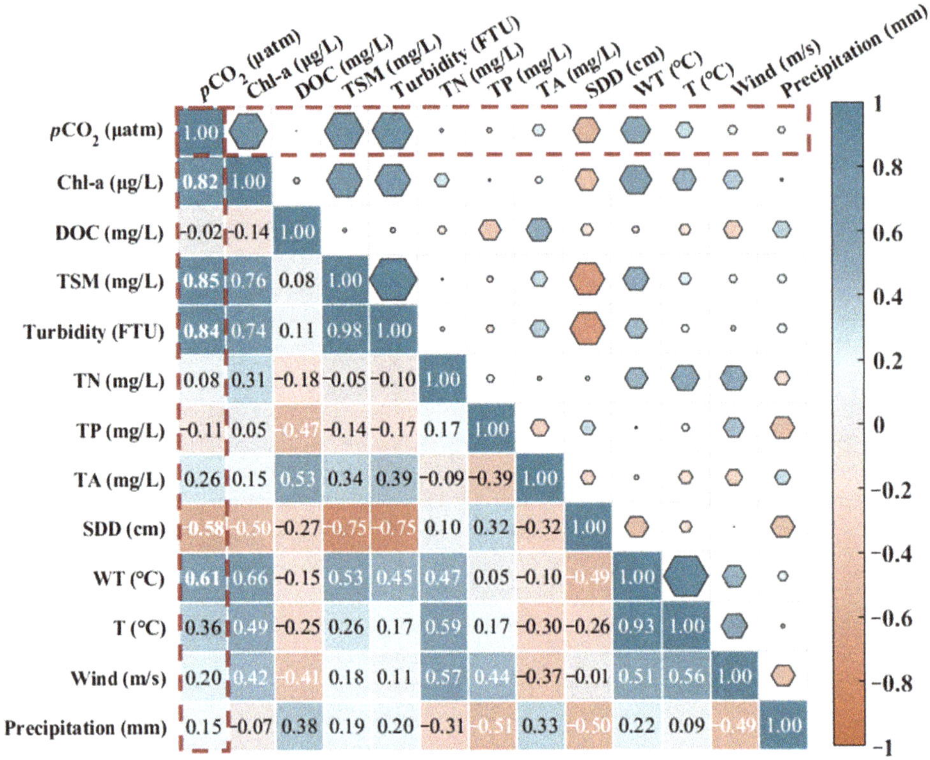

Figure 10. Relationship between pCO_2 values and influencing factors. TN: total nitrogen concentration; TP: total phosphorus concentration; TA: total alkalinity concentration; WT: water temperature; T: air temperature; SDD: Secchi disk depth.

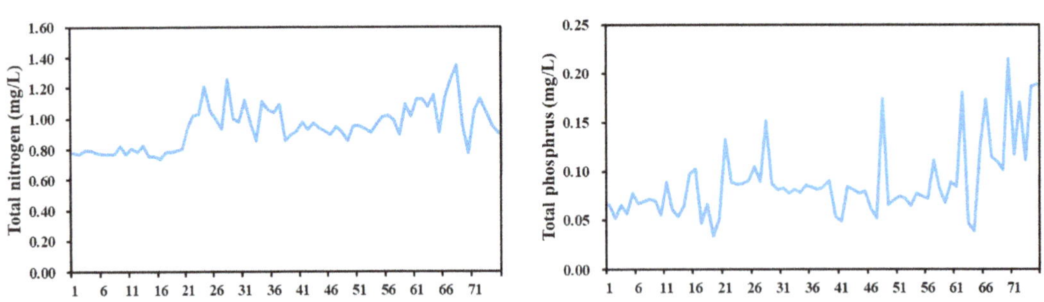

Figure 11. Measured nitrogen and phosphorus concentration at the sampling points.

In addition, several influencing factors (chlorophyll a, total suspended solids, turbidity, Secchi disk depth, water temperature) that showed a high correlation with pCO_2 were analyzed for their changing trends; these factors mainly presented seasonal fluctuations (Figure 12). Among them, the fluctuations in TSM and turbidity were closest to those of pCO_2, with more intense fluctuations, while the changes in chlorophyll and water temperature are relatively gentle, with the values all showing summer > spring > autumn. That is, the values of TSM, turbidity, chlorophyll a, and water temperature are higher in summer, and pCO_2 values are also higher, with the lowest values occurring in autumn. The trend of changes in transparency, with autumn > spring > summer, is opposite to pCO_2 changes: the lowest transparency is found in summer, while pCO_2 is the highest.

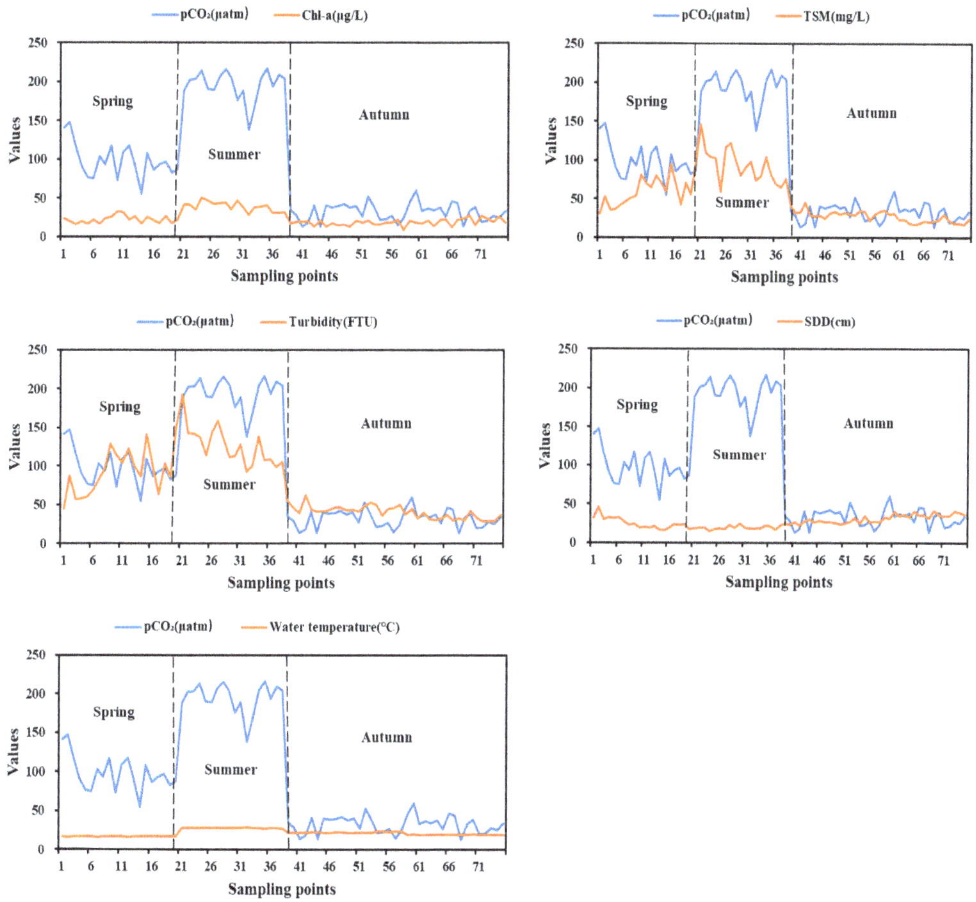

Figure 12. The change trend of influencing factors and pCO_2.

4. Discussion

4.1. Temporal and Spatial Variations of pCO₂ in Chagan Lake

Because Chagan Lake freezes in the winter, we only studied the seasonal distribution of pCO_2 values during the ice-free period of 2021. We found that it exhibited a pattern of high pCO_2 values in summer and low ones in autumn (Figure 3), which was similar to the pCO_2 value distribution at actual sampling points (Figure 4). This distribution was consistent with the seasonal distributions of the pCO_2 values in Beihu Lake and Changchun Park [48], but different from the western part of Chaohu Lake [22] or Taihu Lake [23]. The difference in summer pCO_2 values evident in Figure 4 was attributed to the difference between the actual summer sampling time (18 July) and the sampling times of available satellite images (7 June, 23 June and 22 July). Moreover, the measured water temperature, chlorophyll, TSM, and turbidity of Chagan Lake revealed high values in summer and low values in autumn, and their Pearson correlation coefficients with the pCO_2 values were 0.61, 0.82, 0.85, and 0.84, respectively (Figure 10). Water temperature is one of the main influencing factors on the horizontal distribution of phytoplankton. Phytoplankton and biomass abundance increase with rising water temperature [49,50], leading to increased photosynthesis and decreased carbon dioxide in lakes. However, in the Chagan Lake wetland, although the water temperature in summer is higher than that in autumn, the number, abundance, and biomass of phytoplankton species are lower than they are in autumn [21,32]. This results in higher pCO_2 in summer than in autumn. On the one hand, precipitation increases in summer, and receding water from farmland in the Qianguo irrigation area also flows into Chagan Lake, increasing the water level and diluting phytoplankton and biomass. This leads to a decrease in hydraulic retention time and an increase in water flow speed, significantly affecting the horizontal distribution of phytoplankton [51]. A faster water flow is not conducive to the accumulation of phytoplankton biomass, but simultaneously carries larger amounts of pollutants such as TN and TP, which provide nutrients for phytoplankton growth, manifested as high pigment content [26,50]. On the other hand, as an important fishery industry base in Jilin Province, the Chagan Lake wetland releases hundreds of thousands of catties of fry every spring and autumn. This increases the carbon dioxide concentration in the lake, and as the fish grow, their feed intake also decreases phytoplankton abundance and biomass [50]. This was also visible in the phytoplankton distribution measured at the 2020–2021 sampling points, which was presented by Chang et al. [32].

The pCO_2 values in Chagan Lake in 2021 were high in the southeast region and low in the northwest region (Figure 7b). According to the 2020–2021 phytoplankton sampling analysis of Chang et al. [32], the phytoplankton biomass in the southeast of Chagan Lake was essentially smaller than that in the northwest. It comprised mainly *chlamydomonas* and *chlamydomonas* ovate, the biomass in summer was lower than in autumn, and photosynthesis absorbed less CO_2. This produced a spatial distribution with high pCO_2 values in the southeast region and low pCO_2 values in the northwest region. Compared to the spatial changes between 1990 and 2021 (Figure 7), the spatial distribution was consistent, but the pCO_2 values in 2021 were higher than those in 1990, and the high-value area in southeast China in 2021 was significantly larger than that in 1990, which could be due to the increase in eutrophication in Chagan Lake.

4.2. Chagan Lake as Carbon Source or Sink

In view of the important role and significance of lakes to economic and social development, the ecological environment of lakes has always attracted widespread attention. Since the 18th National Congress of the Communist Party of China, the Party and the state have attached great importance to the cultivation of ecological civilization, and a number of affiliated units have produced five reports. On the basis of these five reports, the Proceedings of the Chinese Academy of Sciences published a series of papers entitled "Progress and Effectiveness of China's Ecological Civilization Construction", which scientifically analyzed the long-term ecological environment changes of the aforementioned typical ecosystems and regions, especially the changes since the 18th National Congress of the Communist Party of China, and formulated suggestions for ecological environmental protection [52,53]. From a single treatment measure to a systematic plan for upstream and downstream and left and right bank coordination, the water quality of Chagan Lake has shown a promising trend. The transparency increased from 0.45 m in 2011 to 0.58 m in 2020; the water quality improved from Class V before 2012 to Class IV in 2020; and the fluoride concentration decreased by 13% year-on-year [53]. Chagan Lake has progressed from "no water" to "water" to "clear water", and the lake area has been stable at 300 km^2 since 2012. The concentration of nitrogen and phosphorus in the water has decreased significantly, with mean nitrogen and phosphorus concentrations of 0.949 mg/L and 0.088 mg/L in 2021, respectively (Figure 11).

The analysis of CO_2 fluxes conducted in this study showed that Chagan Lake, between 1990 (mean = -2.53 ± 0.07 mmol/m^2/d) and 2021 (mean = -2.96 ± 0.13 mmol/m^2/d), has been a weak carbon sink. This may be due to the increase in the nutrient content and temperature of lakes in recent years, leading to the vigorous growth of algae and phytoplankton as well as strong photosynthesis, which absorbs and uses large amounts of CO_2. Furthermore, the flow rate of lake water was slow, and, in such an environment, photosynthesis consumed more CO_2 than the amount of CO_2 produced by phytoplankton respiration. Thus, biophotosynthesis was the main factor contributing to the carbon sink state of Chagan Lake [54,55]. As chlorophyll a concentrations increased, the pCO_2 values increased too (Figure 10), possibly because phytoplankton is an important primary producer in aquatic ecosystems, with fundamental links to the food chain. However, the rates of increase in CO_2 partial pressure in the lake (42.45 $\pm$ 8.50 µatm in 1990; 91.52 $\pm$ 14.31 µatm in 2021) were lower than the rates of increase in atmospheric CO_2 (354.45 ppm in 1990; 416.45 ppm in 2021), keeping Chagan Lake in a carbon sink state. In other words, the degree of eutrophication increased from 1990 to 2021, and the photosynthesis by lake organisms also increased, which in turn promoted carbon dioxide absorption. Eutrophication refers to an abnormal increase in the primary productivity of lake water ecosystems due to excessive intake of nutrients, such as nitrogen and phosphorus, mainly manifested as abnormal reproduction of algae and other plankton [56]. An increase in primary productivity means that water ecosystems fix more CO_2, which in turn reduces CO_2 emissions [22,57].

In addition, in Chagan Lake, the correlations between total suspended matter concentration and turbidity and pCO_2 were high (Figure 10). High turbidity increases the respiration rate of aquatic animals and plants and promotes CO_2 production and release [58]. SDD is negatively correlated with pCO_2, and with decreasing SDD, CO_2 emissions fall and the degree of eutrophication increases.

Among the meteorological factors influencing pCO_2, temperature (R^2 = 0.36) was the strongest, wind speed (R^2 = 0.20) came second, and precipitation (R^2 = 0.15) third, all positively correlated with pCO_2 (Figure 10). However, none of these were dominant factors. When cool lakes warm up due to seasonal or long-term climate changes, they may begin to emit more CO_2 [37]. Wind speed can affect the water–CO_2 flux of CO_2 by affecting the gas transport coefficient, the decomposition of carbon matter in water, and the production of carbon dioxide [54,59]. Precipitation affects exogenous inputs, such as rivers that carry substantial amounts of organic matter and inorganic carbon into lakes, promoting CO_2 emissions [60].

4.3. Uncertainty Analysis

In this paper, a spatiotemporal analysis of the partial pressures and emission fluxes of CO_2 in Chagan Lake was conducted and the factors affecting CO_2 emissions in Chagan Lake were discussed, which provided reference values for studying the contribution of inland lakes to global GHG emissions. However, with the limited measured data available, this study still requires several improvements, as follows:

(1) Although the lake water samples we collected met the basic requirements (including water sample collection depth (0.1–0.5 m), water sample storage temperature (4 °C), etc.), due to the limitations of time and in situ sampling conditions, the number of samples acquired was only modest. This might have had an impact on the accuracy of the models we subsequently constructed using the in situ data. Thus, we will endeavor to collect more in situ samples to improve our models and the accuracy of their predictions.

(2) Because the time span of our data was small, it needs to be increased. Also, the influence of wind speed on CO_2 needs to be more accurately assessed in the analysis of climate impact factors.

5. Conclusions

This study used Landsat 8 satellite images to monitor the pCO_2 and CO_2 flux changes in Chagan Lake. A comparison of the predictive performance of a linear regression model and a machine learning model revealed that the XGBoost model had the highest fitting accuracy (R^2 = 0.94, RMSE = 16.86 μatm), and hence was used to map the pCO_2 variation in Chagan Lake in 2021. The pCO_2 in the surface water of Chagan Lake showed a considerable spatial variability within a range of 0–200 μatm during the ice-free period. In terms of seasonal variation, the highest values were found in the summer, followed by spring and autumn. The CO_2 fluxes calculated from the pCO_2 values ranged from −3.69 to −2.42 mmol/m^2/d, which indicates that Chagan Lake was absorbing CO_2 (i.e., it was a weak carbon sink). Increasing temperature drove the increase in pCO_2, and CO_2 emissions in Chagan Lake, along with the deterioration of the trophic states. Based on the results of this study, actively carrying out eutrophication management, reducing the degree of eutrophication in lakes, and maintaining the reasonable development of fisheries can make important contributions to reducing carbon dioxide emissions in lakes and strengthening the carbon sink function of lakes. The results of this study provide important data and insights needed to implement effective practices for the regulation of the carbon cycle in lakes.

Author Contributions: Conceptualization: R.Z. and Z.W.; methodology: R.Z. and G.L.; software: Q.Y.; validation: C.F.; formal analysis: Y.S.; investigation: Q.Y., S.L. and H.T.; data curation: L.L. and K.S.; writing—original draft preparation: R.Z.; writing—review and editing: Z.W.; visualization and supervision: K.S.; funding acquisition: Z.W., S.L., Y.S., G.L. and K.S. All authors have read and agreed to the published version of the manuscript.

Funding: This research was jointly supported by the Natural Science Foundation of Jilin Province, China (20220203024SF) Zhidan Wen; the National Natural Science Foundation of China (U2243230, 42371390, 42171385, 42201414), Kaishan Song, Yingxin Shang, Ge Liu, Sijia Li; Youth Innovation Promotion Association of Chinese Academy of Sciences, China (2020234) Zhidan Wen; Young Scientist Group Project of Northeast Institute of Geography and Agroecology, Chinese Academy of Sciences (2023QNXZ01) Zhidan Wen; Land Observation Satellite Supporting Platform of National Civil Space Infrastructure Project (CASPLOS-CCSI) Sijia Li, and Postdoctoral Fellowship of Jilin Province of China to Yingxin Shang.

Data Availability Statement: Data are contained within the article.

Acknowledgments: Thanks to USGS and GEE for the Landsat data. Thanks to the NCAR Research Data Archive (RDA) for the FNL data. We also acknowledge the data support from National Climatic Data Center of the National Oceanic and Atmospheric Administration and the Global Monitoring Laboratory (GML) of the National Oceanic and Atmospheric Administration.

Conflicts of Interest: The authors declare no conflict of interest.

References

1. Ballantyne, A.P.; Alden, C.B.; Miller, J.B.; Tans, P.P.; White, J.W.C. Increase in observed net carbon dioxide uptake by land and oceans during the past 50 years. *Nature* **2012**, *488*, 70–72. [CrossRef]
2. IPCC. *Climate Change 2021: The Physical Science Basis. Contribution of Working Group I to the Sixth Assessment Report of the Intergovernmental Panel on Climate Change*; Cambridge University Press: Cambridge, UK; New York, NY, USA, 2021; Volume In Press.
3. Barral, A.; Gomez, B.; Fourel, F.; Daviero-Gomez, V.; Lécuyer, C. CO_2 and temperature decoupling at the million-year scale during the Cretaceous Greenhouse. *Sci. Rep.* **2017**, *7*, 8310. [CrossRef]
4. Davidson, E.A.; Suddick, E.C.; Rice, C.W.; Prokopy, L.S. More Food, Low Pollution (Mo Fo Lo Po): A Grand Challenge for the 21st Century. *J. Environ. Qual.* **2015**, *44*, 305–311. [CrossRef] [PubMed]
5. Linquist, B.A.; Anders, M.M.; Adviento-Borbe, M.A.A.; Chaney, R.L.; Nalley, L.L.; da Rosa, E.F.F.; van Kessel, C. Reducing greenhouse gas emissions, water use, and grain arsenic levels in rice systems. *Glob. Chang. Biol.* **2014**, *21*, 407–417. [CrossRef]
6. Yao, Z.; Zheng, X.; Liu, C.; Lin, S.; Zuo, Q.; Butterbach-Bahl, K. Improving rice production sustainability by reducing water demand and greenhouse gas emissions with biodegradable films. *Sci. Rep.* **2017**, *7*, 39855. [CrossRef] [PubMed]
7. Weyhenmeyer, G.A.; Kosten, S.; Wallin, M.B.; Tranvik, L.J.; Jeppesen, E.; Roland, F. Significant fraction of CO_2 emissions from boreal lakes derived from hydrologic inorganic carbon inputs. *Nat. Geosci.* **2015**, *8*, 933–936. [CrossRef]
8. Verpoorter, C.; Kutser, T.; Seekell, D.A.; Tranvik, L.J. A global inventory of lakes based on high-resolution satellite imagery. *Geophys. Res. Lett.* **2014**, *41*, 6396–6402. [CrossRef]
9. Raymond, P.A.; Hartmann, J.; Lauerwald, R.; Sobek, S.; McDonald, C.; Hoover, M.; Butman, D.; Striegl, R.; Mayorga, E.; Humborg, C.; et al. Global carbon dioxide emissions from inland waters. *Nature* **2013**, *503*, 355–359. [CrossRef] [PubMed]

10. Borges, A.V.; Darchambeau, F.; Teodoru, C.R.; Marwick, T.R.; Tamooh, F.; Geeraert, N.; Omengo, F.O.; Guérin, F.; Lambert, T.; Morana, C.; et al. Globally significant greenhouse-gas emissions from African inland waters. *Nat. Geosci.* **2015**, *8*, 637–642. [CrossRef]
11. Wen, Z.; Shang, Y.; Lyu, L.; Li, S.; Tao, H.; Song, K. A Review of Quantifying pCO_2 in Inland Waters with a Global Perspective: Challenges and Prospects of Implementing Remote Sensing Technology. *Remote Sens.* **2021**, *13*, 4916. [CrossRef]
12. Ciais, P.; Chris, S.; Govindasamy, B.; Bopp, L.; Brovkin, V.; Canadell, J.; Chhabra, A.; Defries, R.; Galloway, J.; Heimann, M. Carbon and other biogeochemical cycles. In *Climate Change 2013: The Physical Science Basis*; Cambridge University Press: Cambridge, UK, 2013; pp. 465–570.
13. Kutser, T.; Verpoorter, C.; Paavel, B.; Tranvik, L.J. Estimating lake carbon fractions from remote sensing data. *Remote Sens. Environ.* **2015**, *157*, 138–146. [CrossRef]
14. Cole, J.J.; Prairie, Y.T.; Caraco, N.F.; McDowell, W.H.; Tranvik, L.J.; Striegl, R.G.; Duarte, C.M.; Kortelainen, P.; Downing, J.A.; Middelburg, J.J.; et al. Plumbing the Global Carbon Cycle: Integrating Inland Waters into the Terrestrial Carbon Budget. *Ecosystems* **2007**, *10*, 172–185. [CrossRef]
15. Holgerson, M.A.; Raymond, P.A. Large contribution to inland water CO_2 and CH_4 emissions from very small ponds. *Nat. Geosci.* **2016**, *9*, 222–226. [CrossRef]
16. Luo, X.; Croft, H.; Chen, J.M.; He, L.; Keenan, T.F. Improved estimates of global terrestrial photosynthesis using information on leaf chlorophyll content. *Glob. Chang. Biol.* **2019**, *25*, 2499–2514. [CrossRef] [PubMed]
17. Xiao, Q.; Xu, X.; Duan, H.; Qi, T.; Qin, B.; Lee, X.; Hu, Z.; Wang, W.; Xiao, W.; Zhang, M. Eutrophic Lake Taihu as a significant CO_2 source during 2000–2015. *Water Res.* **2020**, *170*, 115331. [CrossRef] [PubMed]
18. Huotari, J.; Ojala, A.; Peltomaa, E.; Nordbo, A.; Launiainen, S.; Pumpanen, J.; Rasilo, T.; Hari, P.; Vesala, T. Long-term direct CO_2 flux measurements over a boreal lake: Five years of eddy covariance data. *Geophys. Res. Lett.* **2011**, *38*, L18401. [CrossRef]
19. Reed, D.E.; Dugan, H.A.; Flannery, A.L.; Desai, A.R. Carbon sink and source dynamics of a eutrophic deep lake using multiple flux observations over multiple years. *Limnol. Oceanogr. Lett.* **2018**, *3*, 285–292. [CrossRef]
20. Kling, G.W.; Kipphut, G.W.; Miller, M.C. The flux of CO_2 and CH4 from lakes and rivers in arctic Alaska. *Hydrobiologia* **1992**, *240*, 23–36. [CrossRef]
21. Zhao, M.; Jiao, S.; Liang, H.; Cao, Y.; Zhao, Z.; Zhang, Q.; Yuan, R. Temporal and spatial variation of carbon dioxide partial pressure and exchange flux characteristics in the backwater area of Wanfeng Lake Reservoir. *Environ. Chem.* **2019**, *38*, 1307–1317. [CrossRef]
22. Tianci, Q.I.; Qitao, X.; Yuqing, M.; Hongtao, D. Temporal and spatial variation of carbon dioxide concentration and its exchange fluxes in Lake Chaohu. *J. Lake Sci.* **2019**, *31*, 766–778. [CrossRef]
23. Qi, T.; Xiao, Q.; Cao, Z.; Shen, M.; Ma, J.; Liu, D.; Duan, H. Satellite Estimation of Dissolved Carbon Dioxide Concentrations in China's Lake Taihu. *Environ. Sci. Technol.* **2020**, *54*, 13709–13718. [CrossRef] [PubMed]
24. Zhang, B.; Li, J.; Shen, Q.; Wu, Y.; Zhang, F.; Wang, S.; Yao, Y.; Guo, L.; Yin, Z. Recent research progress on long time series and large scale optical remote sensing of inland water. *Natl. Remote Sens. Bull.* **2021**, *25*, 37–52. [CrossRef]
25. Song, K.; Liu, G.; Wang, Q.; Wen, Z.; Lyu, L.; Du, Y.; Sha, L.; Fang, C. Quantification of lake clarity in China using Landsat OLI imagery data. *Remote Sens. Environ.* **2020**, *243*, 111800. [CrossRef]
26. Wang, J.; Zhao, Q.; Pang, Y.; Hu, K. Research on nutrient pollution load in Lake Taihu, China. *Environ. Sci. Pollut. Res.* **2017**, *24*, 17829–17838. [CrossRef] [PubMed]
27. Shi, K.; Zhang, Y.; Zhou, Y.; Liu, X.; Zhu, G.; Qin, B.; Gao, G. Long-term MODIS observations of cyanobacterial dynamics in Lake Taihu: Responses to nutrient enrichment and meteorological factors. *Sci. Rep.* **2017**, *7*, 40326. [CrossRef] [PubMed]
28. Yin, Z.; Li, J.; Zhang, B.; Liu, Y.; Yan, K.; Gao, M.; Xie, Y.; Zhang, F.; Wang, S. Increase in chlorophyll-a concentration in Lake Taihu from 1984 to 2021 based on Landsat observations. *Sci. Total Environ.* **2023**, *873*, 162168. [CrossRef]
29. Xie, C.; Zhang, X.; Zhuang, L.; Zhu, R.; Guo, J. Analysis of surface temperature variation of lakes in China using MODIS land surface temperature data. *Sci. Rep.* **2022**, *12*, 2415. [CrossRef]
30. Sagan, V.; Peterson, K.; Maimaitijiang, M.; Sidike, P.; Sloan, J.; Greeling, B.; Maalouf, S.; Adams, C. Monitoring inland water quality using remote sensing: Potential and limitations of spectral indices, bio-optical simulations, machine learning, and cloud computing. *Earth-Sci. Rev.* **2020**, *205*, 103187. [CrossRef]
31. Yao, L.; Zhao, X.; Zhou, G.-J.; Liang, R.; Gou, T.; Xia, B.; Li, S.; Liu, C. Seasonal Succession of Phytoplankton Functional Groups and Driving Factors of Cyanobacterial Blooms in a Subtropical Reservoir in South China. *Water* **2020**, *12*, 1167. [CrossRef]
32. Chang, C.; Wang, Y.; Zhao, C.; Liu, Z.; Ma, C. Seasonal succession of phytoplankton functional groups in Chagan Lake and its influencing factors. *Chin. J. Ecol.* **2023**, 1–9. Available online: https://link.cnki.net/urlid/21.1148.Q.20230510.1734.005 (accessed on 6 December 2023).
33. Liu, X.; Zhang, G.; Sun, G.; Wu, Y.; Chen, Y. Assessment of Lake Water Quality and Eutrophication Risk in an Agricultural Irrigation Area: A Case Study of the Chagan Lake in Northeast China. *Water* **2019**, *11*, 2380. [CrossRef]
34. Schindler, D.W.; Carpenter, S.R.; Chapra, S.C.; Hecky, R.E.; Orihel, D.M. Reducing Phosphorus to Curb Lake Eutrophication is a Success. *Environ. Sci. Technol.* **2016**, *50*, 8923–8929. [CrossRef] [PubMed]
35. Cole, J.J.; Caraco, N.F.; Kling, G.W.; Kratz, T.K. Carbon Dioxide Supersaturation in the Surface Waters of Lakes. *Science* **1994**, *265*, 1568–1570. [CrossRef] [PubMed]

36. Lundin, E.J.; Giesler, R.; Persson, A.; Thompson, M.S.; Karlsson, J. Integrating carbon emissions from lakes and streams in a subarctic catchment. *Geophys. Res. Biogeosci.* **2013**, *118*, 1200–1207. [CrossRef]
37. Wen, Z.; Song, K.; Shang, Y.; Fang, C.; Li, L.; Lv, L.; Lv, X.; Chen, L. Carbon dioxide emissions from lakes and reservoirs of China: A regional estimate based on the calculated pCO_2. *Atmos. Environ.* **2017**, *170*, 71–81. [CrossRef]
38. Ma, J. Inversion of Kd(PAR) and Euphotic Zone Depth of Typical Water Bodys in Northeast China with Remote Imagery. Master's Thesis, University of Chinese Academy of Sicences, Shenzhen, China, 2017.
39. Haddout, S.; Priya, K.L.; Hoguane, A.M.; Casila, J.C.C.; Ljubenkov, I. Relationship of salinity, temperature, pH, and transparency to dissolved oxygen in the Bouregreg estuary (Morocco): First results. *Water Pract. Technol.* **2022**, *17*, 2654–2663. [CrossRef]
40. Roy, D.P.; Kovalskyy, V.; Zhang, H.K.; Vermote, E.F.; Yan, L.; Kumar, S.S.; Egorov, A. Characterization of Landsat-7 to Landsat-8 reflective wavelength and normalized difference vegetation index continuity. *Remote Sens. Environ.* **2016**, *185*, 57–70. [CrossRef]
41. Teodoru, C.R.; del Giorgio, P.A.; Prairie, Y.T.; Camire, M. Patterns in pCO_2 in boreal streams and rivers of northern Quebec Canada. *Glob. Biogeochem. Cycles* **2009**, *23*, GB2012. [CrossRef]
42. Cole, J.; Caracao, N. Carbon in catchments: Connecting terrestrial carbon losses with aquatic metabolism. *Mar. Freshw. Res.* **2001**, *52*, 101–110. [CrossRef]
43. Cole, J.J.; Caraco, N.F. Atmospheric exchange of carbon dioxide in a low-wind oligotrophic lake measured by the addition of SF6. *Limnol. Oceanogr.* **1998**, *43*, 647–656. [CrossRef]
44. Crusius, J.; Wanninkhof, R. Gas transfer velocities measured at low wind speed over a lake. *Limnol. Oceanogr.* **2003**, *48*, 1010–1017. [CrossRef]
45. Wanninkhof, R. Relationship between wind speed and gas exchange over the ocean. *J. Geophys. Res. Ocean.* **1992**, *97*, 7373–7382. [CrossRef]
46. Li, S.; Bush, R.T.; Santos, I.R.; Zhang, Q.; Song, K.; Mao, R.; Wen, Z.; Lu, X.X. Large greenhouse gases emissions from China's lakes and reservoirs. *Water Res.* **2018**, *147*, 13–24. [CrossRef]
47. Fang, H.; He, Z.; Chen, X.; Tian, X. Spatial-temporal change of Chagan Lake and response to precipitation change from 1985 to 2018. *China Agric. Inform.* **2020**, *32*, 64–73. [CrossRef]
48. Wen, Z.; Song, K.; Zhao, Y.; Shao, T.; Li, S. Seasonal Variability of Greenhouse Gas Emissions in the Urban Lakes in Changchun, China. *Huanjing Kexue* **2016**, *37*, 102–111. [CrossRef] [PubMed]
49. Lehman, P.W. The influence of climate on phytoplankton community biomass in San Francisco Bay Estuary. *Limnol. Oceanogr.* **2000**, *45*, 580–590. [CrossRef]
50. Li, R.; Zhang, G.; Zhang, L. Multivariate analysis of the relations between phytoplankton assemblages and environmental factors in Chagan Lake Wetland. *Acta Ecol. Sin.* **2014**, *34*, 2663–2673. [CrossRef]
51. Wu, N.; Tang, T.; Zhou, S.; Fu, X.; Jiang, W.; Li, F.; Cai, Q. Influence of cascaded exploitation of small hydropower on phytoplankton in Xiangxi River. *J. Appl. Ecol.* **2007**, *18*, 1091–1096.
52. Zhang, Y.; Qin, B.; Zhu, G.; Song, C.; Deng, J.; Xue, B.; Gong, Z.; Wang, X.; Wu, J.; Shi, K.; et al. Importance and main ecological and environmental problems of lakes in China. *Chin. Sci. Bull.* **2022**, *67*, 3503–3519. [CrossRef]
53. Zhang, G.; Gu, X.; Zhao, T.; Zhang, Y.; Xu, L. Ecological and Environmental Changes and Protection Measures of Lakes in China. *Bull. Chin. Acad. Sci.* **2023**, *38*, 358–364. [CrossRef]
54. Luo, J.C.; Ni, M.F.; Li, S.Y. Water-Air Interface CO_2 Exchange Flux of Typical Lakes in a Mountainous Area of the Western Chongqing and Their Influencing Factors. *Huanjing Kexue* **2019**, *40*, 192–199. [CrossRef]
55. Wang, J.; Zhou, Y.; Zhou, L.; Zhang, Y.; Qin, B.; Spencer, R.G.M.; Brookes, J.D.; Jeppesen, E.; Weyhenmeyer, G.A.; Wu, F. Urbanization in developing countries overrides catchment productivity in fueling inland water CO_2 emissions. *Glob. Chang. Biol.* **2023**, *29*, 1–4. [CrossRef] [PubMed]
56. Xi, S. The Distribution, Environmental Impacts and Preventation Contermeasures of Nitrogen and Phosphate in the Chaohu Lake. Doctoral Dissertation, University of Science and Technology of China, Hefei, China, 2016.
57. Pacheco, F.; Roland, F.; Downing, J. Eutrophication reverses whole-lake carbon budgets. *Inland Waters* **2014**, *4*, 41–48. [CrossRef]
58. Shen, X.; Sun, T.; Su, M.; Dang, Z.; Yang, Z. Short-term response of aquatic ecosystem metabolism to turbidity disturbance in experimental estuarine wetlands. *Ecol. Eng.* **2019**, *136*, 55–61. [CrossRef]
59. Jiang, T.; Guo, J.; Li, Z.; Fang, F.; Bai, L.; Jing, L. Air-water surface greenhouse gas flux in Pengxi river at different operational stages of the Three Gorges Reservoir. *Huanjing Kexue* **2012**, *33*, 1463–1470.
60. Junyu, Z.; Kai, P.; Yuyang, L.; Chaorong, L.; Lei, Z.; Yongqiang, Z.; Yanqing, D. Characteristics and influence factors of carbon dioxide efflux from Lake Hongze under different hydrological scenarios. *J. Lake Sci.* **2022**, *34*, 1347–1358. [CrossRef]

remote sensing

MDPI

Article

Research on the Characteristic Spectral Band Determination for Water Quality Parameters Retrieval Based on Satellite Hyperspectral Data

Xietian Xia [1,2,3], Hui Lu [2,4,*], Zenghui Xu [3], Xiang Li [3] and Yu Tian [1]

1 State Key Laboratory of Urban Water Resource and Environment, Harbin Institute of Technology, Harbin 150090, China; xiaxietian@tju.edu.cn (X.X.)
2 Department of Earth System Science, Ministry of Education Key Laboratory for Earth System Modeling, Institute for Global Change Studies, Tsinghua University, Beijing 100084, China
3 China Construction Power and Environment Engineering Co., Ltd., Nanjing 210012, China
4 Tsinghua University (Department of Earth System Science)—Xi'an Institute of Surveying and Mapping Joint Research Center for Next-Generation Smart Mapping, Beijing 100084, China
* Correspondence: luhui@tsinghua.edu.cn

Abstract: Hyperspectral remote sensing technology has been widely used in water quality monitoring. However, while it provides more detailed spectral information for water quality monitoring, it also gives rise to issues such as data redundancy, complex data processing, and low spatial resolution. In this study, a novel approach was proposed to determine the characteristic spectral band of water quality parameters based on satellite hyperspectral data, aiming to improve data utilization of hyperspectral data and to achieve the same precision monitoring of multispectral data. This paper first introduces the data matching method of satellite hyperspectral data and water quality based on space–time information for guidance in collecting research data. Secondly, the customizable and fixed spectral bands of the existing multispectral camera products were studied and used for the preprocessing of hyperspectral data. Then, the determination approach of characteristic spectral bands of water quality parameters is proposed based on the correlation between the reflectance of different bands and regression modeling. Next, the model performance for retrieval of various water quality parameters was compared between the typical empirical method and artificial neural network (ANN) method of different spectral band sets with different band numbers. Finally, taking the adjusted determination coefficient $\overline{R^2}$ as an evaluation index for the models, the results show that the ANN method has obvious advantages over the empirical method, and band set providing more band options improves the model performance. There is an optimal band number for the characteristic spectral bands of water quality parameters. For permanganate index (COD_{Mn}), dissolved oxygen (DO), and conductivity (EC), the $\overline{R^2}$ of the optimal ANN model with three bands can reach about 0.68, 0.43, and 0.49, respectively, whose mean absolute percentage error (MAPE) values are 14.02%, 16.26%, and 17.52%, respectively. This paper provides technical guidance for efficient utilization of hyperspectral data by determination of characteristic spectral bands, the theoretical basis for customization of multispectral cameras, and the subsequent water quality monitoring through remote sensing using a multispectral drone.

Keywords: water quality retrieval; hyperspectral data; multispectral data; characteristic spectral bands; artificial neural network

1. Introduction

Remote sensing has the advantage of large-scale, all-weather accurate dynamic monitoring, and has been widely used in the water conservancy industry. Remote sensing of the water environment was born, and its development has kept pace with the times [1]. Changes in the composition and concentration of substances in the water often cause changes in the color of the water body [2]. Remote sensing technology of spectral imaging can obtain the color parameters of the water body by obtaining the spectral characteristics of the water body, and then inverse the water quality parameters, so as to realize the water environment monitoring of rivers and lakes [3,4].

Most applications of remote sensing monitoring technology and spectral imaging in the field of water environment monitoring can be summarized in three steps: remote sensing data acquisition, data processing and inversion model construction, and model analysis and application. The available data sources for remote sensing of water quality retrieval are usually multispectral and hyperspectral remote sensors [5] categorized by the spectral resolution of the sensors, carried by the spaceborne, airborne, and portable and ground-based load platforms [6]. Multispectral data available for remote sensing water quality retrieval typically have 3–10 bands. Landsat series data are the most commonly utilized for water quality monitoring, such as TSM, COD, and TP, due to their accessibility and geographic, temporal, and spectral resolution [7,8]. Hyperspectral satellites have multiple bands with about 5–10 nm spectral resolution. For retrieving water quality, hyperspectral data from satellites such as the HuanJing-1 (HJ-1) [9], Gaofen-5 (GF-5) [10], and Ziyuan1-02D (ZY1-02D) [11] have been employed. Higher spectral resolution data have a large number of bands that can be precisely and optimally chosen for developing inversion models of water quality parameters to differentiate the spectral differences in multispectral data, greatly enhancing the accuracy of inversion algorithms [12–14]. Among the four spectral sensor platforms for the water quality monitoring, the portable and ground-based spectrometer [4] is less flexible and more labor-intensive; the airborne spectrometer [5,6] is flexible and has high spatial resolution, but the observation area is small; and the satellite-based spectrometer [8] has low imagery cost and is suitable for large-scale monitoring, but it has the disadvantages of low spatial resolution, poor timeliness, and long revisit cycle.

In terms of model construction for water quality parameter retrieval, it is mainly divided into the empirical method, analytical method (also called bio-optical method [6]), artificial intelligence (AI) method, and combined empirical and analytical methods. The fourth method is called the semi-empirical model or semi-analytical model, while in some studies they were listed together [6]. The empirical method relies only on the statistical relationship between remote sensing data and measured water quality parameters to establish models [15–18]. The principle of the model is simple and the accuracy of the results is high, but the generality of the results is low. The analysis model is based on the principle of water radiative transfer, and the content of each component in the water is calculated from the remote sensing data through the transfer formula [19]. For example, Dekker et al. [20] estimated the water color parameters by building a physical analysis model based on Landsat TM data and the measured intrinsic optical quantity of the water body. Sudduth et al. [12] used airborne hyperspectral data images of the major rivers in Minnesota to establish an analysis model, which was based on the intrinsic optical quantity of the water body and the apparent optical quantity calculated from the spectral data, thereby retrieving the suspended solids concentration of the river and noting that 700 nm is the best band for measuring the suspended solids concentration in the study area. This type of model has good universality and high precision, but the model is difficult to fit. The physical meaning of the spectral index used in the semi-empirical model is clearer than that used in the empirical method. Taking the concentration of chlorophyll-a as an example, the semi-empirical model method is to propose the spectral index related to its concentration based on certain assumptions related to the bio-optical theoretical model [21], and establish a statistical relationship to realize the inversion of the water quality parameters [22,23]. The construction of the analytical model depends on the complex water radiative transfer model, and the existing analytical models almost need to introduce empirical formulas to determine some parameters. Generally, the analytical model with empirical formulas is called the semi-analytical model [24,25], which requires the measured absorption coefficient and other intrinsic optical parameters. The AI model effectively trains a large number of reflectivity and water quality parameters by using an AI algorithm, and automatically learns the nonlinear relationship between the two through the network to realize the prediction of water quality parameters [5,26–28].

The core of modeling various water quality parameters is band selection and band combination [29]. In the past, multispectral data were the main data used for water environment remote sensing, and only a few bands could not accurately obtain the spectral information of different water quality parameters. In recent years, the number of satellites equipped with hyperspectral imagers has gradually increased, such as ZY1-02D [11], GF-5 [30], and the Hyperspectral Precursor and Application Mission (PRISMA) [31]. The on-board hyperspectral camera provides a broad data source for hyperspectral data acquisition, and hyperspectral remote sensing data also provides more bands for model building [32]. The spectral curve for water elements also obtained provides a basis for analyzing the physical meaning of bands, and can help researchers to establish a more accurate semi-empirical model with clearer physical meaning. However, the number of hyperspectral remote sensing bands is usually hundreds. At this stage, the band utilization efficiency of the semi-empirical inversion model is generally low. In general, the establishment of the target water quality parameter inversion model can only be completed with data within four bands, resulting in new problems such as hyperspectral data redundancy and complex processing. Therefore, based on the analysis of spectral characteristics of water elements, the band or band combination can be reasonably selected, and a superior semi-empirical retrieval model of hyperspectral water elements can be constructed to guide the customization of the multispectral camera, so as to realize the same precision inversion of water quality parameters based on the customized multispectral camera. At the same time, there are still some limitations in spaceborne hyperspectral technology, such as the spatial resolution, which creates difficulties in monitoring small lakes, reservoirs, and other small water areas, and there are restrictions on the acquisition of some commercial satellite data. Unmanned aerial vehicles (UAVs) [33] have the advantages of low cost, simple operation, high spatial resolution, and easy realization of scanning and imaging. It is convenient for field operation to patrol the water environment, which can effectively overcome this disadvantage. In summary, for the retrieval of water quality parameters, hyperspectral data have the disadvantages of low spatial resolution, excessive data redundancy, low data utilization, and only a few bands are used in the inversion model. Therefore, it is necessary to determine the method of identifying the characteristic bands of the satellite hyperspectral data inversion model, and realize monitoring at the same accuracy with multispectral data as the hyperspectral data. The follow-up research can be used to guide the effective band customization of the multispectral lens used for UAVs, and realize high-resolution, efficient, flexible, fast and low-cost water environment inspection of small water areas by combining the flexibility of UAVs, the characteristics of efficient data utilization, and the low cost of a multispectral lens.

Li et al. [32] utilized hyperspectral data from the Gaofen-5 satellite and employed machine learning methods to comprehensively characterize the features of the hyperspectral data through the combination of multispectral-scale morphological features. They investigated the relationship between the hyperspectral data and water quality parameters, and established a retrieval model. However, this study did not explore efficient utilization methods for hyperspectral data or redundant data removal techniques. Zheng et al. [11], on the other hand, used hyperspectral data from the ZY1-02D satellite and developed a water quality inversion model using machine learning techniques. They utilized empirical parameters and the ratio between two bands as inputs. The focus of their study was primarily on examining the impact of different machine learning methods and comparing the performance of ZY1-02D satellite hyperspectral data with Sentinel-2 multispectral data. Although both studies employed machine learning (AI) methods to construct water quality parameter inversion models, they did not specifically investigate characteristic spectral bands for water quality parameters. These characteristic bands are often derived through empirical models and have certain limitations.

Therefore, this paper aims to combine the advantages of AI algorithms and the explicit concept of characteristic spectral bands for water quality parameters. Based on hyperspectral data from the ZY1-02D satellite, the objective of this paper is to compare the model performance for retrieving various water quality parameters between the typical empirical method and artificial neural network (ANN) method by using different spectral band sets with different band numbers, so as to provide a guidance approach for the determination of characteristic bands of different water quality parameters. This article first introduces the research area and data sources together with the matching method for quality parameters and satellite hyperspectral data to collect data. Then, the preprocessing methods for water quality data and hyperspectral data are introduced. Then, the method for determining the characteristic spectral bands of water quality parameters is highlighted and derived from the correlation of band reflectance and the regression model methods based on empirical and ANN methods. The Results section introduces the high correlation of the two bands and the reduced computational complexity, along with the results of the empirical method-based model and the ANN-based model. The Discussion section provides recommended values for customizing the characteristic bands and directions for improving the model. Figure 1 is the overall technical flowchart for this work.

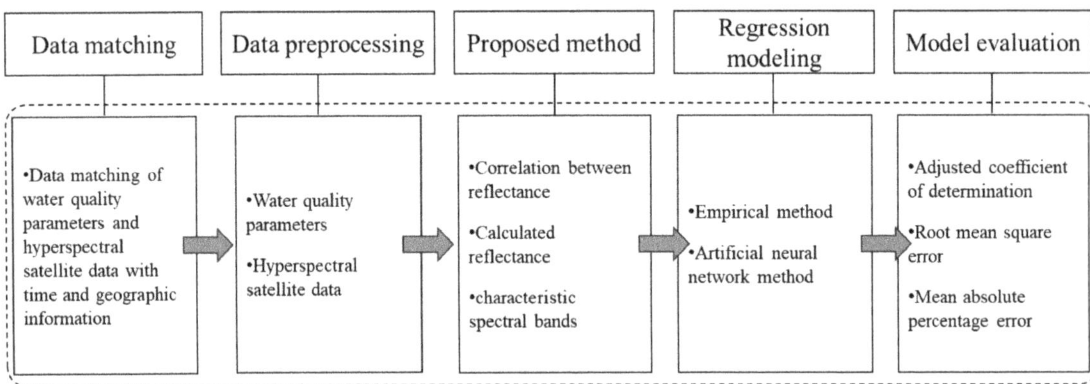

Figure 1. Overall technical flowchart for this work.

2. Materials

2.1. Study Area

As shown in Figure 2, this study selected the areas around Taihu Lake, such as Suzhou, Wuxi, Shanghai, Jiaxing, and Huzhou, as the region to collect water quality and satellite data. The water network in this region is dense; the distribution of National Surface Water Automatic Monitoring Stations (NSWAMS) stations is concentrated and dense, which can effectively improve the utilization rate for the satellite data. Water quality parameters used in this paper were acquired for the period December 2020 to August 2022.

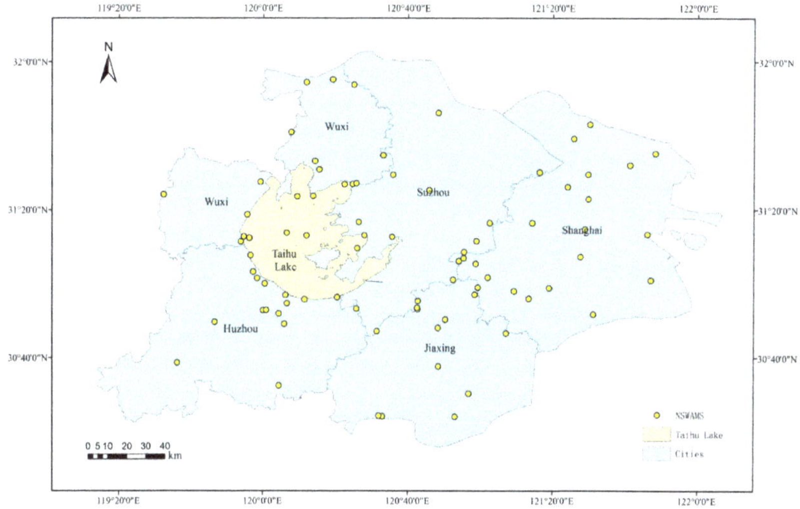

Figure 2. Research area of water quality and satellite data.

2.2. Data Collection

The water quality parameters are from the National Surface Water Automatic Monitoring Real-Time Data Release System of the China Environmental Monitoring Station. The release scope of this system includes data from the National Surface Water Automatic Monitoring Stations (NSWAMS) system, which was built and officially put into operation. From April 2014 to November 2020, NSWAMS included 134 stations. In December 2020, 1506 new stations were added. In 2021, 365 new stations were added. Currently, there are a total of 2005 stations, which can provide sufficient data support for building regression models of satellite hyperspectral data and water quality parameters.

The water parameters released include water temperature, pH, dissolved oxygen (DO), permanganate index (COD_{Mn}), ammonia nitrogen (NH3-N), total phosphorus (TP), total nitrogen (TN), electric conductivity (EC), and turbidity (TUB), a total of 9 monitoring indicators. The data are released every 4 h, which can effectively correspond to the satellite transit at different times. This paper selects 7 water quality parameters as research objects: DO, COD_{Mn}, NH3-N, TP, TN, TUB, and EC.

The hyperspectral satellite data comes from the Natural Resources Satellite Remote Sensing Cloud Service Platform, and are obtained from the hyperspectral camera on the ZY1-02D satellite, the Advanced Hyperspectral Imager (AHSI) sensor [11]. It has been shown that for the AHSI hyperspectral sensor, the average equivalent reflectance for each band in situ R_{rs} and the multispectral sensor Multispectral Imager (MSI) are basically the same [11]. The satellite carries two cameras that can effectively obtain 9-band multispectral data with 115 km width and 166-band hyperspectral data with 60 km width. Among them, the full spectral resolution can reach 2.5 m; the multispectral is 10 m and the hyperspectral is 30 m. The visible near infrared and shortwave infrared spectral resolution of the hyperspectral payload can reach 10 and 20 nm, respectively. The main parameters of the AHSI are shown in Table 1.

Table 1. Main parameters of the ZY1-02D satellite hyperspectral camera.

Spectral Range	Spatial Resolution	Spectral Resolution		Width
		Visible/Near Infrared	SWIR	
0.4–2.5 μm, 166 spectral bands	30 m	10 nm	20 nm	60 km

The hyperspectral data from the ZY1-02D satellite have been radiometrically corrected, bad pixels repaired, and spectrally calibrated. To meet the application requirements of quantitative remote sensing, radiometric calibration, atmospheric correction, and orthorectification must be performed [11], as shown in Figure 3. Among them, orthorectification is performed by using the built-in RPC file and Landsat 8 data after terrain correction as a reference image. One example of the ZY1-02D satellite image products is shown in Figure 4, in which the detailed information was listed including the produce time, longitude and latitude, and so on.

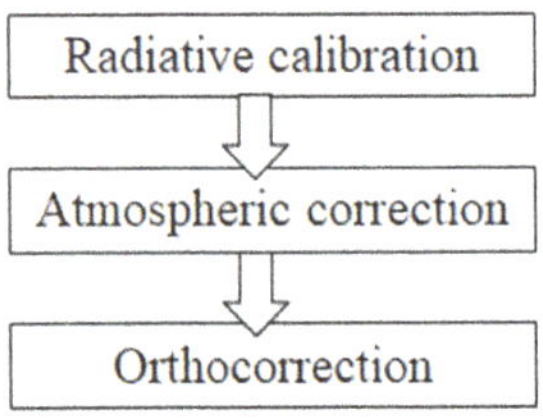

Figure 3. Preprocessing flow of hyperspectral data from satellite ZY1-02D.

Data detail	
SatelliteID:	ZY1E
SensorID:	AHSI
ProduceTime:	2022-08-08 11:00
CloudPercent	0%
Resolution:	30M
OrbitID:	15222
Scene Path/Row	777/131
ProductID:	486535
TopLeft Lon/Lat	120.347802/31.204651
TopRight Lon/Lat	120.972630/31.076574
BottomRight Lon/Lat	120.818736/30.526574
BottomLeft Lon/Lat	120.197424/30.654254

Figure 4. One of the ZY1-02D satellite image products.

3. Method

In this section, we introduce the methodology employed in this paper, which includes data matching, data processing, determination of characteristic spectral bands, regression modeling, and model evaluation. The water quality parameters and hyperspectral data do not possess a direct one-to-one correspondence; the data matching method is to establish a consistent correspondence between water quality data collected at the same geographical location and time and the corresponding satellite hyperspectral data. Preprocessing of the water quality parameters and hyperspectral satellite data is necessary to normalize the distribution of the water quality parameters and ensure the hyperspectral data are processed within the spectral range of interest specific to this paper. The determination of characteristic spectral bands represents a crucial innovation in this research, allowing for efficient selection of relevant bands for different water quality parameters through correlation analysis with a regression model. Various regression modeling methods are presented in this section to evaluate and compare their performance in modeling, and to determine the optimal approach for regression modeling between the water quality parameters and calculated reflectance. Model evaluation is necessary to clearly elaborate the specific parameters used to compare different regression models and to specify the calculation formulas for these comparison metrics.

3.1. Data Matching of Water Quality Parameters and Hyperspectral Satellite Data

The data matching method integrates geometric and temporal information to establish correspondence between water quality parameters and hyperspectral satellite data, enabling the spectral characteristics from the satellite data to represent the water quality parameters.

This paper realizes the heterogeneous data matching of water quality monitoring data and ZY1-02D satellite hyperspectral data of the same NSWAMS based on the location and time, whose principle is shown in Figure 5. Twenty scenes of hyperspectral data with the highest utilization rate were selected as the research data of this paper.

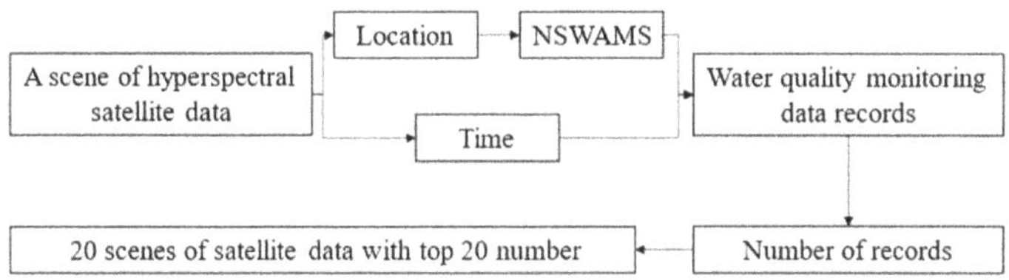

Figure 5. Principle of heterogeneous data matching between hyperspectral satellite data and water quality parameters records.

The specific methods are shown in Figure 6, as follows:

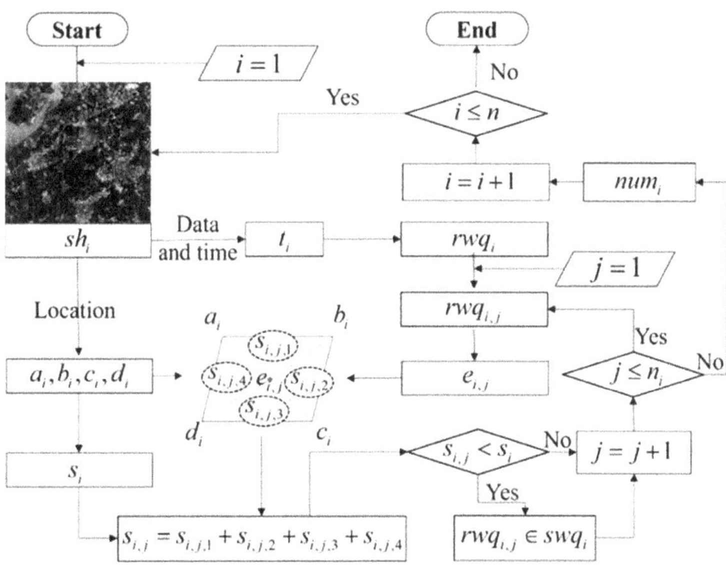

Figure 6. Detailed steps in heterogeneous data matching between hyperspectral satellite data and water quality parameters records.

(1) Data extraction for a range of locations and times. On the Natural Resources Satellite Remote Sensing Cloud Service Platform, the time condition is "December 2020–August 2022", the geographical conditions are Suzhou, Shanghai, Jiaxing, Huzhou, and Wuxi, the satellite sampling conditions are the AHSI sensor of ZY1-02D and 0 cloud amount; a total of 61 scenes were found, of which 8 scenes had low-altitude cloud phenomenon, so 53 scenes were available for selection. Water quality monitoring data were obtained from each NSWAMS in Suzhou, Shanghai, Jiaxing, Huzhou, and Wuxi from December 2020 to August 2022 from the National Surface Water Automatic Monitoring Real-Time Data Release System. Each datum includes the name, time, water quality classification, temperature, pH, DO, COD_{Mn}, NH3-N, TP, TN, EC, and TUB of the NSWAMS, together with latitude and longitude;

(2) Determination if each NSWAMS is within the satellite data. For the *i*th scene of the hyperspectral satellite data sh_i, the sampling date and time is t_i, the four vertices are a_i, b_i, c_i and d_i, whose latitudes and longitudes are (lon_{ai}, lat_{ai}), (lon_{bi}, lat_{bi}), (lon_{ci}, lat_{ci}), and (lon_{di}, $lat_{d}i$). Assuming that the water quality parameter set rwq_i was collected at the same sampling time t_i, and that the number of water quality parameters records in this set is n_i, the latitude and longitude of the NSWAMS $e_{i,j}$ corresponding to the *j*th water quality parameter record $rwq_{i,j}$ is ($lon_{ei,j}$, $lat_{ei,j}$), where $j \in \{1, 2, 3, \ldots, n\}$. The area method is used to determine if this NSWAMS is within the satellite hyperspectral data of this scene. The area of the parallelogram formed by a_i, b_i, c_i, and d_i is s_i, and $e_{i,j}$ forms four triangles with each side of the quadrangle area, whose areas are $s_{i,j,1}$, $s_{i,j,2}$, $s_{i,j,3}$, and $s_{i,j,4}$, respectively. If s_i is less than the sum of the four areas noted, then it proves that the NSWAMS is within the scene data, and these water quality parameter records are collected in the selected set swq_i. Otherwise, if it is higher, it is outside the hyperspectral satellite data of this scene. If the NSWAMS is determined to be within the geographic location of the *i*th satellite data according to the method described above, then the water quality parameter records within the satellite data are collected into the dataset swq_i.

(3) Selection of satellite data with the top 20 water quality parameter records. The number of water quality parameter records in this dataset is calculated, which is the number of NSWAMS in this satellite scene num_i. The number of NSWAMS in each scene of satellite data is calculated using on the method described above, and the numbers of water quality parameter records of all scenes are sorted. The 20 scenes of satellite data with the highest number of NSWAMS are taken for analysis, which are the 20 scenes of satellite hyperspectral data with the highest effective information density;

(4) Extraction of spectral value curve corresponding to each water quality parameter sample. According to the geographic information for the NSWAMSs collected from each scene, ENVI 5.3 software is used to extract the entire spectral value curve for the water body at the corresponding position in the hyperspectral satellite image, the spectral mean of scale 1 is taken as the spectral value [20].

All of the water quality parameter records and the spectral curves for the water body with the same position and time are collected. Using these methods, 188 records of water quality parameters at different times and locations and their corresponding satellite hyperspectral data in time and space were obtained, realizing the matching of 20 scenes of satellite hyperspectral data with the highest effective information density together with their water quality parameters.

3.2. Data Preprocessing

3.2.1. Water Quality Data Preprocessing

According to the Environmental Quality Standards for Surface Water (GB3838-2002), the classification criteria for Class I-V waters with DO, COD_{Mn}, NH3-N, TP, and TN are shown in Table 2.

Table 2. Classification criteria for water quality parameters.

No.	Parameters	Class I	Class II	Class III	Class IV	Class V
1	DO$\geq$	7.5	6	5	3	2
2	$COD_{Mn}\leq$	2	4	6	10	15
3	NH3-N$\leq$	0.15	0.5	1	1.5	2
4	TP$\leq$	0.02	0.1	0.2	0.3	0.4
5	TN$\leq$	0.2	0.5	1	1.5	2

According to these classification criteria, the water classification distribution for the 188 records of various water quality parameters is shown in Figure 7a.

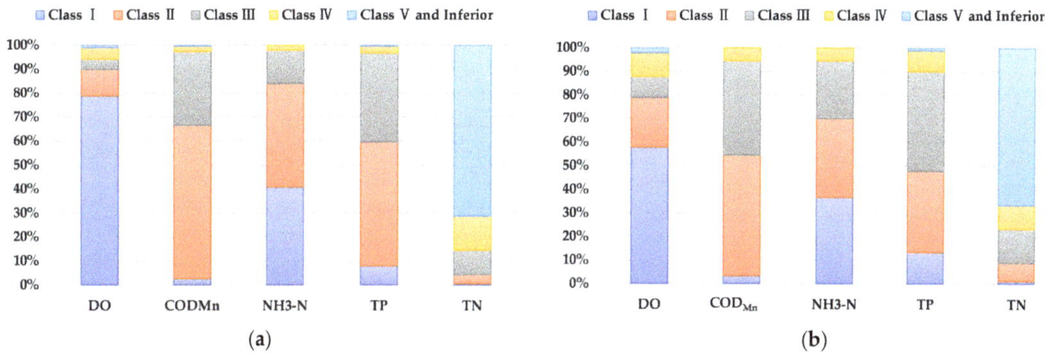

Figure 7. Classifications distribution of water quality parameters: (**a**) original 188 records; (**b**) 90 records after analysis and screening.

It can be seen from Figure 7a that there is a phenomenon of excessive concentration in the distribution of each water quality parameter within a certain interval. For example, the proportion of class I DO is 78.27%, the class II proportion of COD_{Mn} is 63.83%, the sum class I and class II proportion of NH3- N is above 40%, the class II and class III proportion of TP is 51.60% and 37.23%, respectively, and the class V proportion of TN is 53.19%. In order to ensure the accuracy and generalization for the inversion results of various water quality parameters, it is necessary to ensure that each parameter is evenly distributed within each classification interval as much as possible to reduce the phenomenon of distribution concentration. The priority is given to removing data points that simultaneously include Class I DO, Class II COD, Class I or II NH3-N, Class II TP, and Class V and inferior TN. After analysis and screening, 90 records of water quality parameters were selected, and the water classification distribution for each water quality parameter is shown in Figure 7b.

As a result, the water quality monitoring data in relatively concentrated intervals were essentially deleted, resulting in a relatively uniform distribution. Table 3 shows the descriptive statistics of various water quality parameters for the 90 records.

Table 3. Number of different classifications of various water quality parameters after screening.

Statistical Value	DO	COD_{Mn}	NH3-N	TP	TN	EC	TUB
Mean	8.26	3.95	0.35	0.110	2.42	511	80.6
Min	1.32	1.62	0.03	0.005	0.20	155	2.5
Max	15.79	9.46	1.41	0.649	9.25	1020	255.9
Std	2.74	1.43	0.33	0.087	1.79	158	59.1

3.2.2. Hyperspectral Satellite Data Preprocessing

The purpose of this paper is to determine the characteristic spectral bands of different water quality parameters, so as to provide the theoretical basis and application guidance for the band customization of the multispectral camera suitable for drones. The spectral range of interest is from the existing multispectral cameras and customizable multispectral cameras suitable for drones, as follows.

At present, there are many institutions engaged in the development of UAV multispectral imaging equipment, and the multispectral cameras in existing UAVs are mainly used for agricultural crop growth assessment, plant classification, forestry monitoring, etc., as shown in Table 4.

Table 4. Index parameters of existing multispectral cameras for drones.

No.	Institutions	Product Name	Spectral Channel Numbers	Spectral Channel Bands Information
1	Parrot (France)	SEQUOIA	4 multispectral channels + 1 RGB channel	550 nm@40 nm, 660 nm@40 nm, 735 nm@10 nm, 790 nm@40 nm
2	Micasense (USA)	RedEdge-MX	5 multispectral channels	475 nm@32 nm, 560 nm@27 nm, 668 nm@16 nm, 717 nm@12 nm, 842 nm@57 nm
3	Yusense (China)	MS600 Pro	6 multispectral channels	450 nm@35 nm, 555 nm@27 nm, 660 nm@22 nm, 720 nm@10 nm, 750 nm@10 nm, 840 nm@30 nm
4	Yusense (China)	AQ600 Pro	5 multispectral channels + 1 RGB channel	450 nm@35 nm, 555 nm@27 nm, 660 nm@22 nm, 720 nm@10 nm, 840 nm@30 nm
5	DJ (China)	Phantom 4 multispectral version	5 multispectral channels + 1 RGB channel	450 nm@16 nm, 560 nm@16 nm, 650 nm@16 nm, 730 nm@16 nm, 840 nm@26 nm
6	DJ (China)	Mavic3 multispectral version	4 multispectral channels + 1 RGB channel	560 nm@16 nm, 650 nm@16 nm, 730 nm@16 nm, 860 nm@26 nm

The 6 channels and 5 channels of MS600 Pro and AQ600 Pro, respectively, are customizable. The customizable spectral channels include 16 bands with different wavelengths, i.e., 410 nm@35 nm, 450 nm@35 nm, 490 nm@25 nm, 530 nm@27 nm, 555 nm@27 nm, 570 nm@32 nm, 610 nm@30 nm, 650 nm@27 nm, 660 nm@22 nm, 680 nm@25 nm, 720 nm@10 nm, 750 nm@10 nm, 780 nm@13 nm, 800 nm@35 nm, 840 nm@30 nm, and 900 nm@35 nm. The products that realize the integration of the drone body and multispectral camera are the DJ Phantom 4 multispectral version and the DJ Mavic3 multispectral version. The index parameters of the two products are shown in Table 4.

Since the maximum band wavelength of the customized UAV multispectral camera is 900 nm, the 166 bands with the wavelength from 396 nm to 2501 nm are first deleted to 60 bands with the wavelength from 396 to 903 nm. The obtained spectral curves for different collected hyperspectral satellite data are shown in Figure 8. In Figure 8, the y-axis is the remote sensing reflectance and the x-axis is the center wavelength value of the spectral channel bands. It can be seen from Figure 7 that there is a peak in the wavelength range from 550 to 580 nm for the reflectance, which increases sharply between 390 and 580 nm, but some of them are decreased in the wavelength range from 390 to 490 nm. Then, the reflectance increases from 490 to 580 nm, and generally decreases from 580 to 756 nm. There is a small peak in reflectance at 765 nm. For the remaining wavelength range, the regular pattern is not obvious, generally maintaining the level of fluctuation or slightly decreasing.

In this paper, the customizable 16 spectral band combination set is referred to as CM16, and the spectral band combination set of other existing products is referred to as the product name. For example, the 4-multispectral band combination for Parrot multispectral cameras is referred to as the Par set, and so on. The band set name and spectral information of the band set are shown in Table 5.

In order to compare the inversion performance of the above 7 band sets on 7 different water quality parameters and to determine the characteristic spectral bands of each parameter, the following section describes the empirical and artificial neural network (ANN) methods used to fit the reflectance data for single-band, two-band combinations, and three-band combinations from the above 7 band sets with water quality parameters, and to determine the optimal 6 spectral band combinations based on the obtained characteristic spectral bands for each water quality parameter in the customization of a 6-band multispectral camera. To achieve this goal, a total of 26 bands involved in the above 7 band sets were first determined. Then, the reflectance at the corresponding wavelengths of the 26 bands was calculated using the interpolation method. Finally, the remaining bands were deleted, and the preprocessing of the hyperspectral data was completed. The spectral curves for the 26 bands are shown in Figure 9.

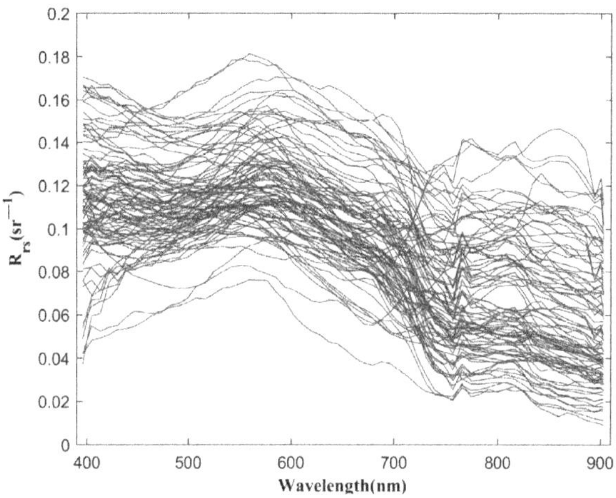

Figure 8. Remote sensing hyperspectral spectral curve with the wavelength from 390 to 900 nm.

Table 5. Information for different multispectral band combination sets.

No.	Band Set Name	Center Wavelengths of Bands in the Band Set (nm)
1	Par	550, 660, 735, 790
2	Mic	475, 560, 668, 717, 842
3	DJ3	560, 650, 730, 860
4	DJ4	450, 560, 650, 730, 840
5	MS600	450, 555, 660, 720, 750, 840
6	AQ600	450, 555, 660, 720, 840
7	CM16	410, 450, 490, 530, 555, 570, 610, 650, 660, 680, 720, 750, 780, 800, 840, 900

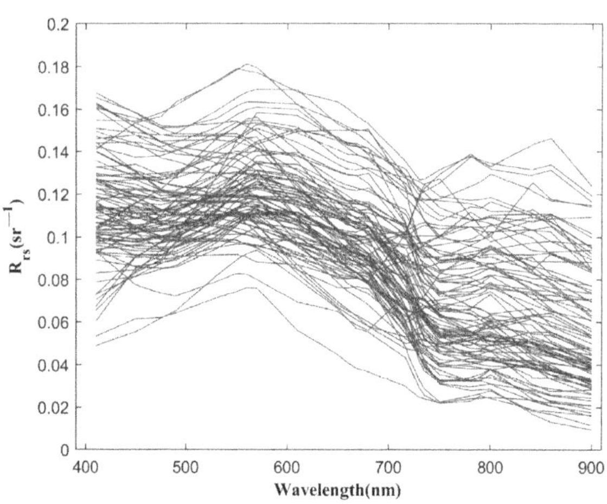

Figure 9. Spectral curves for multispectral data with 26 spectral bands.

3.3. Determination of Characteristic Spectral Bands for Water Quality Parameters Based on the Correlation between Reflectance of Different Bands

From the perspective of water quality parameter measurement, the effective utilization of hyperspectral data leads to selection of the characteristic spectral band combinations for different water quality parameters. By using multiple characteristic bands, accurate inversion of each water quality parameter can be achieved, which can ensure the accuracy of water quality parameter measurement and the simplicity of spectral bands, remove redundant data, improve spectral data processing speed, and achieve efficient utilization of spectral data.

This paper proposes a method to determine the optimal characteristic bands based on the reflectance correlation of different bands, which is shown in Figure 10. For the given band set, the number of bands contained in the band set is n_{bs}. The steps of the approach follow.

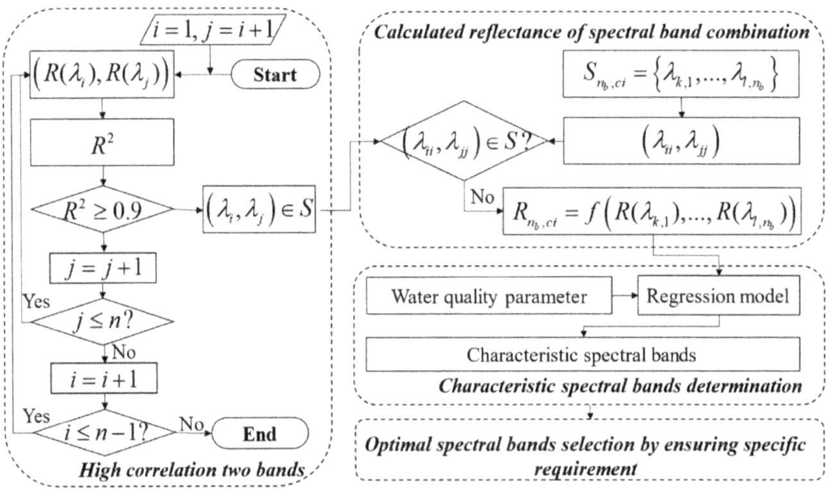

Figure 10. Approach for determining the characteristic spectral bands of water quality parameters.

(1) Determination of a high correlation two-band set. The determination coefficient R^2 [26] between the reflectance data corresponding to each two-band combination in a given band set was calculated by Equation (1).

$$R^2 = 1 - \frac{\sum_{i=1}^{n}\left(R_{rs,i}^{A} - R_{rs,i}^{B}\right)^2}{\sum_{i=1}^{n}\left(R_{rs,i}^{A} - \overline{R}_{rs}^{A}\right)^2} \quad (1)$$

where n is the number of the reflectance data samples, R_{rs}^{A} represents the reflectance data corresponding to band A, and R_{rs}^{B} represents the reflectance data corresponding to band B.

The two bands with the determination coefficient R^2 greater than 0.9 were considered to have the same effect in the same characteristic spectral band combination. Therefore, they cannot appear simultaneously in a characteristic spectral band combination containing two or more bands [34]. The dataset of the two-band combinations with R^2 higher than 0.9 is expressed as S.

The maximum number of spectral bands n_{bmax} contained in a spectral band combination and the number of different spectral band combinations with different numbers of bands could also be determined so that the spectral band combination cannot contain the two highly correlated bands.

(2) Calculated reflectance of the spectral band combination without high correlation between two bands. For the *ci*th spectral band wavelength combination $S_{n_b,ci}$, consisting of the *i*th, *j*th, ..., and *z*th bands wavelength, $(\lambda_{ii}, \lambda_{jj})$ represents arbitrary wavelength combination of two bands from $S_{n_b,ci}$. If $(\lambda_{ii}, \lambda_{jj})$ does not belong to the dataset *S*, then the reflectance data corresponding to the wavelengths in the spectral band combinations $S_{n_b,ci}$ can be used to calculate the combination reflectance $R_{n_b,ci}$ with Equation (2).

$$R_{n_b,ci} = f(R(\lambda_{i,1}), R(\lambda_{j,2}), \ldots, R(\lambda_{z,n_b})) \quad (2)$$

The definition and restriction conditions of the notations are listed in Table 6.

Table 6. Definition and restriction conditions of the notations.

Notations	Definition	Restriction Conditions
$S_{n_b,ci}$	The *ci*th spectral band wavelength combination	$S_{n_b,ci} = \{\lambda_{i,1}, \lambda_{j,2}, \ldots, \lambda_{z,n_b}\}$ $ci \in \{1, 2, 3, \ldots, n_c\}$
$i, j, \ldots, z$	The *i*th, *j*th, ..., and *z*th band wavelength	$i \neq j \neq, \ldots, \neq z$ $i \notin \{1, 2, 3, \ldots, n - n_b + 1\}$ $j \notin \{2, 3, \ldots, n - n_b + 2\}$ $z \notin \{n_b, n_b + 1, n_b + 2, \ldots, n\}$
n_b	The number of bands included in the combination $S_{n_b,ci}$	$n_b \in \{1, 2, 3, \ldots, n_{bmax}\}$
n_c	The number of the maximum value of the combinations with n_b bands	$n_c = C_{n_{bs}}^{n_b}$
$(\lambda_{ii}, \lambda_{jj})$	Arbitrary wavelength combination of two bands from $S_{n_b,ci}$	$\lambda_{ii}, \lambda_{jj} \in S_{n_b,ci}$ $ii \neq jj$
$R_{n_b,ci}$	Calculated reflectance of spectral band combination without high correlation two bands	-

(3) Characteristic spectral bands determination. The combination reflectance $R_{n_b,ci}$ corresponding to the spectral band combinations $S_{n_b,ci}$ containing one to n_{bmax} bands that meet the requirements were traversed to build the regression models with the selected method for inversion with different water quality parameters. The models with the best performance are used to determine the different characteristic spectral band combinations and the number of bands included in the combinations for different water quality parameters. Using the same method mentioned above, other band sets Par, Mic, DJ3, DJ4, MS600, and AQ600 were fitted, and the characteristic spectral band combination for each water quality parameter was selected. The band set that can achieve optimal results was determined by comparing the performance of different band combination models. The characteristic spectral bands of each water quality parameter were summarized within the optimal band set.

(4) Optimal spectral bands selection. By ensuring accurate monitoring of the required water quality parameters while satisfying the overall band quantity requirement, the optimal spectral bands were selected based on the specific monitoring requirements for water quality parameters and the total number of bands that was needed. This approach aimed to achieve precise remote sensing measurements of water quality parameters within the specified number of bands.

3.4. Regression Modeling with the Empirical Method

Considering that empirical models are usually one-band, two-band, and three-band models, this paper adopts a one-, two-, and three-band reflectance index to establish the inversion model for water quality parameters [35]. The reference two-band indexes are to calculate the band ratio (*BR*) [36] and the differential spectral index (*NDSI*) [37] of the reflectance of the two bands. The three-band reference indexes are to calculate the three-band index (*TBI*) [38], the enhanced three-band index (*ETBI*) [39], and the baseline height index (*BH*) [40].

The calculation equation for the single band reflectance data value is expressed as Equation (3).

$$R_{1,ci} = R(\lambda_{i,1}) \tag{3}$$

The equations for the calculated reflectance of the two-band combination are expressed as Equations (4) and (5).

$$R_{2,ci}^{BR} = R(\lambda_{i,1})/R(\lambda_{j,2}) \tag{4}$$

$$R_{2,ci}^{NDSI} = (R(\lambda_{i,1}) - R(\lambda_{j,2}))/(R(\lambda_{i,1}) + R(\lambda_{j,2})) \tag{5}$$

The equations for the calculated reflectance of the three-band combination are expressed as Equations (6)–(8).

$$R_{3,ci}^{TBI} = \left(R(\lambda_{i,1})^{-1} - R(\lambda_{j,2})^{-1}\right)\cdot R(\lambda_{k,3}) \tag{6}$$

$$R_{3,ci}^{ETBI} = \left(R(\lambda_{i,1})^{-1} - R(\lambda_{j,2})^{-1}\right)\cdot\left(R(\lambda_{k,3})^{-1} - R(\lambda_{j,2})^{-1}\right) \tag{7}$$

$$R_{3,ci}^{BH} = R(\lambda_{j,2}) - R(\lambda_{i,1}) - \frac{(\lambda_{j,2} - \lambda_{i,1})\cdot(R(\lambda_{k,3}) - R(\lambda_{i,1}))}{\lambda_{k,3} - \lambda_{i,1}} \tag{8}$$

In this study, the relationship between these different variables and water quality parameters was established using linear least squares regression fitting. In each regression analysis conducted in this section, the water quality parameter of interest was considered as the response variable, such as DO, COD_{Mn}, NH3-N, TP, TN, TUB, and EC. The corresponding variables, including $R_{1,ci}$, $R_{2,ci}^{BR}$, $R_{2,ci}^{NDSI}$, $R_{3,ci}^{TBI}$, $R_{3,ci}^{ETBI}$, and $R_{3,ci}^{BH}$, calculated by Equations (3)–(8), were included as covariates. There was a one-to-one correspondence between the response variable and the respective covariate in each regression analysis.

3.5. Regression Modeling with the ANN Method

The inversion model construction method based on ANN in this study is shown in Figure 11. Firstly, the calculated reflectance and water quality data are divided into a training set and a test set. The training set accounts for 70%, in which the validation set accounts for 10% and the test set accounts for 30%, and the data are preprocessed, which is normalizing the data by mapping the minimum and maximum values to the range [−1, 1]. Next, the ANN model is constructed with 10 hidden neurons and trained using the Levenberg–Marquardt backpropagation algorithm. The network weights, biases, and other parameters are initialized at the beginning and the ANN is trained by accepting the training set as input and undergoing iterative training with forward propagation and loss function. In each iteration, a subset of the training set data is utilized to update the network's weights and biases. The gradient of network parameters is computed using the backpropagation algorithm, and the Levenberg–Marquardt algorithm is applied to optimize the weights and biases. Simultaneously, the network's parameters are adjusted based on the performance metrics of the validation set, assessing the network's generalization ability. To prevent overfitting, training is halted if the network shows no significant improvement in the validation set. After training, the trained ANN is independently evaluated using the testing set. The network's output is computed by passing the testing set samples through the network, and its performance is evaluated by comparing the network's output to the corresponding ground-truth values.

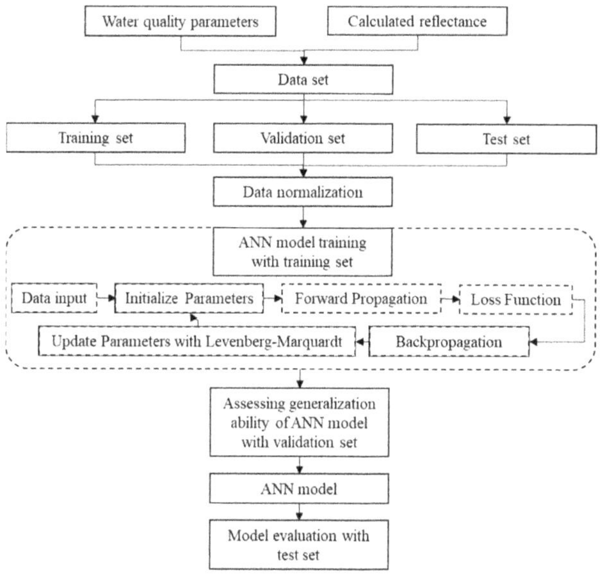

Figure 11. ANN model construction and training method.

The R^2 and root mean square error (RMSE) [26] are used to evaluate the model between the measured water quality parameters and predicted ones with the trained model. The optimal band or band combination of the model evaluation, i.e., the band or band combination model with the best correlation of each water quality parameter, is determined as the characteristic spectral band of the water quality parameter.

For the ANN method [41], the calculation equation of the single-band reflectance data value is expressed as Equation (3). The equations for the calculated reflectance of the two-band combination are expressed as Equation (9).

$$R_{2,ci} = \left(R(\lambda_{i,1}), R(\lambda_{j,2})\right) \tag{9}$$

The equations for the calculated reflectance of the combination with 3-7 band combination are expressed as Equations (10)–(14).

$$R_{3,ci} = \left(R(\lambda_{i,1}), R(\lambda_{j,2}), R(\lambda_{k,3})\right) \tag{10}$$

$$R_{4,ci} = \left(R(\lambda_{i,1}), R(\lambda_{j,2}), R(\lambda_{k,3}), R(\lambda_{l,4})\right) \tag{11}$$

$$R_{5,ci} = \left(R(\lambda_{i,1}), R(\lambda_{j,2}), R(\lambda_{k,3}), R(\lambda_{l,4}), R(\lambda_{m,5})\right) \tag{12}$$

$$R_{6,ci} = \left(R(\lambda_{i,1}), R(\lambda_{j,2}), R(\lambda_{k,3}), R(\lambda_{l,4}), R(\lambda_{m,5}), R(\lambda_{n,6})\right) \tag{13}$$

$$R_{7,ci} = \left(R(\lambda_{i,1}), R(\lambda_{j,2}), R(\lambda_{k,3}), R(\lambda_{l,4}), R(\lambda_{m,5}), R(\lambda_{n,6}), R(\lambda_{o,7})\right) \tag{14}$$

3.6. Model Evaluation

Since the objective of this study was to compare the performance of water quality inversion regression models with varying numbers of spectral bands, the adjusted coefficient of determination $\overline{R^2}$ [42] rather than the raw R^2 was employed to evaluate the models. This was primarily because the former metric provides a truer assessment of model performance by accounting for the influence of the number of bands. As the spectral band is added, if this added band is meaningful, then $\overline{R^2}$ will increase. If the added band is a redundant feature, then $\overline{R^2}$ will decrease.

$\overline{R^2}$ is calculated by Equation (15).

$$\overline{R^2} = 1 - \frac{(1 - R^2)(n - 1)}{n - np - 1} \tag{15}$$

where R^2 is calculated by Equation (16).

$$R^2 = 1 - \frac{\sum_{i=1}^{n}(r_i - p_i)^2}{\sum_{i=1}^{n}(r_i - \bar{r})^2} \tag{16}$$

RMSE is calculated by Equation (17).

$$RMSE = \sqrt{\frac{\sum_{i=1}^{n}(r_i - p_i)^2}{n}} \tag{17}$$

MAPE is calculated by Equation (18).

$$MAPE = \frac{100\%}{n}\sum_{i=1}^{n}\left|\frac{r_i - p_i}{p_i}\right| \tag{18}$$

where n is the number of samples in the dataset, r_i represents the raw measured values, p_i represents the predicted values using the regression models, and np is the number of bands.

4. Results

4.1. Result of High Correlation of Two Bands and Computation Reduction Ratio of CM16 Band Set

The determination coefficient R^2 between the reflectance data corresponding to each two-band combination for the CM16 band set are shown in Figure 12.

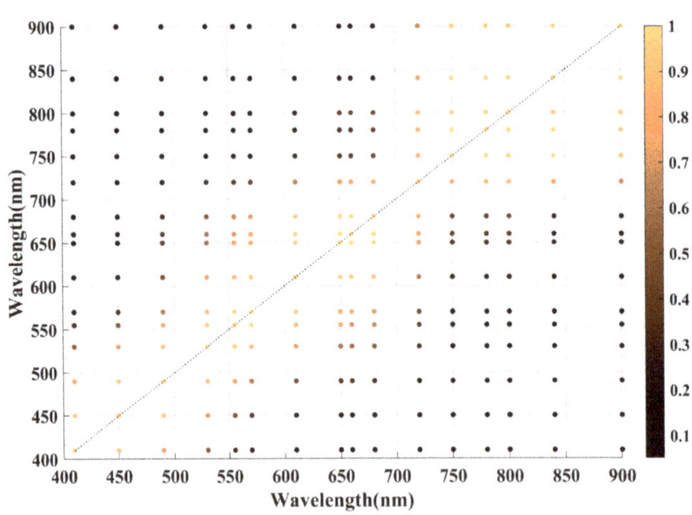

Figure 12. Cloud image of reflectance correlation of 16 spectral bands.

The combinations of the two spectral bands that cannot appear at the same time are shown in the Table 7, with a total of 20 groups.

Table 7. Twenty groups of two-band combinations with high correlation.

No.	Wavelength of Band 1	Wavelength of Band 2	R^2
1	410	450	0.92
2	450	490	0.92
3	530	555	0.94
4	555	570	0.98
5	570	610	0.91
6	610	650	0.95
7	610	660	0.93
8	650	660	1.00
9	650	680	0.98
10	660	680	0.99
11	750	780	0.98
12	750	840	0.96
13	750	800	0.94
14	750	900	0.90
15	780	800	0.98
16	780	840	0.98
17	780	900	0.93
18	800	840	0.95
19	800	900	0.91
20	840	900	0.97

The maximum number of spectral bands included in a spectral band combination and the number of different spectral band combinations with different numbers of bands were calculated, as shown in the Table 8. The reduction proportion refers to the decrease in computational load enabled by the proposed approach. Without this approach, it would be necessary to exhaustively enumerate and evaluate all possible band combinations, which scales exponentially with the number of bands. However, by intelligently pruning away invalid and redundant band combinations before evaluation, our method retains only 3.98% of the complete set of combinations. Thus, the discarded 96.02% of combinations that do not need to be explicitly evaluated lead to the stated reduction in computation. In summary, these percentages quantify the improvement in computational efficiency gained by avoiding brute-force evaluation of all combinations through the selective analysis proposed in this work. For instance, for combinations with seven bands, the number for the enumeration method will be 11,440. However, with the help of the proposed method described in this paper, the number of combinations that meet the requirements is only 25, which means only 0.2% computational load remained and 99.78% of the computational load was reduced.

Table 8. Twenty groups of two-band combinations with high correlation.

Number of Bands	Number of All Combinations	Number of Combinations of This Method	Reduction Proportion
1	16	16	0%
2	120	100	16.67%
3	560	311	44.46%
4	1820	509	72.03%
5	4368	428	90.20%
6	8008	170	97.88%
7	11,440	25	99.78%
8	12,870	0	100.00%
total	39,202	1559	96.02%

As can be seen from the table, the maximum number of spectral bands included in a spectral band combination is 7. Compared with the enumeration method, the method proposed in this paper can reduce the calculation of characteristic spectral band combinations containing only 1–8 bands by 96.02%. Considering that combinations containing 9–16 bands do not need to be calculated, the total calculation can be reduced by 96.02%.

4.2. Result of the Empirical Method with 1, 2, and 3 Bands

Table 9 shows the $\overline{R^2}$, RMSE, and mean absolute percentage error (MAPE) [33] for the linear fitting results of the measured water quality parameters and the calculated reflectance for the one-, two-, and three-band empirical model methods, together with the center wavelengths of the bands corresponding to the model.

Table 9. Optimal performance evaluation of various empirical regression models for different water quality parameters.

Parameters	Calculated Reflectance	$\overline{R^2}$	RMSE	MAPE	λ_1/nm	λ_2/nm	λ_3/nm	Band Set
DO	R_1	0.015	2.653	25.35%	780			CM16
	R_2^{BR}	0.305	2.183	19.75%	610	650		CM16
	R_2^{NDSI}	**0.309**	**2.178**	**19.75%**	**610**	**650**		**CM16**
	R_3^{TBI}	0.288	2.163	19.65%	610	650	680	CM16
	R_3^{ETBI}	0.063	2.482	23.26%	530	555	570	CM16
	R_3^{BH}	0.270	2.192	19.23%	610	650	720	CM16
COD_{Mn}	R_1	0.052	1.358	25.45%	717			Mic
	R_2^{BR}	0.065	1.321	24.41%	530	610		CM16
	R_2^{NDSI}	0.080	1.311	23.99%	450	720		CM16\MS600\AQ600
	R_3^{TBI}	**0.090**	**1.276**	**24.11%**	**530**	**555**	**580**	**CM16**
	R_3^{ETBI}	−0.025	1.355	24.36%	490	680	800	CM16
	R_3^{BH}	0.067	1.292	23.61%	490	800	900	CM16
NH3-N	R_1	**−0.006**	**0.319**	**74.29%**	**610**			**CM16**
	R_2^{BR}	−0.029	0.317	74.20%	490	610		CM16
	R_2^{NDSI}	−0.035	0.318	74.45%	490	610		CM16
	R_3^{TBI}	−0.068	0.316	74.65%	490	610	900	CM16
	R_3^{ETBI}	−0.085	0.318	73.10%	490	610	650	CM16
	R_3^{BH}	−0.046	0.313	70.54%	750	780	900	CM16
TP	R_1	**0.105**	**0.081**	**49.50%**	**720**			**CM16\MS600\AQ600**
	R_2^{BR}	0.088	0.080	47.57%	490	720		CM16
	R_2^{NDSI}	0.087	0.080	47.65%	530	720		CM16
	R_3^{TBI}	0.035	0.080	48.71%	530	720	800	CM16
	R_3^{ETBI}	−0.023	0.083	50.27%	530	650	800	CM16
	R_3^{BH}	0.010	0.081	48.26%	555	660	680	CM16
TN	R_1	0.052	1.698	53.51%	717			Mic
	R_2^{BR}	0.076	1.642	51.44%	530	610		CM16
	R_2^{NDSI}	**0.077**	**1.642**	**51.36%**	**530**	**610**		**CM16**
	R_3^{TBI}	0.041	1.638	51.62%	530	610	750	CM16
	R_3^{ETBI}	−0.030	1.698	52.30%	475	668	717	Mic
	R_3^{BH}	0.047	1.633	51.97%	555	610	680	CM16
EC	R_1	0.060	149.764	22.74%	900			CM16
	R_2^{BR}	0.087	144.655	20.75%	450	730		DJ4
	R_2^{NDSI}	**0.101**	**143.559**	**20.47%**	**450**	**730**		**DJ4**
	R_3^{TBI}	0.038	145.329	21.37%	450	780	800	CM16
	R_3^{ETBI}	−0.067	153.069	22.57%	490	660	800	CM16
	R_3^{BH}	−0.008	148.832	21.61%	490	780	840	CM16
TUB	R_1	0.072	55.462	55.27%	680			CM16
	R_2^{BR}	0.153	51.931	50.10%	555	610		CM16
	R_2^{NDSI}	0.167	51.490	48.94%	530	680		CM16
	R_3^{TBI}	0.126	51.624	49.42%	555	610	660	CM16
	R_3^{ETBI}	−0.011	55.532	54.01%	530	650	800	CM16
	R_3^{BH}	**0.208**	**49.170**	**45.84%**	**570**	**720**	**840**	**CM16**

The empirical model parameters corresponding to each water quality parameter in the table with the best performance are shown in bold in Table 9. Taking $\overline{R^2}$ as the evaluation index, the accuracy of the optimal performance model R_2^{NDSI} for DO is the highest among the different water quality parameters, with R^2 reaching 0.309 and MAPE 19.65%.

This is a three-band model, and the corresponding center wavelengths of the bands are 610, 650, and 680 nm, respectively. This band is combined from the band set CM16. The second is the optimal performance model R_3^{BH} for TUB, which is a three-band model. The corresponding center wavelengths of the bands are 570, 720, and 840 nm, respectively, which are from the band set CM16. Its $\overline{R^2}$ reaches 0.208, but its MAPE is only 45.84%, which has a large relative error. The next were TP, EC, COD_{Mn}, and TN, with $\overline{R^2}$ of 0.105, 0.101, 0.90, and 0.077, respectively. It is worth noting that although the four $\overline{R^2}$ are similar, the MAPE for COD_{Mn} and EC is between 20% and 25%, while the MAPE for TP and TN is between 47% and 52%. NH3-N had the worst performance with $\overline{R^2}$; less than 0 and its optimal performance model was R_1 with MAPE reaching 74.29%. It worth noting that the center wavelengths of the corresponding bands of EC are 450 and 730 nm, respectively, which are from the band set DJ4.

For different empirical models and among the seven water quality parameters, there are two best performance models with three bands, which are R_3^{TBI} and R_3^{BH}, respectively. There are three best performance models with three bands, among which R_2^{BR} accounts for one and $\overline{R^2}$ accounts for two. There are two best performance models with one band. Most of the bands corresponding to the optimal performance model are from CM16, with the exception of the EC model, which is from the DJ4 band set.

In summary, among the band sets of different products, the fitting result of CM16 bands is the best, indicating that, generally, the more bands that can be selected, the better the fitting result. It shows that the empirical models with different band combinations cannot effectively determine the characteristic spectral bands for each water quality parameter, so it is necessary to use the ANN method to carry out the fitting inversion between the calculated reflectance of different band combinations with different band number and each water quality parameter, so as to determine and select the characteristic spectral bands for each water quality parameter.

4.3. Result of the ANN Method

Figure 13 depicts the performance results for the different ANN regression models, employing $\overline{R^2}$ as the evaluation index.

Figure 13 illustrates that the ANN method shows substantially better performance than the empirical method for the regression models of various water quality parameters. Among the optimal ANN regression models for different water quality parameters, the NH3-N model has the poorest performance, as measured by $\overline{R^2}$ of 0.41; nevertheless, this still exceeds the top empirical model (for DO) with an $\overline{R^2}$ of 0.309. The optimal model was for COD_{Mn} with an $\overline{R^2}$ of 0.68, followed by TUB, TP, TN, EC, DO, and NH3-N. The $\overline{R^2}$ values for the optimal TUB and TP ANN models all exceeded 0.6; the $\overline{R^2}$ for the optimal TN ANN model was 0.58, which is a single-band model; and the $\overline{R^2}$ values for all the optimal EC, DO, and NH3-N ANN models were in the range from 0.4 to 0.5. With the exception of TN, the optimal ANN models for the other water quality parameters were all three-band models, whereas the TN ANN model had one band.

For the various band sets, all bands corresponding to the optimal performance ANN models belonged to the CM16 band set. The ANN models for COD_{Mn}, NH3-N, TP, and TUB in the CM16 band set evidenced obvious advantages, with minimum R^2 differences greater than 0.1 compared to the other band sets. The $\overline{R^2}$ differences were all less than 0.1 between the ANN models for DO, TN, and EC in the CM16 band set compared to other band sets. Nevertheless, compared to other band sets, CM16 still demonstrated considerable dominance over other band sets, owing to the ample spectral band options it provides. The existing multispectral lens products are suitable for agriculture and forestry and have large errors in the inversion measurement of water quality parameters.

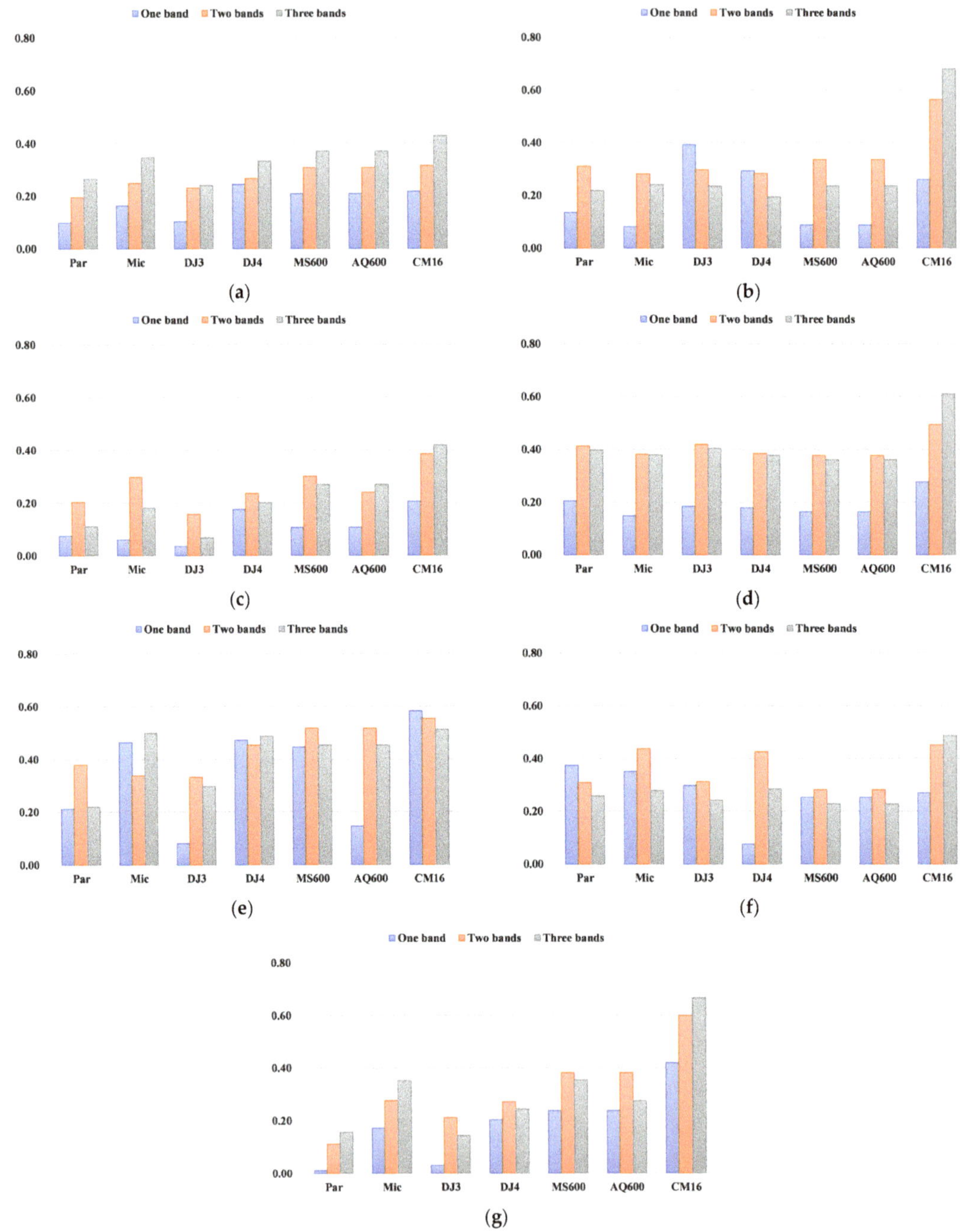

Figure 13. $\overline{R^2}$ for the ANN regression model using different band sets with one, two, and three bands for DO (**a**), COD_{Mn} (**b**), NH3-N (**c**), TP (**d**), TN (**e**), EC (**f**), and TUB (**g**).

For ANN models with different band numbers, most models showed performance that increased with more bands except TN for band set CM16. It is necessary to study the performance of different band number models to determine the optimal number of bands for different water quality parameters.

Figure 14 illustrates the performance of ANN models of water quality parameters with varying numbers of spectral bands. As evidenced in Figure 14, all ANN models of water quality parameters exhibit a consistent pattern in which model accuracy, as measured by $\overline{R^2}$, initially increases with additional bands but subsequently decreases with the exception of TN models. For the ANN regression model of DO and NH3-N, $\overline{R^2}$ values increased monotonically from one to four bands and then decreased monotonically from four to seven bands. For the ANN regression model of CODMn, TP, EC, and NH3-N, $\overline{R^2}$ limbed monotonically from one to three bands but declined monotonically thereafter from three to seven bands. Counterintuitively, the model for TN manifests the greatest performance with one band. Overall, the $\overline{R^2}$ value of the models decreased as the number of bands increased. However, the four-band model and seven-band model had higher $\overline{R^2}$ values than the three-band model and six-band model, respectively.

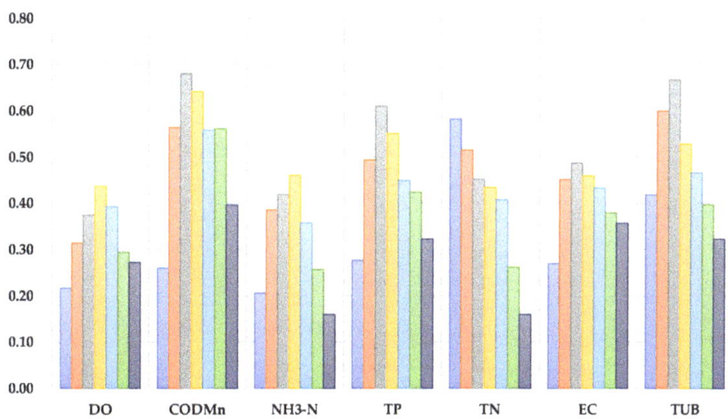

Figure 14. $\overline{R^2}$ for ANN models of different water quality parameters with different numbers of spectral bands.

In summary, despite the increasing availability of information with additional spectral bands, model performance does not improve indefinitely. For most water quality parameters, model efficacy reaches an apex with either three or four bands, beyond which superfluous information degrades predictive accuracy. The anomalous case of TN highlights the idiosyncrasies that can emerge in complex models. It can be extrapolated that a paucity of information precludes achieving optimal model accuracy due to insufficient critical data. Additionally, redundant information introduces random noise into the data, thereby undermining accuracy.

Therefore, there exist an optimal number of characteristic spectral bands for different water quality parameters. The optimal number of bands for DO and NH3-N is four, the optimal number of bands for COD_{Mn}, TP, EC, and TUB is three, and the optimal number of bands for TN is one, because the ANN models with these optimal numbers of bands demonstrated the best performance among the various models.

Figure 15 shows the performance and spectral band information for the optimal ANN models of seven water quality parameters. It can be seen that the ANN model for COD_{Mn} has the best performance among the water quality parameters, with an $\overline{R^2}$ of 0.68 and a MAPE of 14.02%. Though the ANN models for TUB and TP have relatively high $\overline{R^2}$ values of 0.67 and 0.61, respectively, their MAPE values of 36.19% and 28.09% are not low. The $\overline{R^2}$ values of EC and DO ANN regression models are 0.49 and 0.43, respectively, which do not show the advantage among water quality parameters; their MAPE values remain at low levels ranging from 16% to 18%. As with the results of the empirical method and the ANN method with other band sets, the performance of the NH3-N ANN model was the worst, with the lowest $\overline{R^2}$ of 0.54 and the highest MAPE of 65.85%.

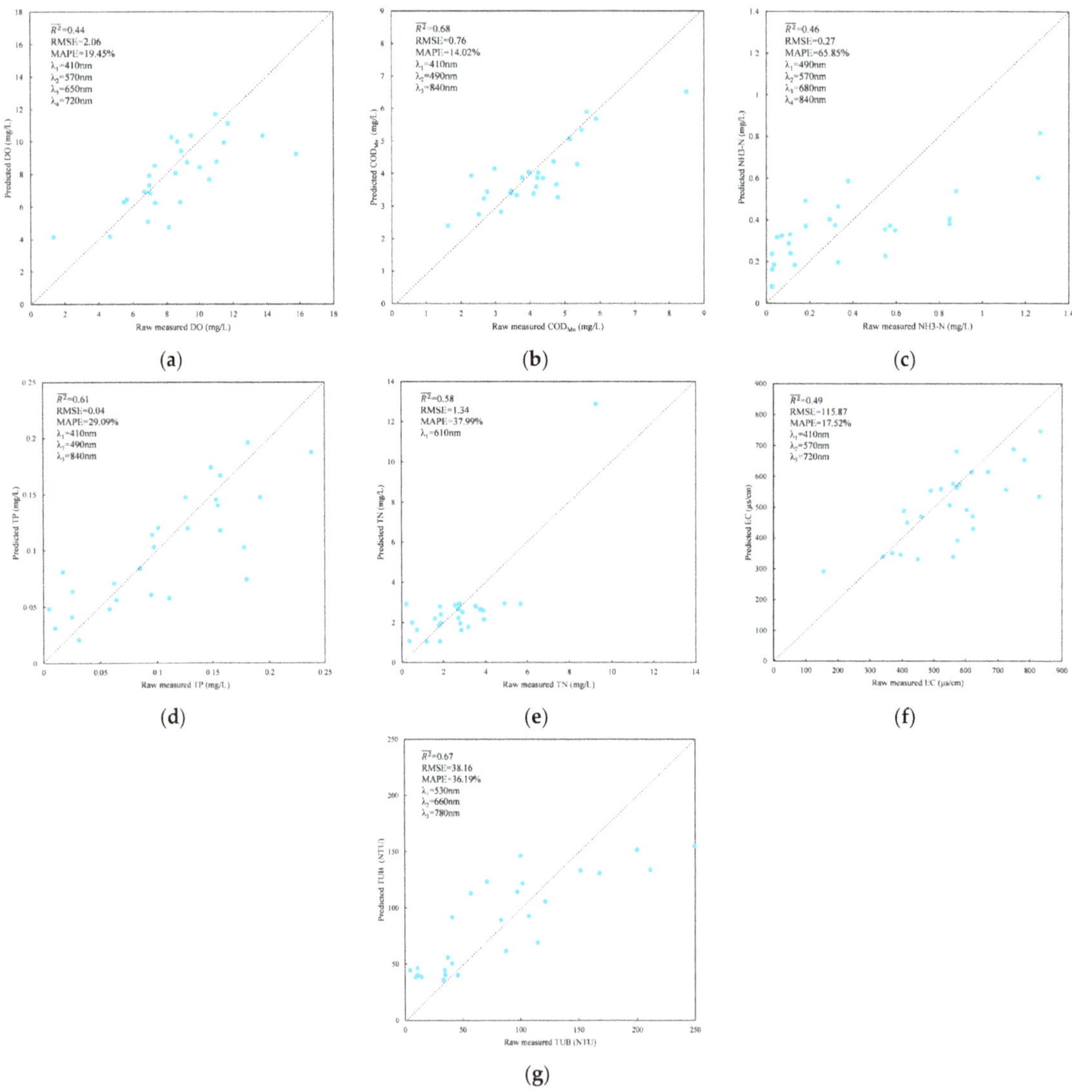

Figure 15. Performance validation and spectral band information for the optimal ANN model of DO (**a**), COD_{Mn} (**b**), NH3-N (**c**), TP (**d**), TN (**e**), EC (**f**), and TUB (**g**).

Furthermore, Figure 15 shows that the characteristic spectral band combinations for COD_{Mn} and TP were the same, namely 410, 490, and 840 nm, which belonged to the set of DO characteristic spectral bands. The DO characteristic spectral bands differed from these two indices only in having an additional 720 nm band. The characteristic spectral bands for EC were 410, 570, and 720 nm, with the 410 and 720 nm bands overlapping with DO and the 570 nm band overlapping with NH3-N. The characteristic spectral bands for

NH3-N were 490, 570, 680, and 840 nm, with the 490 n and 840 nm bands overlapping with DO, COD_{Mn}, and TP. The characteristic spectral band for TN was a single band with a center wavelength of 610 nm, and the characteristic spectral bands for TUB were 530, 660, and 780 nm. The biggest difference from the characteristic spectral bands of them between those of other water quality parameters was that there were no overlapping bands.

5. Discussion

This study also explored the relationship between characteristic spectral bands in the model as the number of bands increased, that is, whether the highly correlated bands in the model with fewer bands would also appear in the model with the best performance with more bands. However, the results did not show that the bands in the model with fewer bands would also appear in the model with the best performance of more bands. Therefore, this content was not discussed in detail.

This paper suggests that the spectral bands with wavelengths of 410, 490, 570, 680, 720, and 840 nm should be used if the researcher needs to customize a six-band multispectral camera to monitor the water quality parameters of DO, COD_{Mn}, NH3-N, TP, and EC and effectively ensure the inversion accuracy of COD_{Mn}, DO, and EC. The inversion accuracy of TN and TUB cannot be ensured. However, due to the common situation that TN exceeds the standard in general water bodies and that TUB is not used to classify water quality, the lack of accurate inversion of TN and TUB will not affect the water quality assessment. If there are other requirements, i.e., the water quality parameters in the research interest of the investigator, the optimal combination of six spectral bands can be determined according to the actual requirements, so as to achieve the optimal inversion result of all these water quality parameters.

The dataset in this paper was obtained from all of the NSWAMS of Suzhou, Shanghai, Jiaxing, Huzhou, and Wuxi in the Changjiang Delta region, covering the data of December 2020, January, November, December 2021, and March and August 2022. The data are diverse and can represent the characteristics of different time periods, and the model is suitable for the inversion of water quality parameters in this region. However, due to the good water quality and low degree of pollution, the dataset used for the training model has fewer class V and inferior class V water quality samples. The performance of the models in water quality inversion and classification is not good. At the same time, the applicability and universality of the model are directly related to the sample dataset during the model training. Therefore, with the progress of this research, the universality of the model can be effectively expanded by adding more regions and more periods of datasets to the model training. In addition, due to different target water quality parameters and different selected bands, there will certainly be different cameras with different spectral band combinations for different purposes, and the bands to be customized should be determined according to the actual needs.

6. Conclusions

This study discussed the determination of the optimal characteristic spectral bands for different water quality parameters with a proposed novel approach, which is based on the correlation between reflectance of different bands and regression modeling with the ANN method. Using fused ZY1-02D hyperspectral images and water quality data from December 2020 to August 2021 around Taihu Lake, the proposed approach was tested. The result showed that it can effectively reduce the computation by 96.02% and quickly find the characteristic bands to improve the modeling efficiency. Comparing different band sets of multispectral cameras, the CM16 band set with 16 bands leads to the best performance, suggesting that more spectral options enable better modeling results. Compared to typical empirical methods, the ANN model shows significant advantages in estimating various water quality parameters. Each parameter has an optimal number of characteristic bands, with model accuracy first increasing and then decreasing as more bands are added, except for TN.

The proposed approach and modeling method provide a new approach for effective spectral characteristic bands and precise estimation of water quality parameters from hyperspectral images. They lay a theoretical foundation for customized multispectral cameras and UAV platforms for water quality monitoring. This study provides a reference for follow-up research to customize multispectral lenses and develop UAV remote sensing techniques for water quality monitoring. The proposed characteristic band approach and modeling method can be extended to other hyperspectral remote sensing studies to address high data dimensionality and improve modeling efficiency.

Author Contributions: Conceptualization, X.X. and H.L.; methodology, X.X.; formal analysis, X.X.; investigation, X.X. and X.L.; resources, H.L., Z.X. and Y.T.; data curation, X.X.; writing—original draft preparation, X.X.; writing—review and editing, X.X. and H.L. All authors have read and agreed to the published version of the manuscript.

Funding: This research was funded by the China Postdoctoral Science Foundation (No. 2022M723535) and the Open Project of State Key Laboratory of Urban Water Resource and Environment, Harbin Institute of Technology (No. SMK202205).

Data Availability Statement: Hyperspectral data from the ZY1-02D satellite can be requested from the website http://sasclouds.com/chinese/normal, accessed on 31 October 2022 and water quality parameters can be obtained from website https://szzdjc.cnemc.cn:8070/GJZ/Business/Publish/Main.html, accessed on 31 October 2022.

Acknowledgments: The authors are grateful to the anonymous reviewers for their constructive comments and suggestions to improve this manuscript.

Conflicts of Interest: Authors Xietian Xia, Zenghui Xu and Xiang Li were employed by the company China Construction Power and Environment Engineering Co., Ltd. The remaining authors declare that the research was conducted in the absence of any commercial or financial relationships that could be construed as a potential conflict of interest.

References

1. Jay, S.; Guillaume, M. A Novel Maximum Likelihood Based Method for Mapping Depth and Water Quality from Hyperspectral Remote-Sensing Data. *Remote Sens. Environ.* **2014**, *147*, 121–132. [CrossRef]
2. Dall'Olmo, G.; Gitelson, A.A.; Rundquist, D.C. Towards a Unified Approach for Remote Estimation of Chlorophyll-a in Both Terrestrial Vegetation and Turbid Productive Waters. *Geophys. Res. Lett.* **2003**, *30*, 1938. [CrossRef]
3. Duan, H.; Ma, R.; Loiselle, S.A.; Shen, Q.; Yin, H.; Zhang, Y. Optical Characterization of Black Water Blooms in Eutrophic Waters. *Sci. Total Environ.* **2014**, *482–483*, 174–183. [CrossRef]
4. Wernand, M.R.; Hommersom, A.; Van Der Woerd, H.J. MERIS-Based Ocean Colour Classification with the Discrete Forel–Ule Scale. *Ocean Sci.* **2013**, *9*, 477–487. [CrossRef]
5. Terentev, A.; Dolzhenko, V.; Fedotov, A.; Eremenko, D. Current State of Hyperspectral Remote Sensing for Early Plant Disease Detection: A Review. *Sensors* **2022**, *22*, 757. [CrossRef]
6. Cao, Q.; Yu, G.; Qiao, Z. Application and Recent Progress of Inland Water Monitoring Using Remote Sensing Techniques. *Environ. Monit. Assess.* **2023**, *195*, 125. [CrossRef] [PubMed]
7. Vakili, T.; Amanollahi, J. Determination of Optically Inactive Water Quality Variables Using Landsat 8 Data: A Case Study in Geshlagh Reservoir Affected by Agricultural Land Use. *J. Clean. Prod.* **2020**, *247*, 119134. [CrossRef]
8. Wang, S.; Garcia, M.; Bauer-Gottwein, P.; Jakobsen, J.; Zarco-Tejada, P.J.; Bandini, F.; Paz, V.S.; Ibrom, A. High Spatial Resolution Monitoring Land Surface Energy, Water and CO_2 Fluxes from an Unmanned Aerial System. *Remote Sens. Environ.* **2019**, *229*, 14–31. [CrossRef]
9. Cao, Y.; Ye, Y.; Zhao, H.; Jiang, Y.; Wang, H.; Shang, Y.; Wang, J. Remote Sensing of Water Quality Based on HJ-1A HSI Imagery with Modified Discrete Binary Particle Swarm Optimization-Partial Least Squares (MDBPSO-PLS) in Inland Waters: A Case in Weishan Lake. *Ecol. Inform.* **2018**, *44*, 21–32. [CrossRef]
10. Wang, Y.; Li, S.; Lin, Y.; Wang, M. Lightweight Deep Neural Network Method for Water Body Extraction from High-Resolution Remote Sensing Images with Multisensors. *Sensors* **2021**, *21*, 7397. [CrossRef]
11. Yang, Z.; Gong, C.; Ji, T.; Hu, Y.; Li, L. Water Quality Retrieval from ZY1-02D Hyperspectral Imagery in Urban Water Bodies and Comparison with Sentinel-2. *Remote Sens.* **2022**, *14*, 5029. [CrossRef]
12. Sudduth, K.A.; Jang, G.-S.; Lerch, R.N.; Sadler, E.J. Long-Term Agroecosystem Research in the Central Mississippi River Basin: Hyperspectral Remote Sensing of Reservoir Water Quality. *J. Environ. Qual.* **2015**, *44*, 71–83. [CrossRef]
13. Kudela, R.M.; Palacios, S.L.; Austerberry, D.C.; Accorsi, E.K.; Guild, L.S.; Torres-Perez, J. Application of Hyperspectral Remote Sensing to Cyanobacterial Blooms in Inland Waters. *Remote Sens. Environ.* **2015**, *167*, 196–205. [CrossRef]

14. Hestir, E.L.; Brando, V.E.; Bresciani, M.; Giardino, C.; Matta, E.; Villa, P.; Dekker, A.G. Measuring Freshwater Aquatic Ecosystems: The Need for a Hyperspectral Global Mapping Satellite Mission. *Remote Sens. Environ.* **2015**, *167*, 181–195. [CrossRef]
15. Lei, S.; Wu, D.; Li, Y.; Wang, Q.; Huang, C.; Liu, G.; Zheng, Z.; Du, C.; Mu, M.; Xu, J.; et al. Remote Sensing Monitoring of the Suspended Particle Size in Hongze Lake Based on GF-1 Data. *Int. J. Remote Sens.* **2019**, *40*, 3179–3203. [CrossRef]
16. Li, Y.; Zhang, Y.; Shi, K.; Zhu, G.; Zhou, Y.; Zhang, Y.; Guo, Y. Monitoring Spatiotemporal Variations in Nutrients in a Large Drinking Water Reservoir and Their Relationships with Hydrological and Meteorological Conditions Based on Landsat 8 Imagery. *Sci. Total Environ.* **2017**, *599–600*, 1705–1717. [CrossRef] [PubMed]
17. Lim, J.; Choi, M. Assessment of Water Quality Based on Landsat 8 Operational Land Imager Associated with Human Activities in Korea. *Environ. Monit. Assess.* **2015**, *187*, 384. [CrossRef] [PubMed]
18. Gholizadeh, M.; Melesse, A.; Reddi, L. A Comprehensive Review on Water Quality Parameters Estimation Using Remote Sensing Techniques. *Sensors* **2016**, *16*, 1298. [CrossRef]
19. Gordon, H.R.; Brown, O.B.; Jacobs, M.M. Computed Relationships between the Inherent and Apparent Optical Properties of a Flat Homogeneous Ocean. *Appl. Opt.* **1975**, *14*, 417–427. [CrossRef] [PubMed]
20. Dekker, A.G.; Vos, R.J.; Peters, S.W.M. Analytical Algorithms for Lake Water TSM Estimation for Retrospective Analyses of TM and SPOT Sensor Data. *Int. J. Remote Sens.* **2002**, *23*, 15–35. [CrossRef]
21. Gordon, H.R.; Morel, A.Y. *Remote Assessment of Ocean Color for Interpretation of Satellite Visible Imagery: A Review*; Lecture Notes on Coastal and Estuarine Studies; American Geophysical Union: Washington, DC, USA, 1983; Volume 4, ISBN 978-0-387-90923-3.
22. Gitelson, A.A.; Schalles, J.F.; Hladik, C.M. Remote Chlorophyll-a Retrieval in Turbid, Productive Estuaries: Chesapeake Bay Case Study. *Remote Sens. Environ.* **2007**, *109*, 464–472. [CrossRef]
23. Le, C.; Hu, C.; Cannizzaro, J.; English, D.; Muller-Karger, F.; Lee, Z. Evaluation of Chlorophyll-a Remote Sensing Algorithms for an Optically Complex Estuary. *Remote Sens. Environ.* **2013**, *129*, 75–89. [CrossRef]
24. Lei, S.; Xu, J.; Li, Y.; Li, L.; Lyu, H.; Liu, G.; Chen, Y.; Lu, C.; Tian, C.; Jiao, W. A Semi-Analytical Algorithm for Deriving the Particle Size Distribution Slope of Turbid Inland Water Based on OLCI Data: A Case Study in Lake Hongze. *Environ. Pollut.* **2021**, *270*, 116288. [CrossRef] [PubMed]
25. Zeng, S.; Lei, S.; Li, Y.; Lyu, H.; Dong, X.; Li, J.; Cai, X. Remote Monitoring of Total Dissolved Phosphorus in Eutrophic Lake Taihu Based on a Novel Algorithm: Implications for Contributing Factors and Lake Management. *Environ. Pollut.* **2022**, *296*, 118740. [CrossRef] [PubMed]
26. Zhang, D.; Zhang, L.; Sun, X.; Gao, Y.; Lan, Z.; Wang, Y.; Zhai, H.; Li, J.; Wang, W.; Chen, M.; et al. A New Method for Calculating Water Quality Parameters by Integrating Space–Ground Hyperspectral Data and Spectral-In Situ Assay Data. *Remote Sens.* **2022**, *14*, 3652. [CrossRef]
27. Doerffer, R.; Schiller, H. The MERIS Case 2 Water Algorithm. *Int. J. Remote Sens.* **2007**, *28*, 517–535. [CrossRef]
28. Pahlevan, N.; Smith, B.; Schalles, J.; Binding, C.; Cao, Z.; Ma, R.; Alikas, K.; Kangro, K.; Gurlin, D.; Hà, N.; et al. Seamless Retrievals of Chlorophyll-a from Sentinel-2 (MSI) and Sentinel-3 (OLCI) in Inland and Coastal Waters: A Machine-Learning Approach. *Remote Sens. Environ.* **2020**, *240*, 111604. [CrossRef]
29. Zhang, D.; Zeng, S.; He, W. Selection and Quantification of Best Water Quality Indicators Using UAV-Mounted Hyperspectral Data: A Case Focusing on a Local River Network in Suzhou City, China. *Sustainability* **2022**, *14*, 16226. [CrossRef]
30. Yin, F.; Wu, M.; Liu, L.; Zhu, Y.; Feng, J.; Yin, C.; Yin, C. Predicting the abundance of copper in soil using reflectance spectroscopy and GF5 hyperspectral imagery. *Int. J. Appl. Earth Obs.* **2021**, *102*, 102420. [CrossRef]
31. Niroumand-Jadidi, M.; Bovolo, F.; Bruzzone, L. Water Quality Retrieval from PRISMA Hyperspectral Images: First Experience in a Turbid Lake and Comparison with Sentinel-2. *Remote Sens.* **2020**, *12*, 3984. [CrossRef]
32. Li, L.; Gu, M.; Gong, C.; Hu, Y.; Wang, X.; Yang, Z.; He, Z. An Advanced Remote Sensing Retrieval Method for Urban Non-Optically Active Water Quality Parameters: An Example from Shanghai. *Sci. Total Environ.* **2023**, *880*, 163389. [CrossRef]
33. Liu, H.; Yu, T.; Hu, B.; Hou, X.; Zhang, Z.; Liu, X.; Liu, J.; Wang, X.; Zhong, J.; Tan, Z.; et al. UAV-Borne Hyperspectral Imaging Remote Sensing System Based on Acousto-Optic Tunable Filter for Water Quality Monitoring. *Remote Sens.* **2021**, *13*, 4069. [CrossRef]
34. Bi, S. *Remote Sensing of Algal Column Integrated Biomass for Inland Waters Based on Soft Classification*; College of Geographical Science: Nanjing, China, 2021.
35. Liu, Y.; Li, J.; Xiao, C.; Zhang, F.; Wang, S. Inland Water Chlorophyll-a Retrieval Based on ZY-1 02D Satellite Hyperspectral Observations. *Natl. Remote Sens. Bull.* **2022**, *26*, 168–178. [CrossRef]
36. Moses, W.J.; Gitelson, A.A.; Berdnikov, S.; Povazhnyy, V. Satellite Estimation of Chlorophyll-a Concentration Using the Red and NIR Bands of MERIS—The Azov Sea Case Study. *IEEE Geosci. Remote Sens. Lett.* **2009**, *6*, 845–849. [CrossRef]
37. Mishra, S.; Mishra, D.R. Normalized Difference Chlorophyll Index: A Novel Model for Remote Estimation of Chlorophyll-a Concentration in Turbid Productive Waters. *Remote Sens. Environ.* **2012**, *117*, 394–406. [CrossRef]
38. Gitelson, A.A.; Dall'Olmo, G.; Moses, W.; Rundquist, D.C.; Barrow, T.; Fisher, T.R.; Gurlin, D.; Holz, J. A Simple Semi-Analytical Model for Remote Estimation of Chlorophyll-a in Turbid Waters: Validation. *Remote Sens. Environ.* **2008**, *112*, 3582–3593. [CrossRef]
39. Yang, W.; Matsushita, B.; Chen, J.; Fukushima, T.; Ma, R. An Enhanced Three-Band Index for Estimating Chlorophyll-a in Turbid Case-II Waters: Case Studies of Lake Kasumigaura, Japan, and Lake Dianchi, China. *IEEE Geosci. Remote Sens. Lett.* **2010**, *7*, 655–659. [CrossRef]

40. Wynne, T.T.; Stumpf, R.P.; Briggs, T.O. Comparing MODIS and MERIS Spectral Shapes for Cyanobacterial Bloom Detection. *Int. J. Remote Sens.* **2013**, *34*, 6668–6678. [CrossRef]
41. Najwa Mohd Rizal, N.; Hayder, G.; Mnzool, M.; Elnaim, B.M.E.; Mohammed, A.O.Y.; Khayyat, M.M. Comparison between Regression Models, Support Vector Machine (SVM), and Artificial Neural Network (ANN) in River Water Quality Prediction. *Processes* **2022**, *10*, 1652. [CrossRef]
42. Piepho, H. An Adjusted Coefficient of Determination for Generalized Linear Mixed Models in One Go. *Biom. J.* **2023**, *65*, 2200290. [CrossRef]

remote sensing

MDPI

Article

Retrievals of Chlorophyll-a from GOCI and GOCI-II Data in Optically Complex Lakes

Yuyu Guo [1,2], Xiaoqi Wei [1,3], Zehui Huang [1,3], Hanhan Li [1,4], Ronghua Ma [1,5], Zhigang Cao [1], Ming Shen [1] and Kun Xue [1,*]

[1] Key Laboratory of Watershed Geographic Sciences, Nanjing Institute of Geography and Limnology, Chinese Academy of Sciences, Nanjing 210008, China
[2] School of Geography, Geomatics and Planning, Jiangsu Normal University, Xuzhou 221116, China
[3] University of Chinese Academy of Sciences, Beijing 100049, China
[4] School of Remote Sensing and Geomatics Engineering, Nanjing University of Information Science and Technology, Nanjing 210044, China
[5] Jiangsu Collaborative Innovation Center of Regional Modern Agriculture & Environmental Protection, Huaiyin Normal University, Huaian 223001, China
* Correspondence: kxue@niglas.ac.cn

Abstract: The chlorophyll-a (Chl*a*) concentration is a key parameter to evaluate the eutrophication conditions of water, which is very important for monitoring algal blooms. Although Geostationary Ocean Color Imager (GOCI) has been widely used in Chl*a* inversion, the consistency of the Rayleigh-corrected reflectance (R_{rc}) of GOCI and GOCI-II sensors still needs to be further evaluated, and a model suitable for lakes with complex optical properties needs to be constructed. The results show that (1) the derived Chl*a* values of the GOCI and GOCI-II synchronous data were relatively consistent and continuous in three lakes in China. (2) The accuracy of the random forest (RF) model (R^2 = 0.84, root mean square error (RMSE) =11.77 μg/L) was higher than that of the empirical model (R^2 = 0.79, RMSE = 12.63 μg/L) based on the alternative floating algae index (AFAI). (3) The interannual variation trend fluctuated, with high Chl*a* levels in Lake Chaohu in 2015 and 2019, while those in Lake Hongze were high in 2013, 2015, and 2022, and those in Lake Taihu reached their peak in 2017 and 2019. There were three types of diurnal variation patterns, namely, near-continuous increase (Class 1), near-continuous decrease (Class 2), and first an increase and then a decrease (Class 3), among which Lake Chaohu and Lake Taihu occupied the highest proportion in Class 3. The results analyzed the temporal and spatial variations of Chla in three lakes for 12 years and provided support for the use of GOCI and GOCI-II data and monitoring of Chl*a* in optical complex inland waters.

Keywords: consistency evaluation; random forest; temporal and spatial variations

1. Introduction

Chlorophyll-a is a common pigment in phytoplankton, and its concentration is an important parameter to describe the nutritional status of water, which is the focus of lake environment quality assessment studies [1]. Traditional water quality monitoring is a time-consuming and labor-intensive task, and it is difficult to monitor the water quality on a large spatial scale [2]. With the development of remote sensing technology and the continuous launch of water color satellites, the differences in the optical properties of natural water caused by phytoplankton pigments, suspended particulate matter (SPM), and colored dissolved organic matter (CDOM) are reflected in remote sensing spectral signals [3]. This provides the possibility for multiscale monitoring in time and space and is very important for lake environment management and ecological protection.

Because of the advantages of easy acquisition and fast image processing, satellites are widely used in the quantitative inversion of Chl*a*. The multispectral satellite data used in inversion include Landsat series data [4], Moderate Resolution Imaging Spectroradiometer

(MODIS) data [5], Sentinel series data [6], Medium Resolution Imaging Spectrometer (MERIS) data [7], GOCI data [8], Visible Infrared Imager Radiometer Suite (VIIRS) data [9], and GF-1 Wide Field of View (WFV) data [10]. The inversion algorithms for Chl*a* mainly include empirical, semi-analytical, and analytical algorithms [11]. In recent years, machine learning models, such as the neural network (NN) [12] algorithm, extreme gradient boosting (XGBoost) algorithm, support vector regression (SVR) model [13], and random forest (RF) model [14], have also been widely used in the inversion of Chl*a*, and a satisfactory performance has been achieved. Among them, the random forest model exhibits the characteristics of high accuracy and high speed [15].

However, influenced by the wind speed, wind direction, precipitation, diurnal variations in algae, and other factors, Chl*a* may change rapidly in a day or even a few hours [16]. Thus, satellites with moderate or low temporal resolution can hardly meet the needs of high-frequency monitoring. In comparison, the GOCI sensor carried by the Communication, Ocean, and Meteorological Satellite (COMS) provides a higher temporal resolution (1 h) and signal-to-noise ratio, which can reduce the effect of missing data due to cloud interference. Previous studies have used GOCI data to estimate Chl*a* in Lake Taihu [17], determine SPM in Lake Hongze [18], and observe cyanobacteria blooms in Lake Taihu [19]. Although the GOCI sensor stopped operating in March 2021, GEO-KOMPSAT-2B (GK-2B) was successfully launched in February 2020, which provides data guarantees for follow-up research. The time coverage and imaging bands of the GOCI-II sensor were increased, making it more suitable for water color research [20]; however, the consistency between the two sensors in estimating Chl*a* must be further confirmed to obtain long-term diurnal observations of Chl*a*.

Lake Chaohu, Lake Hongze, and Lake Taihu are typical lakes with complex optical properties affected by human activities and exhibit serious eutrophication and turbid water quality issues. The aims of this paper were to: (1) evaluate the consistency of GOCI and GOCI-II R_{rc} data and provide a reference for the use of long-term data series; (2) assess the differences between the empirical model and the random forest model and develop a Chl*a* model suitable for lakes with complex optical properties; and (3) produce temporal and spatial distribution products of Chl*a* in Lake Chaohu, Lake Hongze, and Lake Taihu from 2011 to 2022 and analyze the properties of interannual, monthly, and diurnal variations. This paper is the first to evaluate the consistency of GOCI and GOCI-II data in the inversion of Chl*a* in turbid and eutrophic lakes, which can provide a reference for dynamic Chl*a* monitoring.

2. Materials and Methods

2.1. Study Area

Lake Taihu (2338 km^2, 30°56′–31°34′N, 119°54′–120°36′N), Lake Hongze (1577 km^2, 33°06′–33°40′N, 118°10′–118°52′N), and Lake Chaohu (768 km^2, 31°25′–31°43′N, 117°17′–117°52′N) are the third, fourth, and fifth largest freshwater lakes in China, respectively (Figure 1). The mean water depths are 1.9, 3.0, and 1.9 m, and the mean Secchi disk depths are 0.40, 0.27, and 0.27 m in Lake Taihu, Lake Chaohu, and Lake Hongze, respectively [21]. Xukou Bay and the southeastern part of Lake Taihu mostly contain aquatic vegetation, including emergent, floating-leaved, free-floating, and submerged hydrophytes [22]. Lake Hongze is relatively turbid, and the aquatic vegetation in the western part is widely distributed [23]. To ensure the accuracy of Chl*a* data, the growth range of aquatic vegetation in Lake Taihu and Lake Hongze was masked.

2.2. Water Sampling and Measurements

A total of 54 days were selected for field sampling in Lake Chaohu, Lake Hongze, and Lake Taihu from 2013 to 2018. The water samples were filtered through a GF/F glass fiber membrane with an aperture of 47 mm, and the filter membranes were frozen and soaked in an acetone solution with a concentration of 90% for light-proof extraction. The extraction solution was placed in a colorimetric plate, and the concentration of Chl*a* at the sample site was calculated after the absorbance was measured by a UV2700 (Shimadzu, Tokyo, Japan) spectrophotometer [24].

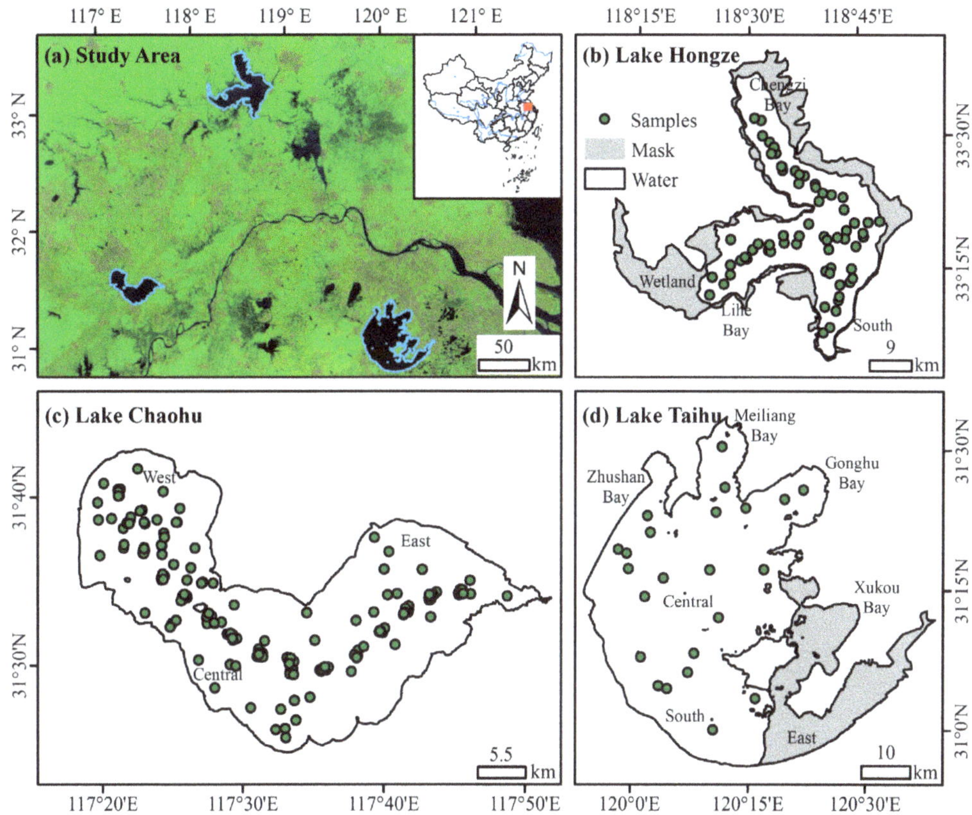

Figure 1. (**a**) Location of Lake Chaohu, Lake Taihu, and Lake Hongze in China. The spatial distributions of samples in (**b**) Lake Hongze, (**c**) Lake Chaohu, and (**d**) Lake Taihu are shown.

2.3. Satellite Image Acquisition and Processing

GOCI data were downloaded from the Korea Ocean Satellite Center (KOSC) (http://kosc.kiost.ac.kr/, accessed on 1 June 2022), covering a target area of 2500 km × 2500 km centered on Korea (36°N, 130°E). The area was observed from May 2011 to March 2021, with a spatial resolution of 500 m and a temporal resolution of 1 h, providing 8 field images from 8:16 to 15:16 (UTC + 8). GOCI L1B data covering the three lakes was processed on the SeaDAS (Version 7.5.3) platform and used to derive R_{rc} by removing Rayleigh scattering signals [25].

The spatial resolution of the GOCI-II sensor has been increased to 250 m, and the number of spectral bands has been increased from 8 to 12 (Figure 2). The satellite transit frequency has been extended to 10 images per day from 7:15 to 16:15 (UTC + 8), which provides higher applicability for monitoring inland lakes. Due to the short online time of GOCI-II data, the SeaDAS platform cannot process first-level data temporarily. Therefore, R_{rc} data processed by the GOCI-II atmospheric correction (G2AC) algorithm was obtained from the official website (https://www.nosc.go.kr/, accessed on 1 April 2023). G2AC was based on the atmospheric correction algorithm of GOCI and attained favorable performance in both clear and turbid water [26].

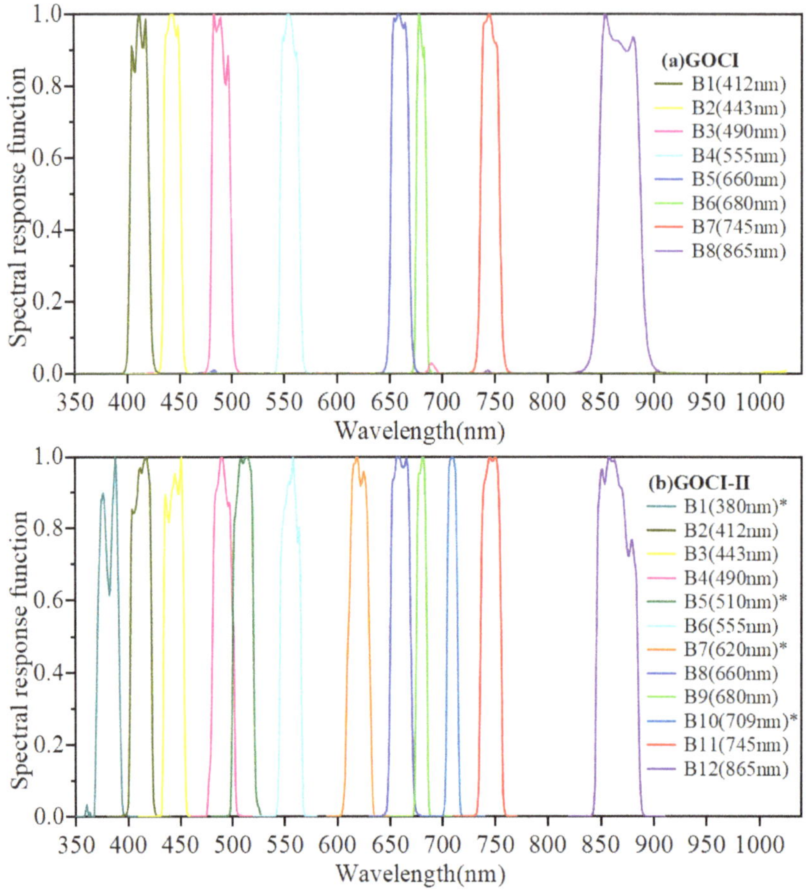

Figure 2. Spectral response functions of (**a**) GOCI and (**b**) GOCI-II in different spectral ranges. Note that the bands marked with * are new bands of GOCI-II.

Cloudless images for this study region were selected from GOCI and GOCI-II data. The number of available images of Lake Chaohu, Lake Hongze, and Lake Taihu from 2011 to 2022 is shown in Figure 3. Inland water does not meet the assumptions of the standard near-infrared atmospheric correction algorithm, and GOCI lacks a shortwave infrared band [27]; therefore, the accuracy of the existing atmospheric correction methods and effective pixel numbers can hardly satisfy the monitoring needs. Therefore, R_{rc} was used in this study to build an inversion model for Chl*a*. In addition, to ensure the uniformity of the inversion results, the resolution of the satellite images was resampled to 250 m. To verify the continuity and availability of the two sensors, GOCI and GOCI-II synchronous data were selected for comparison. From January to March 2021, a total of 215 images (77 scenes of Lake Chaohu, 72 scenes of Lake Hongze, and 66 scenes of Lake Taihu) were derived. Match-up pairs within the three lakes were randomly selected with an interval time of less than 1 min, and samples with a CV > 10% were excluded.

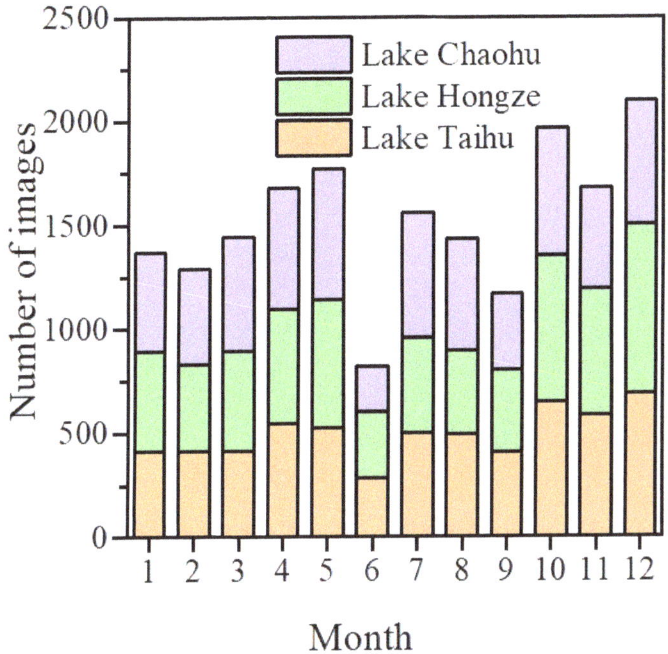

Figure 3. Number of images of Lake Chaohu, Lake Hongze, and Lake Taihu in different months.

2.4. Mask Determination and Match-Up Procedures

Non-water pixels, such as clouds, cloud shadows, sun glint, algal blooms, and aquatic vegetation, affect the accuracy of model development; therefore, it is necessary to confirm the considered mask rules. Combining statistical samples and prior knowledge, when the three conditions of $R_{rc}(490) > 0.14$, $R_{rc}(555) > 0.16$, and $R_{rc}(660) > 0.15$ are satisfied, the pixel can be determined as a cloud or sun glint pixel, while for $R_{rc}(555) < 0.07$, it is a cloud shadow pixel. The threshold of the cyanobacteria and macrophyte index (CMI) was set to 0.02; if the value is lower than this threshold, the pixel is masked as aquatic vegetation [28]. The AFAI has been proven to be effective in identifying algal blooms with GOCI data [19], with a threshold of 0.01 to eliminate algal bloom pixels. Highly turbid water can also lead to high Chl*a* inversion results [29]; thus, combined with the characteristics of the AFAI and adjusted floating algae height (AFAH) [30], when AFAI > −0.01 and AFAH < −0.006 were simultaneously met, the pixel was removed. In addition, considering the influence of the adjacency effect in the nearshore, a one-pixel range along the boundary of the water was masked.

The interval between the satellite transit time and the sampling time was controlled within 0.5 h, and a coefficient of variation (CV) < 10% in a 3 × 3 pixel window was set to reduce the impact of spatial heterogeneity [31]. A total of 327 match-up pairs were finally selected through the above restrictive rules, and the number of sample points in Lake Chaohu, Lake Hongze, and Lake Taihu was 237, 65, and 25, respectively (Figure 1). In addition, in order to effectively analyze the diural variation in Chl*a*, the data were classified according to the curve trend after fitting the quadratic equation. Data with at least six available images in one day were counted for statistical analysis. Moreover, images accounting for less than 30% of the total number of pixels or images with a large difference in the number of pixels on the same day after being masked were eliminated.

2.5. Chla Inversion Algorithms

Many Chl*a* empirical inversion models have been developed with high accuracy for different satellites and water regions (Table 1). By referring to models in previous studies, indices were calculated according to similar GOCI bands, and the performance of various empirical models was comparatively analyzed. Based on model equations commonly used in the literature and the growth trend in training data, three types of equations, including linear, quadratic polynomial, and exponential equations, were selected. Moreover, the random forest model was added to the comparison study, and the model with a higher Chl*a* inversion accuracy was determined for Lake Chaohu, Lake Hongze, and Lake Taihu. Random forest is a highly flexible machine learning approach that constructs independent decision trees based on the bagging concept, and these trees are selected by using a randomized training scheme, which ensures the accuracy of training and judgments [6]. In the modeling process, too many decision trees can reduce efficiency, and too few can reduce accuracy. Thus, the GridSearchCV method was used to adjust the parameters (i.e., n estimators, max_depth, max_features, etc.) to account for the accuracy and efficiency of model operation. In addition, given that Chl*a* is closely related to algal blooms, certain indices, such as the AFAI, index of floating green algae for the GOCI sensor (IGAG), and algal biomass index (ABI), were selected for comparison. Among the 327 match-up pairs, 218 samples were adopted as the training dataset, and 109 samples comprised the validation dataset.

2.6. Model Calibration and Validation

The coefficient of determination (R^2), root mean square error (RMSE) (Equation (1)), mean absolute percentage error (MAPE) (Equation (2)), unbiased percentage difference (UPD) (Equation (3)), and coefficient of variation (CV) (Equation (4)) were selected as metrics for sample screening and model validation, and the inversion accuracy of the Chl*a* model was quantitatively evaluated.

$$\mathrm{RMSE} = \sqrt{\frac{1}{N}\sum_{i=1}^{N}(X_i - Y_i)^2} \tag{1}$$

$$\mathrm{MAPE} = \frac{1}{N}\sum_{i=1}^{N}\frac{|X_i - Y_i|}{X_i} \times 100\% \tag{2}$$

$$\mathrm{UPD} = \frac{1}{N}\sum_{i=1}^{N}\frac{|X_i - Y_i|}{X_i + Y_i} \times 200\% \tag{3}$$

$$\mathrm{CV} = \frac{\sqrt{\sum_{j=1}^{N}(R_j - \overline{R})^2/N}}{\overline{R}} \times 100\% \tag{4}$$

where N is the number of samples, X_i and Y_i are the corresponding values at point i, R_j is the value of j in the 3×3 window and $\overline{R}$ is the mean pixel value within the window.

Table 1. Representative empirical models for estimating Chl*a* using field data and satellite data. Note that the IGAG was used to detect green algae blooms, the AFAI was used to observe diurnal changes in cyanobacteria blooms, and the ABI was proposed to measure the algal biomass.

ID	Model Name	Algorithm Form	Study Area	Data Source	R^2	Reference
1	Four band algorithms (FBA)	$FBA = \frac{1/R_{rs}(662)-1/R_{rs}(693)}{1/R_{rs}(740)-1/R_{rs}(705)}$ $Chla = 0.0097FBA - 0.1268$	Lake Taihu, China	In situ data	0.97	Le (2009) [32]
2	Three band algorithms (TBA)	$TBA = (1/R_{rs}(680) - 1/R_{rs}(745))R_{rs}(865)$ $Chla = 116.04TBA - 51.18$	Lake Taihu, China	GOCI	0.35	Duan (2010) [7]
3	Fluorescence line height (FLH)	$FLH = R_{rc}(680) - R_{rc}(660) - ((R_{rc}(745) - R_{rc}(660)) * \frac{680-660}{745-660})$ $Chla = 101.056FLH^{0.5057}$	Clear and turbid watersaround Korea and Japan	GOCI	0.90	Kim (2016) [33]
4	Band ratio (BR)	$BR = 1.8875+0.8296(1-0.2241^{\lg(R_{rs}(745)/R_{rs}(680))})$ $Chla = 10^{BR}$	Lake Taihu, China	GOCI	0.71	Guo (2020) [17]
5	Normalized difference chlorophyll index (NDCI)	$NDCI = \frac{R_{rs}(708)-R_{rs}(665)}{R_{rs}(708)+R_{rs}(665)}$ $Chla = 14.039+86.115NDCI + 194.325NDCI^2$	4 estuaries and bays, USA	MERIS	0.90	Mishra (2012) [34]
6	Normalized green–red difference index (NGRDI)	$NGRDI = \frac{R_{rc}(560)-R_{rc}(681)}{R_{rc}(560)+R_{rc}(681)}$ $Chla = 0.8724e^{7.0508NGRDI}$	Lake Poyang, China	MERIS	0.70	Feng (2015) [25]
7	Spectral index (SI)	$SI = \frac{e^{R_{rc}(645)} - e^{R_{rc}(859)}}{e^{R_{rc}(645)} + e^{R_{rc}(859)}}$	Lake Taihu, China	MODIS	0.72	Shi (2017) [35]
8	RF	R_{rs}	228 lakes, Global	MODIS	0.51	Cao (2022) [14]
9	Extreme gradient boosting tree (BST)	R_{rc}	67 lakes, China	Landsat 8-Operational Land Imager (OLI)	0.79	Cao (2020) [4]
10	IGAG	$IGAG = \frac{R_{rc}(555)+R_{rc}(660)}{R_{rc}(745)-R_{rc}(660)} + \frac{R_{rc}(745)}{R_{rc}(660)}$	Yellow Sea and East China Sea	GOCI	/	Son (2012) [36]
11	AFAI	$AFAI = R_{rc}(745) - R_{rc}(660) - ((R_{rc}(865) - R_{rc}(660)) * \frac{745-660}{865-660})$	Lake Taihu, China	GOCI	/	Qi (2018) [19]
12	ABI	$ABI = (R_{rs}(645) - R_{rs}(469)) * \frac{555-469}{645-469} - (R_{rs}(859) - R_{rs}(469)) * \frac{555-469}{859-469}$	Lake Chaohu, China	MODIS	/	Hu (2021) [37]

3. Results

3.1. Consistency in R_{rc} of the GOCI and GOCI-II Data

The scatter density plots (Figure 4) for different bands (only six bands were involved in this paper) showed that R_{rc} of the two sensors exhibited favorable consistency ($R^2 > 0.94$, MAPE < 8.82%, and UPD < 9.27%). According to the linear relationship between the GOCI and GOCI-II R_{rc} values in different bands, a linear conversion of GOCI-II R_{rc} values was performed to derive Chl*a* inversion results.

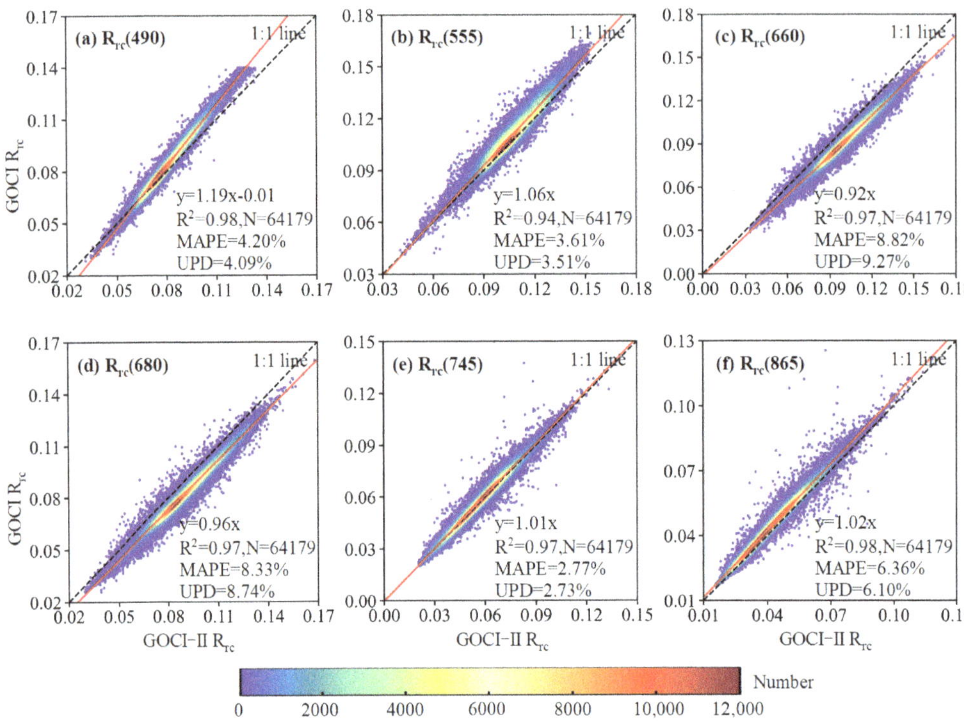

Figure 4. Consistent verification results for different bands based on GOCI and GOCI-II data.

3.2. Establishment and Comparison of Chla Inversion Models

3.2.1. Correlation of Dominant Factors

Based on the 327 match-up pairs, the R_{rc} of each band was extracted, and index factors were calculated (Table 1). The correlation between the index factors and measured Chl*a* was significantly higher than that with the single-band factors (Table 2). Previous studies have shown that blue bands (413, 443, and 490 nm) contribute little to the estimation of Chl*a* in eutrophic and turbid water and are also affected by the uncertainty in atmospheric correction [6]. R^2 between the red and near-infrared bands and Chl*a* was higher than that with the blue and green shortwave bands; thus, longwave bands were used when building the index factors. The index factors with $R^2 > 0.6$ were the AFAI, B7/B5, B7/B6, FLH, and SI (hereinafter referred to as dominant factors), and they were used in the comparison of the empirical model and random forest model. Eight single bands and five dominant factors were selected from the GOCI data as feature variables to build the RF model (recorded as Chl*a*-RF).

Table 2. Correlation between the different factors and measured Chl*a*. The first column provides single-band factors, and the name and formula of the index factors are provided in the second and third columns, respectively, of Table 1. The band order in brackets is the band used from left to right in the formula.

Single-Band Factor	R^2	Index Factor	R^2	Index Factor	R^2
B1	0.13	AFAI (B7, B5, B8)	0.78	FLH (B6, B5, B7)	0.66
B2	0.04	ABI (B5, B3, B4, B8)	0.58	IGAG (B4, B5, B7)	0.47
B3	0.06	B6/B5	0.34	NDCI1 (B6, B5)	0.34
B4	0.08	B7/B5	0.70	NDCI2 (B7, B3)	0.40
B5	0.27	B7/B6	0.74	NGRDI (B4, B6)	0.41
B6	0.29	B8/B7	0.26	SI (B5, B8)	0.68
B7	0.18	FBA1 (B4, B5, B7, B6)	0.32	TBA1 (B5, B4, B6)	0.40
B8	0.21	FBA2 (B5, B6, B8, B7)	0.48	TBA2 (B6, B7, B8)	0.59

3.2.2. Construction and Validation of Empirical Models

The model equations with high R^2 values were quadratic polynomials and exponential equations (Table 3). R^2 of the quadratic polynomial equations was lower than that of the exponential equations (Figure 5). In the range of 0–20 µg/L, training of the quadratic polynomial equations did not conform to the monotonous trend in Chl*a*, whereas the trend in the exponential equation results was more realistic. The R^2 values of the training and validation data indicated the order of AFAI > B7/B6 > B7/B5 > SI > FLH, and the model was best established by using the AFAI (R^2 = 0.79, RMSE = 12.63 µg/L). In summary, for Lake Chaohu, Lake Hongze, and Lake Taihu, the AFAI with an exponential equation (recorded as Chl*a*-AFAI) could better reflect Chl*a*.

Table 3. Correlation between the dominant factors and model equations (R^2). The quadratic polynomial and exponential equations with a higher R^2 for each dominant factor and model equation are noted in bold in Table 3. Among them, the equation with the highest value of R^2 among exponential equations was selected for comparison.

Model Equation	AFAI	B7/B5	B7/B6	FLH	SI
$y = a \times x + b$	0.600	0.487	0.544	0.405	0.443
$y = a \times x^2 + b \times x + c$	**0.764**	**0.668**	**0.675**	**0.485**	**0.589**
$y = a \times \exp(b \times x)$	**0.782**	0.678	0.690	**0.505**	**0.618**
$y = a \times \exp(b \times x) + c$	0.781	**0.694**	**0.696**	0.402	0.425
$y = a \times \exp(b \times x + c)$	0.068	0.677	0.689	0.023	0.087
$y = 100/(1 + \exp(a \times x + b))$	0.655	0.522	0.562	0.452	0.509

3.2.3. Development and Validation of the RF Model

Regarding the single-band factors, the accuracy of the model with B4-B8 (B4, B5, B6, B7, and B8) was higher than that with all eight bands used (Table 4). Then, five dominant factors were added as variables, and the accuracy (R^2 = 0.82) after adding the AFAI was higher than that after adding the other factors. The accuracy was the highest (R^2 = 0.84) after the SI was added, while the accuracy slightly decreased (R^2 = 0.81) when the other factors were added. Therefore, B4-B8, AFAI, and SI were finally selected as variables of the model. The importance of the AFAI and SI was significantly higher than that of the other single-band factors, and the importance of bands B5, B6, and B7 was higher than that of bands B4 and B8, which is relatively consistent with the correlation in Table 2.

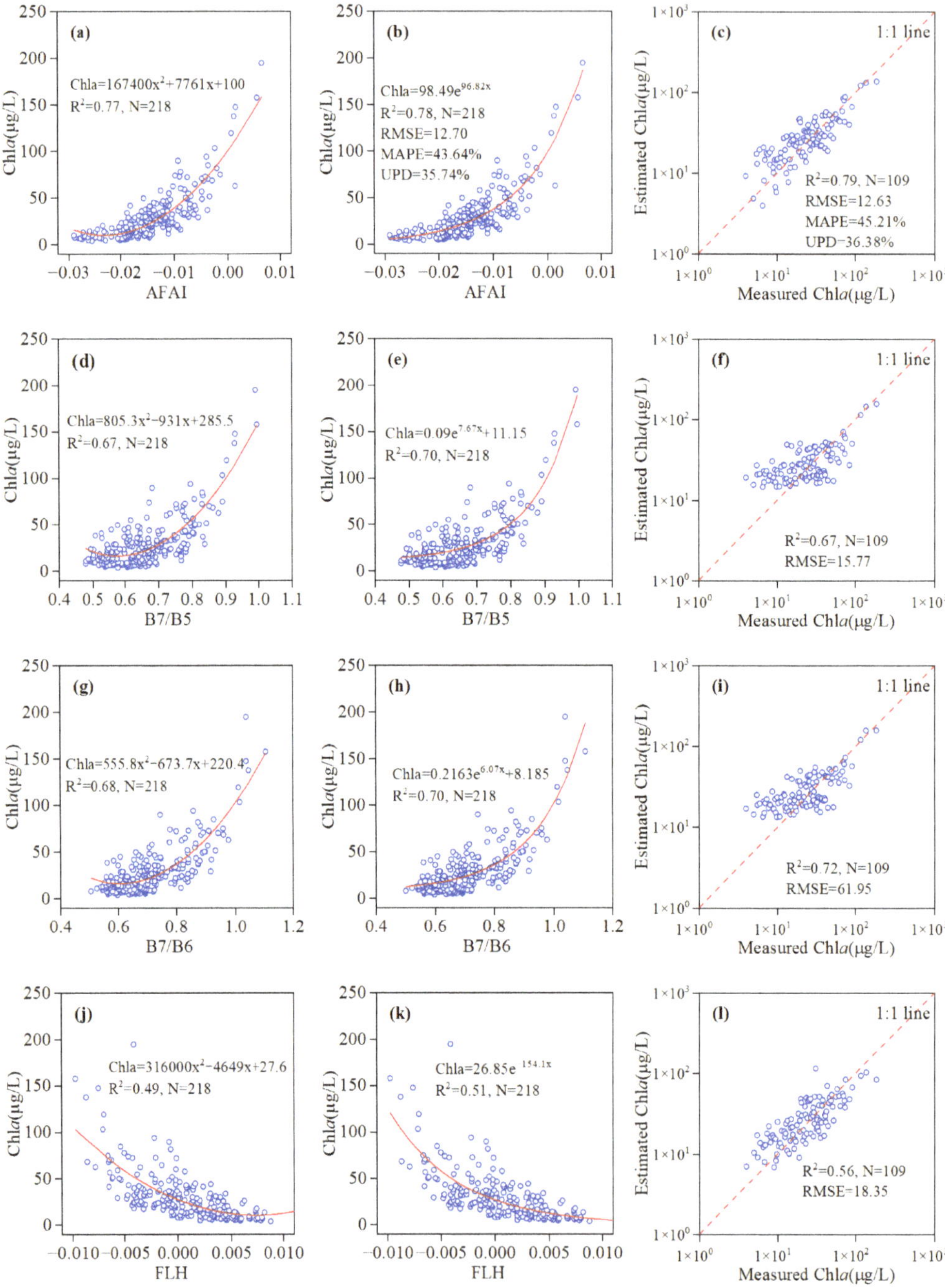

Figure 5. *Cont.*

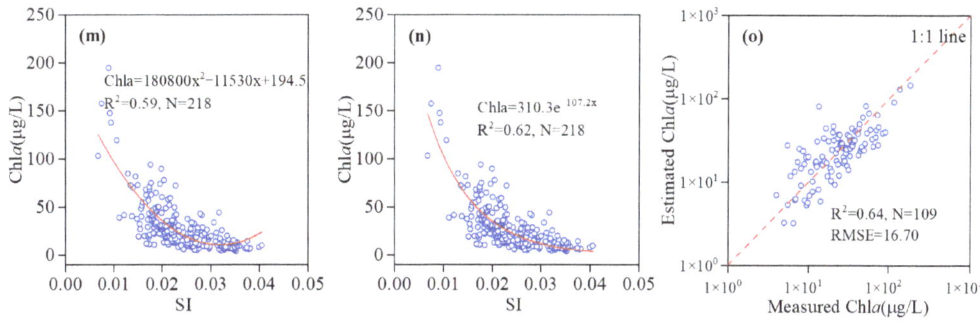

Figure 5. Training and validation of the inversion model for the dominant factor (**a**,**d**,**g**,**j**), and (**m**) show the training results for the quadratic polynomial equation with five factors, namely, the AFAI, B7/B5, B7/B6, FLH, and SI, respectively; (**b**,**e**,**h**,**k**), and (**n**) show the training results for the exponential equation; (**c**,**f**,**i**,**l**), and (**o**) show the validation results for the exponential equation.

Table 4. Training and validation of the RF model with different factors. Note that B1-B8 denotes B1, B2, B3, B4, B5, B6, B7, and B8; similarly, B4-B8 denotes B4, B5, B6, B7, and B8. For each type of comparison, the highest accuracy of training and validation is noted in bold.

ID	Factor	Training (R^2)	Validation (R^2)
1	B1–B8	0.90	0.63
	B4–B8	**0.90**	**0.67**
	B5–B8	0.88	0.66
2	B4–B8, AFAI	**0.92**	**0.82**
	B4–B8, B7/B5	0.91	0.79
	B4–B8, B7/B6	0.92	0.79
	B4–B8, FLH	0.90	0.75
	B4–B8, SI	0.90	0.77
3	B4–B8, AFAI, B7/B5	0.93	0.81
	B4–B8, AFAI, B7/B6	0.93	0.81
	B4–B8, AFAI, FLH	0.93	0.81
	B4–B8, AFAI, SI	**0.94**	**0.84**
4	B4–B8, AFAI, SI, B7/B5	0.92	0.82
	B4–B8, AFAI, SI, B7/B6	0.92	0.82
	B4–B8, AFAI, SI, FLH	0.92	0.83
	B4–B8, AFAI, SI, FLH, B7/B5	0.91	0.81
	B4–B8, AFAI, SI, FLH, B7/B6	0.91	0.82
	All	0.91	0.81

Comparing the Chla-AFAI model (Figure 5b,c) to the Chla-RF model (Figure 6) revealed that the training (R^2 = 0.94, RMSE = 7.18 μg/L, MAPE = 26.26%, UPD = 22.19%) and validation results (R^2 = 0.84, RMSE = 11.77 μg/L, MAPE = 39.02%, UPD = 31.24%) for Chla-RF were obviously better than those for Chla-AFAI. Therefore, Chla-RF was finally selected as the inversion model for Chla in this paper, which attained better applicability for lakes with complex optical properties.

3.3. Temporal and Spatial Variations in Chla in the Three Lakes

Chl*a* in Lake Chaohu (53.06 ± 15.24 μg/L) was higher than that in Lake Taihu (46.53 ± 14.49 μg/L), and that in Lake Hongze was the lowest (30.69 ± 14.97 μg/L). Chl*a* in western Lake Chaohu was higher than that in eastern Lake Chaohu. In Chengzi Bay and Lihe Bay, Chl*a* was significantly higher than that in central Lake Hongze, and Chl*a* in Zhushan Bay, Meiliang Bay, and Gonghu Bay was higher than that in southern and central Lake Taihu.

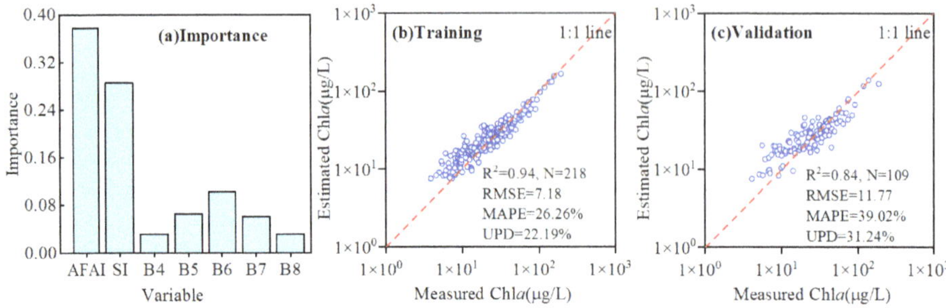

Figure 6. Training and validation of the RF model on Chl*a* estimation: (**a**) importance of seven variables; (**b**) training; and (**c**) validation.

3.3.1. Interannual Variation in Chl*a*

The interannual Chl*a* variation in Lake Chaohu, Lake Hongze, and Lake Taihu from 2011 to 2022 greatly differed (Figure 7); however, the mean Chl*a* still indicated Lake Chaohu > Lake Taihu > Lake Hongze. The mean Chl*a* in Lake Chaohu in 2019 was the highest (63.65 μg/L), while that in 2022 was the lowest (42.99 μg/L). The mean Chl*a* decreased in 2013, reached a subpeak in 2014, and then decreased in the following three years (2015–2017). During the period from 2018 to 2022, Chl*a* in Lake Chaohu reached its peak in 2019 and then showed a slow decline. Chl*a* in western Lake Chaohu was significantly higher than that in eastern Lake Chaohu.

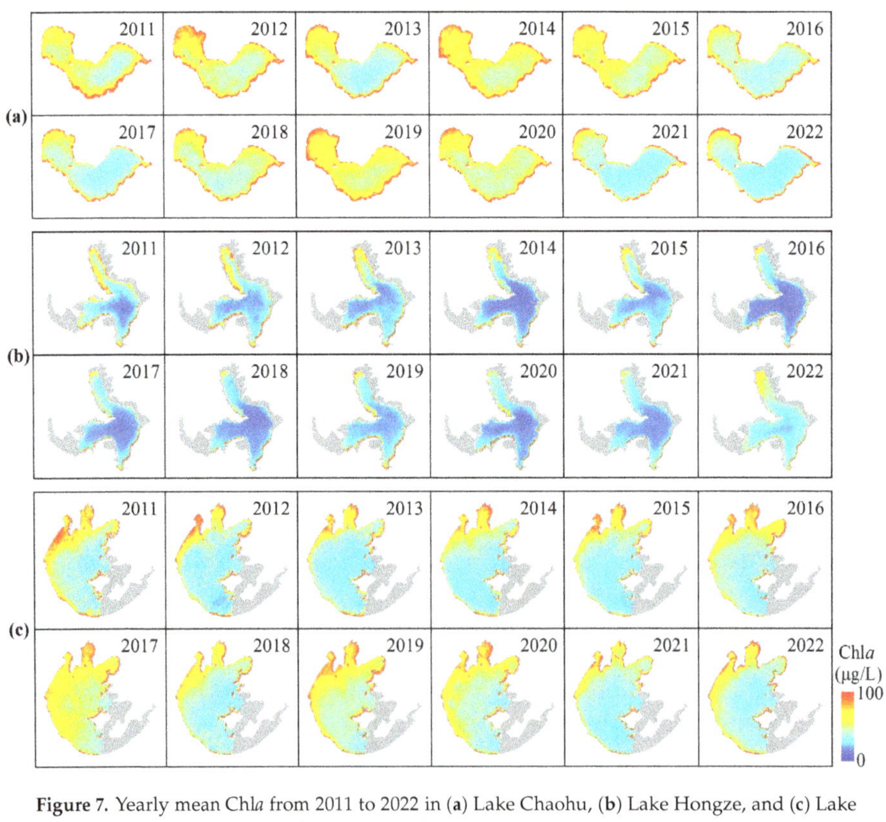

Figure 7. Yearly mean Chl*a* from 2011 to 2022 in (**a**) Lake Chaohu, (**b**) Lake Hongze, and (**c**) Lake Taihu. Note that the GOCI data for 2011 started in May.

Chl*a* in Lake Hongze changed slightly over time and showed a fluctuating decline trend from 2011 to 2019. The mean Chl*a* in 2011 (36.67 µg/L) was the highest, while in 2016 (22.88 µg/L) and 2018 (23.78 µg/L), it was relatively low. The spatial distribution of Chl*a* in Lake Hongze remained consistent over time, with higher Chl*a* values in Chengzi Bay and Lihe Bay and lower Chl*a* values in the central part of the lake.

Chl*a* in Lake Taihu showed significant interannual variation, with the highest mean Chl*a* in 2017 (57.19 µg/L) and the lowest mean Chl*a* in 2013 (40.96 µg/L). The mean Chl*a* gradually increased from 2013 and reached its peak in 2017. From 2018 to 2022, except for a secondary peak in 2019, the mean Chl*a* in the other years showed a fluctuating decline trend. High Chl*a* values were distributed in Meiliang Bay, Zhushan Bay, Gonghu Bay, and other northwest inlet lakes.

3.3.2. Monthly Variation in Chl*a*

The variation in the monthly mean Chl*a* in Lake Chaohu, Lake Hongze, and Lake Taihu from 2011–2022 was derived (Figure 8a,c,e). Chl*a* in summer was higher than that in winter. Lake Chaohu had higher Chl*a* levels in 2015 and 2019, while Lake Hongze had higher Chl*a* levels in 2013, 2015, and 2022, and the values in Lake Taihu peaked in 2017 and 2019. Comparing the daily Chl*a* values calculated by the GOCI and GOCI-II synchronous data (Figure 8b,d,f) revealed that the trend in the three lakes was consistent. The numerical difference was relatively small, which proved the consistency of the two sensors and the practicality of Chl*a*-RF.

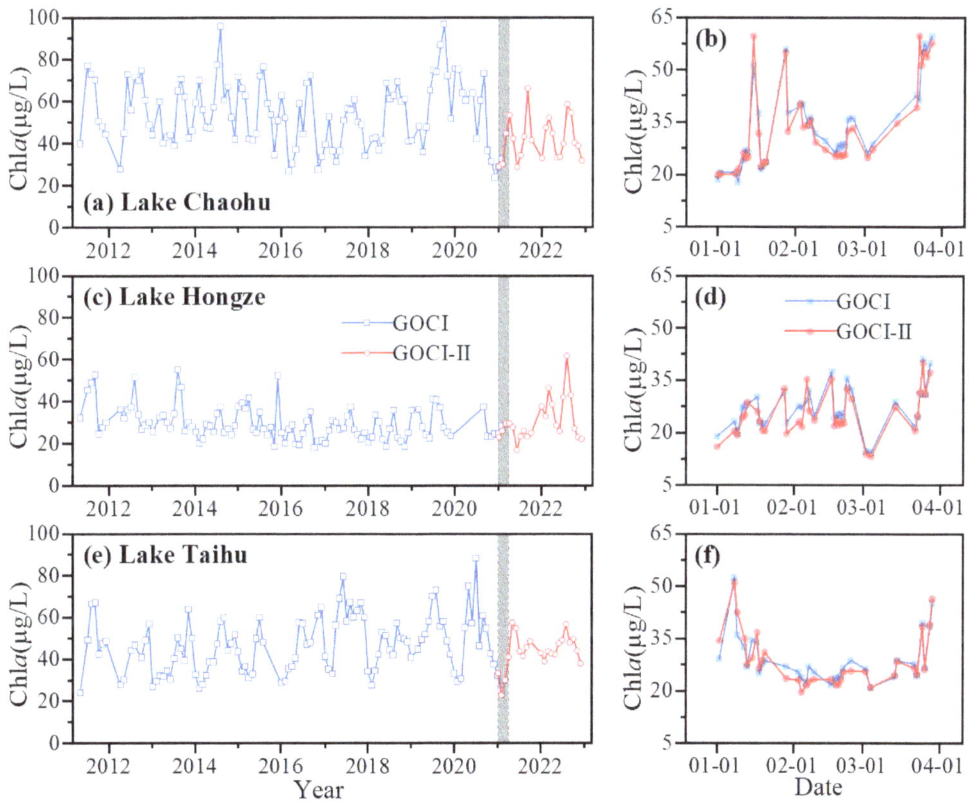

Figure 8. Time series of the monthly mean Chl*a* of GOCI and GOCI-II from 2011 to 2022 in (**a**) Lake Chaohu, (**c**) Lake Hongze, and (**e**) Lake Taihu. Daily mean Chl*a* of GOCI and GOCI-II from January to March 2021 in (**b**) Lake Chaohu, (**d**) Lake Hongze, and (**f**) Lake Taihu.

The spatial distribution of the monthly mean Chl*a* in Lake Chaohu, Lake Hongze, and Lake Taihu was obtained (Figure 9). The mean Chl*a* in Lake Chaohu was higher in September (71.15 μg/L) and August (67.15 μg/L), and lower in April (42.65 μg/L) and December (40.60 μg/L). The overall variation in Chl*a* peaked in September and gradually decreased in October. However, from January to March, Chl*a* in Lake Chaohu was higher than that in the other winter months, especially in February (53.63 μg/L). Chl*a* in the west and east of Lake Chaohu was higher than that in the central lake. The month with the highest Chl*a* value in Lake Hongze was August (40.03 μg/L), followed by March (33.17 μg/L), and the months with the lower values were June (21.56 μg/L) and November (22.32 μg/L). In Chengzi Bay and Lihe Bay, Chl*a* was higher than that in the central lake, especially in the summer from July to September. The mean Chl*a* in Lake Taihu was higher in September (56.34 μg/L) and August (53.45 μg/L), and the mean Chl*a* in February (31.38 μg/L) was the lowest, followed by January (37.03 μg/L) and March (36.19 μg/L). Chl*a* significantly increased from May, reached an annual peak in September, and then gradually decreased, with a slight increase in November (53.10 μg/L) and December (49.41 μg/L). Chl*a* in Zhushan Bay and Meiliang Bay was significantly higher than that in other areas, especially from May to September.

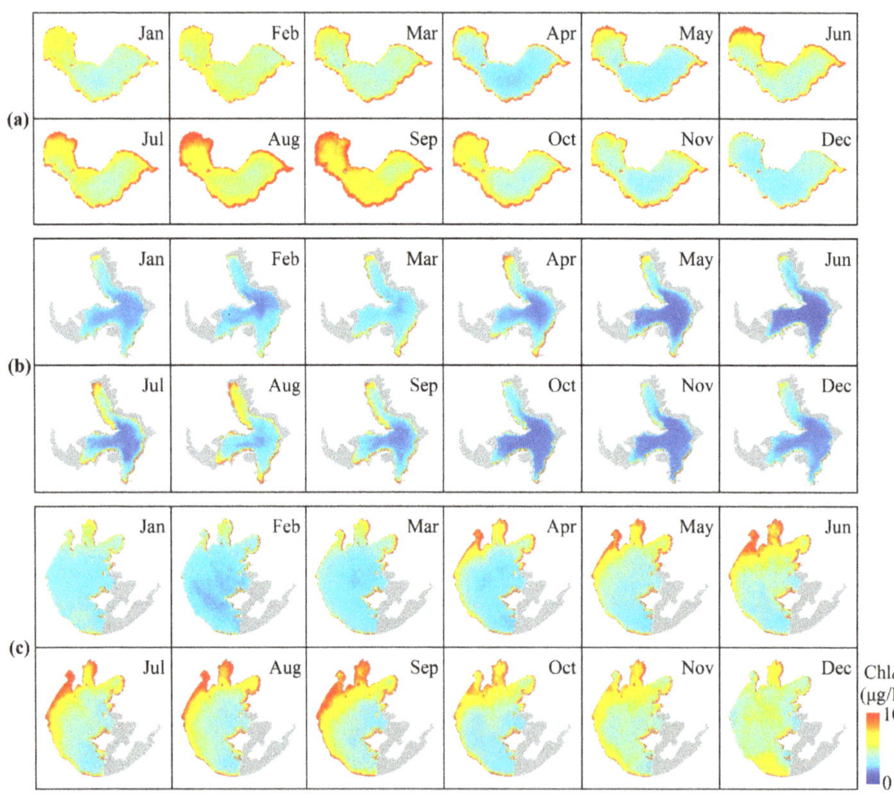

Figure 9. Monthly mean Chl*a* in (**a**) Lake Chaohu, (**b**) Lake Hongze, and (**c**) Lake Taihu.

3.3.3. Diurnal Variation in Chl*a*

On the basis of the selection of effective days (327 days for Lake Chaohu, 265 days for Lake Hongze, and 352 days for Lake Taihu), three typical diurnal variation types were observed (Figure 10), and the images were classified and counted according to the three types (Table 5). During the 8 h period from 8:16 to 15:16, Chl*a* under the first type showed a near-continuous increase (Class 1), Chl*a* under the second type showed a near-continuous decrease (Class 2), and Chl*a* under the third type first increased and then decreased (Class 3). The frequency of the diurnal variation in the three lakes followed the order of Class 3 > Class 2 > Class 1 (Table 5). The occurrence time of Class 3 was concentrated from May to September, while the temporal variation in Class 1 and Class 2 was relatively dispersed. The frequency of Class 3 in Lake Taihu and Lake Chaohu was higher than that of Class 1 or 2, and the frequency of the three types in Lake Hongze was relatively close.

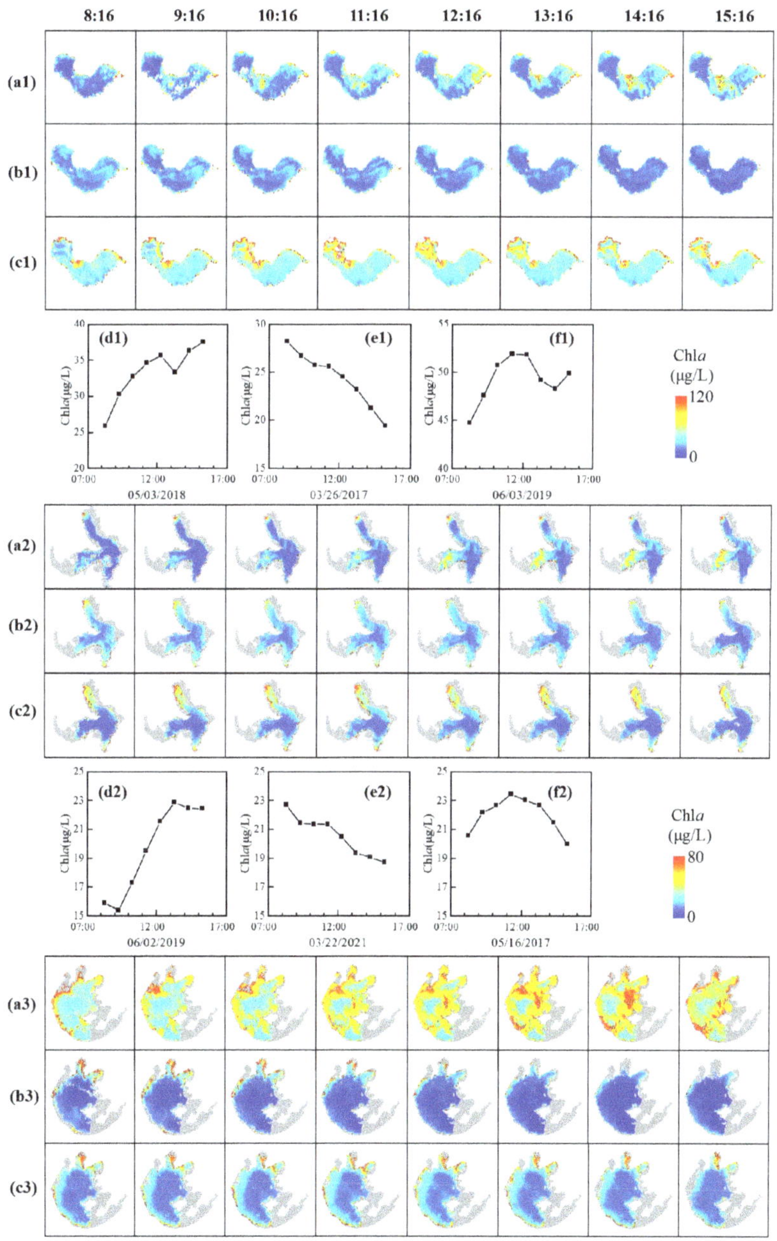

Figure 10. *Cont.*

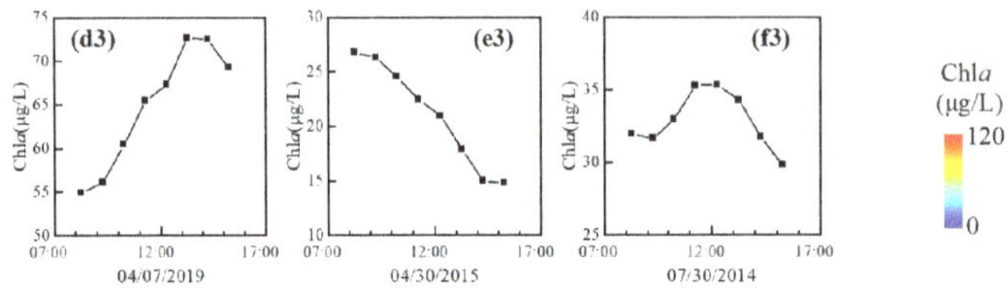

Figure 10. Three typical types of Chl*a* diurnal variation patterns in the three lakes: (**a**) shows images of Class 1 (near-continuous increase), (**b**) shows images of Class 2 (near-continuous decrease), (**c**) shows images of Class 3 (first an increase and then a decrease), and (**d–f**) show corresponding images for three dates. (**a1–f1**) show Lake Chaohu, (**a2–f2**) show Lake Hongze, and (**a3–f3**) show Lake Taihu.

Table 5. Three typical types of diurnal variation patterns in Chl*a*.

Lake	Class	Number of Class	Day of the Year
Lake Chaohu	Class 1	25	228 ± 82
	Class 2	54	163 ± 70
	Class 3	86	211 ± 69
Lake Taihu	Class 1	31	207 ± 99
	Class 2	35	169 ± 73
	Class 3	106	190 ± 69
Lake Hongze	Class 1	26	179 ± 71
	Class 2	28	175 ± 75
	Class 3	43	176 ± 55

4. Discussion

4.1. Performance and Stability of the Proposed Chla Models

To evaluate the stability of the models, different numbers were selected for validation, and the differences between Chla-RF and Chla-AFAI were compared based on R^2, RMSE, MAPE, and UPD (Figure 11). The performance of Chl*a*-RF ($R^2 > 0.94$, RMSE < 8.27 μg/L, MAPE < 24.29%, UPD < 19.60%) was better than that of Chl*a*-AFAI ($R^2 > 0.71$, RMSE < 14.21 μg/L, MAPE < 46.59%, UPD < 37.23%), indicating great advantages in stability and precision. When the sample size was small, the error of both models significantly changed. However, when the number reached 150, the standard deviation decreased, and the models tended to remain stable. In the process of model construction, the accuracy of Chla-RF improved with the increase in sample number after parameter adjustment. However, when the number reached about 240, the increase could only slightly reduce the model error. This is because the sample number is not the only parameter that determines the accuracy of the model; the number of decision trees and max_features considered when constructing the optimal model of the decision tree also have a key influence [6,38]. Therefore, the 218 training and 109 validation samples were relatively reasonable, which not only avoided overfitting problems but also ensured the accuracy of the model and improved its efficiency. Although the inversion principle of an empirical model can be easily defined during model construction, it is not applicable to different regions, and it is sensitive to noise and susceptible to extreme values [39]. Random forest, as a machine learning algorithm, can effectively deal with the complexity of the model, and the nonlinear relationship between variables and Chla can be obtained without prior knowledge [40]. At the same time, the randomness makes the model more accurate and stable, which is also reflected in Figure 11.

Sampling data of 13 October 2022 (N = 11) for Lake Hongze and GOCI-II R_{rc} data of the corresponding time (interval time < 1 h) were used to verify and compare the accuracy of the models. The results (Figure 12) proved that Chl*a*-RF ($R^2 = 0.81$, RMSE = 4.61 μg/L) performed better than Chl*a*-AFAI ($R^2 = 0.37$, RMSE = 10.88 μg/L), and the estimated Chl*a* of Chl*a*-RF was closer to the measured Chl*a* than that of Chl*a*-AFAI. We not only effectively evaluated the accuracy of Chl*a*-RF but also verified the consistency of GOCI-II R_{rc}, providing a guarantee for the continuous use of GOCI series data in Chl*a* inversion.

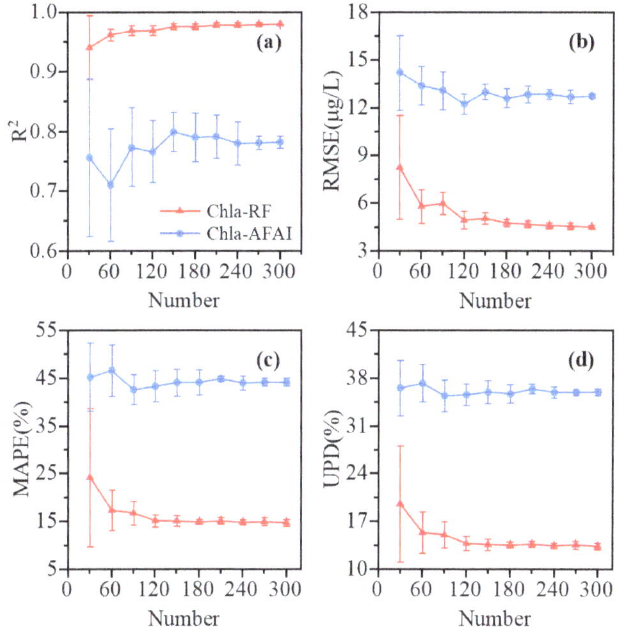

Figure 11. Error of the models with different numbers of samples: (**a**) R^2, (**b**) RMSE, (**c**) MAPE, and (**d**) UPD.

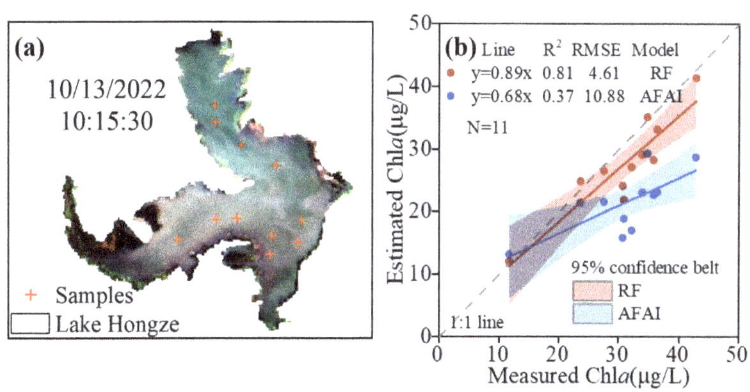

Figure 12. Location of the samples in (**a**) Lake Hongze and (**b**) validation of the models.

4.2. Uncertainties and Limitations in Model Development

The GOCI and GOCI-II data exhibit a high temporal resolution of ≥8 scenes per day, which can capture the rapid variations in Chl*a*. However, previous studies have shown that the use of satellite data with different spatial resolutions affects the results of water color parameter inversion [23]. Compared to the high spatial resolution of Landsat (30 m) and Sentinel 2-MSI (10 m), data with a low resolution of 250 m or even 500 m still yielded a certain error in monitoring the variation in Chl*a* [41]. Although the error could be avoided by the condition of CV < 10%, the impact of the difference in the spatial resolution between GOCI and GOCI-II could not be neglected.

Combined with the single bands and commonly used indices such as the CMI, AFAI, and AFAH, a threshold was determined to remove non-water pixels. Due to the complexity of the image reflectance, the existing threshold value cannot completely remove invalid pixels. The difficulty in distinguishing thin cloud pixels or the loss of data caused by overmasking could reduce the accuracy of Chl*a* inversion [42]. The water pixels near land were often mixed pixels due to the low spatial resolution, which could be easily misidentified as a cyanobacteria bloom or aquatic vegetation, resulting in high Chl*a* values at the lake boundary.

As typical lakes with complex optical properties, Lake Chaohu and Lake Taihu exhibit serious eutrophication, and Lake Hongze is turbid due to perennial sand mining activities [21]. The water situation of Lake Taihu from October to December was complicated, with serious eutrophication and turbidity [2]. Since water with high suspended matter levels produces high reflectance in visible bands, it can be misjudged as water with high Chl*a* levels. Therefore, non-algal bloom but turbid pixels were removed by combining the AFAI and AFAH. Although the Chl*a* inversion accuracy was ensured, this problem could not be completely resolved.

Water signals cannot be detected when algal blooms occur; thus, algal bloom pixels were removed by the AFAI, which could reduce the accuracy of the Chl*a* inversion results. Under the comprehensive effect of wind speed, temperature, and other factors, the algal bloom variation trend mostly first increased and then decreased and often reached a maximum value of bloom area at noon [19]. However, there are still some uncertainties in the classification of the three typical diurnal variation types. In the research process, we also tried to use principal component analysis (PCA) [43] or the *k*-means clustering method [44]; however, due to the influence of data value and trend, the classification results of these methods are quite different from the actual types with low accuracy. In addition, a few images showed a trend of first decreasing and then increasing due to the influence of algal bloom masking in this paper. After the removal of extremely high values, Chl*a* in the lakes showed a decreasing trend; therefore, this variation type was not considered in the analysis of diurnal variation.

4.3. Potential Reasons for the Temporal and Spatial Distributions of Chla

Lake Chaohu, Lake Hongze, and Lake Taihu exhibited similarities and differences in their spatial-temporal variations. There are nine major rivers (33 in total) entering and exiting Lake Chaohu, most of which are distributed in the western part of the lake [45]. Due to the abundance of nutrients at the inlet of the lake, algae often grow explosively; therefore, Chl*a* in the west of Lake Chaohu was higher than that in the east [5]. Chengzi Bay in northern Lake Hongze is relatively closed, which is suitable for phytoplankton growth [46]; therefore, Chl*a* in this area was higher than that in other areas. Chl*a* was still high in Lihe Bay under intense aquatic vegetation growth, such as from July to September [47]. The eastern part of Gonghu Bay, as the inlet of the Water Diversion Project from the Yangtze River to Lake Taihu, also had relatively high Chl*a* levels in the summer. The northwestern part of Lake Taihu is the main source of water pollution, and Chl*a* in Zhushan Bay and Meiliang Bay was always high.

Overall, Chl*a* in summer and autumn was higher than that in spring and winter. This is consistent with the rapid growth of phytoplankton in summer under the influence of various environmental factors such as light and high temperatures. In winter, the water temperature is low, with weak light and low phytoplankton growth; therefore, Chl*a* is relatively low [35]. However, Chl*a* in Lake Chaohu from January to March was higher than that in the winter, especially in February. The reason is that the dominant algae species caused an abnormal increase in Chl*a*, among which Microcystis mostly grows from May to September, while Anabaena blooms in winter [37]. Due to suspended solids in May and June, the water transparency decreased in Lake Hongze. This affected chlorophyll-A photosynthesis, leading to a significant decline in Chl*a* in June [23]. Chl*a* in Lake Taihu slightly increased in November and December; however, a few high-value areas still emerged in the south of Lake Taihu due to the inflow of rivers such as the Tiaoxi River [19].

Lake Chaohu, Lake Hongze, and Lake Taihu showed obvious interannual variations. Due to the increase in the average annual temperature and nutrient salt input, the mean Chl*a* showed a subpeak in 2014 in Lake Chaohu [45], which is consistent with the trend of the risk level of algal blooms proposed by Duan (2017) [5]. The high Chl*a* in Lake Hongze in 2022 was mainly influenced by the Chl*a* levels in March and June of the same year, which increased the overall mean. The mean Chl*a* in Lake Hongze in 2016 and 2018 was low under the influence of a low temperature, lake inflow, precipitation, wind speed, and other factors [47]. Under the effect of factors such as an increasing total phosphorus concentration and temperature and decreasing wind speed, the mean Chl*a* in Lake Taihu from 2013 to 2017 continued to increase. Due to the decreasing water temperature and increasing wind speed [48], the mean Chl*a* in Lake Taihu in 2018 significantly decreased. Influenced by the rising temperature in 2019, the mean Chl*a* in Lake Taihu also increased [49], which is consistent with the variation trend in the frequency of algal blooms proposed by Xue (2022) [41].

5. Conclusions

The GOCI and GOCI-II were consistent in terms of R_{rc} synchronous data, which provided the possibility for long-term use of satellite remote sensing data with high temporal resolution (1 h). The accuracy of the random forest model in Lake Chaohu, Lake Hongze, and Lake Taihu was obviously higher than that of the empirical model, and the accuracy and stability of the inversion of Chl*a* in optically complex lakes were enhanced. This study revealed the temporal and spatial pattern changes in Chl*a* in the three lakes from 2011 to 2022, providing a reference for water environment monitoring and protection. Although random forest requires a large number of samples, it will certainly have a broader application prospect in remote sensing inversion because of its outstanding advantages in dealing with complex nonlinear problems. The spatial resolution of GOCI-II data has been increased to 250 m and combined with high temporal resolution and abundant spectral information, it holds a prominent edge in Chl*a* monitoring, providing an important data source for short-term algal bloom observation and prediction.

Author Contributions: Conceptualization, Y.G. and K.X.; methodology, Y.G.; data curation, Z.H. and M.S.; writing—original draft preparation, Y.G.; writing—review and editing, H.L., Z.C. and K.X.; visualization, X.W.; project administration, K.X.; funding acquisition, R.M. All authors have read and agreed to the published version of the manuscript.

Funding: This research was funded by the National Key R&D Program of China (Grant No. 2022YFF0711603), the National Natural Science Foundation of China (Grant No. 42371371, 42201403, and No. 42071341), Science and Technology Achievement Transformation Foundation of Inner Mongolia Autonomous Region(2021CG0013), the Provincial Natural Science Foundation of Jiangsu of China (Grant No. BK20221159), the National Earth System Science Data Center, and the 14th Five-Year Network Security and Informatization Plan of the Chinese Academy of Sciences (Grant No. WX145XQ06-04, http://data.niglas.ac.cn, accessed on 1 January 2023).

Data Availability Statement: Data available on request.

Acknowledgments: We thank the KOSC and NOSC for providing the GOCI and GOCI-II data and the support from the Lake-Watershed Science Data Center (http://lake.geodata.cn, accessed on 1 January 2023).

Conflicts of Interest: The authors declare no conflict of interest.

References

1. Beck, R.; Zhan, S.; Liu, H.; Tong, S.; Yang, B.; Xu, M.; Ye, Z.; Huang, Y.; Shu, S.; Wu, Q.; et al. Comparison of satellite reflectance algorithms for estimating chlorophyll-a in a temperate reservoir using coincident hyperspectral aircraft imagery and dense coincident surface observations. *Remote Sens. Environ.* **2016**, *178*, 15–30. [CrossRef]
2. Huang, C.; Yang, H.; Zhu, A.; Zhang, M.; Lü, H.; Huang, T.; Zou, J.; Li, Y. Evaluation of the Geostationary Ocean Color Imager (GOCI) to monitor the dynamic characteristics of suspension sediment in Taihu Lake. *Int. J. Remote Sens.* **2015**, *36*, 3859–3874. [CrossRef]
3. Palmer, S.C.J.; Kutser, T.; Hunter, P.D. Remote sensing of inland waters: Challenges, progress and future directions. *Remote Sens. Environ.* **2015**, *157*, 1–8. [CrossRef]
4. Cao, Z.; Ma, R.; Duan, H.; Pahlevan, N.; Melack, J.; Shen, M.; Xue, K. A machine learning approach to estimate chlorophyll-a from Landsat-8 measurements in inland lakes. *Remote Sens. Environ.* **2020**, *248*, 111974. [CrossRef]
5. Duan, H.; Tao, M.; Loiselle, S.A.; Zhao, W.; Cao, Z.; Ma, R.; Tang, X. MODIS observations of cyanobacterial risks in a eutrophic lake: Implications for long-term safety evaluation in drinking-water source. *Water Res.* **2017**, *122*, 455–470. [CrossRef]
6. Shen, M.; Luo, J.; Cao, Z.; Xue, K.; Qi, T.; Ma, J.; Liu, D.; Song, K.; Feng, L.; Duan, H. Random forest: An optimal chlorophyll-a algorithm for optically complex inland water suffering atmospheric correction uncertainties. *J. Hydrol.* **2022**, *615*, 128685. [CrossRef]
7. Duan, H.; Ma, R.; Zhang, Y.; Loiselle, S.A.; Xu, J.; Zhao, C.; Zhou, L.; Shang, L. A new three-band algorithm for estimating chlorophyll concentrations in turbid inland lakes. *Environ. Res. Lett.* **2010**, *5*, 44009. [CrossRef]
8. Huang, X.; Zhu, J.; Han, B.; Jamet, C.; Tian, Z.; Zhao, Y.; Li, J.; Li, T. Evaluation of Four Atmospheric Correction Algorithms for GOCI Images over the Yellow Sea. *Remote Sens.* **2019**, *11*, 1631. [CrossRef]
9. Jiang, G.; Loiselle, S.A.; Yang, D.; Ma, R.; Su, W.; Gao, C. Remote estimation of chlorophyll a concentrations over a wide range of optical conditions based on water classification from VIIRS observations. *Remote Sens. Environ.* **2020**, *241*, 111735. [CrossRef]
10. Zhang, F.; Li, J.; Wang, C.; Wang, S. Estimation of water quality parameters of GF-1 WFV in turbid water based on soft classification. *Natl. Remote Sens. Bull.* **2023**, *27*, 769–779. [CrossRef]
11. Pan, Y.; Guo, Q.; Sun, J. Advances in remote sensing inversion method of chlorophyll a concentration. *Sci. Surv. Mapp.* **2017**, *42*, 43–48. [CrossRef]
12. Kolluru, S.; Tiwari, S.P. Modeling ocean surface chlorophyll-a concentration from ocean color remote sensing reflectance in global waters using machine learning. *Sci. Total Environ.* **2022**, *844*, 157191. [CrossRef] [PubMed]
13. Zhu, X.; Guo, H.; Huang, J.J.; Tian, S.; Xu, W.; Mai, Y. An ensemble machine learning model for water quality estimation in coastal area based on remote sensing imagery. *J. Environ. Manag.* **2022**, *323*, 116187. [CrossRef] [PubMed]
14. Cao, Z.; Shen, M.; Kutser, T.; Liu, M.; Qi, T.; Ma, J.; Ma, R.; Duan, H. What water color parameters could be mapped using MODIS land reflectance products: A global evaluation over coastal and inland waters. *Earth-Sci. Rev.* **2022**, *232*, 104154. [CrossRef]
15. Speiser, J.L.; Miller, M.E.; Tooze, J.; Ip, E. A comparison of random forest variable selection methods for classification prediction modeling. *Expert Syst. Appl.* **2019**, *134*, 93–101. [CrossRef]
16. Li, Y.; Shi, K.; Zhang, Y.; Zhu, G.; Qin, B.; Zhang, Y.; Liu, M.; Zhu, M.; Dong, B.; Guo, Y. Remote sensing of column-integrated chlorophyll a in a large deep-water reservoir. *J. Hydrol.* **2022**, *610*, 127918. [CrossRef]
17. Guo, Y.; Huang, C.; Zhang, Y.; Li, Y.; Chen, W. A Novel Multitemporal Image-Fusion Algorithm: Method and Application to GOCI and Himawari Images for Inland Water Remote Sensing. *IEEE Trans. Geosci. Remote* **2020**, *58*, 4018–4032. [CrossRef]
18. Lei, S.; Xu, J.; Li, Y.; Du, C.; Liu, G.; Zheng, Z.; Xu, Y.; Lyu, H.; Mu, M.; Miao, S.; et al. An approach for retrieval of horizontal and vertical distribution of total suspended matter concentration from GOCI data over Lake Hongze. *Sci. Total Environ.* **2020**, *700*, 134524. [CrossRef]
19. Qi, L.; Hu, C.; Visser, P.M.; Ma, R. Diurnal changes of cyanobacteria blooms in Taihu Lake as derived from GOCI observations. *Limnol. Oceanogr.* **2018**, *63*, 1711–1726. [CrossRef]
20. Prak, M.; Jung, H.C.; Lee, S.; Ahn, J.; Bae, S.; Choi, J. The GOCI-II Early Mission Ocean Color Products in Comparison with the GOCI Toward the Continuity of Chollian Multi-satellite Ocean Color Data. *Korean J. Remote Sens.* **2021**, *37*, 1281–1293. [CrossRef]
21. Xue, K.; Ma, R.; Duan, H.; Shen, M.; Boss, E.; Cao, Z. Inversion of inherent optical properties in optically complex waters using sentinel-3A/OLCI images: A case study using China's three largest freshwater lakes. *Remote Sens. Environ.* **2019**, *225*, 328–346. [CrossRef]
22. Luo, J.; Li, X.; Ma, R.; Li, F.; Duan, H.; Hu, W.; Qin, B.; Huang, W. Applying remote sensing techniques to monitoring seasonal and interannual changes of aquatic vegetation in Taihu Lake, China. *Ecol. Indic.* **2016**, *60*, 503–513. [CrossRef]
23. Cao, Z.; Duan, H.; Shen, M.; Ma, R.; Xue, K.; Liu, D.; Xiao, Q. Using VIIRS/NPP and MODIS/Aqua data to provide a continuous record of suspended particulate matter in a highly turbid inland lake. *Int. J. Appl. Earth Obs.* **2018**, *64*, 256–265. [CrossRef]
24. Jeffrey, S.W.; Humphrey, G.F. New spectrophotometric equations for determining chlorophylls a, b, c1 and c2 in higher plants, algae and natural phytoplankton. *Biochem. Physiol. Pflanz.* **1975**, *167*, 191–194. [CrossRef]
25. Feng, L.; Hu, C.; Han, X.; Chen, X.; Qi, L. Long-Term Distribution Patterns of Chlorophyll-a Concentration in China's Largest Freshwater Lake: MERIS Full-Resolution Observations with a Practical Approach. *Remote Sens.* **2015**, *7*, 275–299. [CrossRef]
26. Ahn, J.; Park, Y.; Ryu, J.; Lee, B.; Oh, I.S. Development of atmospheric correction algorithm for Geostationary Ocean Color Imager (GOCI). *Ocean Sci. J.* **2012**, *47*, 247–259. [CrossRef]

27. Wang, M.; Jiang, L. Atmospheric Correction Using the Information From the Short Blue Band. *IEEE Trans. Geosci. Remote* **2018**, *56*, 6224–6237. [CrossRef]
28. Liang, Q.; Zhang, Y.; Ma, R.; Loiselle, S.; Li, J.; Hu, M. A MODIS-Based Novel Method to Distinguish Surface Cyanobacterial Scums and Aquatic Macrophytes in Lake Taihu. *Remote Sens.* **2017**, *9*, 133. [CrossRef]
29. Xu, J.; Liu, H.; Lin, J.; Lyu, H.; Dong, X.; Li, Y.; Guo, H.; Wang, H. Long-term monitoring particulate composition change in the Great Lakes using MODIS data. *Water Res.* **2022**, *222*, 118932. [CrossRef]
30. Xue, K.; Ma, R.H.; Cao, Z.G.; Hu, M.; Li, J. Applicability evaluation and method selection in detecting cyanobacterial bloom using HY-1C/D CZI data for inland lakes. *Natl. Remote Sens. Bull.* **2023**, *27*, 171–186. [CrossRef]
31. Qi, L.; Hu, C.; Duan, H.; Barnes, B.; Ma, R. An EOF-Based Algorithm to Estimate Chlorophyll a Concentrations in Taihu Lake from MODIS Land-Band Measurements: Implications for Near Real-Time Applications and Forecasting Models. *Remote Sens.* **2014**, *6*, 10694–10715. [CrossRef]
32. Le, C.; Li, Y.; Zha, Y.; Sun, D.; Huang, C.; Lu, H. A four-band semi-analytical model for estimating chlorophyll a in highly turbid lakes: The case of Taihu Lake, China. *Remote Sens. Environ.* **2009**, *113*, 1175–1182. [CrossRef]
33. Kim, W.; Moon, J.; Park, Y.; Ishizaka, J. Evaluation of chlorophyll retrievals from Geostationary Ocean Color Imager (GOCI) for the North-East Asian region. *Remote Sens. Environ.* **2016**, *184*, 482–495. [CrossRef]
34. Mishra, S.; Mishra, D.R. Normalized difference chlorophyll index: A novel model for remote estimation of chlorophyll-a concentration in turbid productive waters. *Remote Sens. Environ.* **2012**, *117*, 394–406. [CrossRef]
35. Shi, K.; Zhang, Y.; Zhou, Y.; Liu, X.; Zhu, G.; Qin, B.; Gao, G. Long-term MODIS observations of cyanobacterial dynamics in Lake Taihu: Responses to nutrient enrichment and meteorological factors. *Sci. Rep.* **2017**, *7*, 40326. [CrossRef]
36. Son, Y.B.; Min, J.; Ryu, J. Detecting massive green algae (Ulva prolifera) blooms in the Yellow Sea and East China Sea using Geostationary Ocean Color Imager (GOCI) data. *Ocean Sci. J.* **2012**, *47*, 359–375. [CrossRef]
37. Hu, M.; Zhang, Y.; Ma, R.; Xue, K.; Cao, Z.; Chu, Q.; Jing, Y. Optimized remote sensing estimation of the lake algal biomass by considering the vertically heterogeneous chlorophyll distribution: Study case in Lake Chaohu of China. *Sci. Total Environ.* **2021**, *771*, 144811. [CrossRef] [PubMed]
38. Yu, X.; Zhao, G.; Chang, C.; Yuan, X.; Wang, Z. Random Forest Classifier in Remote Sensing Information Extraction:A Review of Applications and Future Development. *Remote Sens. Inf.* **2019**, *34*, 8–14. [CrossRef]
39. Wang, J.; Xiaodan, W.; Dujuan, M.; Jianguang, W.; Qing, X. Remote sensing retrieval based on machine learning algorithm: Uncertainty analysis. *Natl. Remote Sens. Bull.* **2023**, *27*, 790–801. [CrossRef]
40. Lary, D.J.; Alavi, A.H.; Gandomi, A.H.; Walker, A.L. Machine learning in geosciences and remote sensing. *Geosci. Front.* **2016**, *7*, 3–10. [CrossRef]
41. Xue, K.; Ma, R.; Cao, Z.; Shen, M.; Hu, M.; Xiong, J. Monitoring Fractional Floating Algae Cover Over Eutrophic Lakes Using Multisensor Satellite Images: MODIS, VIIRS, GOCI, and OLCI. *IEEE Trans. Geosci. Remote* **2022**, *60*, 4211715. [CrossRef]
42. Lin, J.; Huang, T.; Zhao, X.; Chen, Y.; Zhang, Q.; Yuan, Q. Robust Thick Cloud Removal for Multitemporal Remote Sensing Images Using Coupled Tensor Factorization. *IEEE Trans. Geosci. Remote* **2022**, *60*, 5406916. [CrossRef]
43. Kim, S.; Yoon, H. Application of classification coupled with PCA and SMOTE, for obtaining safety factor of landslide based on HRA. *Bull. Eng. Geol. Environ.* **2023**, *82*, 381. [CrossRef]
44. Xue, K.; Ma, R.; Wang, D.; Shen, M. Optical Classification of the Remote Sensing Reflectance and Its Application in Deriving the Specific Phytoplankton Absorption in Optically Complex Lakes. *Remote Sens.* **2019**, *11*, 184. [CrossRef]
45. Yuan, J.; Cao, Z.G.; Ma, J.; Shen, M.; Qi, T.; Duan, H. Remote sensed analysis of spatial and temporal variation in phenology of algal blooms in Lake Chaohu since 1980s. *J. Lake Sci.* **2023**, *35*, 57–72.
46. Liu, G.; Li, Y.; Lv, H.; Mu, M.; Lei, S.; Wen, S.; Bi, S.; Ding, X. Remote Sensing of Chlorophyll-a Concentrations in Lake Hongze Using Long Time Series MERIS Observations. *Environ. Sci.* **2017**, *38*, 3645–3656. [CrossRef]
47. Li, Y.; Zhang, Z.; Cheng, J.; Zou, L.; Zhang, Q.; Zhang, M.; Gong, Z.; Xie, S.; Cai, Y. Water quality change and driving forces of Lake Hongze from 2012 to 2018. *J. Lake Sci.* **2021**, *33*, 715–726. [CrossRef]
48. Wu, D.; Jia, G.; Wu, H. Chlorophyll-a concentration variation characteristics of the algae-dominant and macro-phyte-dominant areas in Lake Taihu and its driving factors, 2007–2019. *J. Lake Sci.* **2021**, *33*, 1364–1375. [CrossRef]
49. Qin, Z.; Ruan, B.; Yang, J.; Wei, Z.; Song, W.; Sun, Q. Long-Term Dynamics of Chlorophyll-a Concentration and Its Response to Human and Natural Factors in Lake Taihu Based on MODIS Data. *Sustainability* **2022**, *14*, 16874. [CrossRef]

remote sensing

MDPI

Article

Spatiotemporal Evolutions of the Suspended Particulate Matter in the Yellow River Estuary, Bohai Sea and Characterized by Gaofen Imagery

Zhifeng Yu [1,2], Jun Zhang [1], Zheyu Chen [1], Yuekai Hu [3], C. K. Shum [4], Chaofei Ma [5], Qingjun Song [5], Xiaohong Yuan [1,2], Ben Wang [1] and Bin Zhou [1,2,6,*]

1 Institute of Remote Sensing and Earth Sciences, School of Information Science and Engineering, Hangzhou Normal University, Hangzhou 311121, China; yu@hznu.edu.cn (Z.Y.); 2019210214028@stu.hznu.edu.cn (J.Z.); 2020213101007@stu.hznu.edu.cn (Z.C.); yuanxh@hznu.edu.cn (X.Y.); 20170056@hznu.edu.cn (B.W.)
2 Zhejiang Provincial Key Laboratory of Urban Wetlands and Regional Change, Hangzhou 311121, China
3 State Key Laboratory of Estuarine and Coastal Research, Shanghai 200241, China
4 Division of Geodetic Science, School of Earth Sciences, The Ohio State University, Columbus, OH 43210, USA; ck.shum@outlook.com
5 National Satellite Ocean Application Service, Beijing 100081, China; mcf@mail.nsoas.org.cn (C.M.); kingdream@mail.nsoas.org.cn (Q.S.)
6 School of Engineering, Hangzhou Normal University, Hangzhou 311121, China
* Correspondence: zhoubin@hznu.edu.cn

Abstract: Suspended particulate matter is a crucial component in estuaries and coastal oceans, and a key parameter for evaluating their water quality. The Bohai Sea, a huge marginal sea covering an expanse of 77,000 km^2 and constantly fed by numerous sediment-laden rivers, has maintained a high level of total suspended particulate matter (TSM). Despite the widespread development and application of TSM retrieval algorithms using commonly available satellite data like Landsat, Sentinel, and MODIS, developing TSM retrieval algorithms for China's Gaofen (GF) series (GF-6 and GF-1) in the Bohai Sea is still a great challenge, mainly due to the limited applicability of empirical algorithms. In this study, 259 in situ measured-TSM samples were collected for algorithm development. The remote sensing reflectance (R_{rs}) curve demonstrates prominent peaks between 550 and 580 nm. Through conversion to remote sensing reflectance, it was found that single-band data had a weak correlation with TSM, reaching a maximum correlation of 0.44. However, by combining bands of band ratio calculations, the correlation was enhanced. Particularly, the blue and green band equivalent R_{rs} ratio had a correlation coefficient of 0.81 with TSM, and the proposed TSM inversion exponential algorithm developed based on this factor obtained an R-squared (R^2) value of 0.76 and a mean relative error (MRE) of 32.24%. Analysis results indicated that: (1) there are spatial variations in the TSM within the Bohai Sea, Laizhou Bay, and the Yellow River estuary, with higher levels near the coast and lower levels in open waters. The Yellow River estuary experiences seasonal fluctuations higher TSM during spring and winter, and lower variations during summer and autumn, and (2) the dynamics of TSM are affected by Yellow River runoff, with increased runoff leads to higher TSM levels and expanded turbid zones. This study proposes a new algorithm to quantify TSM evolutions and distributions in the Bohai Sea and adjacent regions using China's Gaofen imageries.

Keywords: Bohai Sea; total suspended particulate matter (TSM); Gaofen imagery; remote sensing inversion; Yellow River estuary

1. Introduction

Suspended sediment is an important factor that directly affects ocean color, and its concentration and transparency have a large impact on the ocean water column [1–4]. The total suspended particulate matter (TSM) changes rapidly due to ocean dynamics and other factors, such as precipitation, runoff, extreme weather, etc. The Bohai Sea is a vital component of China's continental shelf, and it receives sediment transports from major rivers like the Yellow River, Liaohe River, and Haihe River. Investigating and quantifying the spatiotemporal dynamics of TSM in the Bohai Sea is important for ecological environmental preservation, ocean sustainable development, and for guiding the planning and location selection for coastal environmental projects in the Bohai Sea area [5–7]. The conventional methods for monitoring and analyzing TSM consume a lot of human and material resources, and require certain hydroclimatic conditions, resulting in extensive time and effort. With the integration of remote sensing technology and the widespread utilization of satellite sensing systems including NASA's Landsat, ESA's Sentinel, China's Gaofen (GF), and Korea's GOCI, the abundant data from these diverse satellite sources provide ample resources for advancing ocean water color remote sensing [8–14]. Currently, China's landcover satellite images with a resolution of 2 m and cloud cover of less than 20% have an annual land coverage of 99.8%, and the satellite images with a resolution of 16 m have a coverage of about 99.9%.

With the continuing advancement in remote sensing technology, suspended sediment monitoring methods have evolved to incorporate more mature physical models, empirical models, and semi-empirical models [2,15–17]. Among these methods, analytical models, rooted in rigorous radiative transfer theory, have a great application potential. However, the limitations of accurately determining inherent optical properties such as absorption coefficients and backscattering coefficients for various components in water bodies hinder the widespread use of this approach. Meanwhile, empirical and semi-analytical models rely on data sources that are more readily available and thus have been extensively used.

By analyzing TSM and satellite based spectral data, researchers have identified a strong correlation, which motivates the use of satellite data for TSM inversion [18]. Further, Landsat MSS and TM data have been routinely used to map TSM in the Yellow River estuary and the Gulf of Mexico [19,20]. Subsequent studies introduced hybrid spectral partitioning techniques and semi-analytical models to formulate inversion equations for TSM in diverse regions. Additionally, researchers have investigated temporal variations in the TSM along the western coast of India, considering pre- and post-monsoon TSM variations [21,22]. Empirical algorithms have revealed that the near-infrared (NIR) and thermal bands of the Landsat TM sensor exhibit strong correlations with TSM [23]. Moreover, the blue band/NIR (Near Infrared) band combination has been considered effective in constructing TSM algorithms [24]. However, due to variations in band configurations across different sensors, the selection and combination of bands for constructing TSM algorithms for the Gaofen image data remain challenging.

Along the Chinese coast, studies based on MODIS data have found that TSM variations in the southern Bohai Sea were closely related to winter littoral currents [25]. However, in the Yellow River estuary, the primary cause of the TSM dynamics was the re-suspension of shallow marine sediments [26]. Short-term suspended sediment results derived from GOCI data demonstrated a pronounced correlation between TSM and tides in strong tidal estuaries, including the Hangzhou Bay, especially during the flood and ebb tide phases of the high-tide period [1,4]. A similar pattern was corroborated by Gaofen-4 observations in the Pearl River Estuary [27]. These studies emphasized the divergent driving factors behind TSM variations in different coastal ocean regions. Especially, for the Bohai Sea region, where a comprehensive understanding of the spatiotemporal TSM variations remain elusive.

Here, we propose to employ Gaofen-6 (GF-6) and Gaofen-1 (GF-1) satellite data in combination with the in situ measured TSM and spectral data to identify suitable bands or optimal remote sensing parameters for constructing the TSM algorithm. Finally, this paper proposes a new algorithm to quantify TSM evolutions and distributions in the Bohai Sea and other regions using China's GF-1 and GF-6 images.

2. Study Area and Data Prepare

2.1. Overview of the Study Area

The Bohai Sea is a semi-enclosed inland sea at the mouth of the Yellow River in China, and it is situated in the southwest region (37°15′~38°10′N, 118°10′~119°15′E). It is a vast, shallow, and turbid water body, with an average depth of about 18 m. However, its inherent self-purification capacity is limited. With the rapid growth of industrialization and the impact of human activities, river runoff into the sea has decreased substantially. This reduction impedes the dilution of pollutants, leading to an increase in polluted sediments and severe degradation of the water environment. Meanwhile, the Bohai Sea region has rich marine biocultural, mineral farming, and salt marine resources. The total marine bioculture production alone constitutes about 40.9% of the national mariculture production in the current five-year period.

Within the Bohai Sea, the Yellow River is the primary contributor of sediment, with its estuary (37°59′–37°91′N, 119°09′–119°39′E) serving as the intersection between the Bohai Sea and Laizhou Bay. Most of the water in the Yellow River estuary is less than 10 m deep, and it becomes deeper and deeper as it moves offshore. Notably, the Yellow River is the river with the highest sediment content in the world, annually carrying substantial suspended sediment and various nutrient salts into the sea. The relatively slow flow velocity of the Yellow River can result in material deposition at its mouth.

Over the past two decades, rapid economic development and significant anthropogenic influences in the drainage area have led to increased sewage industrial wastewater discharges, and excessive runoff of agricultural fertilizerse. This has caused severe eutrophication in the seawater near the mouth of the Yellow River, as well as in the waters of Laizhou Bay and Bohai Bay. Apart from the Yellow River, which serves as the primary sediment source for the Bohai Sea, other rivers, such as the Haihe River and Liao River also contribute sediment resources to the Bohai Sea (Figure 1).

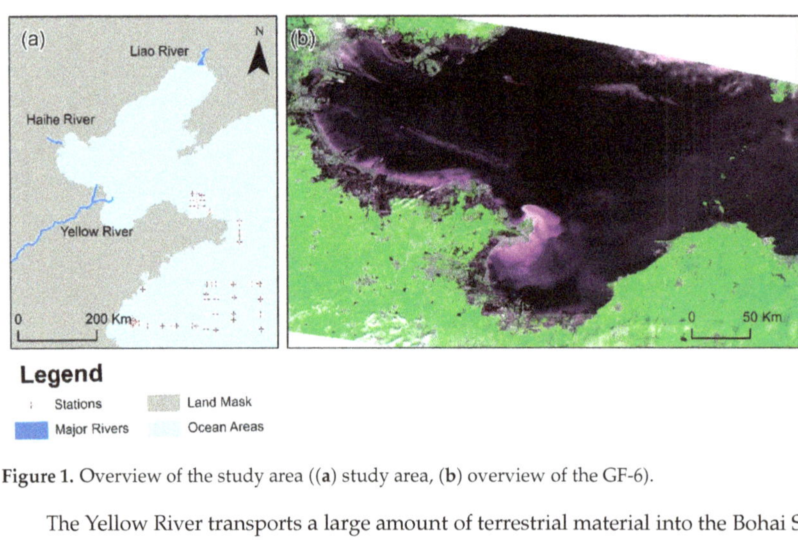

Figure 1. Overview of the study area ((**a**) study area, (**b**) overview of the GF-6).

The Yellow River transports a large amount of terrestrial material into the Bohai Sea every year. As the largest river discharging into the Bohai Sea, the Yellow River contributes over half of the total river flow, and it transports a staggering 420×10^8 m^3 sediment every year. The mouth of the Yellow River constitutes a classic high-turbidity estuary, characterized by an enormous sediment discharge into the sea. Most of the marine sediment from the Yellow River is deposited near the estuary. Thus, this paper aims to develop a TSM algorithm encompassing the entire Bohai Sea, with a specific focus on the Yellow River estuary as a representative area for investigating the seasonal variations in TSM.

2.2. Data Acquisition

GF-6 and GF-1 wide field of view (WFV) image data were utilized in this study (See Table 1 for sensor parameter settings). All the GF data were downloaded from the website of the China Resources Satellite Application Center (http://www.cresda.com, accessed on 4 September 2020, 9 March 2021, 24 July 2021, 27 November 2021 and 9 February 2021). In this study, in situ sampled TSM data were collected from 64 fixed stations and 25 continuous stations, with a total of 249 samples for analysis in 2017. Specifically, the data from 21 stations were simultaneously analyzed to obtain both TSM data and surface spectral reflectance data. The observation equipment and setup are in line with internationally recognized observational standards.

Table 1. Sensor parameter settings.

Sensor	Band	Spectral Range/μm	Resolution/m
GF-1/GF-6 WFV	Band 1 (Blue)	0.45–0.52	16
	Band 2 (Green)	0.52–0.59	
	Band 3 (Red)	0.63–0.69	
	Band 4 (NIR)	0.77–0.89	

2.2.1. Spectral Data

The Field Spec3 spectroscopy instrument with a spectral range of 350–2500 nm is manufactured by ASD (Analytical Spectral Devices, Inc., Boulder, CO, USA). Measurements were conducted following NASA ocean optics specifications, with an observation angle of (40°, 135°). Here, 40° denotes the angle between the instrument probe and the sea surface normal direction, and 135° denotes the angle between the instrument observation plane and the solar incidence plane. According to the above-water method [28], the remote sensing reflectance R_{rs} can be calculated using the following equation:

$$R_{rs} = \frac{\left(L_u - \rho_f \times L_{sky}\right) \times \rho_P}{\pi \times L_P} \quad (1)$$

where L_u, L_{sky}, and L_P represent the measured signals when the spectrometer measures the water, sky, and standard plate, respectively; ρ_f represents the Fresnel reflectance (the magnitude of its value is related to the wind speed).

Each station's data includes records of the respective station name, measurement date and time, wind speed, wind direction, temperature, pressure, cloud cover, sea state, and light variation. After the relative deviation of spectral reflectance for each station was calculated, the spectral data were meticulously screened, and stations with a relative deviation of less than 5% were selected. In this study, after screening the corresponding concentration data, 39 pairs of in situ data or ground-truth measurements were obtained, including both TSM and spectral data.

2.2.2. Processing of the In Situ TSM Data

The Whatman GF/F 0.7 μm glass fiber filter papers were used for water sample filtration. After the water samples were filtered, the membranes were placed in an oven and dried at a constant temperature of 105 °C. The first drying time was 24 h, and the subsequent drying time was only 4 h. After drying, the membranes were cooled in a desiccator for at least 4 h to return to room temperature and then weighed. Each membrane was weighed several times until the difference between two consecutive weightings was within 0.1 mg/L. The membrane was weighed several times until the difference between two consecutive weightings was within 0.1 mg/L. Finally, the concentration of TSM including inorganic and organic suspended matter was measured.

2.2.3. Remote Sensing Data Processing

Considering the spatial offset in the GF satellite imagery, this study initiated a georeferencing process upon data acquisition to pinpoint accurate ground control points, thereby complementing the high-resolution remote sensing data. Meanwhile, radiometric calibration, terrain correction, and other essential procedures were conducted. Following these steps, the FLAASH model for atmospheric correction was implemented, resulting in the derivation of remote sensing reflectance. In this case, both the mean (AVG) and standard deviation (STD) of remote sensing reflectance were computed. Specifically, only the data falling within the range of AVG–3STD to AVG + 3STD were retained, while null data were excluded through median filtering. This preparatory work greatly facilitated subsequent analysis and processing.

3. TSM Algorithm Establishment

3.1. Validation of the Atmospheric Correction

A comparison was conducted between the GF-6 Blue (B1), Green (B2), Red (B3), and NIR (B4) band remote sensing reflectance data and the FLAASH atmospherically corrected remote sensing reflectance data to evaluate the accuracy and suitability of the atmospheric correction method (Figure 2). Some detailed information regarding the validation against in situ data can be found in Hu et al. [1].

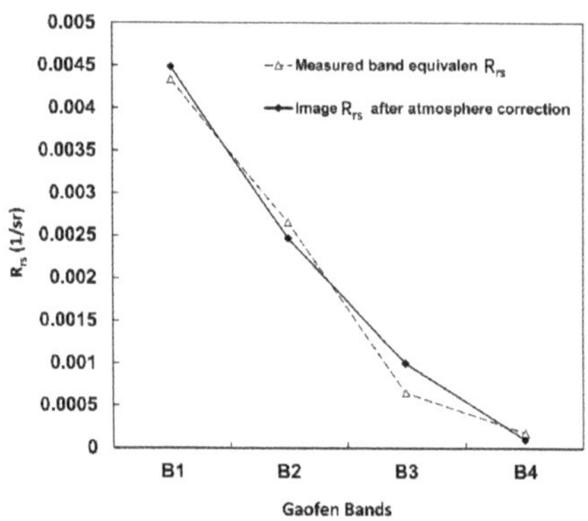

Figure 2. Comparison of atmospheric correction results with the equivalent remote sensing reflectance of the measured waveform for the Gaofen (GF) satellite images.

The results indicate that the FLAASH model yields more accurate outcomes in the B1 and B2 bands, with relative errors of 3.42% and 7.02%, respectively. In contrast, the model performs less effectively in the B3 and B4 bands, with relative errors of 53.32% and 44.64%, respectively. Thus, this study only considered the B1 and B2 bands in the algorithm construction process for subsequent research.

3.2. Characterization of Spectral Data and Equivalent Remote Sensing Reflectance

The measured remote sensing reflectance data and the GF-6/GF-1 band settings for water bodies in Bohai Bay are shown in Figure 3. There is an evident peak around 570 nm, exhibiting a 'broad peak' shape with a tendency to shift toward the red portion of the spectrum, and this is commonly referred to as the 'redshift' phenomenon. The peak reflectance in the visible wavelength range appears around 580 nm, followed by a rapid decline in reflectance beyond the peak wavelength, as wavelengths exceed 600 nm. In the range of 600–700 nm, the reduction in reflectance is relatively gradual, with the reflectance reaching its nadir around 670 nm due to phytoplankton absorption. Generally, spectral reflectance increases with the rise in suspended matter content, and the peak wavelength experiences a red-ward shift.

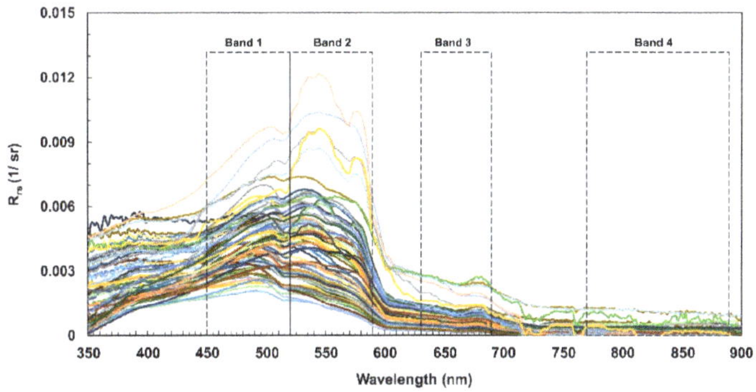

Figure 3. The measured remote sensing reflectance and GF-6/GF-1 band range in the Bohai Sea.

By utilizing the spectral response function, the measured spectral data were converted to GF equivalent remote sensing reflectance for each band. Through this process, 39 data pairs were obtained. The calculation of GF equivalent remote sensing reflectance is shown below:

$$R_{rsi} = \frac{\int_{\lambda_1}^{\lambda_2} S_i(\lambda) R_{rs}(\lambda)\lambda}{\int_{\lambda_1}^{\lambda_2} S_i(\lambda)\lambda} \tag{2}$$

where, R_{rsi} denotes the equivalent remote sensing reflectance, S_i denotes the spectral response function of the adopted image, and R_{rs} denotes the measured remote sensing reflectance at each station. The wavelength λ_1 and λ_2 correspond to the response points on both sides of the response function.

3.3. Bands Sensitive Analysis and TSM Inversion Algorithm

The Pearson correlation coefficient (Equation (3)) was calculated to analyze the correlation of remote sensing reflectance of water bodies corresponding to the TSM data, and the correlation was analyzed to select sensitive wavebands for subsequent modeling.

$$r_{xy} = \frac{\sum_{i=1}^{n} (x_i - \overline{x})(y_i - \overline{y})}{\sqrt{\sum_{i=1}^{n} (x_i - \overline{x})}^{2} \sqrt{\sum_{i=1}^{n} (y_i - \overline{y})^2}} \tag{3}$$

where, r_{xy} denotes the correlation coefficient of remote sensing reflectance; x_i and y_i are two variables, and the three x variables, namely, single band, band difference, and band ratio, are used to derive the most correlated x variable; $\overline{x}$ and $\overline{y}$ represent the means of the two variables, respectively. The closer the value of $|r_{xy}|$ is to 1, the stronger the correlation.

The results (Figure 4) demonstrate that the correlation coefficient increases significantly within the wavelength range of 480–580 nm, with the highest correlation coefficient being observed near 570 nm. Consequently, the correlation coefficient between this measured spectral data and TSM continues to increase across the B1 and B2 bands. Hence, our data modeling efforts focused on these two bands.

The in situ dataset consisting of 39 measurements was utilized to develop the TSM classification algorithm. Two-thirds of this dataset was used to model construction, while the remaining one-third was used for result verification. To establish the inversion algorithms connecting remote sensing factors and TSM, four functional models were developed: exponential, logarithmic, linear, and power functions. Subsequently, the coefficient of determination (R^2), root mean square error (RMSE), and mean relative error (MRE) were computed to evaluate the performance of the TSM algorithm.

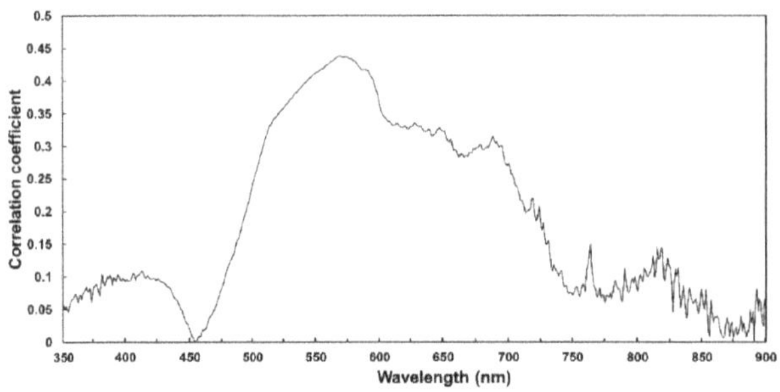

Figure 4. The correlation coefficient between the measured water bodies R_{rs} and TSM.

Specifically, based on the data of the B1 and B2 bands, difference calculations, ratio operations, and combination operations were conducted for the two bands. For example, the blue and green band ratio had a correlation coefficient of 0.81 with TSM. Then, algorithms were constructed using linear, exponential, and power models. Notably, the R^2 values in Table 2 indicated that the exponential model with B2/B1 as the remote sensing factor and the exponential model with (B2–B1) as the remote sensing factor obtained high coefficients of determination, which were 0.76 and 0.75, respectively. Further, the validation dataset was applied to these two algorithms, and the exponential model with B2/B1 as the remote sensing factor outperformed the other two models in both RMSE and MRE metrics. Thus, this algorithm was selected to calculate the TSM in the Bohai Sea in the following equation:

$$\text{TSM} = 0.17 * \text{EXP}(2.93 * \text{X}) \tag{4}$$

where X = R_{rs}(B2)/R_{rs}(B1).

Table 2. Comparison with TSM inversion algorithms.

Models	Remote Sensing Factor (x)	TSM (y) Algorithm	R^2
Linear Function	B2/B1	$y = 7.05 * x - 3.39$	0.71
Exponential Function	B2/B1	$y = 0.17 * EXP(2.93 * x)$	0.76
Power Function	B2/B1	$y = 3.31 * x^{2.43}$	0.74
Linear Function	(B2–B1)/(B2/B1)	$y = 0.014 * x + 3.68$	0.53
Exponential Function	(B2–B1)/(B2/B1)	$y = 3.23 * EXP(0.0059 * x)$	0.60
Polynomial Function	(B2–B1)/(B2/B1)	$y = 10^{-4} * x^2 + 0.03 * x + 3.58$	0.68
Linear Function	B2–B1	$y = 1781.60 * x + 3.63$	0.74
Exponential Function	B2–B1	$y = 3.16 * EXP(741 * x)$	0.75
Polynomial Function	B2–B1	$y = 340,830 * x^2 + 2092.70 * x + 3.45$	0.75

4. Results and Discussion

4.1. The Spatial TSM Distribution in the Bohai Sea

The GF-6 image data used in this study was acquired on 4 September 2020. Before mapping the TSM distribution, the normalized difference water index (NDWI) was employed to distinguish between water bodies and land. A threshold of −0.1 was set for water and land extraction, where regions with NDWI values larger than −0.1 were classified as water bodies, and those with NDWI values smaller than −0.1 were classified as land. Only the regions classified as water bodies were used for TSM mapping.

By using the inversion algorithm, the results in Figure 5 indicate that the Bohai Sea generally exhibits higher TSM concentration near the northern part of the bay as compared to the bay's mouth. Meanwhile, the TSM concentration gradually decreases with the increasing distance from the shore. The TSM concentration in the near-shore waters is the highest, and this is primarily influenced by sediment transport from the

Haihe River, which is the largest contributor to sand transport in Bohai Bay. Moreover, elevated TSM concentrations are particularly noticeable in Bohai Bay, Laizhou Bay, and the Yellow River estuary.

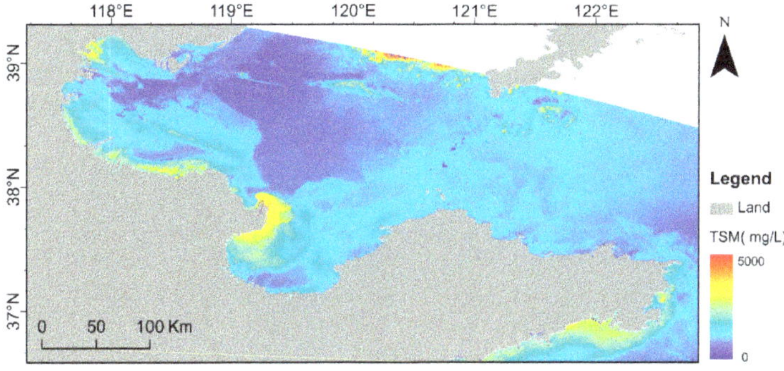

Figure 5. Distribution of TSM in the Bohai Sea.

In Laizhou Bay, the highest TSM is observed in the surface layer of the southwestern sea area, and it is mainly sourced from the Yellow River. The mouth of the Yellow River exhibits the highest TSM, with a gradual decrease toward the middle. This demonstrates that, facilitated by the shallow waters, suspended sediment mainly settles in the near-shore area, making it susceptible to environmental influences. Furthermore, the development of near-shore industries leads to the transport of pollutants by suspended sediment in the water body.

4.2. Quantifying the Spatiotemporal Distribution of TSM in the Yellow River estuary

The Yellow River estuary, a typical high-turbidity estuary in China, was selected as a case study in this paper to further investigate the spatiotemporal distribution of TSM in this region. The retrieval algorithm was adopted to map the TSM in the Yellow River estuary using four high-quality GF-1 satellite images captured in 2021. These images were acquired on 9 March 2021 (spring), 24 July 2021 (summer), 27 November 2021 (fall), and 9 February 2021 (winter). The TSM distribution in the Yellow River estuary was classified based on seasons, and the results from each image were categorized as high, medium, or low (Figure 6).

Analysis shows that there are distinct patterns in the spatial and temporal distribution of TSM in the Yellow River estuary. The dry period of the Yellow River includes January, February, March, November, and December, characterized by minimal runoff and sand transport (Figure 7). In this period, the impact of runoff on suspended sediment near the mouth of the Yellow River is limited. Meanwhile, the flood season includes July, August, and September, with significantly higher water and sand transports from the Yellow River compared to other months. In this period, the residual flow of runoff becomes a crucial factor affecting sediment transport.

The TSM distribution pattern in the Yellow River estuary appears as stripes along the coastline, mainly extending southward into Laizhou Bay. Under various influencing factors, the extent of southward extension varies, and some sediment is also distributed from the estuary to the near-shore sea north of the Xiaoqing River. During spring and autumn, the volume of sand transported from the Yellow River is significantly higher than that in winter, albeit still lower than that during the flood season. The Bohai Sea experiences windy weather during spring and autumn, characterized by variable wind directions. The suspended sediment near the mouth of the Yellow River is the result of the combined action of Yellow River runoff and submarine sediment resuspension. Tides and residual currents play a role in transporting suspended sediment, and the high suspended sediment water range tends to contract toward the shore compared to that in the winter.

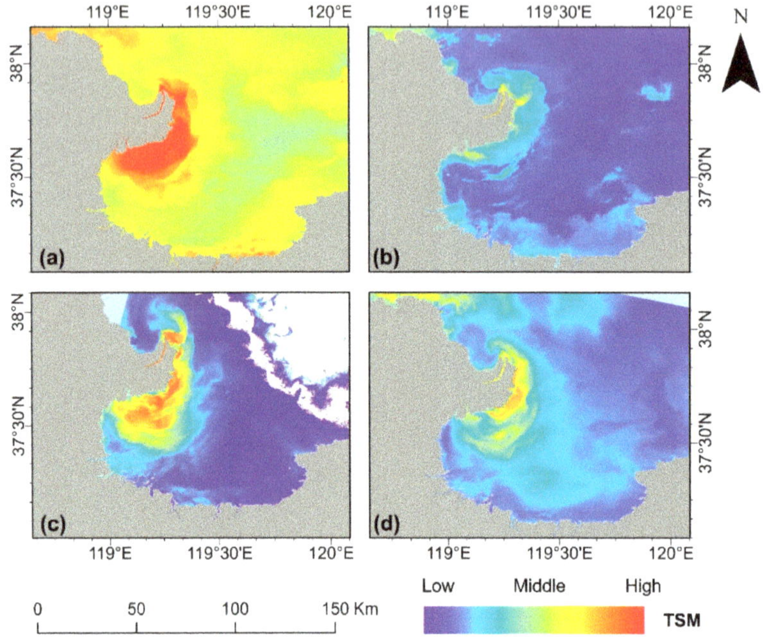

Figure 6. The seasonal variation of TSM in the Yellow River estuary in 2021: (**a**) spring, (**b**) summer, (**c**) fall, (**d**) winter.

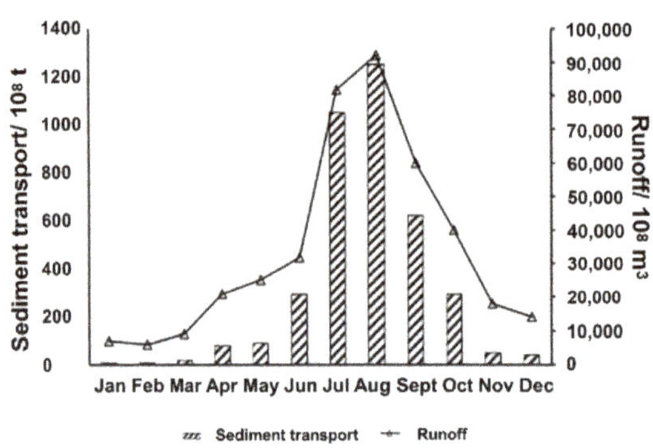

Figure 7. The statistics of incoming water runoff and sediment transport near the mouth of the Yellow River at the Lijin hydrological station in 2020.

In recent years, many scholars have studied the Bohai Sea TSM models and retrieval results using different satellite data. For instance, Li et al. used GOCI images to retrieve the surface suspended sediment concentration in the Bohai Sea [29]. Cui et al. constructed a remote sensing inversion model of TSM using 555 and 670 nm band remote sensing reflectance based on the field-measured or in situ TSM and spectral data in Bohai Bay and Laizhou Bay [30]. Li et al. utilized Landsat 8 Operational Land Imager data to estimate the variation of TSM from 2013 to 2019, and a simple empirical algorithm was developed to map the TSM distribution with an MRE of 33.12% [31]. Compared with different sensor results, the Bohai Sea TSM retrieval distribution in this study is consistent with others. The annual average distribution of TSM suggests that the high turbid area of TSM Is mainly distributed around the Yellow River estuary, northwest of Laizhou Bay and the southern

Bohai Bay; the impact of river runoff is mainly concentrated in the estuary. Meanwhile, the study indicates that the seasonal distribution of TSM varies greatly, and the range and intensity of TSM in winter are higher than those in summer.

4.3. Analysis of Factors Driving TSM Variations in the Yellow River Estuary

4.3.1. Wind, Wind Waves, and Storm Surge

During winter, the Bohai Sea usually experiences prevailing northerly and northwesterly winds with high wind speeds, occasional storm surges, and wave-induced sediment resuspension in shallow nearshore areas. This wave resuspension is an important source of suspended sediment in the sea. The strong north winds, combined with southward waves, generate a southward coastal current in the western delta of Laizhou Bay. This current transports suspended sediment southwestward along the west coast of Laizhou Bay and eventually toward the Bohai Strait. Under the influence of tides, the high-concentration water transported to Laizhou Bay is further transported to distant seas.

Through the generation of rift currents, wind and waves not only disturb shallow sediment but also carry fine sand directly from the shallow sea. This dynamic factor plays a crucial role in sediment movement and transport in the shallow sea. Meanwhile, wind and waves enhance the sea current's ability to carry sediment. The prevailing wave direction in the Yellow River estuary is east to northeast (with a frequency of 29%), and the secondary wave direction is north to northwest (with a frequency of 18%). The strong wave directions include north-northeast and northeast, and secondary strong wave directions include north-northwest and northwest. The wind field affects the northern bank section of the Yellow River, making it more efficient in transporting sand to the west than the eastern bank section to the south. Also, sand transport capacity increases from the upper bay to the mouth of the bay. In the flood season, the sand transport capacity of the northern bank section to the west strengthens, while that of the eastern bank section to the south weakens.

Storm surges are common in Bohai Bay and Laizhou Bay, and they mainly occur in April and September due to a combination of regional wind patterns and geographical factors. These surges are mostly caused by extended periods of high winds, particularly northeasterly winds with speeds ranging from 17.2 to 24.5 m/s. The storm surges lead to a sudden rise in tide levels, typically 2–3 m higher than the normal high tide level. Meanwhile, storm surges have a significant impact on the distribution of suspended sediment, especially the resuspending bottom sediment in shallow nearshore areas.

4.3.2. Human Activities

Human activities mainly affect the distribution and dispersion of incoming sediment through changes in runoff volume and sand transport. Over the past 50 years, the Yellow River has experienced a reduction in marine sediment input, due to both climate change and anthropogenic activities within the basin.

Anthropogenic activities have significantly affected the Yellow River basin through soil and water conservation, irrigation, reservoir construction, etc., leading to a decline in marine sediment. Since the 1950s, comprehensive watershed management and soil conservation efforts have changed sub-basin conditions, reduced erosion intensity, and significantly decreased the amount of sediment reaching the estuary.

The main driving factor for reduced runoff is irrigation diversion. In the water diversion process, irrigation diversion will also divert sediment and reduce the amount of sediment inflow into the sea. When abundant water is diverted in non-abundant periods, the runoff into the sea can be interrupted. Since the 1970s, there have been long-term interruptions in downstream flow, further reducing sediment input.

During the 1990s, the Sanmenxia reservoir in the middle reaches of the Yellow River continued to release muddy water, which changed sediment distribution throughout the year. Although the runoff process remained largely unchanged, this practice changed sediment distribution, particularly during the flood season.

5. Conclusions

In this paper, by using the spectral and measured suspended particulate matter concentration data in the Bohai Sea in 2017, a TSM retrieval algorithm was established based on the Gaofen (GF) satellite images, and the spatiotemporal variations of the TSM in Yellow River estuary in 2021 were analyzed:

(1) We quantified the evolution and spatiotemporal TSM variations in the Bohai Sea using a optimal classification algorithm developed using in situ measured TSM data applied on the GF-6/GF-1 satellite image data. The peak of the measured remote sensing reflectance in the Bohai Sea region appears near the wavelength of 580 nm. Based on the correlation analysis between the GF Band 1, Band 2, Band 3, and Band 4 equivalent remote sensing reflectance and the in situ measured-TSM and the atmospheric correction accuracy evaluation, an exponential model was established by taking the ratio of Band 1 and Band 2 equivalent remote sensing reflectance as the remote sensing factor, and the R^2 value of the model was 0.76. The inversion results suggest that the model can improve the characterization of the spatiotemporal distribution of TSM in the Bohai Sea region using GF images.

(2) The spatiotemporal variations and the pattern distributions of the Yellow River estuary TSM was obvious. High TSM of water bodies was mainly concentrated in Bohai Bay, Laizhou Bay, and the Yellow River estuary near the sea, and the TSM was high near-shore and low offshore. The TSM in the Yellow River estuary sea showed an overall time distribution of being high in the spring and winter and low in summer and autumn.

(3) Yellow River runoff can affect the TSM in the Yellow River estuary. The Yellow River estuary carries a large amount of sediment into the Bohai Sea every year, and the area near the mouth of the Yellow River is affected by the Yellow River runoff; as the flow of the Yellow River runoff rises, the TSM concentration increases, so does the scope of influence. Bohai Bay and Laizhou Bay are less affected by the Yellow River runoff. This is because the sediment carried by the Yellow River runoff enters the Bohai Sea and then falls rapidly and is deposited mostly in the area near the mouth of the Yellow River; however, the deposited sediment is redistributed under the action of wind, tide, waves, and currents, and it can be transported to Bohai Bay, Laizhou Bay and other areas, and even into the Yellow Sea.

Author Contributions: Conceptualization, Z.Y., B.W. and B.Z.; data curation, J.Z., Z.C. and Z.Y.; investigation, C.M. and Q.S.; methodology, X.Y. and B.W.; writing—original draft, Z.Y. and J.Z.; writing—review and editing, C.K.S. and X.Y.; re-writing and re-mapping, Y.H.; funding, Z.Y. and B.Z. All authors have read and agreed to the published version of the manuscript.

Funding: This research was funded by the National Natural Science Foundation of China (#41206169).

Data Availability Statement: Enquiries regarding in situ data availability should be directed to the authors.

Conflicts of Interest: The authors declare no conflict of interest.

References

1. Hu, Y.; Yu, Z.; Zhou, B.; Li, Y.; Yin, S.; He, X.; Peng, X.; Shum, C.K. Tidal-Driven Variation of Suspended Sediment in Hangzhou Bay Based on GOCI Data. *Int. J. Appl. Earth Obs. Geoinf.* **2019**, *82*, 101920. [CrossRef]
2. He, X.; Bai, Y.; Pan, D.; Huang, N.; Dong, X.; Chen, J.; Chen, C.T.A.; Cui, Q. Using Geostationary Satellite Ocean Color Data to Map the Diurnal Dynamics of Suspended Particulate Matter in Coastal Waters. *Remote Sens. Environ.* **2013**, *133*, 225–239. [CrossRef]
3. He, X.; Bai, Y.; Pan, D.; Tang, J.; Wang, D. Atmospheric Correction of Satellite Ocean Color Imagery Using the Ultraviolet Wavelength for Highly Turbid Waters. *Opt. Express* **2012**, *20*, 20754. [CrossRef] [PubMed]
4. Chen, Q.; Zhou, B.; Yu, Z.; Wu, J.; Tang, S. Detection of the Minute Variations of Total Suspended Matter in Strong Tidal Waters Based on GaoFen-4 Satellite Data. *Remote Sens.* **2021**, *13*, 1339. [CrossRef]
5. Zhang, D.; Pang, C.; Liu, Z.; Jiang, J. Winter and Summer Sedimentary Dynamic Process Observations in the Sea Area off Qinhuangdao in the Bohai Sea, China. *Front. Earth Sci.* **2023**, *11*, 1097033. [CrossRef]

6. Zhou, Z.; Bian, C.; Wang, C.; Jiang, W.; Bi, R. Quantitative Assessment on Multiple Timescale Features and Dynamics of Sea Surface Suspended Sediment Concentration Using Remote Sensing Data. *J. Geophys. Res. Oceans* **2017**, *122*, 8739–8752. [CrossRef]
7. Kong, J.L.; Sun, X.M.; Wong, D.W.; Chen, Y.; Yang, J.; Yan, Y.; Wang, L.X. A Semi-Analytical Model for Remote Sensing Retrieval of Suspended Sediment Concentration in the Gulf of Bohai, China. *Remote Sens.* **2015**, *7*, 5373–5397. [CrossRef]
8. Li, P.; Ke, Y.; Bai, J.; Zhang, S.; Chen, M.; Zhou, D. Spatiotemporal Dynamics of Suspended Particulate Matter in the Yellow River Estuary, China during the Past Two Decades Based on Time-Series Landsat and Sentinel-2 Data. *Mar. Pollut. Bull.* **2019**, *149*, 110518. [CrossRef]
9. Liu, H.; Li, Q.; Shi, T.; Hu, S.; Wu, G.; Zhou, Q. Application of Sentinel 2 MSI Images to Retrieve Suspended Particulate Matter Concentrations in Poyang Lake. *Remote Sens.* **2017**, *9*, 761. [CrossRef]
10. Wang, X.; Wen, Z.; Liu, G.; Tao, H.; Song, K. Remote Estimates of Total Suspended Matter in China's Main Estuaries Using Landsat Images and a Weight Random Forest Model. *ISPRS J. Photogramm. Remote Sens.* **2022**, *183*, 94–110. [CrossRef]
11. Zheng, Z.; Li, Y.; Guo, Y.; Xu, Y.; Liu, G.; Du, C. Landsat-Based Long-Term Monitoring of Total Suspended Matter Concentration Pattern Change in the Wet Season for Dongting Lake, China. *Remote Sens.* **2015**, *7*, 13975–13999. [CrossRef]
12. Wang, X.; Song, K.; Liu, G.; Wen, Z.; Shang, Y.; Du, J. Development of Total Suspended Matter Prediction in Waters Using Fractional-Order Derivative Spectra. *J. Environ. Manag.* **2022**, *302*, 113958. [CrossRef] [PubMed]
13. Xu, J.; Gao, C.; Wang, Y. Extraction of Spatial and Temporal Patterns of Concentrations of Chlorophyll-a and Total Suspended Matter in Poyang Lake Using GF-1 Satellite Data. *Remote Sens.* **2020**, *12*, 622. [CrossRef]
14. Cheng, Z.; Wang, X.H.; Paull, D.; Gao, J. Application of the Geostationary Ocean Color Imager to Mapping the Diurnal and Seasonal Variability of Surface Suspended Matter in a Macro-Tidal Estuary. *Remote Sens.* **2016**, *8*, 244. [CrossRef]
15. Sokoletsky, L.; Fang, S.; Yang, X.; Wei, X. Evaluation of Empirical and Semianalytical Spectral Reflectance Models for Surface Suspended Sediment Concentration in the Highly Variable Estuarine and Coastal Waters of East China. *IEEE J. Sel. Top. Appl. Earth Obs. Remote Sens.* **2016**, *9*, 5182–5192. [CrossRef]
16. Teixeira, L.C.; Mariani, P.P.; Pedrollo, O.C.; dos Reis Castro, N.M.; Sari, V. Artificial Neural Network and Fuzzy Inference System Models for Forecasting Suspended Sediment and Turbidity in Basins at Different Scales. *Water Resour. Manag.* **2020**, *34*, 3709–3723. [CrossRef]
17. Hu, Z.; Pan, D.; He, X.; Bai, Y. Diurnal Variability of Turbidity Fronts Observed by Geostationary Satellite Ocean Color Remote Sensing. *Remote Sens.* **2016**, *8*, 147. [CrossRef]
18. Holyer, R.J. Toward Universal Multispectral Suspended Sediment Algorithms. *Remote Sens. Environ.* **1978**, *7*, 323–338. [CrossRef]
19. Aranuvachapun, S.; Walling, D.E. Landsat-MSS Radiance as a Measure of Suspended Sediment in the Lower Yellow River (Hwang Ho). *Remote Sens. Environ.* **1988**, *25*, 145–165. [CrossRef]
20. Jensen, J.R.; Kjerfve, B.; Ramsey, E.W.; Magill, K.E.; Medeiros, C.; Sneed, J.E. Remote Sensing and Numerical Modeling of Suspended Sediment in Laguna de Terminos, Campeche, Mexico. *Remote Sens. Environ.* **1989**, *28*, 33–44. [CrossRef]
21. Sathasivam, S.; Kankara, R.S.; Murugan, U.; Gunasekaran, P.; Palanisamy, T.; James, R.A. Influence of Suspended Sediment Loads on Coastal Hydrodynamics at Vengurla and Ratnagiri Part of the Western Coast of India. *Arab. J. Geosci.* **2021**, *14*, 2231. [CrossRef]
22. Prasad, A.K.; Singh, R.P. Chlorophyll, Calcite, and Suspended Sediment Concentrations in the Bay of Bengal and the Arabian Sea at the River Mouths. *Adv. Space Res.* **2010**, *45*, 61–69. [CrossRef]
23. Baban, S.M.J. The Use of Landsat Imagery to Map Fluvial Sediment Discharge into Coastal Waters. *Mar. Geol.* **1995**, *123*, 263–270. [CrossRef]
24. Duane Nellis, M.; Harrington, J.A.; Wu, J. Remote Sensing of Temporal and Spatial Variations in Pool Size, Suspended Sediment, Turbidity, and Secchi Depth in Tuttle Creek Reservoir, Kansas: 1993. *Geomorphology* **1998**, *21*, 281–293. [CrossRef]
25. Bi, N.; Yang, Z.; Wang, H.; Fan, D.; Sun, X.; Lei, K. Seasonal Variation of Suspended-Sediment Transport through the Southern Bohai Strait. *Estuar. Coast. Shelf Sci.* **2011**, *93*, 239–247. [CrossRef]
26. Liu, K.; Zhang, D.; Xiao, X.; Cui, L.; Zhang, H. Occurrence of Quinotone Antibiotics and Their Impacts on Aquatic Environment in Typical River-Estuary System of Jiaozhou Bay, China. *Ecotoxicol. Environ. Saf.* **2020**, *190*, 109993. [CrossRef]
27. Lu, S.; Deng, R.; Liang, Y.; Xiong, L.; Ai, X.; Qin, Y. Remote Sensing Retrieval of Total Phosphorus in the Pearl River Channels Based on the GF-1 Remote Sensing Data. *Remote Sens.* **2020**, *12*, 1420. [CrossRef]
28. Tang, J.; Tian, G.; Wang, X.; Wang, X.; Song, Q. The Methods of Water Spectra Measurement and Analysis I: Above-Water Method. *J. Remote Sens.* **2004**, *8*, 37–44.
29. Yangdong, L.; Zhenyu, W.; Liang, C. A new method for estimating surface suspended sediment concentration in the Bohai Sea based on GOCI images. *Mar. Lake Bull.* **2020**, *44*, 80–88.
30. Cui, T.W.; Zhang, J.; Wang, K.; Wei, J.W.; Mu, B.; Ma, Y.; Zhu, J.H.; Liu, R.J.; Chen, X.Y. Remote sensing study on the distribution of suspended solids in the Bohai Sea. *J. Oceanogr.* **2009**, *31*, 10–18.
31. Li, J.; Hao, Y.; Zhang, Z.; Li, Z.; Yu, R.; Sun, Y. Analyzing the distribution and variation of Suspended Particulate Matter (SPM) in the Yellow River Estuary (YRE) using Landsat 8 OLI. *Reg. Stud. Mar. Sci.* **2021**, *48*, 102064. [CrossRef]

remote sensing

MDPI

Article

Estimation of Total Phosphorus Concentration in Lakes in the Yangtze-Huaihe Region Based on Sentinel-3/OLCI Images

Xiaoyang Wang [1,2,3], Youyi Jiang [1], Mingliang Jiang [2], Zhigang Cao [2], Xiao Li [1], Ronghua Ma [2,4,5], Ligang Xu [2] and Junfeng Xiong [2,3,*]

1 College of Geomatics, Xi'an University of Science and Technology, Xi'an 710054, China; xywang@stu.xust.edu.cn (X.W.)
2 Key Laboratory of Watershed Geographic Sciences, Nanjing Institute of Geography and Limnology, Chinese Academy of Sciences, Nanjing 210008, China; jiangmingliang18@mails.ucas.ac.cn (M.J.); zgcao@niglas.ac.cn (Z.C.); rhma@niglas.ac.cn (R.M.); lgxu@niglas.ac.cn (L.X.)
3 International Research Center of Big Data for Sustainable Development Goals, Beijing 100094, China
4 University of Chinese Academy of Sciences, Nanjing 211135, China
5 Jiangsu Collaborative Innovation Center of Regional Modern Agriculture and Environmental Protection, Huaiyin Normal University, Huaian 223001, China
* Correspondence: jfxiong@niglas.ac.cn

Abstract: Total phosphorus (TP) concentration is a crucial parameter to assess eutrophication in lakes. As one of the most concentrated regions for freshwater lakes, the Yangtze-Huaihe region plays a significant role in monitoring TP concentrations for the sustainable utilisation of China's water resources. In this study, a TP concentration estimation model suitable for large-sized lake groups was developed using a combination of measured and remote sensing data powered by advanced machine learning algorithms. Compared to traditional empirical models, the model developed in this study demonstrates significant accuracy in fitting (R^2 = 0.53, RMSE = 0.08 mg/L, MAPE = 34.20%). Moreover, the application of this model to lakes in the Yangtze-Huaihe region from 2017 to 2022 has been conducted. The multi-year average TP concentration was 0.18 mg/L. Spatial distribution analyses showed that total phosphorus concentrations were higher in small lakes. In terms of temporal changes, the interannual decreases in total phosphorus concentrations were 0.02 mg/L, 0.01 mg/L, and 0.01 mg/L for small, medium, and large lakes, respectively. We also found that large lakes typically exhibited a "high in spring and summer, low in autumn and winter" pattern until 2020, but transitioned to a "high in summer and autumn, low in spring and winter" pattern after 2020 due to the removal of closed fish nets, which were having a significant impact on the lake ecosystem. Other lakes in the area consistently showed a pattern of "high in spring and summer, low in autumn and winter" during the six-year period. These findings may provide useful references and suggestions for the environmental protection and management of lakes in China.

Keywords: Yangtze-Huaihe lakes; total phosphorus; machine learning; Sentinel-3 OLCI

1. Introduction

The ongoing increase in lake eutrophication has emerged as one of the most significant issues in aquatic environments due to global warming, industrialisation, and intensive agriculture. Approximately 40% of the world's lakes experience severe issues, such as ecosystem and biodiversity deterioration, because of severe eutrophication [1]. Consequently, long-term and extensive monitoring of lake nutrient content is crucial for regulating lake eutrophication.

Currently, various techniques are used to retrieve water quality parameters, including analytical [2], semi-empirical [3], and empirical methods [4]. However, machine learning algorithms have demonstrated unmatched advantages over conventional approaches thanks to the quick development of artificial intelligence technology. By leveraging large datasets, these algorithms can establish complex relationships between water quality parameters and multiple variables, enabling the estimation of crucial parameters, such as the Secchi disk depth (SDD) [5], chlorophyll-*a* (Chl-*a*) [6], total suspended matter (TSM) [7], and chromophoric dissolved organic matter (CDOM) [8], and they have been successfully applied to the long-term monitoring of multiple lakes on a large spatial scale [9–11]. Nonetheless, it is crucial to note that, in addition to these optical characteristic parameters, other non-optical parameters are also closely related to lake eutrophication and play a crucial role in assessing the safety of the water environment, such as total nitrogen (TN), total phosphorus (TP), ammonia nitrogen (NH_3-N), and the permanganate index (COD_{Mn}) [12,13]; however, there are some difficulties in estimating these indices in turbid inland water bodies.

Among these parameters, TP is one of the primary nutrients contributing to lake eutrophication [14]. Extensive research has used remote sensing techniques to construct models for estimating TP concentrations in water bodies. For instance, Baban et al. [15] employed Landsat 5 TM satellite data to establish a regression relationship between remote sensing and in-situ measurements for TP concentration estimation. Wang et al. [16] utilized MODIS images to develop a regression model to estimate TP concentration in Hulun Lake. Additionally, researchers have explored indirect TP estimation by leveraging the relationships between TP and optical water quality parameters, such as TSM, Chl-*a*, and CDOM [17–19]. In recent years, machine learning algorithms have gained prominence in TP concentration estimation, surpassing traditional methods. For instance, García Nieto et al. [20] employed Support Vector Machines (SVM) to construct a TP estimation model for Englishmen Lake in Spain and achieved an impressive coefficient of determination (R^2) of 0.90. Lee et al. [21] applied four different machine learning algorithms to estimate TP concentrations in Euiam Lake, Korea, yielding R^2 values above 0.70. Xiong et al. [22] developed the eXtreme Gradient Boosting (XGBoost) algorithm specifically tailored for TP concentration estimation in Taihu Lake, achieving an R^2 of 0.60 and an RMSE of 0.07 mg/L. Among the various machine-learning techniques, XGBoost, an optimized gradient boosting algorithm, has demonstrated superior performance, self-learning capabilities, and predictive power [23]. Cui et al. [24] identified XGBoost as the optimal choice for estimating Chl-*a* concentration in Nansi Lake using hyperspectral data. Similarly, Hu et al. [25] compared six machine learning methods and concluded that XGBoost was the most effective model for predicting TP concentration in Taihu Lake. The traditional direct inversion method is simple and has good results, but it is not suitable for lakes with complex relationships, while the indirect inversion is detrimental to the model accuracy, so currently these two inversion methods are only applied to a single lake. In contrast, machine learning has higher application ability on a large regional scale, which is not only well proven in parameters with optical characteristics, such as chlorophyll-*a* [6] and Secchi disk depth [26], but also shows good robustness in the inversion of non-optical parameters, such as dissolved CO_2 [27,28]. Although some scholars believe that machine learning has the potential for application in the remote sensing estimation of TP for lakes at a large regional scale [22], a complete proof and application have not been given.

In conclusion, there is a lack of machine learning models that are widely applicable to the estimation of TP concentrations in lake populations. The Yangtze-Huaihe region is recognised as one of the most densely populated freshwater lakes areas in China and represents a prominent region with respect to lake eutrophication issues [29]. Monitoring TP concentrations in the Yangtze-Huaihe region is crucial for effective water management within the region. Therefore, this study aimed to utilise remote sensing techniques and synchronous satellite observations to establish a machine learning-based algorithm for estimating TP concentrations in the Yangtze-Huaihe Lake group. This study further aims to analyze the spatiotemporal variations in the TP and investigate the underlying driving mechanisms, thus providing scientific references for the monitoring and management of lake pollution and eutrophication in the Yangtze-Huaihe region. The objectives of this study were as follows: (1) to develop a machine learning model for estimating TP concentrations in the lakes of the Yangtze-Huaihe region and (2) to analyse the spatiotemporal distribution patterns of TP concentrations in the lakes of the Yangtze-Huaihe region and explore the associated driving mechanisms.

2. Materials and Methods

2.1. Study Area

The study area encompasses the Yangtze-Huaihe region, which lies between the lower reaches of the Yangtze River and the Huai River. It is located at approximately 28°21′–33°40′N, 115°42′–121°1′E and is characterized by a subtropical monsoon climate with mild seasons and significant interannual variations in rainfall. The geographical features of the region are predominantly formed by the sedimentation of the Huai River and the Yangtze River, with higher elevations surrounding the lower-lying central area, where the altitude is less than 10 m. The Yangtze-Huaihe region is one of the areas in China with the highest density of lakes, which is closely associated with the scarcity of water resources in the country [30]. Moreover, the region has a high population density, well-developed freshwater fisheries, and a substantial amount of wastewater discharge. The intensive developmental activities, such as land reclamation and enclosed aquaculture, in the lakes have directly or indirectly contributed to increased nutrient loading [31,32], water quality deterioration, frequent occurrences of cyanobacterial blooms, and severe eutrophication, which pose significant challenges to the sustainable development of the Yangtze-Huaihe region. Therefore, for this study, we have selected shallow lakes within the Yangtze-Huaihe region, with an area larger than 20 km^{2,} as representative lakes for long-term analysis of TP concentrations, as seen in Figure 1. (Please refer to the area of the lake Supplementary Material Table S1 for details).

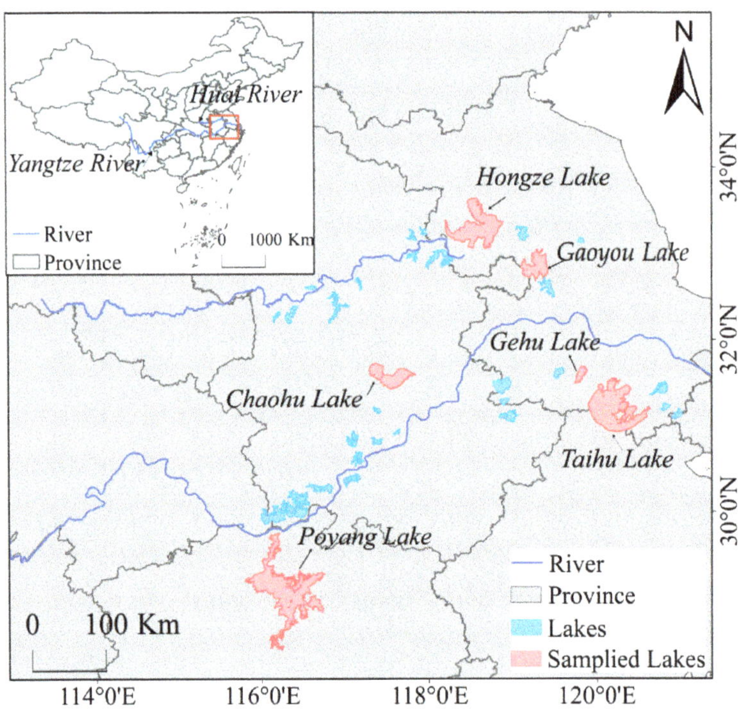

Figure 1. Spatial distribution of 36 lakes (>20 km^2) in Yangtze-Huaihe region (The red box is the Yangtze-Huaihe region).

2.2. Data

2.2.1. In Situ Data

The research team conducted field surveys in six medium to large-sized lakes in the Yangtze-Huaihe region between 2017 and 2022. The sampling points were evenly distributed across the lakes, and specific information is presented in Table 1. 500 mL brown bottles were used to collect surface mixed water samples, which were then immediately transported to the lab with ice packs in light-protected coolers. Filtration experiments were conducted on the same day, followed by refrigeration for chemical analysis. Global positioning system (GPS) devices were used to record the coordinates of the sampling points, and measurement time and weather conditions. TP concentrations were determined using the acid digestion–molybdenum antimony anti-colorimetric method.

Table 1. Summary of Sampling Point Information.

Lake	Sampling Points	TP (mg/L) Mean ± STD	Sampling Time
Taihu Lake	41	0.23 ± 0.16	29 April 2017/30 April 2017/14 November 2019/7 November 2020
Chaohu Lake	24	0.18 ± 0.10	23 August 2017/16 July 2018
Gaoyou Lake	12	0.11 ± 0.04	12 November 2021
Hongze Lake	18	0.10 ± 0.06	14 October 2022
Gehu Lake	8	0.10 ± 0.01	18 July 2018
Poyang Lake	10	0.11 ± 0.05	14 April 2017/12 July 2018

2.2.2. Satellite Data and Preprocessing

The imagery used in this study is obtained from Sentinel-3 images. The Ocean and Land Color Instrument (OLCI) is a multispectral radiometer carried by Sentinel-3A (launched in 2016) and Sentinel-3B (launched in 2018). It has 21 spectral channels in the range of 400–1020 nm, including 16 hydro-chromatic bands and a high signal-to-noise ratio, with a spatial resolution of about 300 m. The data were downloaded from the European Space Agency website (https://scihub.copernicus.eu/dhus/#/home, accessed on 28 June 2022).

Currently, there are various methods for atmospheric correction of water bodies. Among them, the dark spectrum fitting (DSF) atmospheric correction algorithm of Acolite has been proven suitable for the use of Sentinel-3 OLCI imagery in turbid water bodies [33]. Therefore, in this study, the Acolite atmospheric correction algorithm was employed. Subsequently, masking of algal blooms, clouds, and cloud shadows was performed, and terrain correction was conducted using SNAP 8.0.0 software. Spectral reflectance of the sampling points was extracted based on their latitude and longitude coordinates using ENVI 5.3. These data were used for the construction and validation of the TP concentration inversion model.

2.2.3. Auxiliary Data

Meteorological data was incorporated to investigate the influence of natural factors on TP concentration in the Yangtze-Huaihe region. Rainfall data utilized the daily global precipitation dataset provided by NASA (GPM_3IMERGHH_06, https://disc.gsfc.nasa.gov, accessed on 15 February 2023). The spatial resolution of this dataset is $0.1^{\circ} \times 0.1^{\circ}$. Temperature data was obtained from the ERA5-Land dataset (https://cds.climate.copernicus.eu/cdsapp#!/dataset/reanalysis-era5-land-monthly-means?tab=overview, accessed on 15 February 2023) and expressed in Kelvin (K). In this study, the temperature values were converted from Kelvin to Celsius (°C) by subtracting 273.15 from the original data.

2.3. Model Development

2.3.1. Modeling Set Construction

The variable set comprises OLCI image bands and remote sensing indices.

(1) OLCI Image Bands: After performing Acolite atmospheric correction on the OLCI images, 17 bands are generated at wavelengths of 400 nm, 412 nm, 443 nm, 490 nm, 510 nm, 560 nm, 620 nm, 665 nm, 674 nm, 685 nm, 709 nm, 754 nm, 768 nm, 779 nm, 865 nm, 884 nm, and 1016 nm. These bands are denoted as b1 to b17, and transformations such as reciprocal, exponential, square, and square root are applied to each of the 17 bands.

(2) Remote Sensing Indices: Hu et al. [34] developed the Floating Algae Index (FAI) for monitoring algal blooms in water bodies, which have a close relationship with nutrient concentrations [22]. Due to the absence of the Short-Wave Infrared (SWIR) band in OLCI, the Near-Infrared (NIR) band is used as a substitute, resulting in the creation of the Adjusted FAI (AFAI) [35]. Additionally, the Cyanobacteria and Macrophytes Index (CMI) is employed to differentiate cyanobacterial bloom areas and aquatic vegetation areas, while the Turbid Water Index (TWI) is used to assess turbid water bodies [36]. The formulas for calculating these indices are as follows:

$$AFAI = R_{NIR} - [R_{RED} + (R_{NIR2} - R_{RED}) \times (\lambda_{NIR1} - \lambda_{RED})/(\lambda_{NIR2} - \lambda_{RED})] \quad (1)$$

$$CMI = R_{GREEN} - [R_{BLUE} + (R_{SWIR} - R_{BILE}) \times (\lambda_{GREEN} - \lambda_{BLUE})/(\lambda_{SWIR} - \lambda_{BLUE})] \quad (2)$$

$$TWI = R_{RED} - R_{SWIR} \quad (3)$$

Among these, λ_{BLUE}, λ_{GREEN}, λ_{RED}, λ_{NIR1}, λ_{NIR2}, λ_{SWIR} correspond to the center wavelengths of 430 nm, 560 nm, 665 nm, 754 nm, 865 nm, and 1016 nm in OLCI, respectively.

Based on the above, five approaches are constructed using the original bands and transformations, such as reciprocal, exponential, square, and square root, in Table 2. In addition, the AFAI, CMI, and TWI indices are included as variables. In total, each scheme consists of 20 variables, and XGBoost models are built for each approach. Each approach's input variables are marked with an ID (please refer to Supplementary Material Table S2 for details).

Table 2. Approach of input variables.

Approaches	Input Variables
1	AFAI, CMI, TWI, b1~b17
2	AFAI, CMI, TWI, 1/b1~1/b17
3	AFAI, CMI, TWI, e^{b1}~e^{b17}
4	AFAI, CMI, TWI, $(b1)^2$~$(b17)^2$
5	AFAI, CMI, TWI, $\sqrt{b1}$~$\sqrt{b17}$

2.3.2. Algorithm

The XGBoost algorithm was employed to establish a remote sensing algorithm for inversing estimation of phosphorus concentration in the Yangtze-Huaihe region. XGBoost, proposed by Chen et al. [37] in 2016, is an algorithm based on gradient boosting decision trees optimization. The core idea of this algorithm is to utilize gradient information to progressively improve the predictive capability of the model by constructing decision trees. In each iteration, the XGBoost algorithm calculates residuals based on the model's predictions from the previous round and then fits new decision trees to these residuals. Through continuous iterations, new decision trees are added, and their prediction results accumulate. In addition, it controls the model's complexity and prevents overfitting, while improving model runtime speed by introducing regularization terms. In this study, the XGBoost algorithm was implemented using the Scipy library in Python 3.9. To determine the structure of the XGBoost model, multiple hyperparameters need to be adjusted, including learning rate, maximum tree depth, subsample, and regularization. After debugging, the final parameters were set as learning rate of 0.02, max tree depth of 7 layers, subsample of 0.8, and regularization of 0.01, with other parameters set to default values.

Before constructing the algorithm, the importance index of input variables, referred to as IncMSE (increase in mean squared error), needs to be calculated. A higher IncMSE value indicates greater importance of the variable. The accuracy of IncMSE increases with a larger number of random iterations, but at the cost of longer computation time. In this study, 300 iterations were selected. For each scheme, the variables were sorted in descending order based on the calculated IncMSE scores, and then sequentially inputted into the algorithm. The approach with higher accuracy was selected based on model evaluation metrics through cross-validation.

2.4. Model Evaluation

2.4.1. K-Fold Cross-Validation

K-fold cross-validation is widely regarded as a common method for assessing the performance and reliability of machine learning models [38]. It effectively addresses the issue of parameter tuning and prevents information leakage from influencing the model's hyperparameters. The entire dataset is divided into K equally sized subsets, and in each iteration, one subset is used as the validation set while the remaining K-1 subsets are used for training. This process is repeated K times, allowing for model training and evaluation. The average of the evaluation metrics obtained from the K iterations is considered as the final evaluation metric.

2.4.2. Algorithm Accuracy Evaluation

The model accuracy is evaluated using the following metrics: coefficient of determination (R^2), root mean squared error (RMSE), and mean absolute percentage error (MAPE). These metrics provide a comprehensive assessment of the model's performance.

$$R^2 = \frac{\left(\sum (x_i - \overline{x})(y_i - \overline{y})\right)^2}{\left(\sum (x_i - \overline{x})^2 \sum (y_i - \overline{y})^2\right.} \tag{4}$$

$$RMSE = \sqrt{\frac{1}{n-1}\sum_{i-1}^{n}(x_i - y_i)^2} \tag{5}$$

$$MAPE(\%) = \frac{1}{n}\sum_{i-1}^{n}\frac{|y_i - x_i|}{y_i} \times 100\% \tag{6}$$

2.5. Spatial-Temporal Distribution Analysis

A total of 882 cloud-free or partially cloud-free OLCI images were selected from 2017 to 2022 in the Yangtze-Huaihe region (Table 3). The machine learning algorithm was applied to estimate the TP concentration in each lake based on the Yangtze-Huaihe region dataset. Firstly, monthly TP concentrations were calculated for each lake per year and per month. Secondly, the average TP concentrations were determined for each lake across seasons (Spring: from March to May, Summer: from June to August, Autumn: from September to November, Winter: from December to February). Additionally, the mean TP concentration was calculated for each 12′ latitude and longitude grid cell, and the results were visualized. The lakes were classified into four categories based on their surface area: large lakes (>500 km^2), medium lakes (100–500 km^2), medium-small lakes (50–100 km^2), and small lakes (20–50 km^2). The spatial–temporal variations of TP concentration in the Yangtze-Huaihe region were analyzed accordingly.

Table 3. Statistics of OLCI images in the Yangtze-Huaihe region from 2017 to 2022 (unit: scenes).

Year	Jan	Feb	Mar	Apr	May	Jun	Jul	Aug	Sep	Oct	Nov	Dec	Total
2017	16	15	14	17	17	8	10	3	4	12	21	15	152
2018	10	12	17	18	7	11	20	14	17	13	12	14	165
2019	12	1	11	7	4	2	12	23	27	18	28	28	173
2020	7	15	6	16	13	7	1	3	9	10	16	13	116
2021	16	16	12	11	11	7	5	4	16	13	11	14	136
2022	6	8	11	15	10	6	5	18	19	20	4	18	140

3. Results

3.1. Model Accuracy Evaluation

The results of the variable importance index calculations for the five schemes are provided (Supplementary Material Figure S1). Based on the variable importance index, the variables were input into the algorithm in descending order of importance. The training sets of the machine learning algorithms exhibited high fitting accuracy, considering only the accuracy of the validation set. The fitting results were similar for the five scenarios and AFAI, b8, b4, b6, and b11 were selected for all of them, indicating that these five variables are important factors in estimating TP concentrations.

Based on the results of 5-fold cross-validation and scatter plot distributions, as seen in Figure 2, Approach 1 demonstrated higher accuracy and smaller errors (R^2 = 0.53, RMSE = 0.08, MAPE = 34.20%). The scatter plot distribution of approach 1 was more concentrated. Therefore, the AFAI, b8, b4, b6, b11, b17, b14, b2, and CMI were selected as input variables to construct the model for TP concentration estimation.

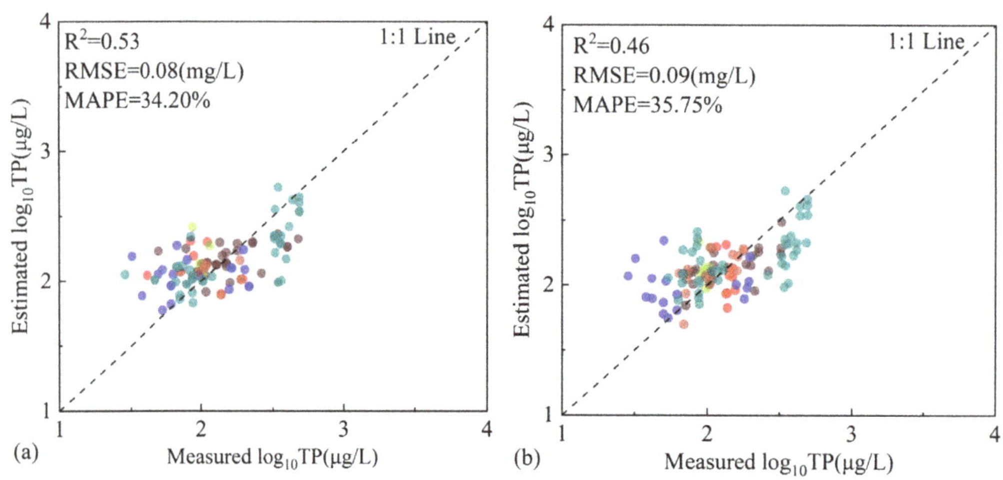

Figure 2. *Cont.*

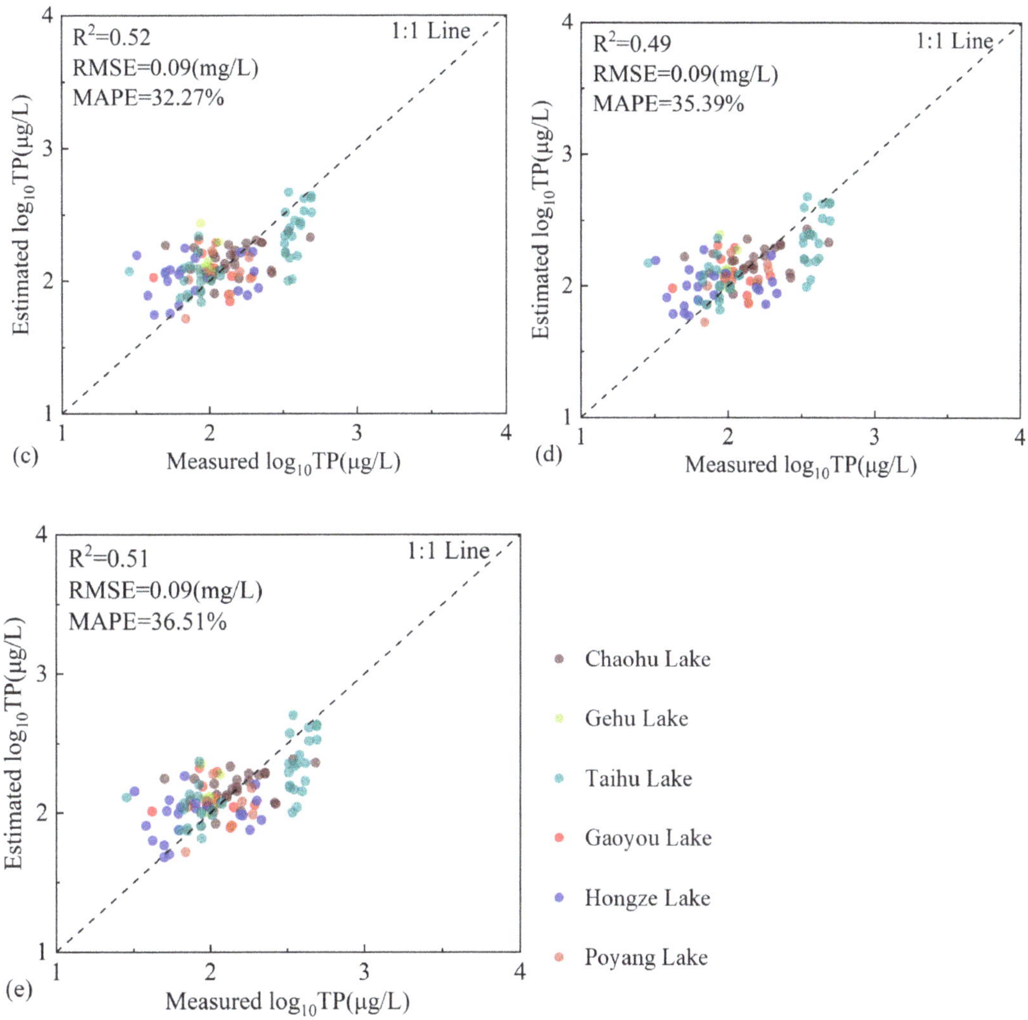

Figure 2. Five-fold cross-validation Results: (**a**) Approach 1, (**b**) Approach 2, (**c**) Approach 3, (**d**) Approach 4 and (**e**) Approach 5.

3.2. Spatial Distribution of TP Concentration in the Yangtze-Huaihe Region

The seasonal variations in TP concentration in the lakes of the Yangtze-Huaihe region during the period 2017–2022 are depicted (Figure 3). The average TP concentrations were relatively higher during the spring (0.21 mg/L) and summer (0.23 mg/L) seasons, while they are lower during autumn (0.19 mg/L) and winter (0.15 mg/L), surpassing the threshold for algal bloom occurrence (0.08 mg/L) [39].

In terms of spatial distribution, smaller lakes near Taihu Lake, Hongze Lake, and Poyang Lake exhibited comparatively higher TP concentrations. The average TP concentration near the lake shore was significantly higher than that in the centre of the lake. Regarding the longitude, the fluctuation trends remained consistent throughout the four seasons. Within the longitude ranges of 117°0′–117°48′E, 118°36′–119°0′E, and 120°0′–120°36′E, which include the western part of Poyang Lake, the smaller lakes between Poyang Lake and Chaohu Lake, and Taihu Lake, the average TP concentrations are relatively lower.

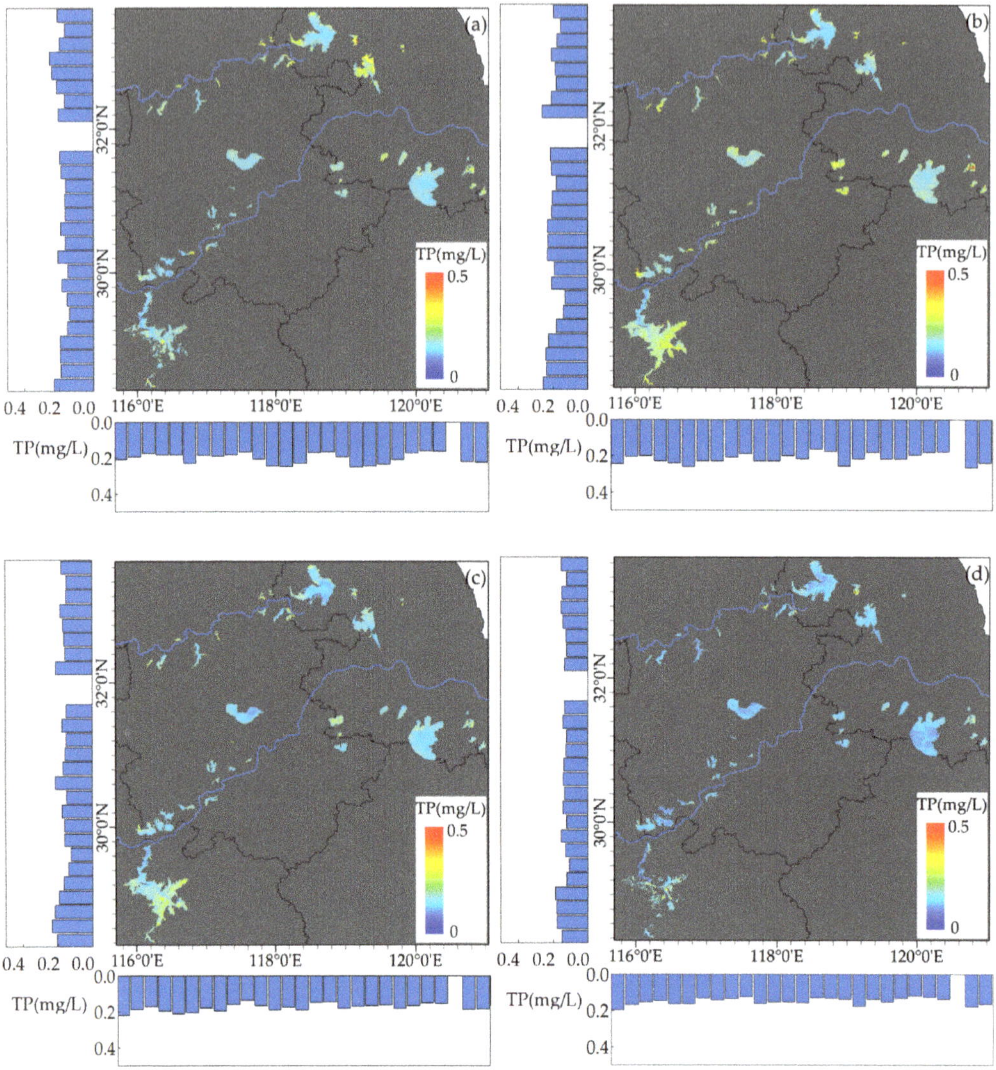

Figure 3. Seasonal mean TP Concentration in Lakes of the Yangtze-Huaihe Region: (**a**) Spring; (**b**) Summer; (**c**) Autumn; (**d**) Winter.

Analyzing the distribution based on latitude, the southern part of Poyang Lake at 29°24′N exhibited higher TP concentrations during spring and summer. Conversely, the northern part of Poyang Lake between 29°36′ and 30°0′N showed lower TP concentrations. Additionally, the northern part of Hongze Lake at 33°48′N consistently displayed higher TP concentrations in all four seasons.

3.3. Temporal Variation of TP Concentration in Lakes of Different Sizes

Figure 4 shows the temporal variation in TP concentrations in lakes of different sizes across the Yangtze-Huaihe region from 2017 to 2022. The large lakes exhibited a distinct pattern, with higher TP concentrations during spring and summer and lower concentrations during autumn and winter from 2017 to 2019. However, from 2020 to 2022, the TP concentrations in large lakes increased during summer and autumn and decreased during spring and winter. Other lake types displayed a consistent pattern over the six-year period, with higher TP concentrations during spring and summer and lower concentrations during autumn and winter.

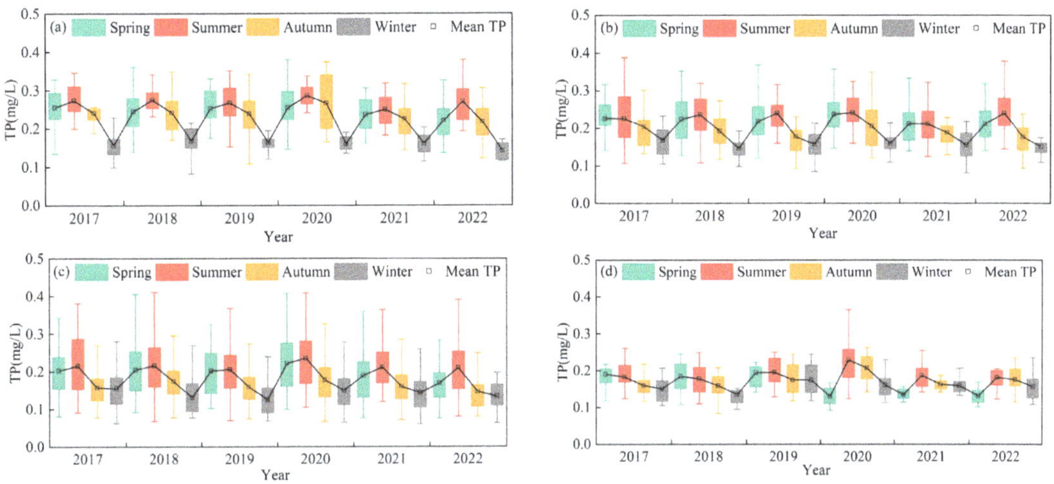

Figure 4. The variation of TP concentration in lakes under different sizes: (**a**) Small-sized lakes (<50 km^2), (**b**) Medium-small sized lakes (50–100 km^2), (**c**) Medium-sized lakes (100–500 km^2), and (**d**) Large-sized lakes (>500 km^2).

Large lakes showed relatively weak and stable seasonal variations, with average TP concentrations ranging from 0.13 to 0.23 mg/L. The highest TP concentration of 0.23 mg/L was recorded during the summer of 2020. Medium and medium-small lakes exhibited larger variation amplitudes than large lakes, and their average TP concentrations during spring and summer were generally higher than 0.20 mg/L. For medium-sized lakes, the highest TP concentration of 0.18 mg/L was observed during autumn 2020, and during winter, TP concentrations remained below 0.15 mg/L. Medium-small lakes showed average TP concentrations ranging from 0.18 to 0.21 mg/L during the autumn and between 0.15 and 0.17 mg/L during the winter. Small lakes displayed the most significant seasonal variation and had higher average TP concentrations, with concentrations above 0.20 mg/L during the spring, summer, and autumn, and between 0.15 and 0.17 mg/L in winter.

Across lakes of different sizes, TP concentrations peaked during the summer of 2020, and TP concentrations during autumn were also higher than those in other years in the same season.

4. Discussion

4.1. Comparison with Other Algorithms

To compare the accuracy of the XGBoost model developed in this study, which is based on both field measurements and OLCI images, we selected two existing empirical models [40,41] that have shown good performance in turbid inland waters, as shown in Table 4. While the existing algorithms achieved high accuracy in specific lakes, with R^2 values above 0.80, their effectiveness in large-sized regions was not satisfactory, with R^2 values below 0.10, RMSE of 0.13 mg/L, and MAPE exceeding 50% (Figure 5). Therefore, these models are unsuitable for TP concentration retrieval in large-sized regions.

Table 4. Algorithm Comparison.

Algorithm Type	Reference	Lake	Factors Used	OLCI Bands	Fitted Model Equation
Empirical Model	Hossen, et al. [40]	Burullus Lake	842/560 nm	b15/b6	$y = 36.201x^2 + 92.078x + 230.74$
	Xu, et al. [41]	Taihu Lake	671/680 nm	b9/b10	$y = 12.8265x^2 - 26.86x + 14.0727$

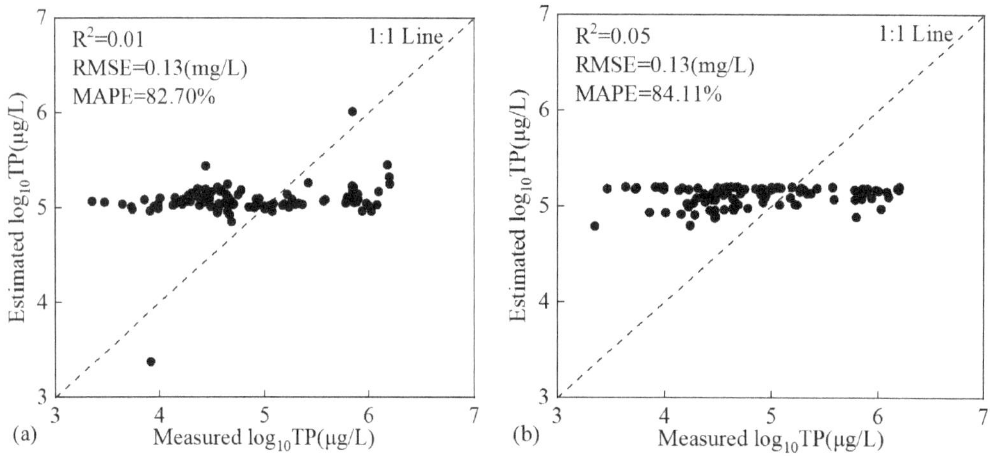

Figure 5. Comparison of algorithms:(**a**) Hossen et al., 2022 [40] (**b**) Xu et al., 2013 [41].

However, machine learning also has certain limitations when it comes to estimating TP concentration. Insufficient sample size in the training dataset can affect the accuracy and generalization ability of the model [42]; additionally, different types of lakes can also influence the model's accuracy [43]. Therefore, cross-validation is employed to divide the dataset into different training and validation sets to minimize the impact of sample size on model performance and enhance its generalization ability [44,45]. Li et al. [46] utilized 67 samples to construct a machine learning model for estimating the concentration of TP in water bodies, while Ding et al. [47] constructed an XGBoost model for estimating TP concentration in Nanyi Lake using 78 samples, both achieving good results. This suggests that, by selecting suitable algorithms and employing appropriate processing methods, machine learning can still yield good validation results even with a limited number of samples. Although the R^2 value of the XGBoost model in this study reached above 0.50, it exhibited differences in estimating results for different types of lakes. Specifically, there was a tendency for overestimation of TP concentration in Gaoyou, Hongze, and Gehu Lake when TP concentrations were low, while there was noticeable underestimation of high TP concentration in Taihu Lake. When comparing the water quality of these lakes, it was found that TSM concentrations were higher in Gaoyou, Hongze and Gehu Lake, while Chl-a concentrations were higher in Taihu Lake (Figure 6), which illustrates that the model in this study tended to overestimate in lakes dominated by non-algal particles, such as Gaoyou, Hongze and Gehu Lake [48], while it underestimated in lakes dominated by phytoplankton, such as Taihu Lake [43]. Currently, few researchers use machine learning to estimate the properties of lakes with different optical characteristics. The number of sampling sites used in this study is limited, so this model has some limitations.

Based on the above findings, no machine-learning model has been specifically developed for TP concentration estimation in lake groups. To enhance the accuracy of the model, future research should focus on increasing the number of sampling points and classifying water bodies based on their optical properties.

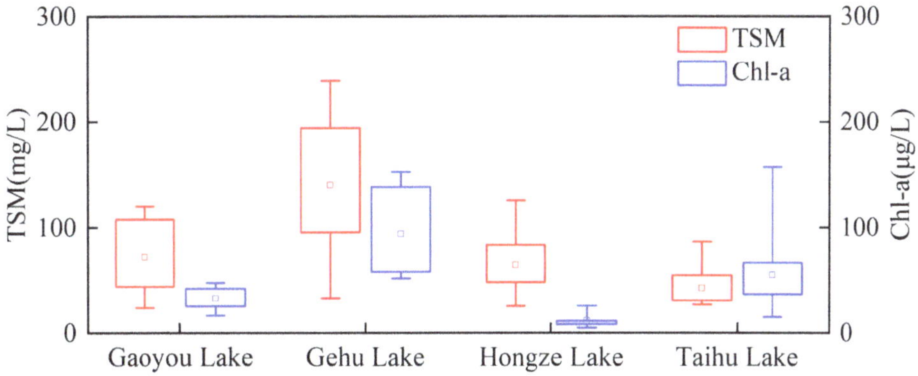

Figure 6. Lake TSM and Chl-*a* concentrations.

4.2. Drivers Analysis

From the perspective of climate change (Figure 7), meteorological factors such as temperature and precipitation directly or indirectly influence nutrient concentrations [49,50]. As the temperature increases, the release rate of nutrients from lake sediments accelerates, leading to an increase in TP concentration in the water. Furthermore, despite the dilution effect of rainfall on lake TP concentration, extreme rainfall events can generate significant external phosphorus loading [51]. During summer, floods are more frequent in the middle and lower reaches of the Yangtze River, while spring experiences higher rainfall and more rainy days compared to autumn [52], indicating that TP concentrations in lakes may be higher during spring and summer. In addition, it should be noted that the TP concentrations in different types of lakes were relatively high during the summer and autumn of 2020. This is attributed to the flooding that occurred in the Yangtze-Huaihe region in July, which resulted in the influx of excessive phosphorus loads into the lakes, directly causing an overall increase in TP concentrations and the accumulation of phosphorus in the sediments, thereby impacting subsequent water quality conditions [53].

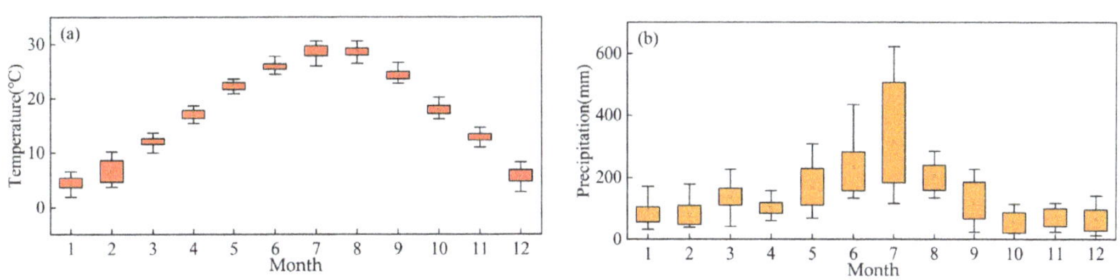

Figure 7. Monthly Scale Climate Variations: (**a**) Temperature; (**b**) Precipitation.

Agricultural pollution can also contribute to the elevated TP concentrations in lakes. In the Yangtze-Huaihe region, the TP concentrations in lakes exhibited a trend of being higher in spring and summer and lower in autumn and winter. This pattern may be related to the double-cropping rice planting system in which fertilisers are applied to the soil during both the spring and summer seasons [54], resulting in a significant inflow of nutrients into lakes [55]. For large lakes, the changing trend in TP concentrations after 2020, characterised by higher levels in summer and autumn and lower levels in spring and winter, is mainly attributed to the time required for the ecological water quality recovery process after the removal of enclosure aquaculture in 2019 [56]. The ecosystem gradually improved after 2020, and a large number of planktonic organisms absorbed phosphorus and organic matter from the water during reproductive growth, leading to a significant decrease in TP concentrations during spring in large lakes. However, although medium- and small-sized lakes have responded positively to the effects of enclosure removal, the presence of large agricultural areas and dense human populations results in a significant discharge of domestic and industrial wastewater [57,58]. The decreasing trend in TP concentrations was not evident in medium and small lakes, highlighting the importance of pollution control and management in the inflow areas of these lakes.

In conclusion, both natural factors and human activities have distinct effects on seasonal and interannual variations in TP concentrations in lakes in the Yangtze-Huaihe region.

4.3. Comparison of TP Concentration Estimation in Typical Lakes

Based on the results of this study, the lakes were categorized into four classes according to their area size, and one representative lake is selected from each category to compare with previously published research results (Figure 8 and Table 5). (1) Xiong et al. [59] analyzed total phosphorus (TP) concentrations in Lake Taihu ranging from 0.14 to 0.17 mg/L in 2017–2019, with an error of only 5% from the average TP concentration during the same period of this study. Shang et al. [60] estimated TP concentrations in Taihu Lake from 2017 to 2020, showing a similar trend to our study, the lower values are due to the fact that TP concentrations are highest in summer, but the Landsat-8 images used by Shang et al. had a longer temporal resolution, possibly not covering periods of high TP concentration, leading to a lower annual average estimation. (2) Qian et al. [61] conducted monthly surveys of TP concentration in Gehu Lake, achieving a correlation of 0.65 with this study. The reasons for the local variations are as follows: (a) There is a discrepancy between the monthly monitoring values and the average values within the monthly interval of the image; (b) Qian et al. used interpolation to calculate TP concentrations in Gehu Lake, and their results may have been influenced by the uneven distribution of sampling data. In contrast, this study employed remote sensing imagery, providing broader continuous spatial coverage. Due to the limitation of spatial resolution, it may be difficult to capture the detailed information for specific localized areas in this study, which leads to local discrepancies compared to the results of Qian et al. (3) The seasonal variation trend for TP concentration in Dianshan Lake observed in the study of Xiong et al. [62] is similar to the trend in this study, but the discrepancies in results are due to: (a) the distribution of sampling points: Xiong et al. analyzed seasonal variations using data from 13 monitoring points in Dianshan Lake, most of which were located near the lakeshore, with high rainfall intensity in spring and summer and a high external phosphorus load [52], possibly leading to overestimations in spring and summer—when the external phosphorus load is lower in autumn and winter, it is likely to be underestimated; (b) the time series of the studies: the time series of this study (2017–2022) differed from that of Xiong et al. (1996–2015). The relevant research indicates that local farmers around Dianshan Lake have extensively converted farmland into orchards in pursuit of higher economic returns. These orchards have a higher application rate of fertilizers [63], resulting in an exacerbation of pollution in Dianshan Lake. Therefore, the differences in water quality during different time periods in the study have led to variations in the estimated values of TP concentration. (4) Gong et al. [64] conducted a water quality survey in Tuohu Lake in 2019, with a TP concentration of 0.35 mg/L in spring, consistent with this study. The inconsistency in autumn and winter is due to Gong et al. conducting only four surveys in select months in 2018. Field surveys can be affected by sudden water quality events, and uneven distribution of remote sensing imagery within seasons can also result in inconsistent statistical results. In addition, Gong et al. established a total of eight sampling points in Taihu Lake, with five of them located near the lake shore. During autumn and winter, the TP concentration is generally lower due to the reduction in external phosphorus loading [65].

Table 5. Comparison of TP Concentrations in Typical Lakes.

Lake Area	Lake	References	Sampled Time	Number of Samples
>500 km^2	Taihu Lake	(Xiong, et al., 2021 [59])	8 March 2017/29 April 2017/ 30 April 2017/30 April 2017	55
		(Shang, et al., 2021 [60])	April 2018/July 2018	28
100–500 km^2	Gehu Lake	(Qian, et al., 2021 [61])	2019 monthly	20
50–100 km^2	Dianshan Lake	(Xiong, et al., 2017 [62])	2005–2015 monthly	13
20–50 km^2	Tuohu Lake	(Gong, et al., 2021 [64])	April 2018/August 2018/November 2018 /February 2019	8

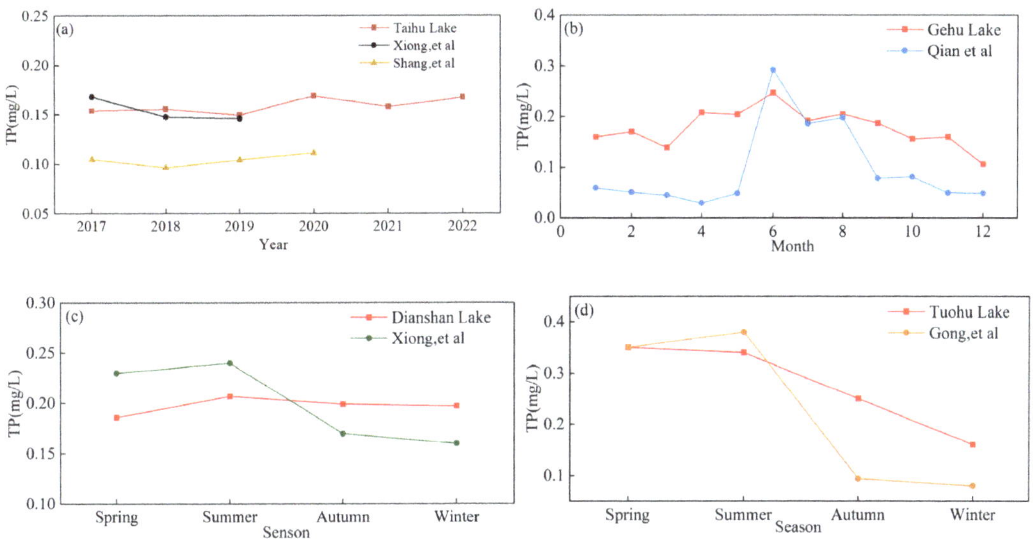

Figure 8. Comparison of Typical Lakes: (**a**) Taihu Lake [59,60]; (**b**) Gehu Lake [61]; (**c**) Dianshan Lake [62]; (**d**) Tuohu Lake [64].

In summary, the TP concentration estimated in this study is generally consistent with the trends observed in previous studies, confirming its scientific validity and reliability. These findings offer invaluable guidance for the monitoring of TP concentration in lakes within the Yangtze-Huaihe region.

5. Conclusions

In this study, a TP concentration estimation model applicable to large-sized regions, using synchronous satellite observations and the XGBoost algorithm, was developed with a focus on the lake group in the Yangtze-Huaihe region. The model utilised nine input variables: AFAI, b8, b4, b6, b11, b17, b14, b2, and CMI. Through cross-validation, the model achieved satisfactory fitting accuracy (R^2 = 0.53, RMSE = 0.08 mg/L, MAPE = 34.20%), outperforming traditional empirical models. The model was applied to OLCI images of lakes in the Yangtze-Huaihe region from 2017 to 2022. The results indicated an interannual decrease in TP concentrations for small, medium, and large lakes, with reductions of 0.02 mg/L, 0.01 mg/L, and 0.01 mg/L, respectively. Furthermore, regarding seasonal variations, the TP concentration in large lakes exhibited two distinct phases: from 2017 to 2019, "high in spring/summer and low in autumn/winter" and "high in summer/autumn and low in spring/winter" during 2020–2022. Other-sized lakes showed a consistent pattern of "high in spring/summer and low in autumn/winter" variations. Additionally, the smaller lakes had higher TP concentrations. The removal of enclosed nets was found to be beneficial for improving water quality in large lakes. However, for medium and small lakes, the discharge of domestic and industrial wastewater in the vicinity resulted in insignificant changes in the TP concentrations. Therefore, when focusing on the ecological environment of large lakes, it is essential to focus on the governance and improvement of water quality in medium and small lakes.

This study demonstrates the potential application of machine learning algorithms for estimating TP concentrations in lakes of the Yangtze-Huaihe region using remote sensing satellite data and in situ measurements. In future research, the establishment of a virtual constellation combining satellites, such as Landsat-TM/ETM/OLI and MODIS, can be considered, in order to further expand the research scale in terms of time and space, enabling comprehensive monitoring and analysis of the global nutrient status of lakes.

Supplementary Materials: The following supporting information can be downloaded at: https://www.mdpi.com/article/10.3390/rs15184487/s1, Figure S1: Importance rank: (a) Original; (b) Inverse; (c) Exponential; (d) Square; (e) Square root; Table S1: Lakes larger than 20 km^2 in the Yangtze-Huaihe region; Table S2: Variables.

Author Contributions: Conceptualization, X.W., Y.J., M.J., Z.C., R.M. and J.X.; Methodology, X.W., Y.J., M.J., Z.C., X.L., L.X. and J.X.; Data curation, X.W., M.J., Z.C. and J.X.; Writing—original draft, X.W., Y.J. and J.X.; Writing—review & editing, X.W., Y.J., M.J., Z.C., X.L., R.M., L.X. and J.X.; Supervision, Y.J., X.L., R.M. and J.X.; Funding acquisition, J.X. All authors have read and agreed to the published version of the manuscript.

Funding: This work was funded by the National Key R&D Program of China (Grant No. 2021YFD1700600), the Natural Science Foundation of Jiangsu Province for Youths (Grant No. BK20221158), the National Natural Science Foundation of China (Grant No. 42201400), the Key Laboratory of Land Satellite Remote Sensing Application, Ministry of Natural Resources of the People's Republic of China (Grant No. KLSMNR-G202215). Data were supported by Lake-Watershed Science Data Center, National Earth System Science Data Center, National Science & Technology Infrastructure of China (http://lake.geodata.cn).

Conflicts of Interest: The authors declare no conflict of interest.

References

1. Ho, L.T.; Goethals, P.L.M. Opportunities and Challenges for the Sustainability of Lakes and Reservoirs in Relation to the Sustainable Development Goals (SDGs). *Water* **2019**, *11*, 1462, Erratum in *Water* **2021**, *13*, 3207. [CrossRef]
2. Roy, S.; Ojha, S.R.; Reddy, N.N.; Samal, R.N.; Das, B.S. Suspended particulate matter and secchi disk depth in the Chilika Lagoon from in situ and remote sensing data: A modified semi-analytical approach. *Int. J. Remote Sens.* **2022**, *43*, 3628–3654. [CrossRef]
3. Hu, Z.F.; Li, D.S.; Guan, D.T. Water quality retrieval and algae inhibition from eutrophic freshwaters with iron-rich substrate based ecological floating beds treatment. *Sci. Total Environ.* **2020**, *712*, 135584. [CrossRef] [PubMed]
4. Dona, C.; Sanchez, J.M.; Caselles, V.; Dominguez, J.A.; Camacho, A. Empirical Relationships for Monitoring Water Quality of Lakes and Reservoirs Through Multispectral Images. *IEEE J. Sel. Top. Appl. Earth Obs. Remote Sens.* **2014**, *7*, 1632–1641. [CrossRef]
5. Maciel, D.A.; Barbosa, C.C.F.; Novo, E.; Flores, R.; Begliomini, F.N. Water clarity in Brazilian water assessed using Sentinel-2 and machine learning methods. *Isprs J. Photogramm. Remote Sens.* **2021**, *182*, 134–152. [CrossRef]
6. Cao, Z.G.; Ma, R.H.; Melack, J.M.; Duan, H.T.; Liu, M.; Kutser, T.; Xue, K.; Shen, M.; Qi, T.C.; Yuan, H.L. Landsat observations of chlorophyll-a variations in Lake Taihu from 1984 to 2019. *Int. J. Appl. Earth Obs. Geoinf.* **2022**, *106*, 102642. [CrossRef]
7. Liu, X.Y.; Zhang, Z.; Jiang, T.; Li, X.H.; Li, Y.Y. Evaluation of the Effectiveness of Multiple Machine Learning Methods in Remote Sensing Quantitative Retrieval of Suspended Matter Concentrations: A Case Study of Nansi Lake in North China. *J. Spectrosc.* **2021**, *2021*, 5957376. [CrossRef]
8. Ruescas, A.B.; Hieronymi, M.; Mateo-Garcia, G.; Koponen, S.; Kallio, K.; Camps-Valls, G. Machine Learning Regression Approaches for Colored Dissolved Organic Matter (CDOM) Retrieval with S2-MSI and S3-OLCI Simulated Data. *Remote Sens.* **2018**, *10*, 786. [CrossRef]
9. Somasundaram, D.; Zhang, F.F.; Ediriweera, S.; Wang, S.L.; Yin, Z.Y.; Li, J.S.; Zhang, B. Patterns, Trends and Drivers of Water Transparency in Sri Lanka Using Landsat 8 Observations and Google Earth Engine. *Remote Sens.* **2021**, *13*, 2193. [CrossRef]
10. Balasubramanian, S.V.; Pahlevan, N.; Smith, B.; Binding, C.; Schalles, J.; Loisel, H.; Gurlin, D.; Greb, S.; Alikas, K.; Randla, M.; et al. Robust algorithm for estimating total suspended solids (TSS) in inland and nearshore coastal waters. *Remote Sens. Environ.* **2020**, *246*, 111768. [CrossRef]
11. Sun, X.; Zhang, Y.L.; Zhang, Y.B.; Shi, K.; Zhou, Y.Q.; Li, N. Machine Learning Algorithms for Chromophoric Dissolved Organic Matter (CDOM) Estimation Based on Landsat 8 Images. *Remote Sens.* **2021**, *13*, 3560. [CrossRef]

12. Niu, C.; Tan, K.; Jia, X.P.; Wang, X. Deep learning based regression for optically inactive inland water quality parameter estimation using airborne hyperspectral imagery. *Environ. Pollut.* **2021**, *286*, 117534. [CrossRef] [PubMed]
13. Sagan, V.; Peterson, K.T.; Maimaitijiang, M.; Sidike, P.; Sloan, J.; Greeling, B.A.; Maalouf, S.; Adams, C. Monitoring inland water quality using remote sensing: Potential and limitations of spectral indices, bio-optical simulations, machine learning, and cloud computing. *Earth-Sci. Rev.* **2020**, *205*, 103187. [CrossRef]
14. Hecky, R.E.; Kilham, P. Nutrient limitation of phytoplankton in fresh-water and marine environments—A review of recent-evidence on the effects of enrichment. *Limnol. Oceanogr.* **1988**, *33*, 796–822. [CrossRef]
15. Baban, S.M.J. Detecting water-quality parameters in the norfolk broads, uk, using landsat imagery. *Int. J. Remote Sens.* **1993**, *14*, 1247–1267. [CrossRef]
16. Wang, L.Y.; Li, C.Y.; Sun, B. Determination total phosphorus concentration and evaluation eutrophication of Hulun Lake with MODIS data. *Chin. J. Environ. Eng.* **2014**, *8*, 5527–5534.
17. Chen, J.; Quan, W.T. Using Landsat/TM Imagery to Estimate Nitrogen and Phosphorus Concentration in Taihu Lake, China. *IEEE J. Sel. Top. Appl. Earth Obs. Remote Sens.* **2012**, *5*, 273–280. [CrossRef]
18. Zhang, L.S.; Zhang, L.F.; Cen, Y.; Wang, S.; Zhang, Y.; Huang, Y.; Sultan, M.; Tong, Q.X. Prediction of Total Phosphorus Concentration in Macrophytic Lakes Using Chlorophyll-Sensitive Bands: A Case Study of Lake Baiyangdian. *Remote Sens.* **2022**, *14*, 3077. [CrossRef]
19. Liu, Y.; Jiang, H. Retrieval of Total Phosphorus Concentration in the Surface Waters of Poyang Lake Based on Remote Sensing and Analysis of Its Spatial-Temporal Characteristics. *J. Nat. Resour.* **2013**, *28*, 2169–2177. [CrossRef]
20. Nieto, P.J.G.; Garcia-Gonzalo, E.; Fernandez, J.R.A.; Muinz, C.D. Water eutrophication assessment relied on various machine learning techniques: A case study in the Englishmen Lake (Northern Spain). *Ecol. Model.* **2019**, *404*, 91–102. [CrossRef]
21. Lee, H.W.; Kim, M.; Son, H.W.; Min, B.; Choi, J.H. Machine-learning-based water quality management of river with serial impoundments in the Republic of Korea. *J. Hydrol. -Reg. Stud.* **2022**, *41*, 101069. [CrossRef]
22. Xiong, J.F.; Lin, C.; Cao, Z.G.; Hu, M.Q.; Xue, K.; Chen, X.; Ma, R.H. Development of remote sensing algorithm for total phosphorus concentration in eutrophic lakes: Conventional or machine learning? *Water Res.* **2022**, *215*, 118213. [CrossRef] [PubMed]
23. Lundberg, S.M.; Erion, G.; Chen, H.; DeGrave, A.; Prutkin, J.M.; Nair, B.; Katz, R.; Himmelfarb, J.; Bansal, N.; Lee, S.-I. From local explanations to global understanding with explainable AI for trees. *Nat. Mach. Intell.* **2020**, *2*, 56–67. [CrossRef] [PubMed]
24. Cui, Y.; Meng, F.; Fu, P.J.; Yang, X.Y.; Zhang, Y.X.; Liu, P.D. Application of hyperspectral analysis of chlorophyll a concentration inversion in Nansi Lake. *Ecol. Inform.* **2021**, *64*, 101360. [CrossRef]
25. Hu, Y.R.; Du, W.J.; Yang, C.; Wang, Y.; Huang, T.Y.; Xu, X.Y.; Li, W.W. Source identification and prediction of nitrogen and phosphorus pollution of Lake Taihu by an ensemble machine learning technique. *Front. Environ. Sci. Eng.* **2023**, *17*, 55. [CrossRef]
26. Zhang, Y.B.; Shi, K.; Sun, X.; Zhang, Y.L.; Li, N.; Wang, W.J.; Zhou, Y.Q.; Zhi, W.; Liu, M.L.; Li, Y.; et al. Improving remote sensing estimation of Secchi disk depth for global lakes and reservoirs using machine learning methods. *Giscience Remote Sens.* **2022**, *59*, 1367–1383. [CrossRef]
27. Chen, S.; Hu, C.; Barnes, B.B.; Wanninkhof, R.; Cai, W.-J.; Barbero, L.; Pierrot, D. A machine learning approach to estimate surface ocean pCO(2) from satellite measurements. *Remote Sens. Environ.* **2019**, *228*, 203–226. [CrossRef]
28. Jang, E.; Im, J.; Park, G.H.; Park, Y.G. Estimation of Fugacity of Carbon Dioxide in the East Sea Using In Situ Measurements and Geostationary Ocean Color Imager Satellite Data. *Remote Sens.* **2017**, *9*, 821. [CrossRef]
29. Qin, B.Q.; Gao, G.; Zhu, G.W.; Zhang, Y.L.; Song, Y.Z.; Tang, X.M.; Xu, H.; Deng, J.M. Eutrophication of lakes and their ecosystem response. *Chin. Sci. Bull.* **2013**, *58*, 855–864. [CrossRef]
30. Hu, M.; Ma, R.; Xiong, J.; Wang, M.; Cao, Z.; Xue, K. Eutrophication state in the Eastern China based on Landsat 35-year observations. *Remote Sens. Environ.* **2022**, *277*, 113057. [CrossRef]
31. Zhao, Y.; Li, T.; Huang, B.; Feng, Y.; Lei, M.; Zhuo, H.; Wu, L. Evolution characteristics and driving factors of water quality and eutrophication of typical lakes in the middle reaches of the Yangtze River from 2016 to 2020. *J. Lake Sci.* **2022**, *34*, 1441–1451.
32. Yang, G.; Ma, R.; Zhang, L.; Jiang, J.; Yao, S.; Zhang, M.; Zeng, H. Lake status major problems and protection strategy in China. *J. Lake Sci.* **2010**, *22*, 799–810.
33. Vanhellemont, Q.; Ruddick, K. Atmospheric correction of Sentinel-3/OLCI data for mapping of suspended particulate matter and chlorophyll-a concentration in Belgian turbid coastal waters. *Remote Sens. Environ.* **2021**, *256*, 112284. [CrossRef]
34. Hu, C.; Lee, Z.; Ma, R.; Yu, K.; Li, D.; Shang, S. Moderate Resolution Imaging Spectroradiometer (MODIS) observations of cyanobacteria blooms in Taihu Lake, China. *J. Geophys. Res.-Ocean.* **2010**, *115*, 5511. [CrossRef]
35. Qi, L.; Hu, C.M.; Visser, P.M.; Ma, R.H. Diurnal changes of cyanobacteria blooms in Taihu Lake as derived from GOCI observations. *Limnol. Oceanogr.* **2018**, *63*, 1711–1726. [CrossRef]
36. Liang, Q.C.; Zhang, Y.C.; Ma, R.H.; Loiselle, S.; Li, J.; Hu, M.Q. A MODIS-Based Novel Method to Distinguish Surface Cyanobacterial Scums and Aquatic Macrophytes in Lake Taihu. *Remote Sens.* **2017**, *9*, 133. [CrossRef]
37. Chen, T.; Guestrin, C.; Assoc Comp, M. XGBoost: A Scalable Tree Boosting System. In Proceedings of the 22nd ACM SIGKDD International Conference on Knowledge Discovery and Data Mining (KDD), San Francisco, CA, USA, 13–17 August 2016; pp. 785–794.

38. Brooks, W.; Corsi, S.; Fienen, M.; Carvin, R. Predicting recreational water quality advisories: A comparison of statistical methods. *Environ. Model. Softw.* **2016**, *76*, 81–94. [CrossRef]
39. Xu, H.; Paerl, H.W.; Zhu, G.W.; Qin, B.Q.; Hall, N.S.; Zhu, M.Y. Long-term nutrient trends and harmful cyanobacterial bloom potential in hypertrophic Lake Taihu, China. *Hydrobiologia* **2017**, *787*, 229–242. [CrossRef]
40. Hossen, H.; Mahmod, W.E.; Negm, A.; Nakamura, T. Assessing Water Quality Parameters in Burullus Lake Using Sentinel-2 Satellite Images. *Water Resour.* **2022**, *49*, 321–331. [CrossRef]
41. Xu, L.J.; Huang, C.C.; Li, Y.M.; Chen, X. Deriving Concentration of TN,TP based on Hyper Spectral Reflectivity. *Remote Sens. Technol. Appl.* **2013**, *28*, 681–688.
42. Holloway, J.; Mengersen, K. Statistical Machine Learning Methods and Remote Sensing for Sustainable Development Goals: A Review. *Remote Sens.* **2018**, *10*, 1365. [CrossRef]
43. Du, C.G.; Wang, Q.; Li, Y.M.; Lyu, H.; Zhu, L.; Zheng, Z.B.; Wen, S.; Liu, G.; Guo, Y.L. Estimation of total phosphorus concentration using a water classification method in inland water. *Int. J. Appl. Earth Obs. Geoinf.* **2018**, *71*, 29–42. [CrossRef]
44. Vu, H.L.; Ng, K.T.W.; Richer, A.; An, C. Analysis of input set characteristics and variances on k-fold cross validation for a Recurrent Neural Network model on waste disposal rate estimation. *J. Environ. Manag.* **2022**, *311*, 114869. [CrossRef]
45. Hadjisolomou, E.; Stefanidis, K.; Herodotou, H.; Michaelides, M.; Papatheodorou, G.; Papastergiadou, E. Modelling Freshwater Eutrophication with Limited Limnological Data Using Artificial Neural Networks. *Water* **2021**, *13*, 1590. [CrossRef]
46. Li, N.; Zhang, Y.L.; Shi, K.; Zhang, Y.B.; Sun, X.; Wang, W.J.; Qian, H.M.; Yang, H.Y.; Niu, Y.K. Real-Time and Continuous Tracking of Total Phosphorus Using a Ground-Based Hyperspectral Proximal Sensing System. *Remote Sens.* **2023**, *15*, 507. [CrossRef]
47. Ding, L.; Qi, C.C.; Li, G.; Zhang, W.Q. TP Concentration Inversion and Pollution Sources in Nanyi Lake Based on Landsat 8 Data and InVEST Model. *Sustainability* **2023**, *15*, 9678. [CrossRef]
48. Lei, S.H.; Xu, J.; Li, Y.M.; Li, L.; Lyu, H.; Liu, G.; Chen, Y.; Lu, C.Y.; Tian, C.; Jiao, W.Z. A semi-analytical algorithm for deriving the particle size distribution slope of turbid inland water based on OLCI data: A case study in Lake Hongze. *Environ. Pollut.* **2021**, *270*, 116288. [CrossRef]
49. Janatian, N.; Olli, K.; Cremona, F.; Laas, A.; Noges, P. Atmospheric stilling offsets the benefits from reduced nutrient loading in a large shallow lake. *Limnol. Oceanogr.* **2020**, *65*, 717–731. [CrossRef]
50. Xiong, J.; Xue, K.; Li, J.; Hu, M.; Li, J.; Wang, X.; Lin, C.; Ma, R.; Chen, L. Vertical distribution analysis and total mass estimation of nitrogen and phosphorus in large shallow lakes. *J. Environ. Manag.* **2023**, *344*, 118465. [CrossRef]
51. Huang, J.C.; Zhang, Y.J.; Bing, H.J.; Peng, J.; Dong, F.F.; Gao, J.F.; Arhonditsis, G.B. Characterizing the river water quality in China: Recent progress and on-going challenges. *Water Res.* **2021**, *201*, 117309. [CrossRef]
52. Zhang, Y.X.; Hu, Z.H. Variation Characteristics of Rainstorms in Middle-Lower Reaches of the Yangtze River in Recent 60 Years. *Clim. Environ. Res.* **2019**, *24*, 755–768. [CrossRef]
53. Zhu, W.; Hu, S.; Feng, G.; Zhang, Y.; Zhai, S.; Xue, Z.; Zhao, S.; Chen, H.; Wang, R. Effects of great floods on phosphorus in shallow lakes: A case study of Lake Taihu in 2016. *J. Lake Sci.* **2020**, *32*, 325–336. [CrossRef]
54. Wang, L.C.; Hoang, D.V.; Liou, Y.A. Quantifying the Impacts of the 2020 Flood on Crop Production and Food Security in the Middle Reaches of the Yangtze River, China. *Remote Sens.* **2022**, *14*, 3140. [CrossRef]
55. Tong, S.T.Y.; Chen, W.L. Modeling the relationship between land use and surface water quality. *J. Environ. Manag.* **2002**, *66*, 377–393. [CrossRef]
56. Luo, J.H.; Pu, R.L.; Ma, R.H.; Wang, X.L.; Lai, X.J.; Mao, Z.G.; Zhang, L.; Peng, Z.L.; Sun, Z. Mapping Long-Term Spatiotemporal Dynamics of Pen Aquaculture in a Shallow Lake: Less Aquaculture Coming along Better Water Quality. *Remote Sens.* **2020**, *12*, 1866. [CrossRef]
57. Liao, R.; Wu, X.; Wang, Z.; Wang, B.; Ke, F.; Han, C.; Zhou, Q.; Ren, J. Spatial distribution and pollution assessment on the main nutrients and heavy metals in sediments of Lake Gehu, Taihu Basin after removing the aquaculture net. *J. Lake Sci.* **2021**, *33*, 1436–1447. [CrossRef]
58. Zeng, S.; Du, C.G.; Li, Y.M.; Lyu, H.; Dong, X.Z.; Lei, S.H.; Li, J.D.; Wang, H.J. Monitoring the particulate phosphorus concentration of inland waters on the Yangtze Plain and understanding its relationship with driving factors based on OLCI data. *Sci. Total Environ.* **2022**, *809*, 151992. [CrossRef]
59. Xiong, J.F. *Remote Sensing Estimation of Phosphorus Concentration in Lake Taihu and Its Response to Non-Point Source Phosphorus Load. Nanjing Institute of Geography and Limnology*; Chinese Academy of Sciences: Beijing, China, 2021.
60. Shang, W.; Jin, S.G.; He, Y.; Zhang, Y.Y.; Li, J. Spatial-Temporal Variations of Total Nitrogen and Phosphorus in Poyang, Dongting and Taihu Lakes from Landsat-8 Data. *Water* **2021**, *13*, 1704. [CrossRef]
61. Qian, W.H.; Gao, Y.X.; Zhang, Y.M.; Yang, F.; Zhang, Z.W.; Zhao, Y.; Liu, Y.; Xu, D.J. Analysis of characteristics and reasons of temporal and spatial changes of Gehu Lake water quality based on multivariate statistical analysis. *Water Resour. Hydropower Eng.* **2021**, *52*, 116–128. [CrossRef]
62. Xiong, G.H.; Wang, G.C.; Wang, D.Q.; Yang, W.L.; Chen, Y.Y.; Chen, Z.L. Spatio-Temporal Distribution of Total Nitrogen and Phosphorus in Dianshan Lake, China: The External Loading and Self-Purification Capability. *Sustainability* **2017**, *9*, 500. [CrossRef]
63. Zhang, M.S.; Li, W.; Zhang, W.G.; Li, Y.T.; Li, J.Y.; Gao, Y. Agricultural land-use change exacerbates the dissemination of antibiotic resistance genes via surface runoffs in Lake Tai Basin, China. *Ecotoxicol. Environ. Saf.* **2021**, *220*, 112328. [CrossRef] [PubMed]

64. Gong, W.J. Evaluation of water quality and eutrophication status of Tuo Lake in Wuhe County, Anhui Province. *Anhui Archit.* **2021**, *28*, 101–103. [CrossRef]
65. Li, X.F.; Xu, W.Z.; Song, S.; Sun, J. Sources and spatiotemporal distribution characteristics of nitrogen and phosphorus loads in the Haihe River Basin, China. *Mar. Pollut. Bull.* **2023**, *189*, 114756. [CrossRef] [PubMed]

remote sensing

MDPI

Article

Combined Retrievals of Turbidity from Sentinel-2A/B and Landsat-8/9 in the Taihu Lake through Machine Learning

Zhe Yang [1,2], Cailan Gong [1,*], Zhihua Lu [3], Enuo Wu [4], Hongyan Huai [4], Yong Hu [1], Lan Li [1] and Lei Dong [1,2]

1 Key Laboratory of Infrared System Detection and Imaging Technologies, Shanghai Institute of Technical Physics, Chinese Academy of Sciences, Shanghai 200083, China; yangzhe@mail.sitp.ac.cn (Z.Y.); huyong@mail.sitp.ac.cn (Y.H.); lilan@mail.sitp.ac.cn (L.L.); donglei20@mail.ucas.ac.cn (L.D.)
2 University of Chinese Academy of Sciences, Beijing 100049, China
3 Shanghai Academy of Environmental Sciences, Shanghai 200233, China
4 Shanghai Environment Monitoring Center, Shanghai 200235, China
* Correspondence: gcl@mail.sitp.ac.cn

Abstract: Lakes play a crucial role in the earth's ecosystems and human activities. While turbidity is not a direct biochemical indicator of lake water quality, it is relatively easy to measure and indicates trophic status and lake health. Although ocean color satellites have been widely used to monitor water color parameters, their coarse spatial resolution makes it hard to capture the fine spatial variability of turbidity in lakes. The combination of Sentinel-2 and Landsat provides an opportunity to monitor lake turbidity with high spatial and temporal resolution. This study aims to generate consistent turbidity products in Taihu Lake from 2018 to 2022 using the Multispectral Instrument (MSI) on board Sentinel-2A/B and the Operational Land Imager (OLI) on board Landsat-8/9. We first tested the performance of three atmospheric correction methods to retrieve consistent reflectance from MSI and OLI images. We found that the Rayleigh correction and a subtraction of the SWIR band from Rayleigh-corrected reflectance can generate the most consistent reflectance (the coefficient of determination (R^2) > 0.84, the mean absolution percentage error (MAPE) < 7%, the median error (ME) < 0.0035, and slope > 0.92). Machine learning models outperformed an existing semi-analytical retrieval algorithm in retrieving turbidity (MSI: R^2 = 0.92, MAPE = 18.78%, and OLI: R^2 = 0.93, MAPE = 16.20%). The consistency of turbidity from the same-day MSI and OLI images was also satisfactory (N = 3110 and MAPE = 26.48%). The distribution of turbidity exhibited obvious spatial and seasonal variability in Taihu Lake from 2018 to 2022. The results show the potential of MSI and OLI when combined to monitor inland lake water quality.

Keywords: turbidity; remote sensing; water quality; Sentinel-2A/B; Landsat-8/9

1. Introduction

Lakes are widely distributed on the earth's surface and are critical for the survival and development of human society. Lakes not only provide valuable freshwater resources but also have various ecological and environmental functions [1]. However, human activities and climate change have caused serious water quality problems in numerous lakes throughout the world, which directly affect human well-being [2]. For this reason, monitoring lake water quality parameters has important socioeconomic significance. Turbidity is closely associated with the nutritional status and primary productivity of the water body [3]. As an optical scattering indicator of water turbidity, turbidity is measured by detecting the scattering intensity of water and used as a rough indicator of fine suspended matter in water [4]. Although turbidity is one of the simplest measures of optical characteristics and pollution degree of a water body, traditional field sampling methods are difficult to capture the overall spatial and temporal changes of lake turbidity due to a limited number of points. In comparison, satellite observation has been widely used to provide long-term and large-scale water quality monitoring [5].

Generally, ocean color satellites, such as the Sea-Viewing Wide Field-of-View Sensor (SeaWiFS), Moderate Resolution Imaging Spectroradiometer (MODIS), and Medium Resolution Imaging Spectrometer (MERIS), have been successfully employed to analyze the turbidity distribution in coastal waters and inland lakes [6–8]. However, the coarse spatial resolution (250–1000 m) of ocean color instruments limits their use at small scales due to their insufficient ability to capture spatial details in turbidity. While moderate spatial resolution missions (10–100 m) were initially designed primarily for land applications, some researchers have carried out research using them to monitor the turbidity of water. For example, Goodin et al. [9] found that the SPOT-HRV2 red band has a good correlation with turbidity and mapped turbidity in the Tuttle Creek Reservoir in Kansas, USA. Bustamante et al. [10] predicted water turbidity from Landsat band 3 using generalized additive models. Zhou et al. [11] used a long-time series of Landsat images to retrieve urban water turbidity and analyze factors driving turbidity changes. Rodríguez-López et al. [12] combined in situ measurement of Secchi disk depth with Landsat images to develop empirical turbidity models in Araucanian lakes.

Recently, the new generation of sensors with several enhancements, such as the Multispectral Instrument (MSI) on board Sentinel-2A/B and the Operational Land Imager (OLI) on board Landsat-8/9, have become appealing instruments for inland water monitoring [13,14]. The OLI has nine spectral bands in the visible, near-infrared (NIR), and shortwave infrared (SWIR) with a spatial resolution of 30 m. While the revisit period of Landsat-8 is 16 days, it was improved to 8 days when Landsat-9 became operational in 2021. Compared to the OLI, the MSI has three additional red-edge bands, a spatial resolution of 10–60 m, and a 5-day revisit time. Although the OLI and MSI have finer spatial resolution than ocean color sensors, their low temporal resolution makes it challenging to frequently monitor quickly changing inland lakes.

Recent publications have demonstrated the capability of combining the MSI and OLI data for water quality parameter monitoring. MSI and OLI have similar spatial resolution and radiometric performances, and their combination can improve the revisit time to 2–3 days, which is comparable to the Ocean and Land Color Instrument (OLCI) on board Sentinel-3 and MODIS. For example, Ciancia et al. [15] trained and validated a Sentinel-2A MSI-based total suspended matter (TSM) model, then generated a MSI-OLI merged model by calibrating OLI data. Benjamin et al. [16] mapped water clarity in some lakes through harmonized images from Landsat-8 and Sentinel-2. Pahlevan et al. [17] assessed extensive Landsat-8 and Sentinel-2A/B aquatic products, including the top-of-atmosphere (TOA) reflectance, the remote-sensing reflectance (R_{rs}), and water quality parameters. Nonetheless, several challenges remain in combing multiple moderate spatial resolution satellite missions to generate consistent turbidity products in inland lakes.

A mass of related research and field observation data has proved that the visible and near-infrared reflectance are elevated with increasing turbidity [18–20]. The accuracy of turbidity retrievals depends on the performance of atmospheric correction (AC) methods to recover R_{rs} from at-sensor measurements. However, the common AC methods are not yet available for optically complex inland waters, and few attempts have been made to evaluate the performance of several specialized AC methods [21–23]. Thus, AC is one of the biggest challenges in creating consistent MSI-OLI turbidity products in inland waters. A second major challenge is to develop a specialized algorithm to retrieve turbidity in inland lakes from MSI and OLI Level-2 data. The number of MSI and OLI near-simultaneous nadir overpasses (n-SNO) is insufficient considering cloud coverage, which requires that the field data used for MSI and OLI modeling, respectively, be long-term stable. However, in inland lakes, the use of platforms for stationary water quality measurements is still incipient [24]. Regarding the estimation algorithm of turbidity, most models are empirical and rely on a strong correlation between turbidity and derived reflectance [25]. In fact, turbidity is often influenced by several matters such as chlorophyll-a (Chl-a), TSM, and colored dissolved organic matter (CDOM), so the relationship between turbidity and reflectance is not obvious when the optical or biochemical properties of water change [26]. Due to their good data mining capabilities, machine learning algorithms provide an alternative way to describe the nonlinear relationship between objective and feature variables.

The goal of this article is to generate consistent turbidity products in Taihu Lake in China from 2018 to 2022 using Sentinel-2A/B and Landsat-8/9. To achieve this goal, we began by calculating the agreement of reflectance derived from MSI and OLI to evaluate the performance of three AC methods. We then trained and calibrated machine learning models for retrieving turbidity using MSI and OLI datasets, respectively. Finally, we assessed the consistency of turbidity products and mapped turbidity in Taihu Lake from MSI and OLI images from 2018 to 2022. The overarching methodological framework is shown in Figure 1. Specifically, this study is innovative in the following three aspects:

(1) Testing the performance of three aquatic AC methods to retrieve consistent reflectance products;
(2) Developing a machine learning model for turbidity retrieval in Taihu Lake;
(3) Proposing a method for evaluating the consistency of MSI and OLI products in inland water bodies.

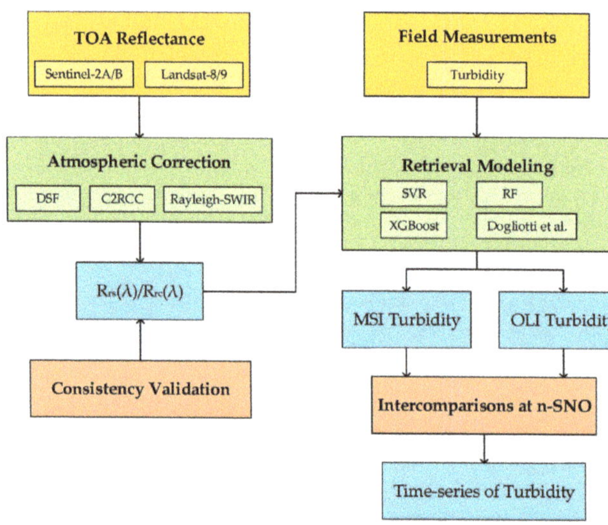

Figure 1. Flow chart of this study, yellow rectangles represent the input, green rectangles represent the model, blue rectangles represent the output, and orange rectangles represent the validation.

2. Materials and Methods

2.1. Study Area

Taihu Lake is located in the Lower Reaches of the Yangtze River and is the third largest freshwater lake in China with a water surface area of 2338 km^2 (between 30°55′–31°33′N and 119°52′–120°36′E, Figure 2). The lake's average water depth is 1.9 m, and its maximum water depth is 3 m [27]. At present, the maximum length of Taihu Lake is 68 km, the average width is 35.7 km, and the total length of the shoreline is more than 393 km. The rapid economic development of the cities around Taihu Lake has led to the deterioration of water quality, which has an adverse impact on people's drinking water. Although the government has taken some measures to improve water quality, algal blooms are still frequently observed [28]. Taihu Lake is characterized by varying turbidity due to inputs from rivers and sediment resuspension [29]. In this study, East Taihu is masked due to aquatic vegetation coverage.

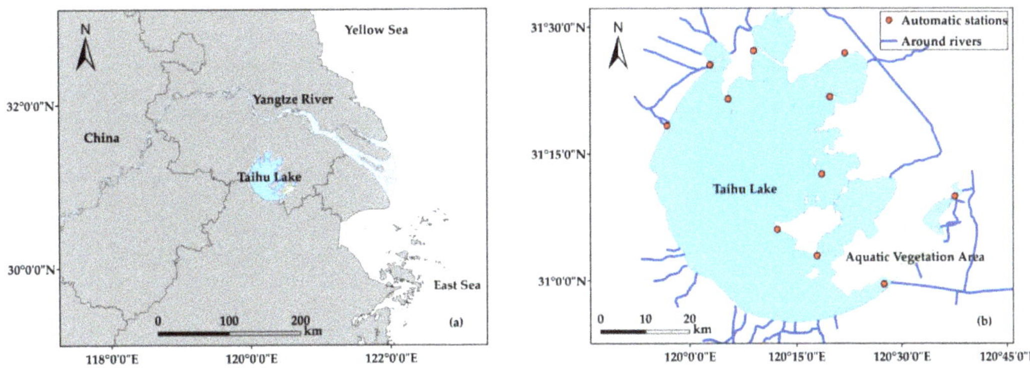

Figure 2. Location of Taihu Lake in Eastern China (**a**) and 11 automatic water quality stations (**b**).

2.2. Field Data

The turbidity data coincident with satellite overpasses was collected from the daily measurement data of automatic water quality monitoring stations in Taihu Lake. In this study, 11 stations distributed throughout the lake were considered (Figure 2). In each station, turbidity was measured by nephelometry [30]. The unit of measurement is nephelometric turbidity units (NTU). Specifically, a water sample is collected into the sample pool by the pump and mixed with a 400 NTU standard turbidity solution. The standard solution was prepared by mixing $C_6H_{12}N_4$, $N_2H_6SO_4$, and distilled water. A beam of light from an LED light with a wavelength of 860 nm passes through the sample pool. The intensity of scattered light is measured by a sensor positioned perpendicular to the emitted light. The turbidity of the water sample has a certain proportional relationship with the intensity of scattered light.

2.3. Satellite Data Acquisition and Processing

All available Level-1C Sentinel-2A/B and Level-1T Landsat-8/9 images from 2018 to January 2023 in Taihu Lake were downloaded from the Copernicus Open Access Hub (https://scihub.copernicus.eu/ accessed on 1 February 2023) and the United States Geological Survey (https://earthexploer.usgs.gov/ accessed on 1 February 2023), respectively(Figure 3). Details of MSI and OLI are given in Table 1. Ultimately, 76 MSI and 42 OLI data scenes with less than 30% cloud cover were obtained. Note that all images are TOA products. In this study, we tested three mature AC methods: the Case-2 Regional Coast Color (C2RCC) processor, the Dark Spectrum Fitting (DSF) algorithm, and Rayleigh Correction.

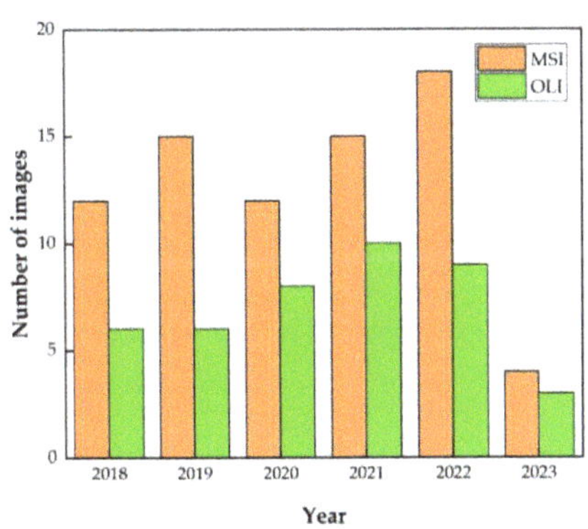

Figure 3. Temporal distributions of the MSI and OLI images from 2018 to 2023.

Table 1. Band setting and SNRs of Sentinel-2A/B MSI and Landsat-8/9 OLI.

Sentinel-2A/B MSI			Landsat-8/9 OLI		
Center Wavelength (nm)	Resolution (m)	Signal-to-Noise Ratio	Center Wavelength (nm)	Resolution (m)	Signal-to-Noise Ratio
443/442	60	1367	443	30	332
493/492	10	206	482	30	381
560/559	10	235	561	30	256
665	10	218	655/654	30	134
704	20	243			
741/740	20	212			
783/780	20	220			
833	10	213			
865/864	20	155	865	30	92
2202/2186	20	165	2201	30	40

The C2RCC processor is based on neural networks and relies on a large database of simulated water-leaving reflectance and related TOA radiance [31]. We used C2RCC on the Sentinel Applications Platform (SNAP, version 9.0) software and set the salinity to 12 PSU. The DSF algorithm assumes that the atmosphere is homogeneous and estimates atmospheric path reflectance from the lowest TOA reflectance of multiple targets in all bands [32]. This algorithm has been integrated into the ACOLITE software, and the version we used is 20210802. The outputs of the C2RCC processor and the DSF algorithm were both R_{rs} (sr^{-1}). Rayleigh correction has been validated as an alternative AC method for application in inland waters in various studies [33–35]. Rayleigh correction is also available on the ACOLITE software. The outputs of the Rayleigh correction were Rayleigh-corrected reflectance (R_{rc}, dimensionless) instead of R_{rs}. After correcting the absorption of water vapor and ozone as well as Rayleigh scattering, based on the assumption that the signals in the SWIR are dominated by aerosol signals, all R_{rc} between visible and NIR are subtracted by R_{rc_swir} to get $R_{rc}^{'}$ [36]. We call this sample the aerosol correction of R_{rc} the Rayleigh-SWIR.

2.4. Satellite Data to In Situ Match-Ups

In general, match-ups between satellite and field data used a time window of ±3 h [37]. The overpass times for MSI and OLI at Taihu Lake are 10:35 a.m. and 10:30 a.m. Each water quality monitoring station measures turbidity at 4:00 a.m., 12:00 a.m., and 20:00 p.m., respectively. We collected field measurements at 12:00 a.m., which satisfied a time difference of <3 h. Moreover, we excluded the measurements when the stations were covered by clouds, shadows, and algae blooms. The value of turbidity greater than 300 NTU was also excluded because high turbidity was rare. Spatial windows (3 × 3 pixels) centered on stations were applied to extract the mean value of images as match-ups. A total of 193 MSI match-ups and 101 OLI match-ups were obtained to train and validate turbidity retrieval models. Table 2 and Figure 4 show the descriptive statistics and distributions of match-ups.

Table 2. Statistics of turbidity from eleven stations matched with satellites. Std denotes the standard deviation.

Sensor	Number	Turbidity (NTU)		
		Range	Mean	Std
MSI	193	6–216	50	39
OLI	101	8–208	47	34

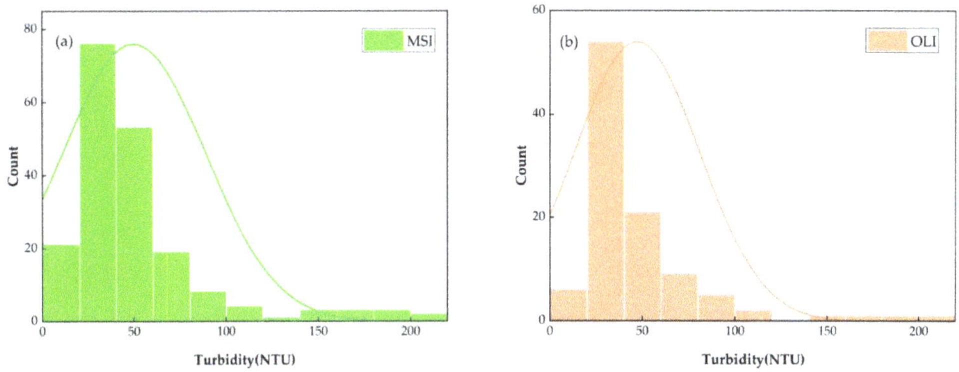

Figure 4. Distributions of MSI-matched turbidity (**a**) and OLI-matched turbidity (**b**) in this study.

2.5. Retrieval Model Development

Dogliotti et al. [38] constructed a single-band semi-analytical turbidity retrieval algorithm using 645 nm and 895 nm bands. The algorithm is suitable for turbidity retrieval in the range of 1–1000 Formazin Nephelometric Units (FNU) in coastal and estuarine waters. Both NTU and FNU use Formazin's primary standard and can be converted to each other. However, turbid lakes have more complicated optical properties than coastal and estuarine waters, resulting in difficulty retrieving turbidity using a single band [39].

In recent years, machine learning has been used to estimate water quality parameters from multi-bands [40–42]. In this study, we test three machine learning models using reflectance as inputs: eXtreme Gradient Boosting (XGBoost) [43], support vector regression (SVR) [44], and random forest (RF) [45]. SVR can solve the non-linear problems in low-dimensional feature space by seeking a linear function in high-dimensional feature space. RF establishes multiple decision trees and outputs the average predicted value for each tree. XGBoost predicts the sum of scores in multiple regression trees [46].

The inputs to the models for MSI and OLI were nine and five spectral bands from visible to NIR, respectively. We separated all match-ups into two subsets: 70% MSI (n = 135) and OLI (n = 70) samples were used to train models, and the remaining 30% MSI (n = 58) and OLI (n = 31) samples were used to test the performance of models. The hyperparameters of each model were determined by cross-validation grid search in Scikit-Learn of Python.

2.6. Intercomparisons at n-SNO

The similar overpass time of Sentinel-2A/B and Landsat-8/9 at the same location allows us to assess the consistency of the intercomparisons at n-SNO events [47]. We performed the consistency assessment for reflectance and turbidity products, and the reflectance intercomparisons were carried out for the MSI and OLI's similar spectral bands. Due to similarities in atmospheric and aquatic conditions, we could choose the optimal AC method according to the consistency of the image-derived reflectance and address the input differences in turbidity retrieval. The comparison between synchronous MSI and OLI image-derived turbidity determines whether they can generate combined turbidity products in Taihu Lake.

A total of 7 n-SNO events in all of the scenes we acquired. In order to minimize uncertainties in the intercomparisons, we first generated random points with 1 km intervals in Taihu Lake, so that each scene has 468 points. Then the MSI data were resampled to 30 m for consistency with the OLI data. The values of reflectance and turbidity were taken from 3 × 3 spatial windows centered on random points.

2.7. Performance Metrics

Common metrics such as the coefficient of determination (R^2), mean absolution percentage error (MAPE), and root mean square error (RMSE) (Equations (1)–(3)) were used to access the performance of models and the consistency of aquatic products [48].

$$R^2 = 1 - \frac{\sum_{i=1}^{n}(y_i - x_i)^2}{\sum_{i=1}^{n}(x_i - \overline{x})^2}, \tag{1}$$

$$\text{MAPE} = \frac{1}{n}\sum_{i=1}^{n}\frac{|y_i - x_i|}{x_i} \times 100\%, \tag{2}$$

$$\text{RMSE} = \sqrt{\frac{\sum_{i=1}^{n}(y_i - x_i)^2}{n}}, \tag{3}$$

where n is the number of points. For the performance evaluation, x_i and y_i are the retrieved values and measured values, respectively; for the consistency analyses, x_i and y_i are the MSI and OLI image-derived values, respectively. We also used the median error (ME) and slope (Equations (4) and (5)) to access consistency in MSI and OLI products due to the presence of outliers and noise [49].

$$\text{ME} = Median(y_i - x_i), \tag{4}$$

$$\text{Slope} = \frac{y_i - 1}{x_i}. \tag{5}$$

3. Results

3.1. Performance of AC Algorithms

The statistical metrics for intercomparisons of R_{rs} after DSF and C2RCC as well as R'_{rc} after Rayleigh-SWIR are summarized in Table 3. Since the ranges of reflectance estimated by the three AC methods are different, RMSE cannot be used as a metric of consistency assessment but can represent the performance in different bands. With the values of R^2, MAPE, ME, and slope, it is possible to evaluate the performance of each AC method based on the consistency of reflectance products. Overall, different AC methods performed differently on various bands, and all AC methods performed worst in NIR bands.

Among the three AC methods, the consistency of R'_{rc} derived from MSI and OLI images by Rayleigh-SWIR performance was better than R_{rs} derived by DSF and C2RCC, particularly in the band of 865 nm. All the visible and NIR bands of MSI and OLI intercomparisons, R'_{rc} indicate relative consistency ($R^2 > 0.84$, MAPE < 7%, ME < 0.0035, and Slope > 0.92). R_{rs} derived by C2RCC presented the worst consistency (average MAPE = 45.76%), possibly due to C2RCC underestimated R_{rs} of turbid water.

The scatterplots show good consistency between the MSI-derived R'_{rc} through the Rayleigh-SWIR method and the corresponding OLI-derived values (Figure 5). Most intercomparisons are evenly distributed around line 1:1. However, R'_{rc} greater than 0.06 at 865 nm depicted a large deviation, which is possibly related to slight algal blooms. In particular, R'_{rc} (665) and R'_{rc} (655) agree to within <6%, which is extremely beneficial for generating consistent turbidity products, as the red band contributes the most to turbidity retrieval.

Table 3. Statistical metrics for MSI-OLI intercomparisons of reflectance after AC.

DSF					
Central Wavelength (nm)	443	492/483	560/561	665/655	865
R^2	0.65	0.69	0.49	0.79	0.61
MAPE	20.47%	13.38%	12.37%	20.47%	81.70%
RMSE (sr^{-1})	0.0046	0.0043	0.0055	0.0054	0.0069

Table 3. *Cont.*

			DSF		
ME (sr^{-1})	0.0019	0.0014	0.0026	0.0036	0.0022
Slope	1.10	0.98	0.72	0.91	0.86
			C2RCC		
Central Wavelength (nm)	443	492/483	560/561	665/655	865
R^2	0.84	0.83	0.75	0.76	0.77
MAPE	22.58%	27.78%	50.49%	57.43%	70.50%
RMSE (sr^{-1})	0.0014	0.0021	0.0055	0.0044	0.0010
ME (sr^{-1})	−0.0005	−0.0011	−0.0033	−0.0017	−0.0007
Slope	0.88	0.81	0.55	0.60	0.47
			Rayleigh-SWIR		
Central Wavelength (nm)	443	492/483	560/561	665/655	865
R^2	0.84	0.89	0.90	0.96	0.88
MAPE	4.53%	3.81%	3.19%	5.76%	6.19%
RMSE	0.0056	0.0049	0.0050	0.0056	0.0052
ME	−0.0016	0.0009	0.0004	0.0035	-3×10^{-5}
Slope	0.92	0.92	0.97	0.95	0.98

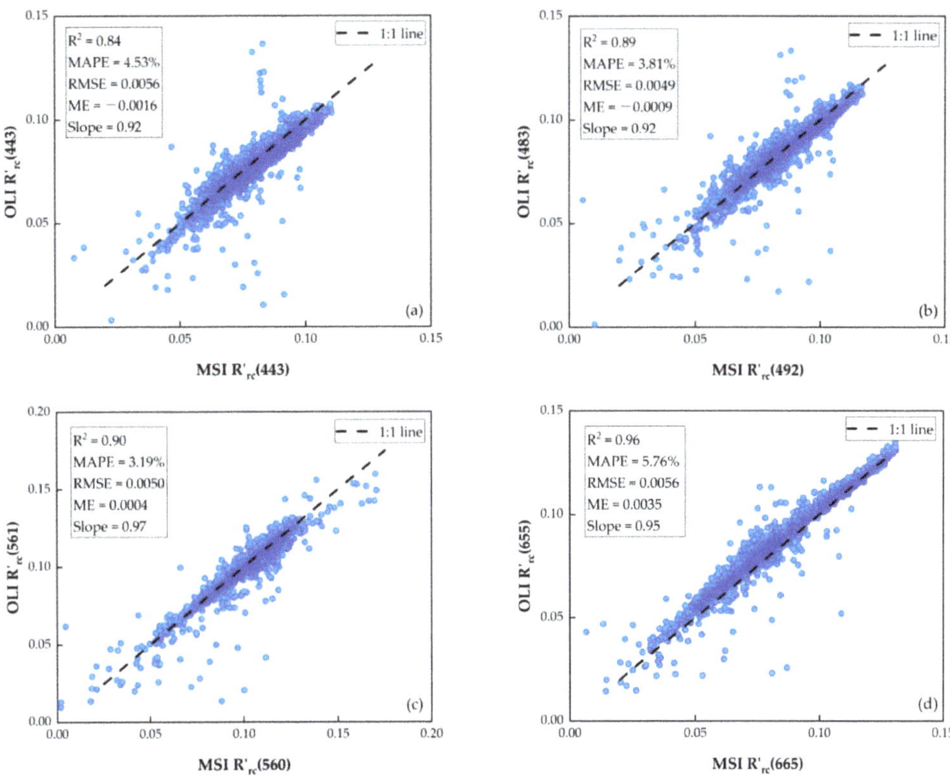

Figure 5. *Cont.*

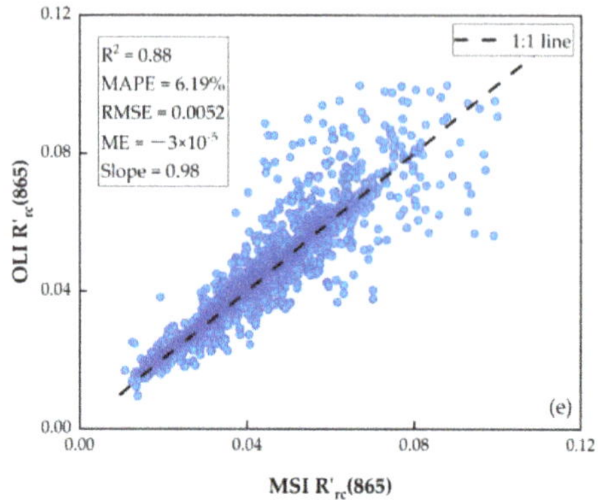

Figure 5. The intercomparisons (n = 3260) of R'_{rc} products from MSI and OLI at n-SNO, (**a**–**e**) represent the comparison of five similar bands of MSI and OLI, respectively.

3.2. Performance of Algorithms in Turbidity Retrieval

The mean values of R'_{rc} of MSI and OLI bands for turbidity at different values are shown in Figure 6. Overall, the average R'_{rc} was proportional to turbidity, and the signals from turbidity changes were well captured by MSI- and OLI-derived R'_{rc} through the Rayleigh-SWIR method. Each R'_{rc} at MSI and OLI visible bands had significant correlations with turbidity, and the difference of R'_{rc} gradually reduced at the NIR bands. The R'_{rc} curve shape was similar in each turbidity range, and the peak of the curse was at 560 nm.

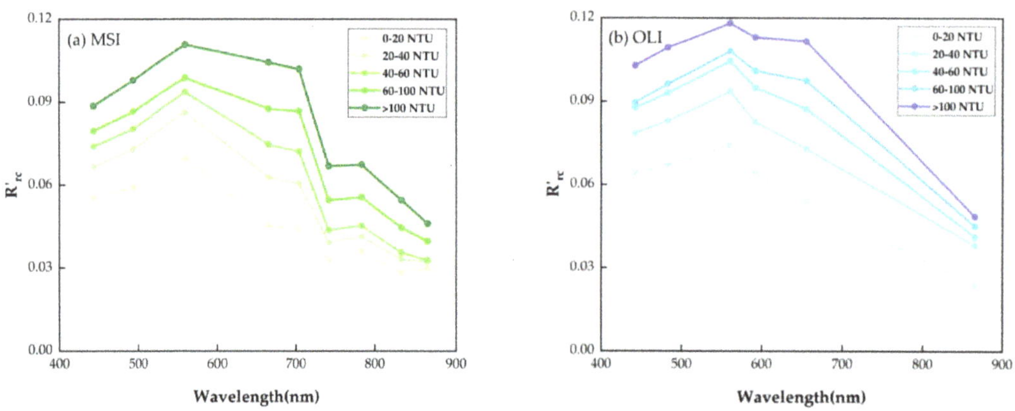

Figure 6. Average R'_{rc} at MSI (**a**) and OLI (**b**) bands in different turbidity ranges.

According to the statistical metrics calculated on the validation data, XGBoost, SVR, and RF had similar performance for turbidity retrievals, although RF had a slightly worse performance (Figure 7a–c). Machine learning models outperformed the semi-analytical algorithm built by Dogliotti et al. (Figure 7d) from R'_{rc}, particularly for high turbidity values. Among them, XGBoost had the best performance for MSI (R^2 = 0.92, MAPE = 18.78%, RMSE = 10.23 NTU), and SVR had the best performance for OLI (R^2 = 0.93, MAPE = 16.20%, RMSE = 9.53 NTU).

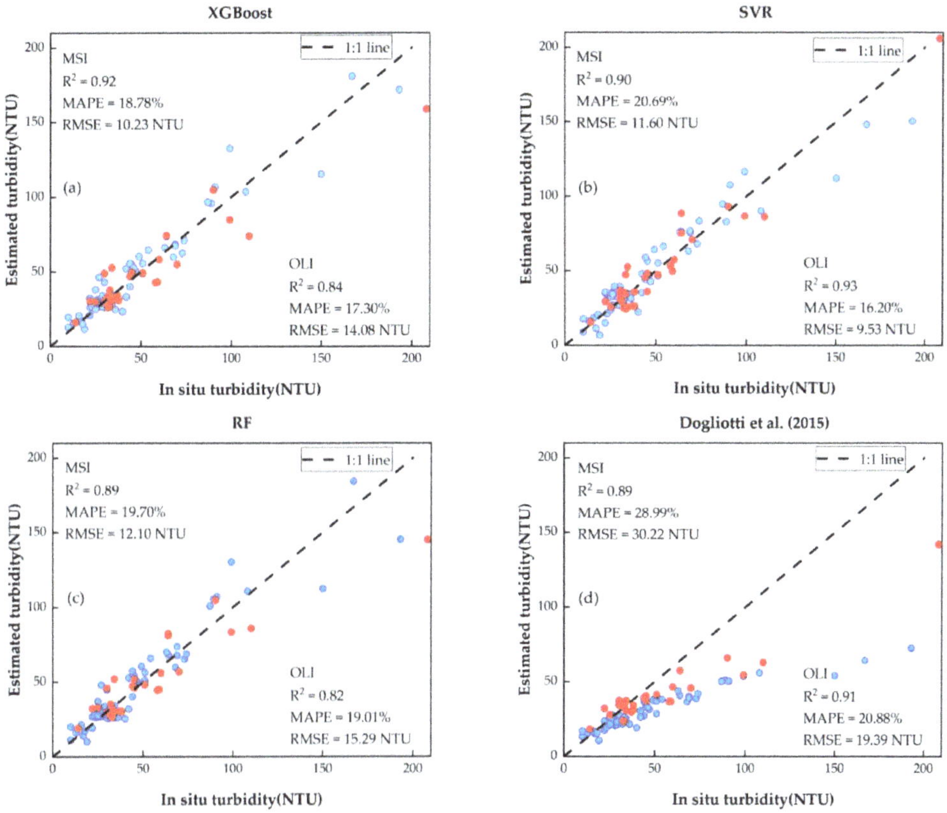

Figure 7. Performance evaluation of turbidity retrieval using XGBoost (**a**), SVR (**b**), RF (**c**), and Dogliotti et al. (**d**). Blue circles represent MSI, and red circles represent OLI.

3.3. Comparison of Turbidity Products between MSI and OLI

Examples of turbidity distributions from the same-day MSI and OLI images were mapped in Figure 8. The range of the color bar was retained consistently for a better comparison. XGBoost-derived turbidity from MSI $R^{'}_{rc}$ exhibited consistent spatial distributions with SVR-derived turbidity from synchronous OLI $R^{'}_{rc}$ in Taihu Lake. MSI-derived turbidity was highly consistent with OLI-derived turbidity in clean lake water, such as in the north and east of Taihu Lake. However, MSI-derived turbidity was slightly higher than OLI-derived turbidity in turbid lake water, such as in the south and west of Taihu Lake.

For all n-SNO of MSI and OLI images, the difference between intercomparisons of MSI- and OLI-derived turbidity was relatively small (R^2 = 0.88, MAPE = 26.48%, RMSE = 34.45 NTU, ME = 15.7 NTU, Slope = 0.68). The scatterplots (Figure 9) show that MSI- and OLI-derived turbidity (<75 NTU) were evenly distributed along the 1:1 line. Despite MSI-derived turbidity being higher than OLI-derived turbidity (>75 NTU), they still maintain a certain correlation.

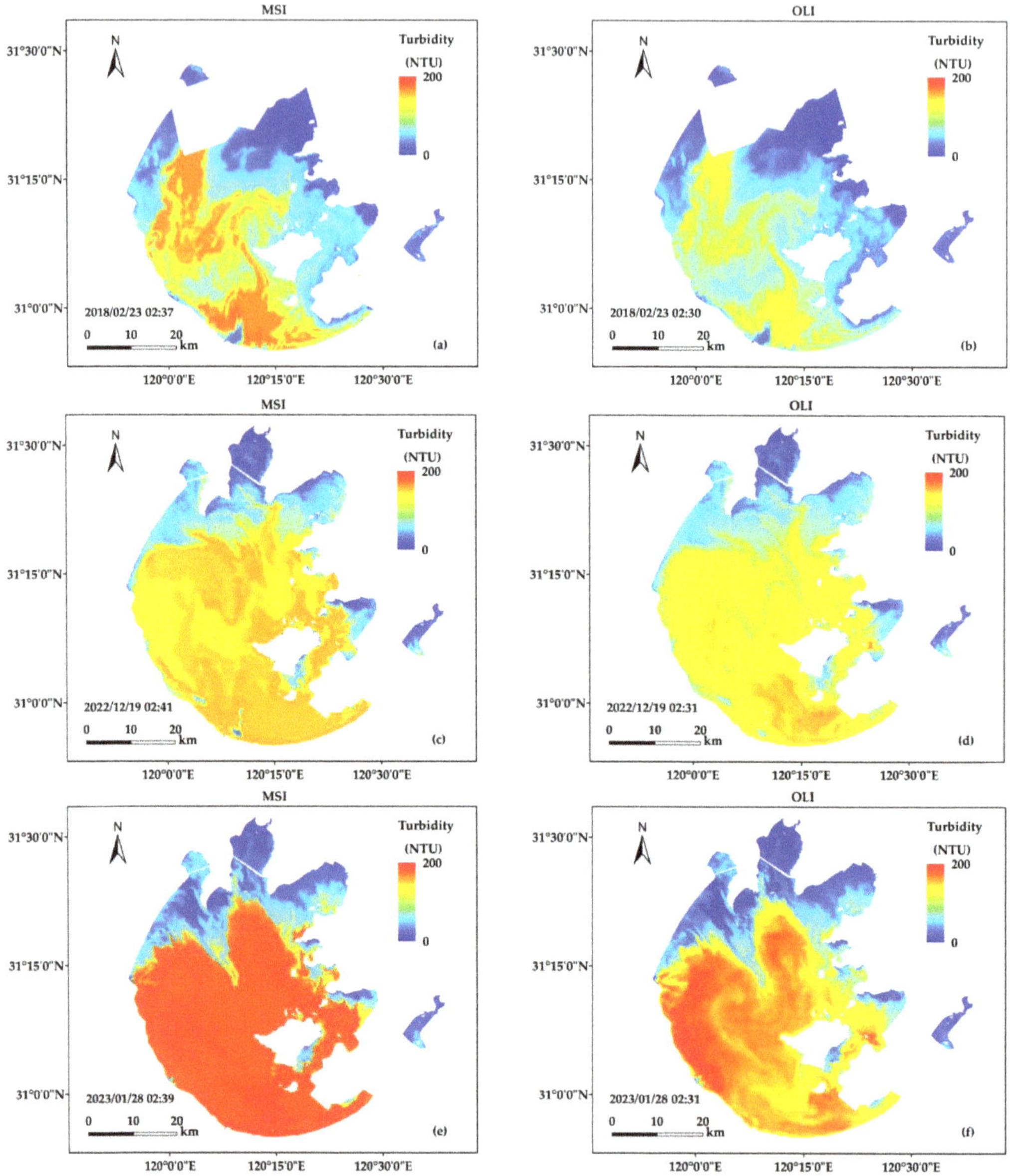

Figure 8. Comparison of the same-day MSI and OLI images derived turbidity. Part of the images on 23 February 2018 were masked due to cloud cover. (**a**,**b**) are comparisons on 23 February 2018, (**c**,**d**) are comparisons on 19 December 2022, and (**e**,**f**) are comparisons on 28 January 2023.

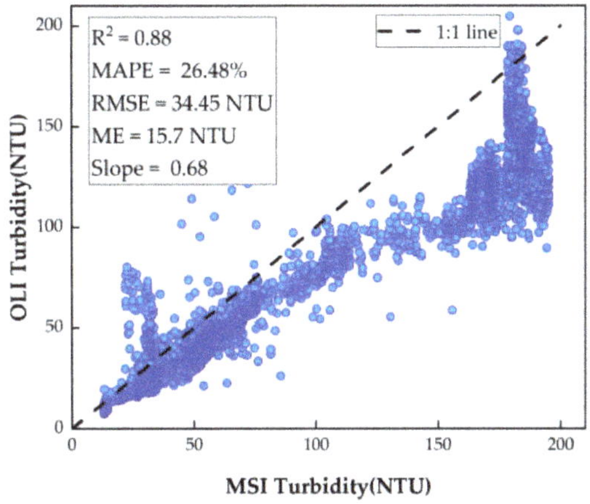

Figure 9. Intercomparisons (n = 3110) of turbidity products from MSI and OLI at n-SNO.

4. Discussion

4.1. Strengths and Limitations of AC

Ideally, in situ R_{rs} should be used to test the performance of different AC methods. However, it is difficult to collect sufficient in situ data to test the ability of AC methods to estimate accurate and consistent reflectance. The Rayleigh-SWIR method outperformed other methods in deriving consistent reflectance products from MSI and OLI images in Taihu Lake. The errors of AC generally come from aerosol corrections, while R_{rc} has been proven accurate without an aerosol correction [50]. Aerosol correction of R_{rc} is the subtraction of the SWIR band from $R_{rc}(\lambda)$, which does not change the spectral shape of R'_{rc}. DSF algorithm needs dark pixels to identify aerosol models. For Taihu Lake, located in the urban agglomeration, insufficient dark pixels may lead to the failure of aerosol correction [51]. The accuracy of C2RCC is influenced by the input of the model, such as bio-optical parameters, and several studies have also reported the inadequacy of the C2RCC algorithm in retrieving R_{rs} in highly turbid water [52].

However, SWIR signals over water are influenced by nearby land pixels and underwater vegetation, resulting in negative retrieval of R'_{rc} in nearshore and shallow water. In addition, the subtraction of SWIR from all bands ignores the spatial and spectral variability of the aerosol, so R'_{rc} may retain residual aerosol signals or overcorrect the aerosol signals [53].

4.2. Sources of Uncertainty

Although the consistency of MSI- and OLI-derived turbidity indicated that MSI and OLI can combine to generate turbidity time series in Taihu Lake, the difference in the calibrated results is still 26%. First, the MSI and OLI's similar bands have different spectral specifications and radiometric performances [54]. For example, the MSI and OLI red channels have significant differences in center wavelength and spectral responses, and new instruments always have a better Signal-to-Noise Ratio than the older ones. Since we did not calibrate the original TOA reflectance, this instrument-induced deviation is about 1–3% and will translate to further products [55]. In addition, MSI- and OLI-derived R'_{rc} using the Rayleigh-SWIR method had an average 4.7% difference, which would translate into retrieval model development and turbidity retrieval.

Machine learning models had good performance for turbidity retrievals from MSI and OLI in the optically complex Taihu Lake. Although the changes in turbidity are closely related to spectral bands, the complex water composition of inland water makes turbidity retrieval challenging. Machine learning algorithms can capture non-linearities between all bands and turbidity in the presence of residual atmospheric signals [56]. However, the applicability of machine learning models depends on the range of input data [57]. MSI and OLI match-ups are insufficient in high turbidity, which results in uncertainties. In particular, the performance of the OLI model decreases with the increase in turbidity due to fewer match-ups. In addition, although the match-ups were collected from MSI and OLI images in all seasons, the rainy summer climate in the lower reaches of the Yangtze River made images in the summer insufficient. Finally, precipitation and wind can cause rapid changes in water quality [58], and the uncertainty between field measurements and image-derived increased the deviation of turbidity products between MSI and OLI.

4.3. Time-Series Analyses

The annual mean turbidity from MSI and OLI images is shown in Figure 10. MSI images were resampled to 30 m to compute the mean turbidity with OLI images. The distribution of turbidity exhibited obvious spatial and interannual changes in Taihu Lake from 2018 to 2022. Five bays in the north and east of Taihu Lake had lower turbidity, and the turbidity of the open area in the north and west of Taihu Lake was high (mean turbidity above 100 NTU). This was mainly due to the resuspension of sediment by wind in open areas [59]. Among the five bays, the turbidity values of the two eastern bays were the lowest (mean turbidity below 30 NTU), and the mean turbidity values of the three northern bays ranged from 50 to 100 NTU. Similar to the results of Yin et al. [27], Secchi disk depth (Z_{sd}) was low in the open area and high in the five bays, which also confirms the high correlation between turbidity and Z_{sd}.

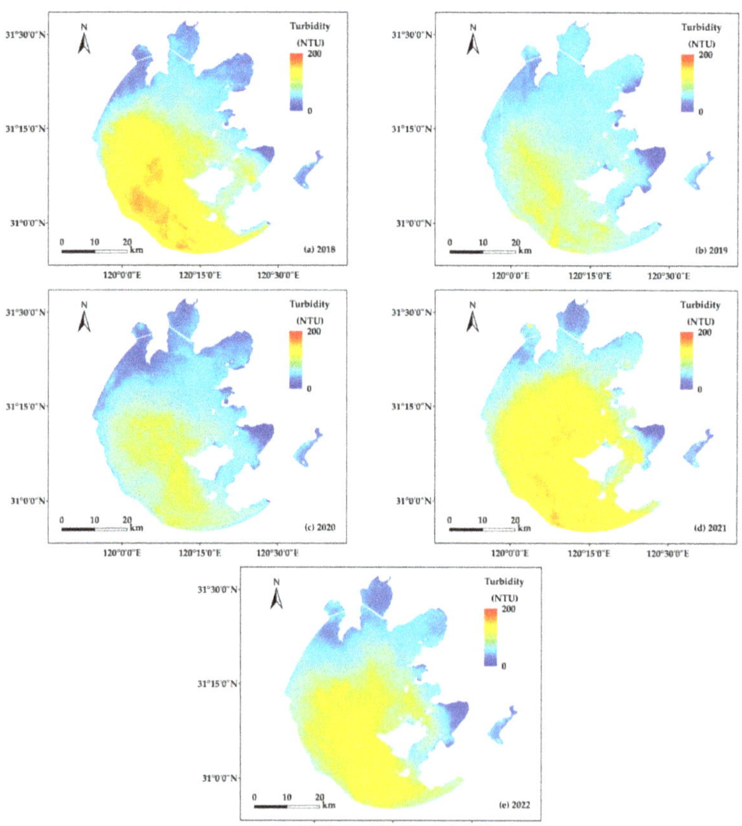

Figure 10. Mean turbidity distributions derived from MSI and OLI images in Taihu Lake from 2018 to 2022 (**a**–**e**).

The interannual changes in turbidity in different sub-regions of the lake varied. The turbidity trend in the open area varied greatly from year to year. The turbidity of the five bays is basically about 50 NTU. In contrast, turbidity in open areas peaked in 2018 and 2021, with both mean turbidities of 150 NTU. Yin et al. [27] also indicated that Z_{sd} showed fluctuations over time. We need more available images from past years to summarize turbidity trends in open areas.

Moreover, the distributions of seasonal mean turbidity were also mapped in Figure 11, with winter mean turbidity included in the available images for 2023. The distribution of turbidity exhibited obvious seasonal characteristics in Taihu Lake. Turbidity was highest in the winter and lowest in the summer. Observations used in statistics for seasonal averages of turbidity included more than 20 images, indicating that seasonal mean turbidity can reflect the actual trends of turbidity.

Daily in situ and MSI- and OLI-derived turbidity time series at one of the stations in Taihu Lake from 2018 to 2023 are illustrated in Figure 12. This station is located in the center of the lake and has the most complete daily in situ observations. In situ data also indicated that turbidity in winter was higher than in other seasons at the central lake. MSI- and OLI-derived turbidity had a consistent temporal trend with in situ turbidity. Although image-derived turbidity still deviated from in situ turbidity, the combination of MSI and OLI increased the number of available observations of dynamical inland lakes. In addition, the time series of the other two stations are illustrated in Figure A1.

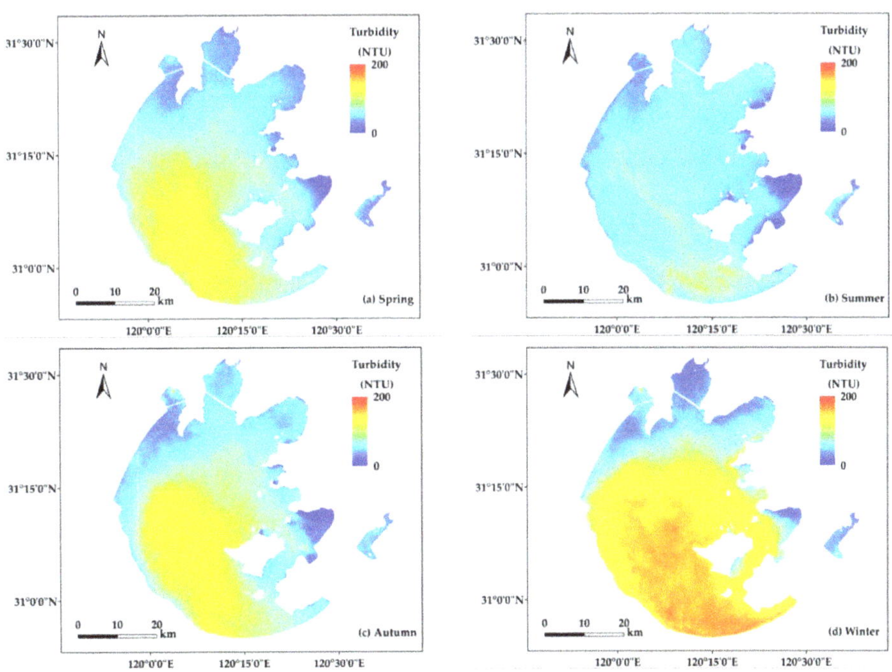

Figure 11. Seasonal mean turbidity in the spring (**a**), summer (**b**), autumn (**c**), and winter (**d**) in Taihu Lake from 2018 to 2023.

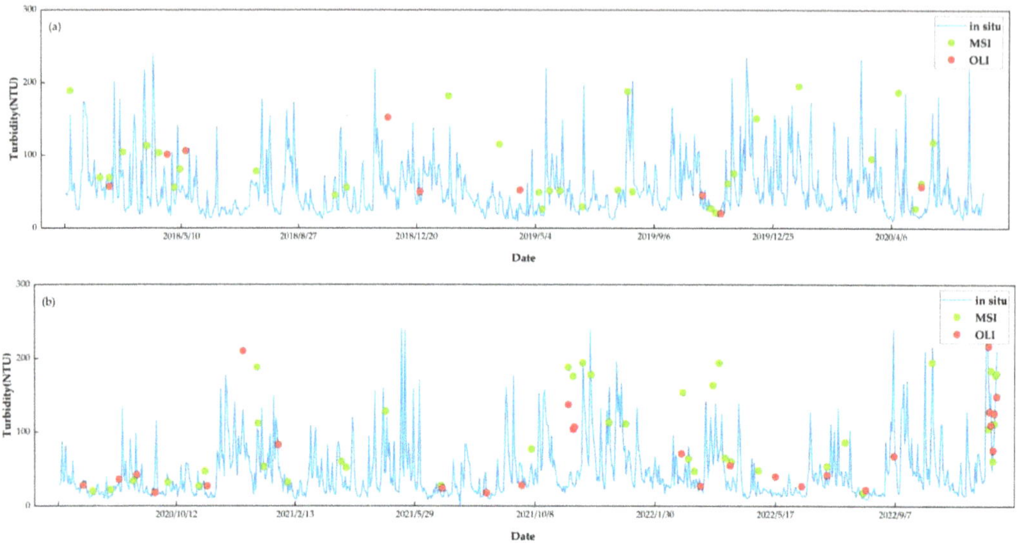

Figure 12. Time-series of in situ and MSI- and OLI-derived turbidity for one station from 2018 to 2023 in Taihu Lake. The date of (**a**) is from 1 January 2018 to 30 June 2020 and the date of (**b**) is from 1 July 2020 to 28 February 2023.

5. Conclusions

In this study, we generated consistent turbidity products in Taihu Lake from 2018 to 2022 using Sentinel-2A/B MSI and Landsat-8/9 OLI. We tested the performance of three aquatic AC methods to retrieve consistent reflectance products. The R'_{rc} derived from MSI and OLI images by Rayleigh-SWIR were most consistent. Machine learning models outperformed an existing semi-analytical retrieval algorithm, where XGboost and SVR had the best performance for MSI and OLI turbidity retrievals, respectively. Machine learning models were selected to retrieve turbidity from MSI and OLI R'_{rc} images. The consistency of turbidity retrievals was validated with the intercomparisons from the same-day MSI and OLI images. Finally, we generate the annual and seasonal mean turbidity from all MSI and OLI images. The time series indicated that MSI and OLI had high potential for combining to monitor lake water quality.

However, some uncertainties in our study should be resolved in the future. Firstly, Rayleigh-SWIR is not a perfect AC method; the low signal-to-noise ratios of MSI and OLI in SWIR bands affect the accuracy of correcting aerosol signals. Methods for estimating aerosol signals in R_{rc} need some improvement. Then, we need more MSI and OLI images of other inland lakes to verify the generality of the process and more in situ turbidity data to reduce the differences between models in high turbidity.

Author Contributions: Conceptualization, C.G. and Y.H.; methodology, Z.Y. and L.L.; formal analysis, Z.Y. and Z.L.; resources, C.G. and Y.H.; data curation, L.D. and E.W.; writing—original draft preparation, Z.Y.; writing—review and editing, Z.Y. and H.H. All authors have read and agreed to the published version of the manuscript.

Funding: This research was funded by Shanghai 2021 "Science and Technology Innovation Action Plan" Social Development Science and Technology Research Project (21DZ1202500), the Jiangsu Provincial Water Conservancy Science and Technology Project (No. 2020068), and the Science and Technology Project of Shanghai Municipal Water Bureau (Shanghai Branch 2021-10, Shanghai Branch 2018-07).

Data Availability Statement: Not applicable.

Acknowledgments: We are also thankful to all anonymous reviewers for their constructive comments provided on the study.

Conflicts of Interest: The authors declare no conflict of interest.

Appendix A

Figure A1. Time-series of in situ and MSI- and OLI-derived turbidity for two stations from 2018 to 2023 in Taihu Lake. (**a**,**b**) map a station in the open area of the lake, (**c**,**d**) map a station in the bay of the lake.

References

1. Adrian, R.; O'Reilly, C.M.; Zagarese, H.; Baines, S.B.; Hessen, D.O.; Keller, W.; Livingstone, D.M.; Sommaruga, R.; Straile, D.; Van Donk, E.; et al. Lakes as sentinels of climate change. *Limnol. Oceanogr.* **2009**, *54*, 2283–2297. [PubMed]
2. Jane, S.F.; Hansen, G.J.A.; Kraemer, B.M.; Leavitt, P.R.; Mincer, J.L.; North, R.L.; Pilla, R.M.; Stetler, J.T.; Williamson, C.E.; Woolway, R.I.; et al. Widespread deoxygenation of temperate lakes. *Nature* **2021**, *594*, 66–70. [PubMed]
3. Ma, Y.; Song, K.; Wen, Z.; Liu, G.; Shang, Y.; Lyu, L.; Du, J.; Yang, Q.; Li, S.; Tao, H.; et al. Remote Sensing of Turbidity for Lakes in Northeast China Using Sentinel-2 Images with Machine Learning Algorithms. *IEEE J. Sel. Top. Appl. Earth Obs. Remote Sens.* **2021**, *14*, 9132–9146.
4. Lee, Z.; Shang, S.; Hu, C.; Du, K.; Weidemann, A.; Hou, W.; Lin, J.; Lin, G. Secchi disk depth: A new theory and mechanistic model for underwater visibility. *Remote Sens. Environ.* **2015**, *169*, 139–149.
5. Shi, K.; Zhang, Y.; Zhu, G.; Liu, X.; Zhou, Y.; Xu, H.; Qin, B.; Liu, G.; Li, Y. Long-term remote monitoring of total suspended matter concentration in Lake Taihu using 250m MODIS-Aqua data. *Remote Sens. Environ.* **2015**, *164*, 43–56.
6. Nechad, B.; Ruddick, K.G.; Neukermans, G. Calibration and validation of a generic multisensor algorithm for mapping of turbidity in coastal waters. *Proc. SPIE* **2009**, *7473*, 74730H.
7. Hou, X.; Feng, L.; Duan, H.; Chen, X.; Sun, D.; Shi, K. Fifteen-year monitoring of the turbidity dynamics in large lakes and reservoirs in the middle and lower basin of the Yangtze River, China. *Remote Sens. Environ.* **2017**, *190*, 107–121.
8. Potes, M.; Costa, M.J.; Salgado, R. Satellite remote sensing of water turbidity in Alqueva reservoir and implications on lake modelling. *Hydrol. Earth Syst. Sci.* **2012**, *16*, 1623–1633.
9. Goodin, D.G.; Harrington, J.A., Jr.; Nellis, M.D.; Rundquist, D.C. Mapping reservoir turbidity patterns using SPOT-HRV data. *Geovarto Int.* **1996**, *11*, 71–78.
10. Bustamante, J.; Pacios, F.; Díaz-Delgado, R.; Aragonés, D. Predictive models of turbidity and water depth in the Doñana marshes using Landsat TM and ETM+ images. *J. Environ. Manag.* **2009**, *90*, 2219–2225.
11. Zhou, Q.; Wang, J.; Tian, L.; Feng, L.; Li, J.; Xing, Q. Remotely sensed water turbidity dynamics and its potential driving factors in Wuhan, an urbanizing city of China. *J. Hydrol.* **2021**, *593*, 125893.
12. Rodríguez-López, L.; Duran-Llacer, I.; González-Rodríguez, L.; Cardenas, R.; Urrutia, R. Retrieving Water Turbidity in Araucanian Lakes (South-Central Chile) Based on Multispectral Landsat Imagery. *Remote Sens.* **2021**, *13*, 3133.
13. He, Y.; Jin, S.; Shang, W. Water Quality Variability and Related Factors along the Yangtze River Using Landsat-8. *Remote Sens.* **2021**, *13*, 2241.
14. Gao, Z.; Shen, Q.; Wang, X.; Peng, H.; Yao, Y.; Wang, M.; Wang, L.; Wang, R.; Shi, J.; Shi, D.; et al. Spatiotemporal Distribution of Total Suspended Matter Concentration in Changdang Lake Based on In Situ Hyperspectral Data and Sentinel-2 Images. *Remote Sens.* **2021**, *13*, 4230.
15. Ciancia, E.; Campanelli, A.; Lacava, T.; Palombo, A.; Pascucci, S.; Pergola, N.; Pignatti, S.; Satriano, V.; Tramutoli, V. Modeling and Multi-Temporal Characterization of Total Suspended Matter by the Combined Use of Sentinel 2-MSI and Landsat 8-OLI Data: The Pertusillo Lake Case Study (Italy). *Remote Sens.* **2020**, *12*, 2147.
16. Page, B.P.; Olmanson, L.G.; Mishra, D.R. A harmonized image processing workflow using Sentinel-2/MSI and Landsat-8/OLI for mapping water clarity in optically variable lake systems. *Remote Sens. Environ.* **2019**, *231*, 111284.
17. Pahlevan, N.; Chittimalli, S.K.; Balasubramanian, S.V.; Vellucci, V. Sentinel-2/Landsat-8 product consistency and implications for monitoring aquatic systems. *Remote Sens. Environ.* **2019**, *220*, 19–29.
18. Novoa, S.; Doxaran, D.; Ody, A.; Vanhellemont, Q.; Lafon, V.; Lubac, B.; Gernez, P. Atmospheric Corrections and Multi-Conditional Algorithm for Multi-Sensor Remote Sensing of Suspended Particulate Matter in Low-to-High Turbidity Levels Coastal Waters. *Remote Sens.* **2017**, *9*, 61.
19. Warren, M.A.; Simis, S.G.H.; Selmes, N. Complementary water quality observations from high and medium resolution Sentinel sensors by aligning chlorophyll-a and turbidity algorithms. *Remote Sens. Environ.* **2021**, *265*, 112651. [PubMed]
20. Caballero, I.; Stumpf, R.P.; Meredith, A. Preliminary Assessment of Turbidity and Chlorophyll Impact on Bathymetry Derived from Sentinel-2A and Sentinel-3A Satellites in South Florida. *Remote Sens.* **2019**, *11*, 645.
21. Wang, D.; Ma, R.; Xue, K.; Loiselle, S.A. The Assessment of Landsat-8 OLI Atmospheric Correction Algorithms for Inland Waters. *Remote Sens.* **2019**, *11*, 169. [CrossRef]
22. Warren, M.A.; Simis, S.G.H.; Martinez-Vicente, V.; Poser, K.; Bresciani, M.; Alikas, K.; Spyrakos, E.; Giardino, C.; Ansper, A. Assessment of atmospheric correction algorithms for the Sentinel-2A MultiSpectral Imager over coastal and inland waters. *Remote Sens. Environ.* **2019**, *225*, 267–289. [CrossRef]
23. Zhang, M.; Hu, C. Evaluation of Remote Sensing Reflectance Derived From the Sentinel-2 Multispectral Instrument Observations Using POLYMER Atmospheric Correction. *IEEE Trans. Geosci. Remote Sens.* **2020**, *58*, 5764–5771. [CrossRef]
24. Kuhn, C.; Valerio, A.M.; Ward, N.; Loken, L.; Sawakuchi, H.O.; Kampel, M.; Richey, J.; Stadler, P.; Crawford, J.; Striegl, R.; et al. Performance of Landsat-8 and Sentinel-2 surface reflectance products for river remote sensing retrievals of chlorophyll-a and turbidity. *Remote Sens. Environ.* **2019**, *224*, 104–118. [CrossRef]
25. Cao, Z.; Ma, R.; Liu, M.; Duan, H.; Xiao, Q.; Xue, K.; Shen, M. Harmonized Chlorophyll-a Retrievals in Inland Lakes From Landsat-8/9 and Sentinel 2A/B Virtual Constellation Through Machine Learning. *IEEE Trans. Geosci. Remote Sens.* **2022**, *60*, 1–16. [CrossRef]

26. Güttler, F.N.; Niculescu, S.; Gohin, F. Turbidity retrieval and monitoring of Danube Delta waters using multi-sensor optical remote sensing data: An integrated view from the delta plain lakes to the western—Northwestern Black Sea coastal zone. *Remote Sens. Environ.* **2013**, *132*, 86–101. [CrossRef]
27. Yin, Z.; Li, J.; Liu, Y.; Xie, Y.; Zhang, F.; Wang, S.; Sun, X.; Zhang, B. Water clarity changes in Lake Taihu over 36 years based on Landsat TM and OLI observations. *Int. J. Appl. Earth Obs.* **2021**, *102*, 102457. [CrossRef]
28. Lyu, H.; Yang, Z.; Shi, L.; Li, Y.; Guo, H.; Zhong, S.; Miao, S.; Bi, S.; Li, Y. A Novel Algorithm to Estimate Phytoplankton Carbon Concentration in Inland Lakes Using Sentinel-3 OLCI Images. *IEEE Trans. Geosci. Remote Sens.* **2020**, *58*, 6512–6523. [CrossRef]
29. Liu, D.; Yu, S.; Xiao, Q.; Qi, T.; Duan, H. Satellite estimation of dissolved organic carbon in eutrophic Lake Taihu, China. *Remote Sens. Environ.* **2021**, *264*, 112572. [CrossRef]
30. Telesnicki, G.J.; Goldberg, W.M. Comparison of turbidity measurement by nephelometry and transmissometry and its relevance to water quality standards. *B Mar. Sci.* **1995**, *57*, 540–547.
31. Brockmann, C.; Doerffer, R.; Peters, M.; Kerstin, S.; Embacher, S.; Ruescas, A. Evolution of the C2RCC neural network for Sentinel 2 and 3 for the retrieval of ocean colour products in normal and extreme optically complex waters. In Proceedings of the Living Planet Symposium 2016, Prague, Czech Republic, 9–13 May 2016; pp. 9–13.
32. Vanhellemont, Q.; Ruddick, K. Atmospheric correction of metre-scale optical satellite data for inland and coastal water applications. *Remote Sens. Environ.* **2018**, *216*, 586–597. [CrossRef]
33. Duan, H.; Feng, L.; Ma, R.; Zhang, Y.; Loiselle, S.A. Variability of particulate organic carbon in inland waters observed from MODIS Aqua imagery. *Environ. Res.* **2014**, *9*, 084011. [CrossRef]
34. Feng, L.; Hu, C.; Han, X.; Chen, X.; Qi, L. Long-Term Distribution Patterns of Chlorophyll-a Concentration in China's Largest Freshwater Lake: MERIS Full-Resolution Observations with a Practical Approach. *Remote Sens.* **2015**, *7*, 275–299. [CrossRef]
35. Zhang, Y.; Ma, R.; Duan, H.; Loiselle, S.; Zhang, M.; Xu, J. A novel MODIS algorithm to estimate chlorophyll a concentration in eutrophic turbid lakes. *Ecol. Ind.* **2016**, *69*, 138–151. [CrossRef]
36. Feng, L.; Hou, X.; Li, J.; Zheng, Y. Exploring the potential of Rayleigh-corrected reflectance in coastal and inland water applications: A simple aerosol correction method and its merits. *ISPRS J. Photogramm. Remote Sens.* **2018**, *146*, 52–64. [CrossRef]
37. Bailey, S.W.; Werdell, P.J. A multi-sensor approach for the on-orbit validation of ocean color satellite data products. *Remote Sens. Environ.* **2006**, *102*, 12–23. [CrossRef]
38. Dogliotti, A.I.; Ruddick, K.G.; Nechad, B.; Doxaran, D.; Knaeps, E. A single algorithm to retrieve turbidity from remotely-sensed data in all coastal and estuarine waters. *Remote Sens. Environ.* **2015**, *156*, 157–168. [CrossRef]
39. Cao, Z.; Ma, R.; Duan, H.; Pahlevan, N.; Melack, J.; Shen, M.; Xue, K. A machine learning approach to estimate chlorophyll-a from Landsat-8 measurements in inland lakes. *Remote Sens. Environ.* **2020**, *248*, 111974. [CrossRef]
40. Shi, J.; Shen, Q.; Yao, Y.; Li, J.; Chen, F.; Wang, R.; Xu, W.; Gao, Z.; Wang, L.; Zhou, Y. Estimation of Chlorophyll-a Concen-trations in Small Water Bodies: Comparison of Fused Gaofen-6 and Sentinel-2 Sensors. *Remote Sens.* **2022**, *14*, 229. [CrossRef]
41. Hafeez, S.; Wong, M.S.; Ho, H.C.; Nazeer, M.; Nichol, J.E.; Abbas, S.; Tang, D.; Lee, K.-H.; Pun, L. Comparison of Machine Learning Algorithms for Retrieval of Water Quality Indicators in Case-II Waters: A Case Study of Hong Kong. *Remote Sens.* **2019**, *11*, 617. [CrossRef]
42. Guo, H.; Huang, J.J.; Chen, B.; Guo, X.; Singh, V.P. A Machine Learning-Based Strategy for Estimating Non-Optically Active Water Quality Parameters Using Sentinel-2 Imagery. *Int. J. Remote Sens.* **2020**, *42*, 1841–1866. [CrossRef]
43. Chen, T.; Guestrin, C. XGBoost: A Scalable Tree Boosting System. In Proceedings of the KDD'16: 22nd ACM SIGKDD International Conference on Knowledge Discovery and Data Mining, New York, NY, USA, 13–17 August 2016; pp. 785–794.
44. Mountrakis, G.; Im, J.; Ogole, C. Support vector machines in remote sensing: A review. *ISPRS J. Photogramm. Remote Sens.* **2011**, *66*, 247–259. [CrossRef]
45. Huangfu, K.; Li, J.; Zhang, X.; Zhang, J.; Cui, H.; Sun, Q. Remote Estimation of Water Quality Parameters of Medium- and Small-Sized Inland Rivers Using Sentinel-2 Imagery. *Water* **2020**, *12*, 3124. [CrossRef]
46. Yang, Z.; Gong, C.; Ji, T.; Hu, Y.; Li, L. Water Quality Retrieval from ZY1-02D Hyperspectral Imagery in Urban Water Bodies and Comparison with Sentinel-2. *Remote Sens.* **2022**, *14*, 5029. [CrossRef]
47. Hafeez, S.; Wong, M.S.; Abbas, S.; Asim, M. Evaluating Landsat-8 and Sentinel-2 Data Consistency for High Spatiotemporal Inland and Coastal Water Quality Monitoring. *Remote Sens.* **2022**, *14*, 3155. [CrossRef]
48. Seegers, B.N.; Stumpf, R.P.; Schaeffer, B.; Loftin, K.A.; Werdell, P.J. Performance metrics for the assessment of satellite data products: An ocean color case study. *Opt. Express* **2018**, *26*, 7404–7422. [CrossRef]
49. Pahlevan, N.; Smith, B.; Alikas, K.; Anstee, J.; Barbosa, C.; Binding, C.; Bresciani, M.; Cremella, B.; Giardino, C.; Gurlin, D.; et al. Simultaneous retrieval of selected optical water quality indicators from Landsat-8, Sentinel-2, and Sentinel-3. *Remote Sens. Environ.* **2022**, *270*, 112860. [CrossRef]
50. Cao, Z.; Ma, R.; Pahlevan, N.; Liu, M.; Melack, J.M.; Duan, H.; Xue, K.; Shen, M. Evaluating and optimizing VIIRS retrievals of chlorophyll-a and suspended particulate matter in turbid lakes using a machine learning approach. *IEEE Trans. Geosci. Remote Sens.* **2022**, *60*, 1–17. [CrossRef]
51. Vanhellemont, Q.; Ruddick, K. Turbid wakes associated with offshore wind turbines observed with Landsat 8. *Remote Sens. Environ.* **2014**, *145*, 105–115. [CrossRef]
52. Sent, G.; Biguino, B.; Favareto, L.; Cruz, J.; Sá, C.; Dogliotti, A.I.; Palma, C.; Brotas, V.; Brito, A.C. Deriving Water Quality Parameters Using Sentinel-2 Imagery: A Case Study in the Sado Estuary, Portugal. *Remote Sens.* **2021**, *13*, 1043. [CrossRef]

53. Feng, L.; Hu, C. Land adjacency effects on MODIS Aqua top-of-atmosphere radiance in the shortwave infrared: Statistical assessment and correction. *J. Geophys. Res.* **2017**, *122*, 4802–4818. [CrossRef]
54. Buma, W.G.; Lee, S.-I. Evaluation of Sentinel-2 and Landsat 8 Images for Estimating Chlorophyll-a Concentrations in Lake Chad, Africa. *Remote Sens.* **2020**, *12*, 2437. [CrossRef]
55. Pahlevan, N.; Smith, B.; Binding, C.; O'Donnell, D.M. Spectral band adjustments for remote sensing reflectance spectra in coastal/inland waters. *Opt. Express* **2017**, *25*, 28650–28667. [CrossRef]
56. Hu, C.; Feng, L.; Guan, Q. A Machine Learning Approach to Estimate Surface Chlorophyll a Concentrations in Global Oceans From Satellite Measurements. *IEEE Trans. Geosci. Remote Sens.* **2021**, *59*, 4590–4607. [CrossRef]
57. Pahlevan, N.; Smith, B.; Schalles, J.; Binding, C.; Cao, Z.; Ma, R.; Alikas, K.; Kangro, K.; Gurlin, D.; Nguyen, H.; et al. Seamless retrievals of chlorophyll-a from Sentinel-2 (MSI) and Sentinel-3 (OLCI) in inland and coastal waters: A machine-learning approach. *Remote Sens. Environ.* **2020**, *240*, 111604. [CrossRef]
58. Xu, G.; Li, P.; Lu, K.; Tantai, Z.; Zhang, J.; Ren, Z.; Wang, X.; Yu, K.; Shi, P.; Cheng, Y. Seasonal changes in water quality and its main influencing factors in the Dan River basin. *CATENA* **2018**, *173*, 131–140. [CrossRef]
59. Feng, L.; Hou, X.; Zheng, Y. Monitoring and understanding the water transparency changes of fifty large lakes on the Yangtze Plain based on long-term MODIS observations. *Remote Sens. Environ.* **2019**, *221*, 675–686. [CrossRef]

remote sensing MDPI

Article

Retrieval of Chla Concentrations in Lake Xingkai Using OLCI Images

Li Fu [1,3], Yaming Zhou [2], Ge Liu [1], Kaishan Song [1,3], Hui Tao [1], Fangrui Zhao [1], Sijia Li [1], Shuqiong Shi [4] and Yingxin Shang [1,*]

1 Northeast Institute of Geography and Agroecology, Chinese Academy of Sciences, Changchun 130102, China; 2010140213@stu.lcu.edu.cn (L.F.); liuge@iga.ac.cn (G.L.); songkaishan@iga.ac.cn (K.S.); taohui@iga.ac.cn (H.T.); zhaofr226@nenu.edu.cn (F.Z.); lisijia@iga.ac.cn (S.L.)
2 Center for Satellite Application on Ecology and Environment, Ministry of Ecology and Environment of the People's Republic of China, Beijing 100029, China; zhouyaming@secmep.cn
3 School of Geography and Environment, Liaocheng University, Liaocheng 252059, China
4 Ministry of Ecology and Environment of China, Beijing 100029, China; shi.shuqiong@mee.gov.cn
* Correspondence: shangyingxin@iga.ac.cn

Abstract: Lake Xingkai is a large turbid lake composed of two parts, Small Lake Xingkai and Big Lake Xingkai, on the border between Russia and China, where it represents a vital source of water, fishing, water transport, recreation, and tourism. Chlorophyll-a (Chl*a*) is a prominent phytoplankton pigment and a proxy for phytoplankton biomass, reflecting the trophic status of waters. Regularly monitoring Chl*a* concentrations is vital for issuing timely warnings of this lake's eutrophication. Owing to its higher spatial and temporal coverages, remote sensing can provide a synoptic complement to traditional measurement methods by targeting the optical Chl*a* absorption signals, especially for the lakes that lack regular in situ sampling cruises, like Lake Xingkai. This study calibrated and validated several commonly used remote sensing Chl*a* retrieval algorithms (including the two-band ratio, three-band method, four-band method, and baseline methods) by applying them to Sentinel-3 Ocean and Land Colour Instrument (OLCI) images in Lake Xingkai. Among these algorithms, the four-band model (FBA), which removes the absorption signal of detritus and colored dissolved organic matter, was the best-performing model with an R^2 of 0.64 and a mean absolute percentage difference of 38.26%. With the FBA model applied to OLCI images, the monthly and spatial distributions of Chl*a* in Lake Xingkai were studied from 2016 to 2022. The results showed that over the seven years, the Chl*a* concentrations in Small Lake Xingkai were higher than in Big Lake Xingkai. Unlike other eutrophic lakes in China (e.g., Lake Taihu and Lake Chaohu), Lake Xingkai did not display a stable seasonal Chl*a* variation pattern. We also found uncertainties and limitations of the Chl*a* algorithm models when using a larger satellite zenith angle or applying it to an algal bloom area. Recent increases in anthropogenic nutrient loading, water clarity, and warming temperatures may lead to rising phytoplankton biomass in Lake Xingkai, and the results of this study can be applied for the satellite-based monitoring of its water quality.

Keywords: Lake Xingkai; Chl*a*; remote sensing; OLCI; high solar zenith angle

1. Introduction

The chlorophyll-a concentrations (Chl*a*) are the most important parameter to assess the trophic status of water; Chl*a* represents an easy and simple proxy for phytoplankton biomass and has been used for decades as a key variable to assess external loading levels of phosphorus for eutrophication [1]. Regularly monitoring Chl*a* concentrations is essential for assessing and monitoring cyanobacterial blooms and implementing efficient management strategies, avoiding further deterioration of water quality [2]. Traditional in situ sampling methods are expensive and time-consuming, and therefore direct Chl*a* concentration measurements are unavailable for many lakes [3,4]. The Chl*a* concentration and eutrophication, and increased productivity typically result in variations in the optical properties of water; therefore, remote sensing approaches have been employed for water trophic state assessment; using satellite remote sensing to estimate Chl*a* concentration offers multiple advantages and allows for identifying large-scale and long-term dynamics [5,6].

It is acknowledged that algorithms for Chl*a* retrieval for oceans generally return inaccurate results in turbid lakes due to the significant difference in the optical properties of these two types of water bodies [7,8]. Clear ocean water optical properties are determined by phytoplankton and related substances [9]. For turbid lakes, the optical properties are strongly influenced by suspended sediment or colored-dissolved organic matter [10]. Accurately estimating Chl*a* concentration from remotely sensed data is particularly challenging in turbid, productive waters. Until now, a variety of algorithms have been developed for the retrieval of Chl*a* in turbid waters, including the two-band ratio method [11], the three-band algorithm [5], the Fluorescence Line Height algorithm (FLH) [12], the MERIS Maximum Chlorophyll Index (MCI) [13], and Quasi-Analytical Algorithm and inherent optical property (IOP) inversion-based algorithms [14–18].

Since the NASA Coastal Zone Color Scanner (CZCS) was launched in 1978, ocean-color satellite instruments have provided temporal and spatial distributions of Chl*a* [19]. In the past decades, significant advances have been made in applying remotely sensed data, and various remote sensing sensors have been used to derive Chl*a* products, such as MODIS (Moderate Resolution Imaging Spectroradiometer), SeaWiFS (Sea-viewing Wide Field-of-view Sensor), and MERIS (Medium Resolution Imaging Spectrometer). A recently introduced sensor for monitoring turbid lakes is the Ocean and Land Color Instrument (OLCI) onboard the Sentinel-3 satellite. OLCI includes 21 spectral bands, improved signal-to-noise ratio (SNR), cameras for mitigation of sunglint, a global spatial resolution of 300 m, daily revisit times, and enhanced coverage. Recently, many researchers have attempted to map Chl*a* concentration based on OLCI images in various turbid lakes [6,20–23].

In this study, we attempted to apply the OLCI sensors to estimate Chl*a* concentrations in Lake Xingkai, one of the largest but less well-known freshwater lakes in Northeast Asia, where the recent gradual expansion of agriculture, industry, urbanization, and land-use change, has led to water quality degradation and development of algal blooms [24]. Three specific objectives were pursued in this study: (1) performance evaluation of different Chl*a* models in the turbid waters of Lake Xingkai; (2) retrieval of the OLCI Chl*a* products of Lake Xingkai to show the variations of phytoplankton biomass between 2016 and 2022; and (3) assessment of the uncertainties and limitations of Chl*a* algorithm models when using a larger satellite zenith angle or applied to the algal bloom area.

2. Materials and Methods

2.1. Study Area

Lake Xingkai (44°32″–45°21″N, 131°58″–132°51″E), located on the border between China and Russia, is the largest freshwater lake in Northeast Asia [25] and covers an area of 4380 km^2 (Figure 1). A natural sand bar separates Lake Xingkai into two parts: Small Lake Xingkai and Big Lake Xingkai. The former is wholly located in China. The latter is located across the China–Russia border and most of its water is in Russia. The lake basin has a temperate continental monsoon climate, with less spring and autumn precipitation, more summer, and a long dry winter. The lake freezes in December and thaws in late April of the subsequent year [26]. It is fed by 26 rivers (9 in China and 17 in Russia) and has only one outflowing river in the northeast, the Songacha River, which flows into the Ussuri River [27]. For China and Russia, Lake Xingkai represents an essential food supply base and has tremendous economic and social value for flood control, irrigation, agriculture, and tourism, and plays a particular ecological function in regulating climate and improving water quality. Research on its water quality can provide data support for local environmental management and sustainable development. In this study, Big Lake Xingkai was divided into seven sub-regions, including the Northeast Coast, North Coast, Northwest Coast, West Coast, Open Waters, South Coast, and East Coast (Figure 1).

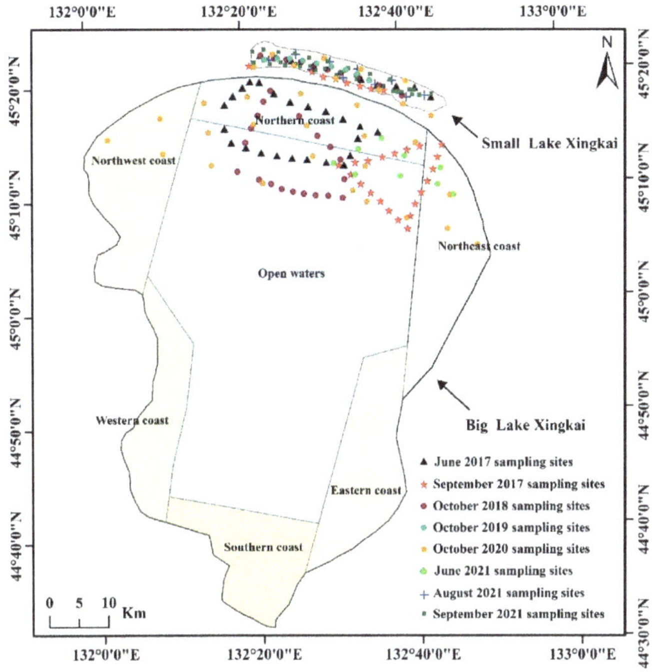

Figure 1. Location, sub-regions, and sampling cruises of Lake Xingkai.

2.2. Field Sampling and Laboratory Analysis

Over the course of eight field experiments on Lake Xingkai, 260 samples were collected between 2017 and 2021 (Figure 1). Surface water (depth ~30 cm) was collected using a standard two-liter polyethylene water-sampling instrument and then transferred to cold, dark coolers in the laboratory for analysis within 24 h. The absorption coefficient of a total particular matter, subsequently separated into phytoplankton ($a_{ph}(\lambda)$) and detrital components ($a_d(\lambda)$) through bleaching with 80% hot ethanol, was measured following the quantitative filter technique approach [28]. The absorption coefficient of the dissolved organic matter ($a_g(\lambda)$) was determined spectrophotometrically with reference to ultrapure water using filtration of the water sample over a 0.2 µm membrane filter [28]. The sum of $a_d(\lambda)$ and $a_g(\lambda)$ was denoted as $a_{dg}(\lambda)$. The suspended particulate matter concentrations (SPM) were gravimetrically determined from samples that were dried at 105 °C overnight. SPM was differentiated into suspended particular inorganic matter and suspended particular organic matter by burning the organic matter from the filters at 450 °C for four hours [29]. The Chla concentrations were filtered on Whatman GF/F fiberglass filters and extracted with ethanol (80%) at 80 °C. The Chla concentration was determined spectrophotometrically [30].

The measurements were taken in Lake Xingkai for the samples collected between the local hours of 8:00 and 15:00 using a FieldSpec spectroradiometer (Analytical Spectral Devices, Inc.). The radiance of the water surface, sky, and a 30% gray board was measured over the ship's deck. The $R_{rs}(\lambda)$ were extracted using the following Equation:

$$R_{rs}(\lambda) = \frac{(L_{sw} - rL_{sky})\rho_p}{L_p \pi} \quad (1)$$

2.3. Meteorological Data

Daily wind speed (m/s) and temperature (°C) in August of each year between 2016 and 2021 were downloaded from the Jixi meteorological station near Lake Xingkai.

2.4. Candidate Chla Algorithms for Lake Xingkai

Six proposed Chl*a* algorithms, the two-band ratio method (BR), the three-band algorithm (TBA), a four-band algorithm (FBA), which is a simplified expression of the TC2 algorithm [17], the Fluorescence Line Height algorithm (FLH), the MERIS Maximum Chlorophyll Index (MCI), and the maximum peak-height (MPH) algorithm [31] were evaluated in Lake Xingkai. In the model construction of turbid inland waters, one of the most important steps was the extraction of $a_{ph}(\lambda)$ from $a_{nw}(\lambda)$ (removing $a_{dg}(\lambda)$). In the model assumption, the BR method ignored the ($a_{dg}(665)$, TBA considered $a_{dg}(709)$ nearly close to $a_{dg}(665)$, FBA thought of $a_{dg}(665)$ was approximately equal to $0.17a_{dg}(560) + 0.83a_{dg}(709)$. FLH, MCI, and MPH attempt to remove $a_{dg}(\lambda)$ using baseline methods. The calculation steps of these algorithms were as follows:

$$BR = R_{rs}(709)/R_{rs}(665) \quad (2)$$

$$TBA = (1/R_{rs}(665) - 1/R_{rs}(709)) * R_{rs}(754) \quad (3)$$

$$FBA = (1/R_{rs}(665) - \frac{0.17}{R_{rs}(560)} - \frac{0.83}{R_{rs}(709)}) * R_{rs}(754) \quad (4)$$

$$FLH = R_{rs}(681) - R_{rs}(665) - (R_{rs}(709) - R_{rs}(665)) * \frac{681 - 665}{709 - 665} \quad (5)$$

$$MCI = R_{rs}(709) - R_{rs}(665) - (R_{rs}(754) - R_{rs}(665)) * \frac{709 - 665}{754 - 665} \quad (6)$$

$$MPH = R_{rs}(\lambda_{max}) - R_{rs}(665) - (R_{rs}(885) - R_{rs}(665)) * (\lambda_{max} - 665)/(885 - 665) \quad (7)$$

In Equation (7), $R_{rs}(\lambda_{max})$ and λ_{max} are, respectively, the magnitude and position of the highest $R_{rs}(\lambda_{max})$ value from OLCI bands at 681, 709, and 754 nm.

2.5. Atmospheric Correction of OLCI Imagery

In this study, the process of atmospheric correction of OLCI imagery (denoted as Rud) was as follows.

The OLCI cloud-free subset radiance images in Lake Xingkai from May to October between 2016 and 2022 were downloaded from NASA's Ocean Biology Processing Group (OBPG) (https://oceancolor.gsfc.nasa.gov/, 1 January 2020.).

In the first step, the Rayleigh corrected reflectance ($R_{rc}(\lambda)$) was calculated with the SeaDAS package (version 8.2.0) OLCI products with Equation (8):

$$R_{rc}(\lambda) = \pi/F_0(\lambda)/cos\theta * (L_t(\lambda)/t_{g_{sol}}(\lambda)/t_{g_{sen}}(\lambda) - L_r(\lambda) - TL_g(\lambda)) \quad (8)$$

where $L_t(\lambda)$ represents the top-of-atmosphere (TOA) radiance, $L_r(\lambda)$ represents the Rayleigh radiance, $TL_g(\lambda)$ represents the TOA glint radiance, θ represents the solar zenith angle, $F_0(\lambda)$ represents the mean solar flux, $t_{g_{sol}}(\lambda)$ represents gaseous transmittance from the sun to the surface, and $t_{g_{sen}}(\lambda)$ represents gaseous transmittance from surface to sun.

The second step was the extraction of the lake water region. The lake boundary was obtained using one cloud-free OLCI image observed on 10 September 2018, considering this lake's minimal water level variations. The lake boundary shapefile was drawn on a standard false color display by visual interpretation using ArcGIS 10.2. Then, the shapefile was converted to a raster image. A two-pixel buffer inward of the water boundary was removed to avoid the influence of adjacent land and substrates on nearby offshore water pixels.

The third step was obtaining the OLCI retrieved products. A near-infrared (NIR) bright pixel atmospheric correction was used in this study. The shape model in NIR bands proposed by Ruddick [32] was used to parametrize (Equation (9)).

$$R_{rs}(\lambda) = R_{rs}(\lambda_0) * \frac{a_w(\lambda_0)}{a_w(\lambda)} * \left(\frac{\lambda}{\lambda_0}\right)^{-\eta} \tag{9}$$

The reference band was 754 nm, and the parameterized bands were 779, 865, 885, and 1012 nm. The a_w represents the absorption by pure water obtained from the OLCI "msl12_sensor_info.dat" in SeaDAS 8.0.2 (ocssw/share/olci, Table S1 in Supplementary Materials). Applying Equation (9) to Equation (8), the simulated $R_{rc}(\lambda)$ can be expressed as (denoted as $R^*_{rc}(\lambda)$):

$$R^*_{rc}(\lambda) = \frac{\rho_a * t(\lambda_0) - \rho_a * t(\lambda) * \frac{a_w(\lambda_0)}{a_w(\lambda)} * \left(\frac{\lambda}{\lambda_0}\right)^{-\eta} + R_{rc}(\lambda_0) * t(\lambda) * \frac{a_w(\lambda_0)}{a_w(\lambda)} * \left(\frac{\lambda}{\lambda_0}\right)^{-\eta}}{t(\lambda_0)} \tag{10}$$

where η represents a shape parameter that depends on the size of the particles, and ρ_a represents the aerosol reflectance. In this study, the aerosol was assumed to be "white" (i.e., the aerosol reflectance was assumed to be wave-independent). The total diffuse atmosphere transmittance $t(\lambda)$ was calculated using the solar zenith angle (θ), the satellite zenith angle (θ_v), the Rayleigh optical thickness ($\tau_r(\lambda)$) and the ozone optical thickness ($\tau_{O_z}(\lambda)$) [33] as:

$$t(\lambda) = exp\left(\left(-\frac{\tau_r(\lambda)}{2} + \tau_{O_z}(\lambda)\right)/cos\theta\right) * exp\left(\left(-\frac{\tau_r(\lambda)}{2} + \tau_{O_z}(\lambda)\right)/cos\theta_v\right) \tag{11}$$

The $\tau_r(\lambda)$ of each OLCI band in the standard atmospheric pressure (1013.25 hPa, denoted as $\tau_r^s(\lambda)$) was obtained from the OLCI Rayleigh lookup table in SeaDAS 8.2.0 (ocssw/share/olci). For each pixel in one specific OLCI image over Lake Xingkai, the $\tau_r(\lambda)$ was calculated as:

$$\tau_r(\lambda) = \tau_r^s(\lambda) * a_p/1013.25 \tag{12}$$

where a_p represents the National Centers for Environmental Prediction (NCEP) atmospheric pressure product in hPa obtained from the l2 gen process of SeaDAS 8.2.0. The $\tau_{O_z}(\lambda)$ was calculated by multiplying the ozone concentration (Ozone) in cm obtained from the SeaDAS l2 gen process by the ozone absorption coefficients (OAC) in matm-1*cm^{-1} (Table S2 in the Supplementary Materials [34]):

$$\tau_{O_z}(\lambda) = ozone * OAC \tag{13}$$

The fourth step was the identification of ρ_a. For each OLCI water pixel, the value of ρ_a was set between 0 and the maximum of $R_{rc}(NIR)$ of this pixel ($R_{rc}^{max}(NIR)$) with a step of $R_{rc}^{max}(NIR)/100$; the value of η was set between −4 and +4 with a step of 0.1. For each combination of ρ_a and η, we can obtain a group of $R^*_{rc}(NIR)$ at 779, 865, and 885 nm. The identification of ρ_a was done with the minimum Euclidean distance (MED) between $R^*_{rc}(NIR)$ and $R_{rc}(NIR)$:

$$\text{MED} = \sqrt{(R^*_{rc}(779) - R_{rc}(779))^2 + (R^*_{rc}(865) - R_{rc}(865))^2 + (R^*_{rc}(885) - R_{rc}(885))^2 + (R^*_{rc}(1012) - R_{rc}(1012))^2} \tag{14}$$

The fifth step was the retrieval of $R_{rs}(\lambda)$. Applying the determined ρ_a from Equation (14) and $t(\lambda)$ estimated from Equation (11), the $R_{rs}(\lambda)$ of each band of the OLCI water pixels were obtained with Equation (15):

$$R_{rs}(\lambda) = (R_{rc}(\lambda) - \rho_a)/t(\lambda)/\pi \tag{15}$$

2.6. OLCI $R_{rs}(\lambda)$ Quality Control in Lake Xingkai

The valid OLCI $R_{rs}(\lambda)$ pixels were determined to be barely affected by clouds, cloud shadows, adjacency effects, sun glint, and higher sun zenith angle or viewing zenith angle. Here, we removed the invalid OLCI $R_{rs}(\lambda)$ pixels as follows.

The first step was masking thick clouds. The pixels that simultaneously satisfied $R_{rc}(412) > 0.25$, $R_{rc}(442) > 0.25$, and $R_{rc}(490) > 0.25$ were regarded as thick cloud pixels and eliminated.

The second step was masking high solar zenith angles (Solz) or satellite zenith angles (Senz). In this study, strict high-angle limitations were used, and pixels were removed if Solz > 55° or Senz > 55°.

The third step was removing non-water pixels by the normalized difference water index (NDWI) calculated with Equation (16). By visual interpretation, the NDWI of the water pixels covered by thin clouds or submerged aquatic vegetation, influenced by the adjacency efforts, often was <0.3. With the threshold of NDWI $\geq$ 0.3, we can remove most of these non-water pixels.

$$\text{NDWI} = (R_{rc}(560) - R_{rc}(1012))/(R_{rc}(560) + R_{rc}(1012)) \tag{16}$$

The fourth step was the removal of non-water pixels by visual inspection. Water pixels were eliminated in the following situations: (1) submerged aquatic vegetation influenced the water pixels, and (2) undetected clouds or cloud shadows still covered the water pixels.

2.7. Production of Matchups (Synchronous Satellite $R_{rs}(\lambda)$ and In Situ Chla) for Chla Model Development

The time window between the valid OLCI $R_{rs}(\lambda)$ and in situ samples was set as <1 day. If the number of the valid OLCI $R_{rs}(\lambda)$ pixels within a 3 $\times$ 3 window centered at the location of each in situ sample was nine, and the coefficient of variation (CV) of the valid spectra was <0.15, the median $R_{rs}(\lambda)$ of these valid pixels was calculated as the OLCI $R_{rs}(\lambda)$ of this sample. Based on these criteria, a total of 119 matchups (synchronous OLCI $R_{rs}(\lambda)$ and in situ Chl*a*) were selected. We evenly divided the matchup data into two parts: one part of 60 samples was used to calibrate the candidate models for estimating Chl*a* from OLCI image data, and the other part comprising 59 samples, was used to validate these models.

2.8. Accuracy Assessment

We used five evaluation indicators to evaluate the accuracy of these Chl*a* models, including coefficient of determination (R^2), relative difference (RD), mean normalized difference (MND), mean absolute percentage difference (MAPD), and root-mean-square deviation (RMSD):

$$R^2 = \frac{\sum_{i=1}^{N}\left(\text{Value}_{\text{predicted},i} - \overline{\text{Value}_{\text{measured}}}\right)^2}{\sum_{i=1}^{N}\left(\text{Value}_{\text{measured},i} - \overline{\text{Value}_{\text{measured}}}\right)^2} \tag{17}$$

$$\text{RD} = \left|\frac{\text{Value}_{\text{measured},i} - \text{Value}_{\text{predicted},i}}{\text{Value}_{\text{measured},i}}\right| * 100\% \tag{18}$$

$$\text{MND} = \frac{1}{N}\sum_{i=1}^{N}\frac{\text{Value}_{\text{measured},i} - \text{Value}_{\text{predicted},i}}{\text{Value}_{\text{measured},i}} * 100\% \tag{19}$$

$$\text{MAPD} = \frac{1}{N}\sum_{i=1}^{N}\left|\frac{\text{Value}_{\text{measured},i} - \text{Value}_{\text{predicted},i}}{\text{Value}_{\text{measured},i}}\right| * 100\% \tag{20}$$

$$\text{RMSD} = \sqrt{\frac{\sum_{i=1}^{N}\left(\text{Value}_{\text{measured,i}} - \text{Value}_{\text{predicted,i}}\right)^2}{N}} \tag{21}$$

where $\text{Value}_{\text{measured,i}}$ is the reference value from the in situ measurements, $\text{Value}_{\text{predicted,i}}$ represents the value derived by the algorithms; $\text{Value}_{\text{measured}}$ is the mean value of all the $\text{Value}_{\text{predicted,i}}$, and N is the number of observations.

3. Results

3.1. Calibration and Validation of Chla Algorithms

In this study, the atmospheric correction performance of Rud was validated using the synchronous in situ $R_{rs}(\lambda)$ (Figures S8 and S9, and Table S3). For the bands, including the 665, 681, 709, 754, and 885 nm that were used for the Chl*a* algorithms, the mean MAPDs, MNDs, and RMSDs were 16.44%, −6.8%, and 0.0052 sr^{-1}, respectively. Rud outperformed the commonly used C2RCC algorithm in Lake Xingkai. Thus, Rud was used to produce the long-time series OLCI $R_{rs}(\lambda)$ products in this Lake. The calibration dataset with in situ Chl*a* concentrations ranging from 0.4 to 13.59 mg m^{-3} was used to establish the expression of the BR, TBA, MCI, FLH, MPH, and FBA algorithms (Figure 2, Equations (2)–(7)). Different regression methods (exponential, linear, logarithmic, second-order polynomial, and power) were tested, and the method with the highest R^2 was selected for each algorithm as follows:

$$y = 51.152 \times BR^2 - 35.672 \times BR - 4.9263 \tag{22}$$

$$y = 143.12 \times TBA^2 + 117.45 \times TBA + 10.375 \tag{23}$$

$$y = 371998 \times MCI^2 - 2976.2 \times MCI + 10.568 \tag{24}$$

$$y = -42853 \times FLH^2 - 3608.6 \times FLH + 6.5238 \tag{25}$$

$$y = 960594 \times MPH^2 - 3585.9 \times MPH + 7.5487 \tag{26}$$

$$y = 507.02 \times FBA^2 + 169.7 \times FBA + 10.313 \tag{27}$$

For each algorithm, a second-order polynomial regression function had the highest R^2 compared with other regression function types. Previous studies [35–37] agreed that a second-order polynomial regression function is a good choice in the construction of water-quality parameters retrieving models for turbid waters with algae (case-II waters). Thus, all these models were calibrated using the second-order polynomial regression type. Figure 3 shows strong and similar relationships between each reflectance index of BR, TBA, and FBA (R^2 = 0.64, 0.58, and 0.67, respectively). The relationships between the reflectance indices of MCI and MPH were weaker (R^2 = 0.04 and 0.25).

The performance of all the algorithms (Equations (22)–(27)) was validated using the validation dataset (Figure 3 and Table 1). A positive MND value indicates an overestimation of the Chl*a* concentrations by these algorithms. Regarding R^2, FBA showed the best performance (R^2 = 0.64), and FLH, BR, and TBA followed closely (R^2 = 0.60, 0.57, and 0.56, respectively). The estimated Chl*a* concentrations of MCI and MPH were weakly correlated with in situ Chl*a* concentrations; thus, these algorithms were not used for the rest of the analysis. The MAPD and RMSD of the remaining four algorithms showed a similar performance to R^2. FBA had the lowest value (MAPD = 38.26% and RMSD = 1.66 mg m^{-3}), and the other three algorithms followed (FLH: MAPD = 40.94%, and 1.68 mg m^{-3}, BR: 41.49%, and 1.85 mg m^{-3}, and TBA: 43.02%, and 1.85 mg m^{-3}). The results indicated that FBA was the best-performing algorithm compared to the other five. We counted the contribution of $a_{ph}(\lambda)$ to the total non-water absorption ($a_{nw}(\lambda)$), the sum of $a_{ph}(\lambda)$, $a_g(\lambda)$, and $a_d(\lambda)$) at 665 nm for the samples in Lake Xingkai (Figure 4). Although located near the phytoplankton peak, there were few samples in which the phytoplankton absorption was dominated. Thus, FBA, which had a reasonable assumption in removing $a_g(665)$ and $a_d(665)$, was selected as the algorithm to retrieve Chl*a* concentrations in Lake Xingkai using OLCI images from June to October between 2016 and 2022.

Table 1. Performance of different algorithms in Lake Xingkai.

Algorithms	R^2	MAPD (%)	MND (%)	RMSD (mg m^{-3})
BR	0.57	41.49	24.82	1.85
TBA	0.56	43.02	27.11	1.85
MCI	0	49.52	2.12	2.98
FLH	0.60	40.94	27.41	1.68
MPH	0.34	54.65	35.58	2.14
FBA	0.64	38.26	24.60	1.66

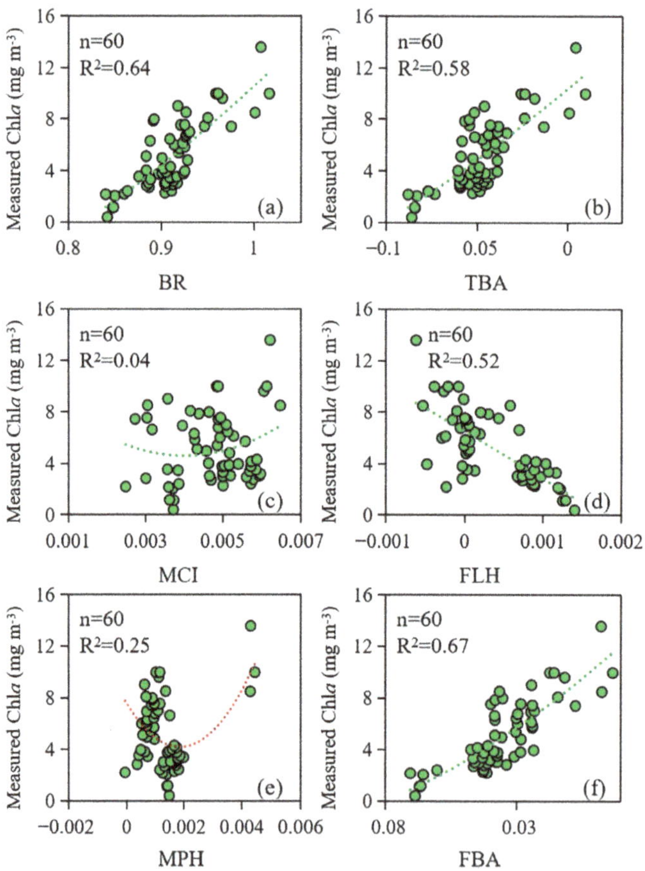

Figure 2. Indices of the models versus measured Chl*a* concentrations, (**a**) BR, (**b**) TBA, (**c**) MCI, (**d**) FLH, (**e**) MPH, and (**f**) FBA. The red dashed lines are the linear regression lines.

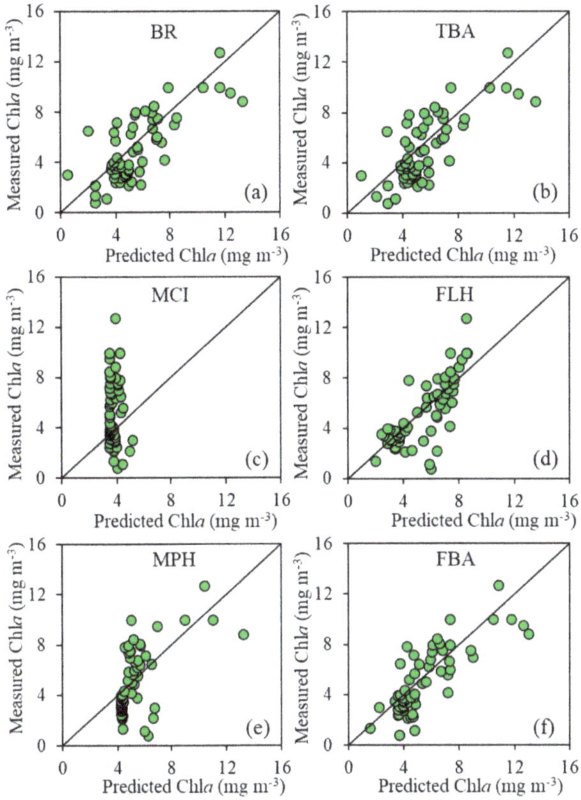

Figure 3. Comparison of measured and predicted Chl*a* concentrations by (**a**) BR, (**b**) TBA, (**c**) MCI, (**d**) FLH, (**e**) MPH, and (**f**) FBA. The solid lines are the 1:1 lines.

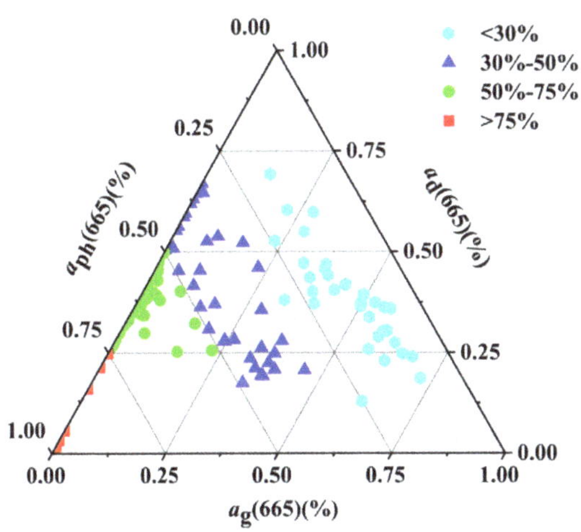

Figure 4. Contribution of $a_{ph}(665)$, $a_g(665)$, and $a_d(665)$ to $a_{nw}(665)$. Cyan, blue, green, and red represent the sample ratios of $a_{ph}(665)$: $a_{nw}(665)$ in the range of <30%, 30–50%, 50–75%, and ≥75%, respectively.

3.2. Spatial and Temporal Variations of Chla Concentrations in Lake Xingkai

The OLCI-retrieved Chl*a* products by the FBA algorithm from May to October between 2016 and 2022 were used to evaluate the phytoplankton biomass variation in this Lake (the monthly mean Chl*a* maps in 2022 are shown in Figure 5, and 2016, 2017, 2018, 2019, 2020, and 2021 are shown in Figures S1–S6). Big Lake Xingkai and Small Lake Xingkai generally had relatively low Chl*a* concentrations (with a long-term annual mean Chl*a* concentration of 4.74 and 10.98 mg m^{-3}) compared with other lakes with frequent blooms in China (e.g., Lake Taihu [1,38], Lake Chaohu [39], and Lake Dianchi [40]). The trophic status of Small Lake Xingkai was higher than that of Big Lake Xingkai. The maximum monthly mean value of Small Lake Xingkai (21.75 mg m^{-3}) appeared in August 2017, while that of Big Lake Xingkai (8.23 mg m^{-3}) appeared in May 2019. The minimum monthly mean value of Small Lake Xingkai (6.50 mg m^{-3}) appeared in May 2018, and that of Big Lake Xingkai (2.56 mg m^{-3}) appeared in September 2016. Overall, the Chla spatial difference of Big Lake Xingkai was not noticeable, except during specific months with relatively high values (August 2016 and August 2017) in the South Coast sub-region. The South Coast was the sub-region with the highest Chl*a* concentrations, ranging from 2.67 mg m^{-3} in October 2020 to 12.97 mg m^{-3} in August 2016, with a long-term annual mean of 5.86 mg m^{-3}. The Open Waters was the sub-region with the lowest Chl*a* concentrations ranging from 2.09 mg m^{-3} in September 2016 to 8.16 mg m^{-3} in May 2019, with a long-term annual mean of 4.37 mg m^{-3}. The long-term Chl*a* annual mean of the rest of the sub-regions, North Coast, Northwest Coast, East Coast, Northeast Coast, and West Coast, were 4.84, 5.04, 5.09, 5.2, and 5.42 mg m^{-3}, respectively.

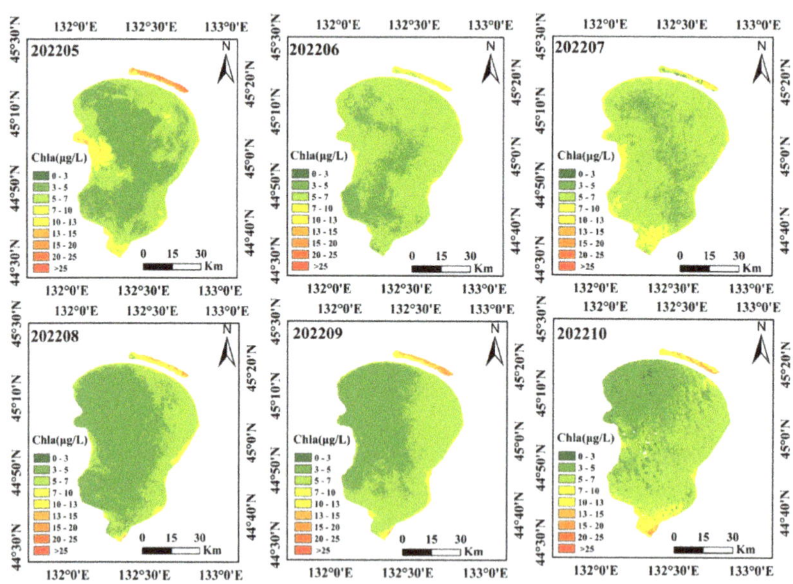

Figure 5. Monthly mean Chl*a* distributions of Lake Xingkai in 2022.

Unlike other Chinese lakes with frequent algal blooms, which have relatively stable seasonal Chl*a* variation patterns (e.g., Lake Taihu and Lake Chaohu often have relatively higher values in summer and autumn and relatively lower values in winter and spring) [41–43], the seasonal Chl*a* variation trends in Lake Xingkai varied across years (Figure 6). The minimum (maximum) monthly mean Chl*a* value in 2016, 2017, 2018, 2019, 2020, 2021, and 2022 were in July (May), July (August), May (June), October (June), June (September), October (June), July (May) in Small Lake Xingkai, and in September (August), July (May), September (May), September (May), October (June), October (May), and August (July) in Big Lake Xingkai.

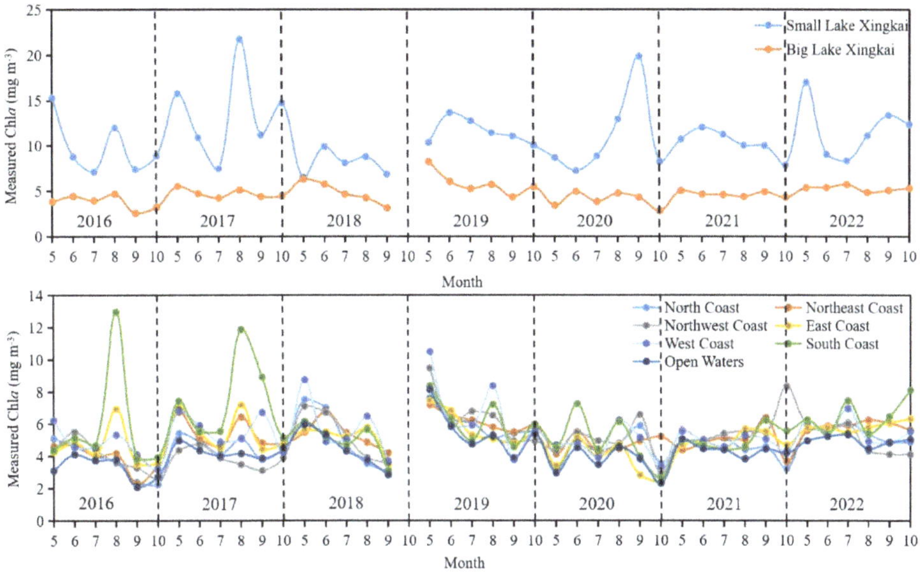

Figure 6. Monthly mean Chl*a* concentration of Lake Xingkai between 2016 and 2022.

4. Discussion

4.1. Reduction of Chla Model Accuracy Due to High Sun Angle

We only retained the sampling with both sun zenith angle (Solz) and satellite zenith angle (Senz) < 55° for model calibration and validation. The threshold was stricter than the one proposed by Bailey and Werdell [44] for NASA's ocean watercolor products with Solz < 70° and Senz < 60°. This study selected a stricter Solz and Senz threshold because we found an observable decrease in the performance of atmospheric correction when Solz > 55° in Lake Xingkai. The correlation between in situ Chl*a* concentrations and the spectral index of the well-performing models in Lake Xingkai decreased when only the data with Solz > 55° were added (Figure 7). The R^2 of FBA, TBA BR, and FLH decreased from 0.67, 0.58, 0.60, and 0.56 to 0.40, 0.23, 0.19, and 0.46, respectively.

Some studies have shown that a high Solz or Senz angle will increase the residual errors in satellite $R_{rs}(\lambda)$ products due to many factors, including higher atmospheric multiple scattering effects and underestimation of the Rayleigh contribution [44,45]. The atmospheric correction methods for waters originated from oceans and were then applied or improved to inland waters. The aerosol models in SeaDAS, the software package for processing ocean color data of NASA, were built based on analyzing aerosol properties of eight open seas and three coastal AERONET stations [46]. Aerosols over land often tend to have more absorbent particles and larger aerosol optical thickness than those over sea [47–49]. Using aerosol models designed for coastal waters and oceans to perform the atmospheric correction of inland waters may have uncertainties due to the different optical properties of aerosols between land and sea, especially when the Solz were high, like >55° as in Lake Xingkai.

The OLCI $R_{rs}(\lambda)$ products of other inland lakes of the global middle and high latitudes also may use this strict threshold to ensure the $R_{rs}(\lambda)$ products with good quality, but it will reduce the number of valid OLCI images. In this study, we select five regions (the longitude and latitude ranges were 1°, with center latitude at 0°N, 10°N, 20°N, 30°N, 40°N, 50°N, 60°N, 70°N, 80°N and with the same longitude of 120°E) to analyze the OLCI solar observation differences over a whole year (2021). All the available OLCI images of these nine regions were downloaded and processed to obtain each image's solar observation data for 2021. Without consideration of clouds, sunglint, ice, or other certifications, the solar observation information for different months is shown in Figure 8. The 0°N and 10°N regions did not have days with Solz > 55° for the entire year. The rate of high Solz angle images increased with latitude. At latitudes ≥ 40°N, most days in October and November (when many lakes are not yet frozen in the northern hemisphere) with Solz > 55° were not used to retrieve Chl*a* concentrations, resulting in the absence of Chl*a* inversion products this month. This indicates that in autumn, valid data of lakes located in the mid or high-latitude ranges may be less than those in the lower latitudes. For example, in October 2018, Lake Xingkai had seven scenes without clouds and obvious sunglint. However, because the Solz of all these images was >55°, they were not used to retrieve Chl*a* concentrations, resulting in the absence of Chl*a* inversion products in this month (Figure 5 and Figure S3). We also counted the Senz angle distribution at different months for regions at different latitudes (Figure S7). Unlike the Solz, Senz did not vary with latitude. The Senz of most images was <55°.

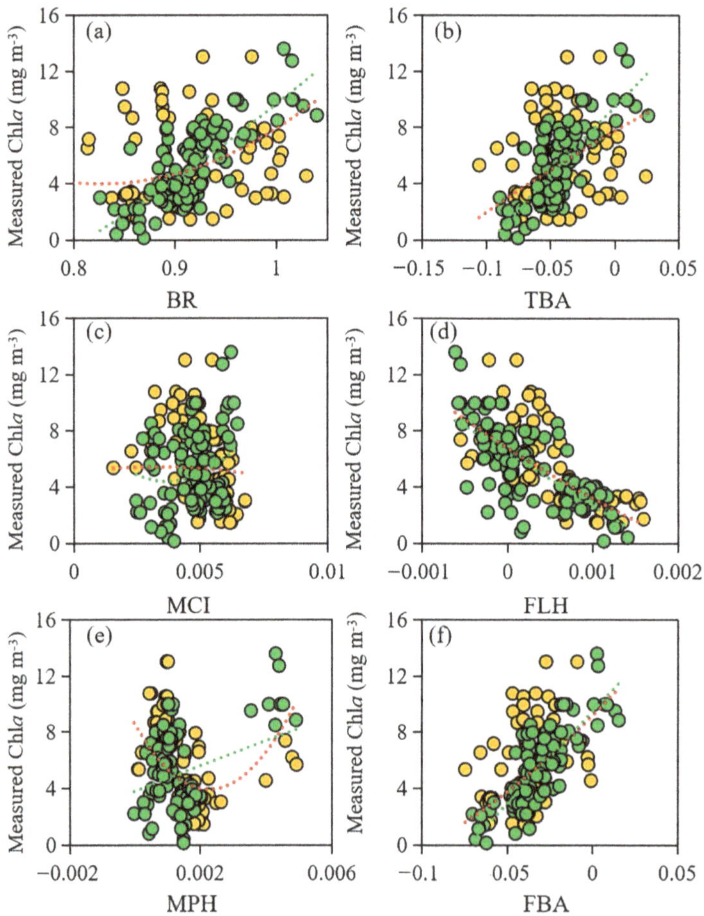

Figure 7. Scatter plots of spectral index versus measured Chl*a* concentrations, (**a**) BR, (**b**) TBA, (**c**) MCI, (**d**) FLH, (**e**) MPH, and (**f**) FBA. The yellow circles are all data, while the green circles are data with Solz < 55°. The red dashed line is the regression line of all the data, and the green dashed line is the regression line of the data with Solz < 55°.

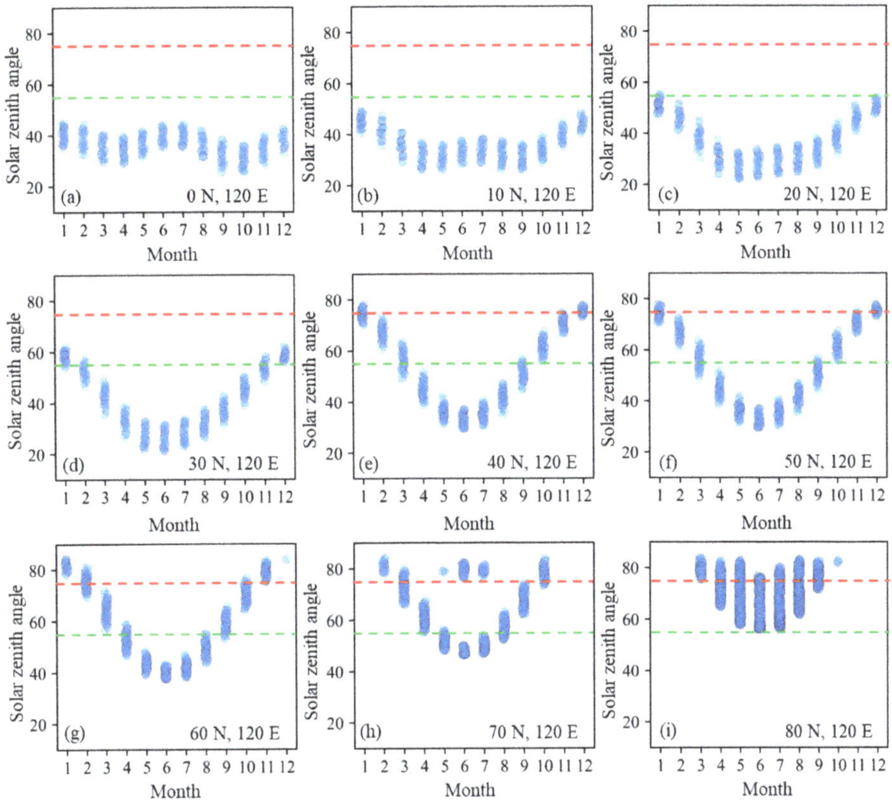

Figure 8. Monthly Solz distribution for regions at different latitudes, (**a**) 0°N, (**b**) 10°N, (**c**) 20°N, (**d**) 30°N, (**e**) 40°N, (**f**) 50°N, (**g**) 60°N, (**h**) 70°N, (**i**) 80°N. The green line is 55°, and the red line is 70°.

4.2. Limitations of Using Retrieved Chla Products to Evaluate the Trophic Status of the Lake

Unfortunately, the Chl*a* product used in this study did not cover areas with dense algal blooms, which are the most intuitive manifestation of lake eutrophication. There are many reasons why previously proposed Chl*a* algorithms or products are not suitable for algal blooms. Algae floating on the water's surface have higher NIR reflectance than water. The assumptions of most atmospheric correction and Chl*a* retrieval algorithms are, therefore, invalid in this situation. Additionally, the NDWI of dense algal blooms is often negative, and therefore, a positive NDWI threshold for extracting water pixels would also remove the algal bloom pixels [23,41]. The algal bloom area usually had large Chl*a* concentrations, >10 mg m^{-3} [50,51], >100 mg m^{-3} [20], or even >1000 mg m^{-3} [52]. Masking the algal bloom area would underestimate the lake's trophic status when using the retrieved Chl*a* products (Figure 9). Thus, many studies used both the retrieved Chl*a* concentrations and the extracted bloom areas to assess the trophic status of a lake [21,42]. It is not intuitive to use two indicators with different meanings. Many oligotrophic and mesotrophic lakes also have algal blooms [53,54]. Conversely, the waters classified as eutrophic (e.g., Chl*a* > 10 mg m^{-3}) by the trophic state index maybe not present algal blooms.

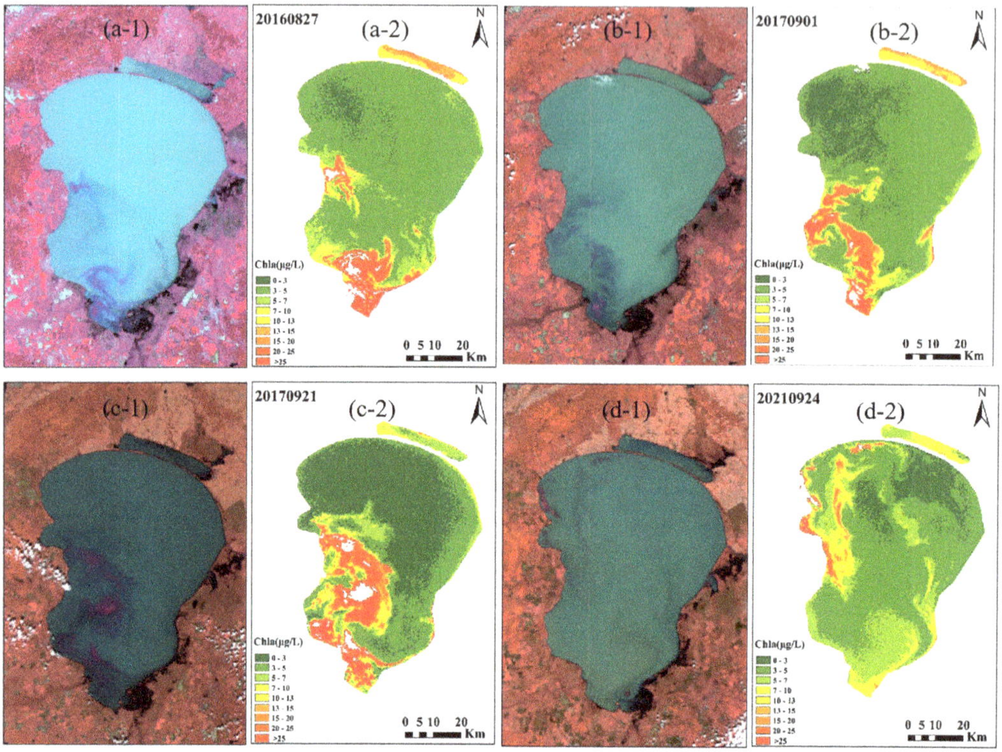

Figure 9. Standard false-color images (x-1) and corresponding Chl*a* products (x-2) were observed on 27 August 2016 (**a**), 1 September 2017 (**b**), 21 September 2017 (**c**), and 24 September 2021 (**d**), respectively.

Only using satellite-retrieved Chl*a* concentrations to measure the trophic status of water bodies will be clear. Few algorithms have been proposed for the Chl*a* estimation of algal blooms. Many studies showed that the MPH algorithm performed well for algal blooms in some African lakes [31,55,56]. The relatively worse performance of MPH compared with FLH, BR, TBA, and FBA for waters with no algal bloom in Lake Xingkai indicated that we may need hybrid algorithms to retrieve Chl*a* concentrations across blooming and non-blooming waters. It is vital to match synchronous satellite images with in situ samples when constructing atmospheric correction methods and Chl*a* retrieval models for waters with algal blooms. Our experience indicated that obtaining accurate matches for blooming waters by ship is difficult. The waves caused by the boat will affect the algal bloom floating state during sampling (Figure 10). Additionally, the commonly used time window for inland lakes ($\leq$3 h, $\leq$1 day, or $\leq$3 days) may be invalid. Therefore, we may need other data sampling methods, such as buoys, drones, and ground-based spectrometers, to collect samples for model calibration and validation.

Figure 10. Changing of the algal bloom floating state by the waves of the ship was observed by drones in Lake Hulun, China (**a**), and cameras in Lake Taipingchi, China (**b**).

4.3. Risk of Increased Trophic Status of Lake Xingkai Due to Increased Phytoplankton Biomass

In the past fifty years, the Lake Xingkai basin has become an important granary for China and Russia. Many wastelands, natural wetlands, woodlands, river beaches, etc., in the area have been reclaimed as farmland [57,58], creating a problem of agricultural non-point source pollution. The nutrient concentration in this lake is also very high. The measured total phosphorous of all our samples of Lake Xingkai was 0.112 mg L^{-1}, which is similar to those of other highly eutrophic lakes in China, have the following values: Lake Taihu (0.081 mg L^{-1}), Lake Chaohu (0.108 mg L^{-1}), and Lake Dianchi (TP 0.135 mg L^{-1}) [59]. Sufficient nutrients provide the primary conditions for the rapid growth of algae in Lake Xingkai.

We also gauged the possible effects of increased water clarity and air temperature on the phytoplankton biomass of Lake Xingkai. The in situ Chl*a* concentrations significantly correlated with the in situ measured SDD. It may be presumed that the change in water clarity may be a factor that affects the phytoplankton biomass in Lake Xingkai. Results from Lake Chagan, another turbid lake in the same region of Northeast China, showed that the total suspended matter (SPM, which had a high correlation with SDD in Lake Xingkai) is the main controlling factor limiting phytoplankton biomass [60]. The water clarity in Lake Xingkai has been increasing in the last few decades using long-time series of Landsat images (Figure 11) [61]. Between 1984 and 2018, the SDD of Small Lake Xingkai and Big Lake Xingkai showed an increasing trend, with a mean rate of 0.2728 and 0.1840 cm per year, respectively. Increased water clarity may result in more light entering the water and promote the growth of phytoplankton, increasing the risk of algal blooms in this lake.

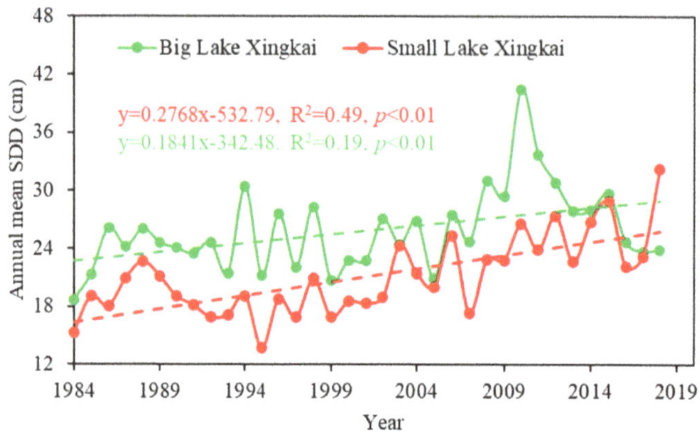

Figure 11. Annual mean SDD retrieved from Landsat images in Lake Xingkai from 1984 to 2018.

Northeast China has been one of the most rapidly warming regions of China and East Asia [62]. Figure 12 shows the annual average air temperature between May and October (the non-frozen period) observed at the Jixi meteorological station near Lake Xingkai from 1980 to 2020. In the past forty years, the annual average air temperature in the Jixi has shown an increasing trend, with a mean rate of 0.0251 °C per year. Like greater water clarity, warmer air temperatures also promote the growth of phytoplankton, increasing the trophic status of this lake.

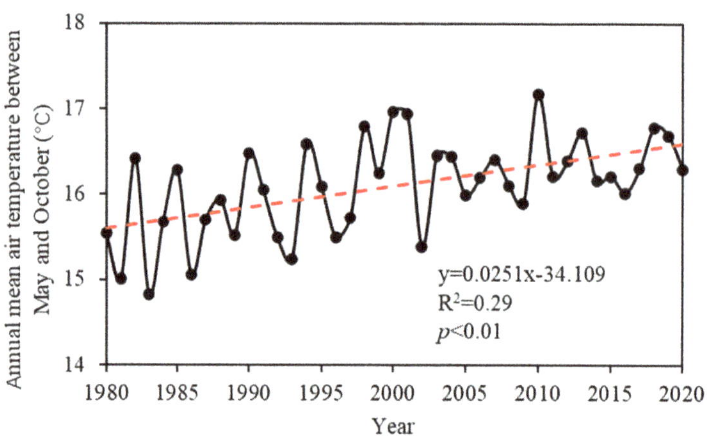

Figure 12. Annual mean air temperature between May and October was observed at the Jixi meteorological station near Lake Xingkai from 1980 to 2020.

5. Conclusions

In this study, we compared six Chl*a*-retrieved models in Lake Xingkai, a typical turbid but rarely noticed mid-high latitude lake in Northeast Asia. The best-performed algorithm, the FBA algorithm, had a reasonable assumption in removing $a_g(665)$ and $a_d(665)$ was selected as the algorithm to retrieve Chl*a* concentrations in Lake Xingkai using OLCI images from June to October between 2016 and 2022. Big Lake Xingkai and Small Lake Xingkai generally had relatively low Chl*a* concentrations (with a long-term annual mean Chl*a* concentration of 4.74 and 10.98 mg m^{-3}).

We also gauged the uncertainties and limitations of using satellite images to monitor Chl*a* concentrations of mid-high latitude lakes by taking Lake Xingkai as an example. We found a large decrease in accuracy when the Chl*a* retrieval models were applied to the images with Solz > 55°. Meanwhile, we also gauged the limitations of remote-sensing Chl*a* products for evaluating the trophic status of lakes due to the absence of areas with algal blooms. We suggest using hybrid algorithms to retrieve Chl*a* concentrations across non-algal bloom and bloom waters.

We also found uncertainties and limitations of the Chl*a* algorithm models when using a larger satellite zenith angle or applying it to an algal bloom area. Recent increases in anthropogenic nutrient loading, water clarity, and warming temperatures may lead to rising phytoplankton biomass in Lake Xingkai, and the results of this study can be applied for the satellite-based monitoring of its water quality.

Supplementary Materials: The following supporting information can be downloaded at: https://www.mdpi.com/article/10.3390/rs15153809/s1, Figure S1. Monthly mean Chl*a* distributions in Lake Xingkai in 2016. Figure S2. Monthly mean Chla distributions in Lake Xingkai in 2017. Figure S3. Monthly mean Chla distributions in Lake Xingkai in 2018. Figure S4. Monthly mean Chla distributions in Lake Xingkai in 2019. Figure S5. Monthly mean Chla distributions in Lake Xingkai in 2020. Figure S6. Monthly mean Chla distributions in Lake Xingkai in 2021. Figure S7. Satellite angle distribution in different months for regions at different latitudes. The green line is 55°, and the red line is 60°. Figure S8. The comparison of in situ $R_{rs}(\lambda)$ and predicted by Rud for 560, 665, 681, 709, 754, and 885 nm. Figure S9. The comparison of in situ $R_{rs}(\lambda)$ and predicted by C2RCC for 560, 665, 681, 709, 754, and 885 nm. Table S1. Absorption and backscatter coefficients of pure water in each OLCI band. Table S2. The absorption coefficient of ozone (ACO) in different bands is used to calculate ozone optical depth. Note that one atm-cm equals DU/1000; DU stands for Dobson Unit. For example, for a columnar amount of ozone of 300 DU, at 615 nm, =300/1000 × 0.1162 = 0.03486. Table S3. The atmospheric correction performance of Rud in Lake Xingkai in different bands with the C2RCC algorithm as a comparison in the bracket.

Author Contributions: L.F.: Conceptualization, data curation, formal analysis, methodology, visualization, roles/writing—original draft. Y.S.: Funding acquisition, project administration, supervision, writing. K.S.: review and editing. G.L.: Software, investigation. S.S.: Software, visualization. F.Z.: investigation, resources. H.T. & S.L.: investigation, resources. Y.Z.: Software, resources. All authors have read and agreed to the published version of the manuscript.

Funding: The research was jointly supported by the High Resolution Earth Observation System Major Project Funding (30-Y60B01-9003-22/23), the National Key Research and Development Program of China (2021YFB3901101), the National Natural Science Foundation of China (No. 42171385), and the Youth Innovation Promotion Association of Chinese Academy of Sciences, China (2022228).

Data Availability Statement: Not applicable.

Conflicts of Interest: The authors declare no conflict of interest.

References

1. Wang, S.; Li, J.; Zhang, B.; Spyrakos, E.; Tyler, A.N.; Shen, Q.; Zhang, F.; Kuster, T.; Lehmann, M.K.; Wu, Y.; et al. Trophic state assessment of global inland waters using a MODIS-derived Forel-Ule index. *Remote Sens. Environ.* **2018**, *217*, 444–460. [CrossRef]
2. Le, C.; Hu, C.; Cannizzaro, J.; English, D.; Muller-Karger, F.; Lee, Z. Evaluation of chlorophyll-a remote sensing algorithms for an optically complex estuary. *Remote Sens. Environ.* **2013**, *129*, 75–89. [CrossRef]
3. Li, S.; Song, K.; Wang, S.; Liu, G.; Wen, Z.; Shang, Y.; Lyu, L.; Chen, F.; Xu, S.; Tao, H.; et al. Quantification of chlorophyll-a in typical lakes across China using Sentinel-2 MSI imagery with machine learning algorithm. *Sci. Total Environ.* **2021**, *778*, 146271. [CrossRef] [PubMed]
4. Song, K.; Wang, Q.; Liu, G.; Jacinthe, P.-A.; Li, S.; Tao, H.; Du, Y.; Wen, Z.; Wang, X.; Guo, W.; et al. A unified model for high resolution mapping of global lake (>1 ha) clarity using Landsat imagery data. *Sci. Total Environ.* **2022**, *810*, 151188. [CrossRef] [PubMed]
5. Shi, K.; Zhang, Y.; Song, K.; Liu, M.; Zhou, Y.; Zhang, M.; Li, Y.; Zhu, G.; Qin, B. A semi-analytical approach for remote sensing of trophic state in inland waters: Bio-optical mechanism and application. *Remote Sens. Environ.* **2019**, *232*, 111349. [CrossRef]

6. Pahlevan, N.; Smith, B.; Schalles, J.; Binding, C.; Cao, Z.; Ma, R.; Alikas, K.; Kangro, K.; Gurlin, D.; Hà, N.; et al. Seamless retrievals of chlorophyll-a from Sentinel-2 (MSI) and Sentinel-3 (OLCI) in inland and coastal waters: A machine-learning approach. *Remote Sens. Environ.* **2020**, *240*, 111604. [CrossRef]
7. Le, C.; Li, Y.; Zha, Y.; Sun, D.; Huang, C.; Zhang, H. Remote estimation of chlorophyll a in optically complex waters based on optical classification. *Remote Sens. Environ.* **2011**, *115*, 725–737. [CrossRef]
8. Lubac, B.; Loisel, H. Variability and classification of remote sensing reflectance spectra in the eastern English Channel and southern North Sea. *Remote Sens. Environ.* **2007**, *110*, 45–58. [CrossRef]
9. Prieur, L.; Sathyendranath, S. An optical classification of coastal and oceanic waters based on the specific spectral absorption curves of phytoplankton pigments, dissolved organic matter, and other particulate materials. *Limnol. Oceanogr.* **1981**, *26*, 671–689. [CrossRef]
10. Shi, K.; Li, Y.; Li, L.; Lu, H.; Song, K.; Liu, Z.; Xu, Y.; Li, Z. Remote chlorophyll-a estimates for inland waters based on a cluster-based classification. *Sci. Total Environ.* **2013**, *444*, 1–15. [CrossRef]
11. Gurlin, D.; Gitelson, A.A.; Moses, W.J. Remote estimation of chl-a concentration in turbid productive waters—Return to a simple two-band NIR-red model? *Remote Sens. Environ.* **2011**, *115*, 3479–3490. [CrossRef]
12. Gower, J.; Doerffer, R.; Borstad, G. Interpretation of the 685nm peak in water-leaving radiance spectra in terms of fluorescence, absorption and scattering, and its observation by MERIS. *Int. J. Remote Sens.* **1999**, *20*, 1771–1786. [CrossRef]
13. Gower, J.; King, S.; Borstad, G.; Brown, L. Detection of intense plankton blooms using the 709 nm band of the MERIS imaging spectrometer. *Int. J. Remote Sens.* **2005**, *26*, 2005–2012. [CrossRef]
14. Le, C.; Li, Y.; Zha, Y.; Sun, D.; Yin, B. Validation of a quasi-analytical algorithm for highly turbid eutrophic water of Meiliang Bay in Taihu Lake, China. *IEEE Trans. Geosci. Remote Sens.* **2009**, *47*, 2492–2500. [CrossRef]
15. Lee, Z.; Carder, K.; Arnone, R.A. Deriving inherent optical properties from water color: A multiband quasi-analytical algorithm for optically deep waters. *Appl. Opt.* **2002**, *41*, 5755–5772. [CrossRef]
16. Lee, Z.; Shang, S.; Qi, L.; Yan, J.; Lin, G. A semi-analytical scheme to estimate Secchi-disk depth from Landsat-8 measurements. *Remote Sens. Environ.* **2016**, *177*, 101–106. [CrossRef]
17. Liu, G.; Li, L.; Song, K.; Li, Y.; Lyu, H.; Wen, Z.; Fang, C.; Bi, S.; Sun, X.; Wang, Z.; et al. An OLCI-based algorithm for semi-empirically partitioning absorption coefficient and estimating chlorophyll a concentration in various turbid case-2 waters. *Remote Sens. Environ.* **2020**, *239*, 111648. [CrossRef]
18. Xue, K.; Ma, R.; Duan, H.; Shen, M.; Boss, E.; Cao, Z. Inversion of inherent optical properties in optically complex waters using sentinel-3A/OLCI images: A case study using China's three largest freshwater lakes. *Remote Sens. Environ.* **2019**, *225*, 328–346. [CrossRef]
19. O'Reilly, J.E.; Werdell, P.J. Chlorophyll algorithms for ocean color sensors-OC4, OC5 & OC6. *Remote Sens. Environ.* **2019**, *229*, 32–47. [CrossRef]
20. Bi, S.; Li, Y.; Liu, G.; Song, K.; Xu, J.; Dong, X.; Cai, X.; Mu, M.; Miao, S.; Lyu, H. Assessment of algorithms for estimating chlorophyll-a concentration in inland waters: A round-robin scoring method based on the optically fuzzy clustering. *IEEE Trans. Geosci. Remote Sens.* **2021**, *60*, 4200717. [CrossRef]
21. Guan, Q.; Feng, L.; Hou, X.J.; Schurgers, G.; Zheng, Y.; Tang, J. Eutrophication changes in fifty large lakes on the Yangtze Plain of China derived from MERIS and OLCI observations. *Remote Sens. Environ.* **2020**, *246*, 111890. [CrossRef]
22. Kravitz, J.; Matthews, M.; Bernard, S.; Griffith, D. Application of Sentinel 3 OLCI for chl-a retrieval over small inland water targets: Successes and challenges. *Remote Sens. Environ.* **2020**, *237*, 111562. [CrossRef]
23. Shen, M.; Duan, H.; Cao, Z.; Xue, K.; Qi, T.; Ma, J.; Liu, D.; Song, K.; Huang, C.; Song, X.; et al. Sentinel-3 OLCI observations of water clarity in large lakes in eastern China: Implications for SDG 6.3. 2 evaluation. *Remote Sens. Environ.* **2020**, *247*, 111950. [CrossRef]
24. Song, K.; Fang, C.; Jacinthe, P.-A.; Wen, Z.; Liu, G.; Xu, X.; Shang, Y.; Lyu, L. Climatic versus anthropogenic controls of decadal trends (1983–2017) in algal blooms in lakes and reservoirs across China. *Environ. Sci. Technol.* **2021**, *55*, 2929–2938. [CrossRef] [PubMed]
25. Long, H.; Shen, J. Sandy beach ridges from Xingkai Lake (NE Asia): Timing and response to palaeoclimate. *Quat. Int.* **2017**, *430*, 21–31. [CrossRef]
26. Yu, S.; Li, X.; Wen, B.; Chen, G.; Hartleyc, A.; Jiang, M.; Li, X. Characterization of water quality in Xiao Xingkai Lake: Implications for trophic status and management. *Chin. Geogra. Sci.* **2021**, *31*, 558–570. [CrossRef]
27. Yang, Q.; Huang, X.; Wen, Z.; Shang, Y.; Wang, X.; Fang, C.; Song, K. Evaluating the spatial distribution and source of phthalate esters in the surface water of Xingkai Lake, China during summer. *J. Gt. Lakes Res.* **2021**, *47*, 437–446. [CrossRef]
28. Sun, D.; Li, Y.; Le, C.; Shi, K.; Huang, C.; Gong, S.; Yin, B. A semi-analytical approach for detecting suspended particulate composition in complex turbid inland waters (China). *Remote Sens. Environ.* **2013**, *134*, 92–99. [CrossRef]
29. Shi, K.; Zhang, Y.; Zhu, G.; Liu, X.; Zhou, Y.; Xu, H.; Qin, B.; Liu, G.; Li, Y. Long-term remote monitoring of total suspended matter concentration in Lake Taihu using 250 m MODIS-Aqua data. *Remote Sens. Environ.* **2015**, *164*, 43–56. [CrossRef]
30. Simis, S.G.; Peters, S.W.; Gons, H.J. Remote sensing of the cyanobacterial pigment phycocyanin in turbid inland water. *Limnol. Oceanogr.* **2005**, *50*, 237–245. [CrossRef]
31. Matthews, M.W.; Bernard, S.; Robertson, L. An algorithm for detecting trophic status (chlorophyll-a), cyanobacterial-dominance, surface scums and floating vegetation in inland and coastal waters. *Remote Sens. Environ.* **2012**, *124*, 637–652. [CrossRef]

32. Ruddick, K.G.; De Cauwer, V.; Park, Y.-J.; Moore, G. Seaborne measurements of near infrared water-leaving reflectance: The similarity spectrum for turbid waters. *Limnol. Oceanogr.* **2006**, *51*, 1167–1179. [CrossRef]
33. Wang, M.; Gordon, H. A simple, moderately accurate, atmospheric correction algorithm for SeaWiFS. *Remote Sens. Environ.* **1994**, *50*, 231–239. [CrossRef]
34. Koontz, A.; Flynn, C.; Hodges, G.; Michalsky, J.; Barnard, J. *Aerosol Optical Depth Value-Added Product*; US Department of Energy: Washington, DC, USA, 2013; Volume 32. [CrossRef]
35. Chavula, G.; Brezonik, P.; Thenkabail, P.; Johnson, T.; Bauer, M. Estimating chlorophyll concentration in Lake Malawi from MODIS satellite imagery. *Phys. Chem. Earth Parts A/B/C* **2009**, *34*, 755–760. [CrossRef]
36. Hou, X.; Feng, L.; Duan, H.; Chen, X.; Sun, D.; Shi, K. Fifteen-year monitoring of the turbidity dynamics in large lakes and reservoirs in the middle and lower basin of the Yangtze River, China. *Remote Sens. Environ.* **2017**, *190*, 107–121. [CrossRef]
37. Petus, C.; Chust, G.; Gohin, F.; Doxaran, D.; Froidefond, J.-M.; Sagarminaga, Y. Estimating turbidity and total suspended matter in the Adour River plume (South Bay of Biscay) using MODIS 250-m imagery. *Cont. Shelf Res.* **2010**, *30*, 379–392. [CrossRef]
38. Qin, B.; Paerl, H.; Brookes, J.; Liu, J.; Jeppesen, E.; Zhu, G.; Zhang, Y.; Xu, H.; Shi, K.; Deng, J.; et al. Why Lake Taihu continues to be plagued with cyanobacterial blooms through 10 years (2007–2017) efforts. *Sci. Bull.* **2019**, 64. [CrossRef]
39. Li, J.; Zhang, Y.; Ma, R.; Duan, H.; Loiselle, S.; Xue, K.; Liang, Q. Satellite-based estimation of column-integrated algal biomass in nonalgae bloom conditions: A case study of Lake Chaohu, China. *IEEE J. Sel. Top. Appl. Earth Obs. Remote Sens.* **2016**, *10*, 450–462. [CrossRef]
40. Mu, M.; Wu, C.; Li, Y.; Lyu, H.; Fang, S.; Yan, X.; Liu, G.; Zheng, Z.; Du, C.; Bi, S.; et al. Long-term observation of cyanobacteria blooms using multi-source satellite images: A case study on a cloudy and rainy lake. *Environ. Sci. Pollut. Res.* **2019**, *26*, 11012–11028. [CrossRef]
41. Cao, Z.; Ma, R.; Melack, J.; Duan, H.; Liu, M.; Kutser, T.; Xue, K.; Shen, M.; Qi, T.; Yuan, H. Landsat observations of chlorophyll-a variations in Lake Taihu from 1984 to 2019. *Int. J. Appl. Earth Observ. Geoinf.* **2022**, *106*, 102642. [CrossRef]
42. Shi, K.; Zhang, Y.; Zhou, Y.; Liu, X.; Zhu, G.; Qin, B.; Gao, G. Long-term MODIS observations of cyanobacterial dynamics in Lake Taihu: Responses to nutrient enrichment and meteorological factors. *Sci. Rep.* **2017**, *7*, 1–16. [CrossRef] [PubMed]
43. Xun, S.; Zhai, W.; Fan, W. MODIS in monitoring the chlorophyll-a concentrations of Chaohu Lake. *J. Appl. Meteor. Sci.* **2009**, *20*, 95–101. [CrossRef]
44. Bailey, S.W.; Werdell, P.J. A multi-sensor approach for the on-orbit validation of ocean color satellite data products. *Remote Sens. Environ.* **2006**, *102*, 12–23. [CrossRef]
45. Chen, J.; He, X.; Quan, W.; Ma, L.; Jia, M.; Pan, D. A statistical analysis of residual errors in satellite remote sensing reflectance data from oligotrophic open oceans. *IEEE Trans. Geosci. Remote Sens.* **2021**, *60*, 4203912. [CrossRef]
46. Ahmad, Z.; Franz, B.A.; McClain, C.R.; Kwiatkowska, E.J.; Werdell, J.; Shettle, E.P.; Holben, B.N. New aerosol models for the retrieval of aerosol optical thickness and normalized water-leaving radiances from the SeaWiFS and MODIS sensors over coastal regions and open oceans. *Appl. Opt.* **2010**, *49*, 5545–5560. [CrossRef]
47. Kaufman, Y.; Tanré, D.; Remer, L.A.; Vermote, E.; Chu, A.; Holben, B. Operational remote sensing of tropospheric aerosol over land from EOS moderate resolution imaging spectroradiometer. *J. Geophys. Res.* **1997**, *102*, 17051–17067. [CrossRef]
48. Liu, G.; Li, Y.; Lyu, H.; Wang, S.; Du, C.; Huang, C. An improved land target-based atmospheric correction method for Lake Taihu. *IEEE J. Sel. Top. Appl. Earth Observ. Remote Sens.* **2015**, *9*, 793–803. [CrossRef]
49. Lynch, P.; Reid, J.S.; Westphal, D.L.; Zhang, J.; Hogan, T.F.; Hyer, E.J.; Curtis, C.A.; Hegg, D.A.; Shi, Y.; Campbell, J.R.; et al. An 11-year global gridded aerosol optical thickness reanalysis (v1. 0) for atmospheric and climate sciences. *Geosci. Model Dev.* **2016**, *9*, 1489–1522. [CrossRef]
50. Huang, J.; Zhang, Y.; Arhonditsis, G.B.; Gao, J.; Chen, Q.; Peng, J. The magnitude and drivers of harmful algal blooms in China's lakes and reservoirs: A national-scale characterization. *Water Res.* **2020**, *181*, 115902. [CrossRef]
51. Kang, J.; Xie, Y.; Lin, Y.; Wang, Y. Algal Bloom, Succession, and Drawdown of Silicate in the Chukchi Sea in Summer 2010. *Ecosystems* **2022**, *25*, 320–336. [CrossRef]
52. O'Shea, R.E.; Pahlevan, N.; Smith, B.; Bresciani, M.; Egerton, T.; Giardino, C.; Li, L.; Moore, T.; Ruiz-Verdu, A.; Ruberg, S.; et al. Advancing cyanobacteria biomass estimation from hyperspectral observations: Demonstrations with HICO and PRISMA imagery. *Remote Sens. Environ.* **2021**, *266*, 112693. [CrossRef]
53. Reinl, K.L.; Brookes, J.D.; Carey, C.C.; Harris, T.D.; Ibelings, B.W.; Morales-Williams, A.M.; De Senerpont Domis, L.N.; Atkins, K.S.; Isles, P.D.; Mesman, J.; et al. Cyanobacterial blooms in oligotrophic lakes: Shifting the high-nutrient paradigm. *Freshwater Biol.* **2021**, *66*, 1846–1859. [CrossRef]
54. Zhang, M.; Zhang, Y.; Deng, J.; Liu, M.; Zhou, Y.; Zhang, Y.; Shi, K.; Jiang, C. High-resolution temporal detection of cyanobacterial blooms in a deep and oligotrophic lake by high-frequency buoy data. *Environ. Res.* **2022**, *203*, 111848. [CrossRef] [PubMed]
55. Matthews, M.W.; Odermatt, D. Improved algorithm for routine monitoring of cyanobacteria and eutrophication in inland and near-coastal waters. *Remote Sens. Environ.* **2015**, *156*, 374–382. [CrossRef]
56. Salem, S.I.; Strand, M.H.; Higa, H.; Kim, H.; Kazuhiro, K.; Oki, K.; Oki, T. Evaluation of MERIS chlorophyll-a retrieval processors in a complex turbid lake Kasumigaura over a 10-year mission. *Remote Sens.* **2017**, *9*, 1022. [CrossRef]
57. Dong, J.; Xiao, X.; Menarguez, M.; Zhang, G.; Qin, Y.; Thau, D.; Biradar, C.; Moore, B. Mapping paddy rice planting area in northeastern Asia with Landsat 8 images, phenology-based algorithm and Google Earth Engine. *Remote Sens. Environ.* **2016**, *185*, 142–154. [CrossRef]

58. Mao, D.; Tian, Y.; Wang, Z.; Jia, M.; Du, J.; Song, C. Wetland changes in the Amur River Basin: Differing trends and proximate causes on the Chinese and Russian sides. *J. Environ. Manag.* **2021**, *280*, 111670. [CrossRef]
59. Tong, Y.; Wang, M.; Peñuelas, J.; Liu, Y. Improvement in municipal wastewater treatment alters lake nitrogen to phosphorus ratios in populated regions. *Proc. Natl. Acad. Sci. USA* **2020**, *117*, 11566–11572. [CrossRef]
60. Liu, X.; Chen, L.; Zhang, G.; Zhang, J.; Wu, Y.; Ju, H. Spatiotemporal dynamics of succession and growth limitation of phytoplankton for nutrients and light in a large shallow lake. *Water Res.* **2021**, *194*, 116910. [CrossRef]
61. Fang, C.; Jacinthe, P.A.; Song, C.; Zhang, C.; Song, K. Climate-driven variations in suspended particulate matter dominate water clarity in shallow lakes. *Opt. Express.* **2022**, *30*, 4028–4045. [CrossRef]
62. Ren, G.; Ding, Y.; Zhao, Z.; Zheng, J.; Wu, T.; Tang, G.; Xu, Y. Recent progress in studies of climate change in China. *Adv. Atmos. Sci.* **2012**, *29*, 958–977. [CrossRef]

remote sensing

MDPI

Article

Mapping Irish Water Bodies: Comparison of Platforms, Indices and Water Body Type

Minyan Zhao [1] and Fiachra O'Loughlin [1,2,*]

[1] Dooge Centre for Water Resources Research, School of Civil Engineering, University College Dublin, D04 V1W8 Dublin, Ireland
[2] UCD Earth Institute, University College Dublin, D04 V1W8 Dublin, Ireland
* Correspondence: fiachra.oloughlin@ucd.ie

Abstract: Accurate monitoring of water bodies is essential for the management and regulation of water resources. Traditional methods for measuring water quality are always time-consuming and expensive; furthermore, it can be very difficult capture the full spatiotemporal variations across regions. Many studies have shown the possibility of remote-sensing-based water monitoring work in many areas, especially for water quality monitoring. However, the use of optical remotely sensed imagery depends on several factors, including weather, quality of images and the size of water bodies. Hence, in this study, the feasibility of optical remote sensing for water quality monitoring in the Republic of Ireland was investigated. To assess the value of remote sensing for water quality monitoring, it is critical to know how well water bodies and the existing in situ monitoring stations are mapped. In this study, two satellite platforms (Sentinel-2 MSI and Landsat-8 OLI) and four indices for separating water and land pixel (Normalized Difference Vegetation Index—NDVI; Normalized Difference Water Index—NDWI; Modified Normalized Difference Water Index—MNDWI; and Automated Water Extraction Index—AWEI) have been used to create water masks for two scenarios. In the first scenario (Scenario 1), we included all pixels classified as water, while for the second scenario (Scenario 2) accounts for potential land contamination and only used water pixels that were completed surround by other water pixels. The water masks for the different scenarios and combinations of platforms and indices were then compared with the existing water quality monitoring station and to the shapefile of the river network, lakes and coastal and transitional water bodies. We found that both platforms had potential for water quality monitoring in the Republic of Ireland, with Sentinel-2 outperforming Landsat due to its finer spatial resolution. Overall, Sentinel-2 was able to map ~25% of the existing monitoring station, while Landsat-8 could only map ~21%. These percentages were heavily impacted by the large number of river monitoring stations that were difficult to map with either satellite due to their location on smaller rivers. Our results showed the importance of testing several indices. No index performed the best across the different platforms. $AWEI_{nsh}$ (Automated Water Extraction Index—no shadow) and Sentinel-2 outperformed all other combinations and was able to map over 80% of the area of all non-river water bodies across the Republic of Ireland. While MNDWI was the best index for Landsat-8, it was the worst performer for Sentinel-2. This study showed that optical remote sensing has potential for water monitoring in the Republic of Ireland, especially for larger rivers, lakes and transitional and coastal water bodies.

Keywords: water resources monitoring; Republic of Ireland; remote sensing; Landsat-8; Sentinel-2

1. Introduction

It is critical that we sustain and improve our water resources as they are inseparable from the ecosystems and are necessary for life [1]. The pressures on water resources are increasing globally with ever-growing demands for freshwater, over-abstraction of water resources, pollution and climate change [2]. One of the most effective controls to prevent water pollution is to identify the potential sources promptly [3]; therefore, it is crucial to obtain timely access to water information, such as water quality and water quantity.

Under the European Union's Water Framework Directive (WFD), which came into force in 2000 [4], and the National Hydrometric Monitoring Program [5], information on water quality and quantity across Ireland is collected periodically via conventional methods [6]; for example, water flow in some locations is measured via the use of weirs or handheld mechanical current meter, while water quality parameters are derived from laboratory analyses undertaken on water samples collected at each monitoring station. This approach provides precise information on water at discrete sampling points; however, it cannot grasp the spatiotemporal aspects of water quality that occur in large water bodies [7]. Across the Republic of Ireland, there are 4829 water bodies designated by the WFD; however, only approximately two-thirds of these are included in the national WFD monitoring program, and this does not include all water bodies. This can be broken down to its constituent parts, with 74% of river bodies, 6.8% of lakes and 40% of coastal and transitional waters currently monitored [6].

Remote sensing (RS), with its frequent revisit time and large spatial coverage, is an alternative approach to facilitating water monitoring [8]. The use of RS in water quality monitoring started in the early 1970s [9], when the correlation between solar radiation reflectance and sediment concentration was investigated [10]. Today, RS is used in many aspects of water resources including flooding estimation, water erosion assessment, water area detection, the phenology of water cycle, water quality derivation, etc. [11]. In Ireland, very few studies have investigated the use of remote sensing for water resources (e.g., [12–14]). Agarwal et al. [12] and Karki et al. [13], as part of the Remote Sensing of Irish Surface Waters Project (INFER), used remote sensing to estimate the chlorophyll-a content, turbidity, and algal coverage in some regional Irish transitional and coastal (TraC) areas. However, the feasibility of RS at national scale for water quality monitoring has not been established.

To establish if remote sensing has potential in water quality monitoring in Ireland, it is first necessary to see if the spatial coverage is sufficient. One way to test this is via the generation of water masks, which separate land and water pixels. The spectral, temporal, radiometric and spatial resolutions of different satellite platforms are the determining factors in this. Therefore, the choice of RS platforms is important. For example, MODIS (Moderate Resolution Imaging Spectroradiometer), onboard the Terra and Aqua spacecrafts, has the advantage of high temporal resolution (1–2 times/day), broad spectral range (0.4–14.4 μm) and long tracking history (1999–present) [14]. It has been utilized in tracking surface water evolutions [15–17]. However, its disadvantage is its spatial resolution, which ranges from 250 to 1000 m, restricting application only to large areas [14]. The deployment of different high-resolution platforms offers great possibilities in detecting smaller water bodies. IKONOS, one of the first high-spatial resolution satellites, has been used to map lake areas in Myanmar [18]. However, the presence of shadows on IKONOS images makes water detection harder [19]. In recent years, high resolution data are becoming more easily accessible, with more than 1600 CubeSats currently orbiting the Earth's surfaces. These CubeSat have approximately daily revisit times and their 3-m spatial resolution is much higher than many of the other high-resolution products available [20]. CubeSats have been used to monitor lake dynamics in Arctic-Boreal [21], in Yukon Flats, US [22] and in the Tocantins–Araguaia hydrographic region, Brazil [23]. However, the relatively low radiometric quality and inter-sensors inconsistencies of CubeSats have posed limitations in water monitoring [24]. Two platforms that have been widely used in aquatic science are the Landsat missions and Sentinel-2. These platforms, with their 30- and 20-m resolutions, respectively, are freely available and have the ideal spectral range for aquatic science [25,26]. Some of the uses of Landsat include the generation of global river widths [27]; multi-temporal global water masks [28]; and water masks and river widths in Congo [29] and North America [30]. Sentinel-2, despite being a relatively newer platform has also had a range of applications from coastal shoreline mapping [31] to surface water mapping across France [32] to mapping the tidal flat around the Bohai and Yellow Seas [33]. In some cases, both Landsat and Sentinel-2 have been combined—e.g., Pardo-Pascual et al. [34]—for mapping the shoreline of natural beaches.

To generate water masks from remote sensing, there are various approaches available. Yang et al. [32] has classified these approaches into two groups: (1) sample-based methods; and (2) rule-based methods. Sample-based methods classify the water pixels based on the supervised classification at pixel and object levels within the same groups [35]. The accuracy of this method depends on prior reliable knowledge, which is usually acquired at a high cost. Rule-based approaches, on the other hand, are faster and more convenient to use than sample-based methods and are therefore more widely used. Rule-based methods normally rely on proven algorithms based on comparing the surface reflectance of spectral bands [32]. While there is a large selection of water spectral indices, the Normalized Difference Water index (NDWI) was the first index proposed for water extraction. Though it suffers from noise in build-up areas, which affects accuracy, the Modified Normalized Difference Water Index (MNDWI) was created to address this issue [1]. However, it cannot distinguish between water and snow [36]. To further improve water mapping accuracy, the Automated Water Extraction Index (AWEI) was then created, which includes two indices capable of shadow and non-shadow scenarios, respectively [37]. While the above are the most commonly used water indices, many more have been created to address other shortcomings. For example, the Water Index2015 (WI2015) [38] was developed to improve the performance of classifying water pixels via Landsat. The Water Ratio Index (WRI), High Resolution Water Index (HRWI) and Enhanced Water Index (EWI) were developed for urban areas and cities and have been shown to perform better than other more widely used water indices [39–41]. Recently, to improve upon the MNDWI, the Contrast Difference Water Index (CDWI) and the Shadow Difference Water Index (SDWI) were created for built-up and non-built-up areas, respectively, and the BDWI (Background Difference Water Index) was designed for complex background [42]. Other indices, such as the NDVI (Normalized Difference Vegetation Index) and EVI (Enhanced Vegetation Index), which were developed for extracting land pixel areas, have been used in distinguishing water from land [43,44]. Memon et al. [45] compared NDWI with two other indices—Red and Short-Wave Infra-Red (RSWIR) and MNDWI in Pakistan—and showed the tendencies of NDWI to underestimate the inundated area. Acharya, Subedi and Lee [43] compared the same indices for Nepal and found that NDVI and NDWI showed better results than MNDWI and AWEI for pure water pixels. Liu et al. [46] and Du et al. [47] compared various water indices (NDWI, MNDWI, AWEI and WI2015) in Guangzhou and the Venice coastland, respectively, and demonstrated that pan-sharpened WI2015 and MNDWI provide more accurate results than the other indices. The AWEI was compared with both MNDWI and Maximum Likelihood (ML) for Denmark, Switzerland, Ethiopia, South Africa and New Zealand, and the results showed that the accuracy of the AWEI was much higher than the other two indices [37]. Furthermore, Sivanpillai et al. [48] showed that the MNDWI outperformed NDWI in calculating pre- and post-flood inundation extents in the US. In addition to the use of these indices, machine learning methods are also becoming popular for the extraction of water bodies. These machine learning methods include regression trees, neural networks and support vector machines, and several studies have compared these methods [49–51].

The overall objective of this study is to help assess the value of remote sensing optical imagery for water quality monitoring of water bodies across the Republic of Ireland. To answer this objective, it is critical to know the extent to which remote sensing can map the different water bodies, and three sub-objectives will be addressed:

1. What percentages of water bodies are mapped by the different remote sensing platforms? What is the difference between using Landsat-8 or Sentinel-2?
2. Which is the best water index for detecting water pixels across Ireland? Does this vary by water body type?

3. How well does remote sensing map the existing in situ monitoring points? This is critical for the calibration of water quality estimates from remote sensing.

2. Study Area and Methodology

2.1. Study Area

In this study, we focus on the Republic of Ireland, which is in the northwest of Europe (Figure 1). The Republic of Ireland covers five-sixths of Ireland's surface, with a total area of 70,273 km^2, and the remainder is Northern Ireland [52]. The Republic of Ireland has over 12,200 lakes; however, only 812 have been designated as lake water bodies by the WFD [53]. These lake water bodies are predominately small (66% < 1 ha) and mostly in Western Ireland. Furthermore, Ireland has more than 74,000 km of river channels [54], where first- and second-order streams account for 95.8% of the total (https://gis.epa.ie/, accessed on 22 July 2023).

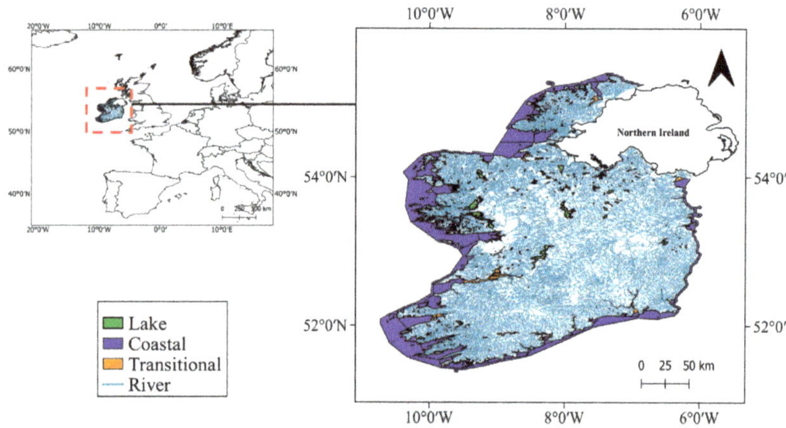

Figure 1. Location map showing the Republic of Ireland and its river, lakes and coastal and transitional water bodies.

Water quality is monitored in the Republic of Ireland by the Irish Environmental Agency (EPA) [55] to ensure that it adheres to European legislation such as the WFD and directives pertaining to nitrates, habitats, groundwater, drinking water and bathing water [56]. The Republic of Ireland is currently in the third cycle (2022–2027) of the WFD directive [57]. However, at the end of the second WFD cycle, the EPA reported that while almost half of Ireland's rivers and lakes (50% and 69%, respectively) meet or exceed the mandatory 'good ecological status', 64% of estuaries failed to meet the WFD's mandatory 'good ecological status' standard (https://www.epa.ie/publications/monitoring{-}{-}assessment/freshwater{-}{-}marine/).

Under the third WFD cycle, the EPA has broken down the Republic of Ireland into 46 catchments and 583 sub-catchment units for reporting purposes. There are a total of 4842 water bodies, with 3–15 water bodies in each sub-catchment [58]. The total area of transitional, coastal and lake water bodies monitored in this cycle are 842.37 km^2, 13,642.07 km^2 and 1200.31 km^2, respectively. The total number of monitoring points is 14,097, with 13,706 of them monitoring surface water bodies. The physical, chemical and biological changes in each water body are monitored with the sampling frequency depending on parameter and water types. For example, chemical oxygen demand (COD) is sampled an average of 80 times per month nationally, while the sampling time for nitrogen is 229 times per month nationally.

2.2. Earth Observation Platforms

Two satellites missions with publicly available data—Sentinel-2A/B multi-spectral instrument (MSI) and Landsat-8 Operational Land Imager (OLI)—were used in this study. Sentinel-2A/B are two identical Earth Observation satellites launched in June 2015 and March 2017 by the European Space Agency as part of the Copernicus Project (GMES). These satellites have spectral bands that span from the visible (VIS) to the short wave infrared (SWIR) (458–2280 nm) at high spatial resolution (10 m to 60 m), with a revisit time of 5 days at the equator [59].

Landsat-8 was launched by the U.S. Geological Survey (USGS) in February 2013 and carries a two-sensor payload (Operational Land Imager (OLI) and the Thermal Infrared Sensor (TIRS)) with a revisit time of 16 days [60]. Landsat-8 OLI was used, and it has a spectral range of Landsat-8 OLI (435–2294 nm) with a spatial resolution of 30 m. Detailed band information of each platform is listed in Table 1.

Table 1. Sentinel-2A/B MSI and Landsat-8 OLI band characteristics.

Sentinel-2A/B MSI			Landsat-8 OLI		
Band	**Wavelength Range (nm)**	**Resolution (m)**	**Band**	**Wavelength Range (nm)**	**Resolution (m)**
Band 1: Coastal aerosol	442.2–442.7	60	Band 1: Coastal/Aerosol	435–451	30
Band 2: Blue	492.1–492.4	10	Band 2: Blue	452–512	
Band 3: Green	559.0–559.8	10	Band 3: Green	533–590	
Band 4: Red	664.6–664.9	10	Band 4: Red	636–673	
Band 5—Vegetation red edge	703.8–704.1	20	Band 5: Near Infrared Red (NIR)	851–879	
Band 6—Vegetation red edge	739.1–740.5	20	Band 6: Shortwave Infrared 1 (SWIR1)	1566–1651	
Band 7—Vegetation red edge	779.7–782.8	20	Band 7: Shortwave Infrared 2 (SWIR2)	2107–2294	
Band 8: Near Infrared Red (NIR)	832.8–832.9	10	Band 8: Panchromatic	500–680	15
Band 8A—Narrow Near Infrared Red	864.0–864.7	20	Band 9: Cirrus	1360–1390	30
Band 9—Water vapor	943.2–945.1	60			
Band 10—SWIR—Cirrus	1373.5–1376.9	60			
Band 11: Shortwave Infrared (SWIR1)	1613.7–1610.4	20			
Band 12: Shortwave Infrared (SWIR2)	2185.7–2202.4	20			

2.3. Methodology

The method used in this research is depicted in Figure 2. In the following sections, the different stages of the research will be explained as follows: (a) two datasets have been pre-processed in the Google Earth Engine (GEE); the pre-process includes image selection, band selection and the creation of image catalogs; (b) to fit different resolutions of Sentinel-2, pan-sharpening was used in upscaling the 20-m band of Sentinel-2 to 10-m; (c) the calculation of the water indices: NDVI, NDWI, MNDWI and AWEI; and (d) the comparison of the maps from satellites with EPA datasets.

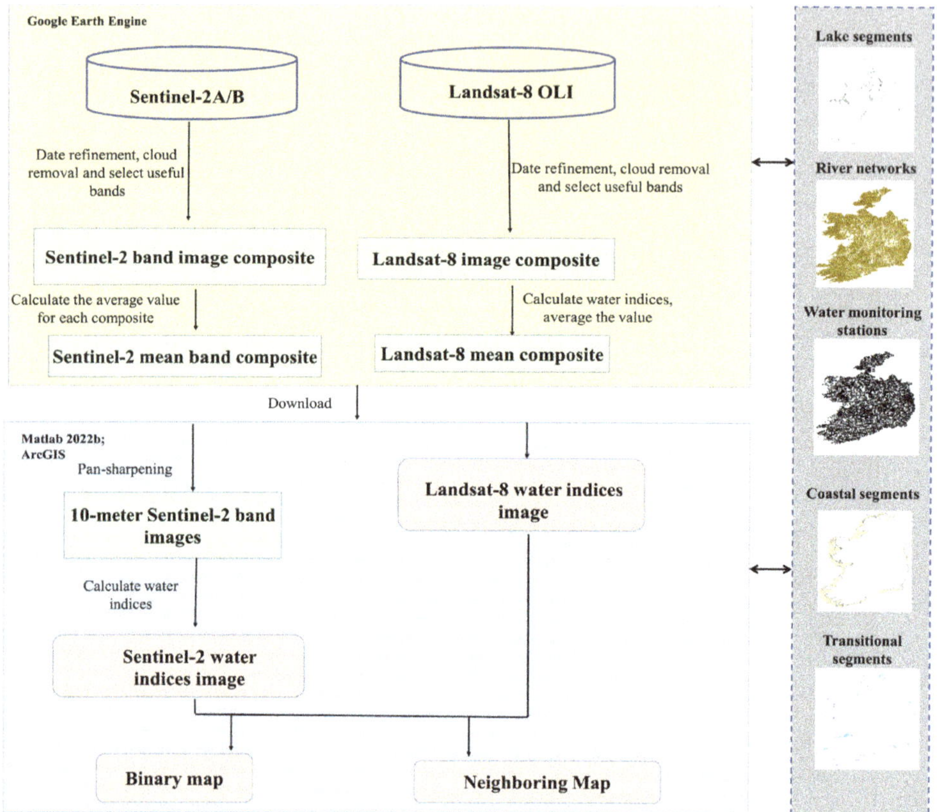

Figure 2. Methodology flowchart. Input data are indicated by cylinders, while output data and processing steps are shown in orange and green, respectively. Maps used in comparison are shown in the gray frame. Remotely sensed imagery from January 2018 to December 2019 was used.

2.3.1. Pre-Processing

Sentinel-2A/B MSI and Landsat-8 OLI, from January 2018 to December 2019, were initially processed on the Google Earth Engine (GEE) platform (https://code.earthengine.google.com/). GEE is a cloud-based geospatial analysis platform used for geographical computing and analysis, which allows academic users free access to the process and to analyze large amounts of data as they need [61], and the time period was chosen to ensure that a cloud free mosaic was able to be produce across the entire Republic of Ireland.

The Sentinel-2A/B level 2A and Landsat-8 Tier 1 datasets were used in this study. Level 2A is bottom-of-atmosphere (BOA) dataset, which are produced from level-1C data (top of atmosphere) via Sen2Cor [62], while Landsat-8 Collection 2 Tier 1 is the highest quality radiometric and positional data available.

The first step in pre-processing was to select images with cloud coverage less than 20%; then, cloud pixels were removed from Sentinel-2A/B and Landsat-8 imagery using the QA60 band and CFMask, respectively. The QA60 band is a bitmask band which contains cloud mask information, while CFMask is the C implementation of function of Mask, initially created by Zhu [63]. It is currently utilized by USGS to identify clouds, cloud shadows, cirrus, snow and ice during OLI data processing [64]. Its accuracy has been tested and compared with the other cloud detection algorithms (See5, Automated Cloud Cover Assessment (ACCA), Artificial Thermal-Automated Cloud Cover Algorithm (AT-ACCA), Fixed Temperature-Automated Cloud Cover Algorithm (FT-ACCA)) [65]. Results showed that CFMask outperforms other methods, producing the highest accuracy for cloud shadow results and working without geographic restrictions. While CFMask is a spectral characteristic-based multi-pass algorithm which used decision trees to produce two potential cloud masks, cloud mask was produced by combining these two potential cloud masks together [63].

2.3.2. Panchromatic Sharpening

An additional processing step was required for Sentinel-2 data to ensure that the spatial resolution of all bands in generating water masks are identical. Sentinel's SWIR1 and SWIR2 (Table 1) bands have a spatial resolution of 20 m, while Bands 2, 3, 4 and 8 are all at 10 m. Therefore, to obtain more explicit spatial information from Sentinel-2, panchromatic sharpening or pan-sharpening was necessary to upscale SWIR1 and SWIR2, and the overview of the process is shown in Figure 3.

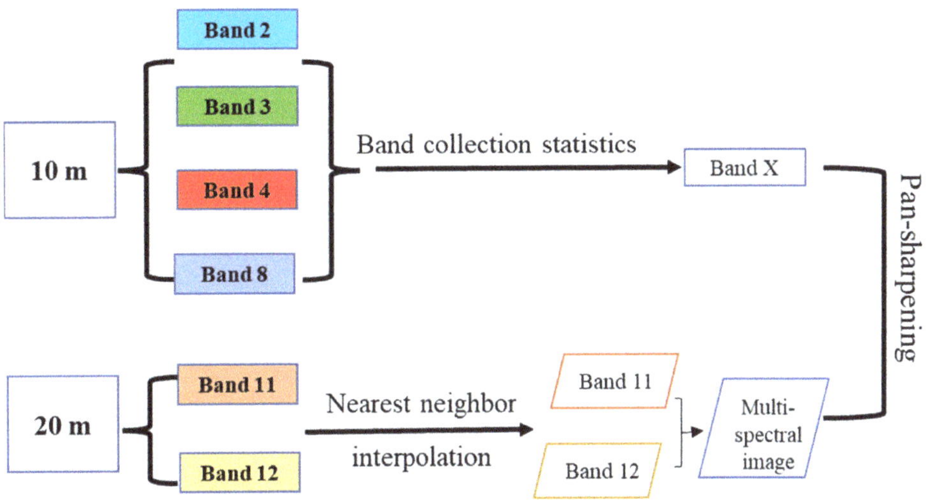

Figure 3. Pan-sharpening flowchart for Sentinel-2 data.

Band 11 and Band 12 have first been resampled to 10 m via nearest neighbor interpolation because nearest neighbor interpolation reserves the properties of the original dataset and is highly efficient compared to other resampling techniques over large areas. Additionally, previous research has shown that the selection of different resampling methods will not have a significant effect on the final visual appearance of fused images. Pan-sharpening is a technique that combines high-resolution detail from a panchromatic band with the lower-resolution color information of coarse bands, and it can help obtain a higher resolution map, which is crucial in some studies [66]. First, the lower resolution bands were needed to resample to obtain the same resolution with high-resolution band [67]. Many pan-sharpening approaches have been used and compared in extracting water masks [68,69]. One of the most common methods is the High Pass Filter (HPF) [70]. The HPF uses a filter to add information from the high-resolution image to the low-resolution image. This filter has a specified weight, and there are many kinds of soothing filters, including the Box (used in this study), Gaussian, and Laplacian. Du et al. [47] tested different image pan-sharpening approaches to generate the water masks at the Venice coastland, Italy, and showed that HPF was the most accurate method.

2.3.3. Water Mask Creation

The water indices from Landsat-8 were created directly on GEE, while Matlab2022a was used for Sentinel-2 to overcome the computation limit of GEE and the need for pan-sharpening Sentinel-2 images prior to water mask creation. For both Landsat-8 and Sentinel-2, water masks were created for four different indices: NDVI, NDWI, MNDWI and AWEI ((1)–(6)).

$$\text{NDVI} = \frac{\text{NIR} - \text{Red}}{\text{NIR} + \text{Red}} \tag{1}$$

$$\text{NDWI} = \frac{\text{Green} - \text{NIR}}{\text{Green} + \text{NIR}} \tag{2}$$

$$\text{MNDWI1} = \frac{\text{Green} - \text{SWIR1}}{\text{Green} + \text{SWIR1}} \tag{3}$$

$$\text{MNDWI2} = \frac{\text{Green} - \text{SWIR1}}{\text{Green} + \text{SWIR2}} \tag{4}$$

$$\text{AWEI}_{\text{sh}} = \text{Blue} + 2.5 \times \text{Green} - 1.5 \times (\text{NIR} + \text{SWIR1}) - 0.25 \times \text{SWIR2} \tag{5}$$

$$\text{AWEI}_{\text{nsh}} = 4 \times (\text{Green} - \text{SWIR1}) - (0.25 \times \text{NIR} + 2.75 \times \text{SWIR2}). \tag{6}$$

For each of these indices', binary maps, where 1 is water and 0 is land, were produced using the Bottom Valley threshold approach. This approach locates the lowest point between the two peaks, which is deemed the threshold for separating water and land pixels. The Otsu threshold approach [71] was also tested; however, this method resulted in the incorrect classification of pixels as land or water. Previous studies have noted this same issue with Ostu when the amplitudes of the bimodal are largely different, which causes the Ostu threshold to become biased [69,72]. Figure 4 shows the different thresholds calculated using the Otsu and the Bottom Valley approaches and shows the bias in the Ostu threshold.

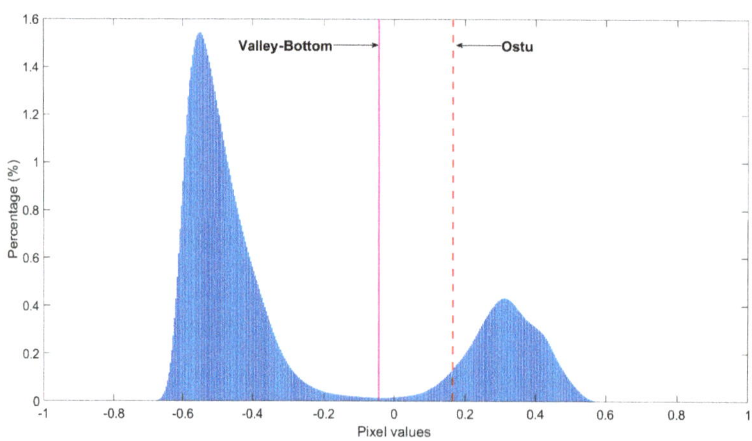

Figure 4. Comparison of Ostu and Bottom Valley approaches for land–water thresholding (Landsat-8 and NDWI used).

2.3.4. Comparison of Binary Water Masks

Two sets of binary maps were created for two different scenarios. Scenario 1 ignores many potential sources of errors, including boundary effects such as land contamination at the edge of water pixels [73], and assumes that if a pixel is classified as water, it is water. However, Scenario 2 aims to account for these boundary effects by only classifying a pixel as water if it classified as a water pixel itself and is also surrounded by other classified water pixels. This approach was previously applied to remove the boundary effect on ICESat [74].

Scenario 1 and Scenario 2 for the detection of water were compared with datasets available from the Irish EPA to investigate the spatial coverage of water bodies by the two remote sensing platforms and the impact of using different indices in distinguishing water land pixels. Comparison datasets include national water monitoring stations, shapefile of river net-routes, lake, coastal and transitional water bodies, and all were downloaded from the EPA website (https://gis.epa.ie/EPAMaps/). These datasets are for third-cycle WFD and are shown in Figure 2 gray frame.

The national water monitoring station map includes all 9736 rivers, 2883 lakes, 603 transitional, 484 coastal and 391 groundwater monitoring stations (although the groundwater stations were not applicable to this study). These monitoring stations are used for the WFD to periodically monitor water quality (e.g., concentration of chlorophyll-a and phosphorus). Therefore, it is important that remote sensing imagery is able to map these locations so that relationships between in situ measurements and remotely sensed optical properties can be developed. The river shapefile comprises over 100,000 polylines with the stream order information. The lakes, transitional and coastal maps comprise 810, 108 and 194 polygons, respectively. Since this study focused on the Republic of Ireland, the water bodies which lie entirely in Northern Ireland have been removed manually.

The binary maps for each scenario and for every index used for each platform were compared with the EPA data, and the following percentages were calculated:

- Percentage of water cells which overlay water monitoring stations;
- The percentage of lakes mapped and their areas;
- The percentage of rivers mapped and their stream order;
- Percentage of coastal and transitional areas mapped.

3. Results

3.1. Water Masks

Two regions in the Republic of Ireland were chosen to show the detailed results of the impact of using the different platforms and indices (Figure 5). The first region, as shown in Figure 5b, is Lough Derg. Lough Derg is the third-largest lake in the Republic of Ireland and is in the Lower Shannon River basin. This area is predominately rural. The second region is centered around Dublin City and Bay (Figure 5c). Dublin is the capital city of the Republic of Ireland, is heavily urbanized and is located on the east coast of Ireland.

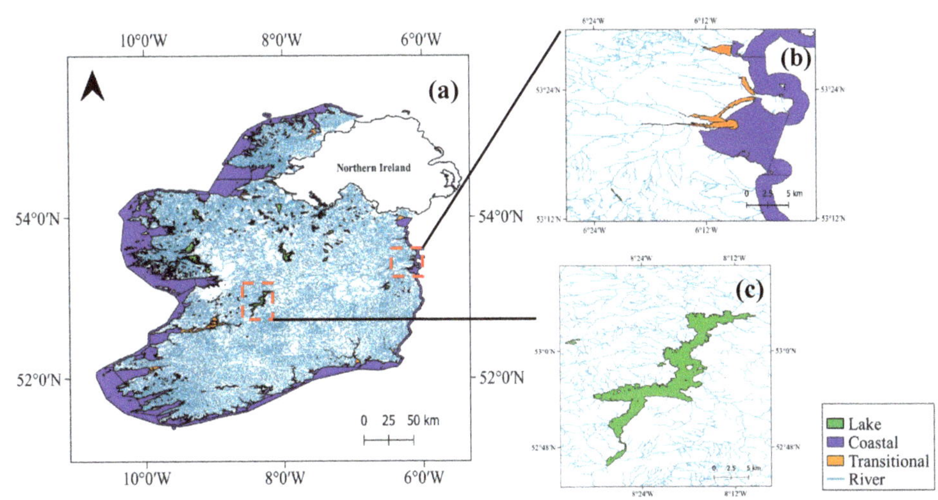

Figure 5. Location of the regions used in detailed analysis. Central panel (**a**) shows the locations of the two regions: Dublin Bay (right-up panel (**b**)) and Lough Derg (right-down panel (**c**)).

The water masks of Lough Derg and Dublin Bay mapped via Landsat-8 are depicted in Figure 6, corresponding to Scenario 1 and Scenario 2, while Figure 7 depicts those derived from Sentinel-2. The shape of Lough Derg was clearly mapped by all indices and platforms and for both scenarios, as was the river meander just south of Lough Derg, except for MNDWI2 via Sentinel-2. Although Sentinel-2 did map more details due to its higher spatial resolution, as expected, more of the river network was mapped in the Sentinel-2 water masks compared with those derived from Landsat. However, all indices did not perform equally across the platforms. By comparing Figures 6 and 7, it is clear that for Landsat, MNDWI1 was able to map more of the river network compared with the other indices, while for Sentinel-2, $AWEI_{sh}$ outperformed the other indices. The results for Dublin Bay region showed that both transitional and coastal areas were obvious and similar for all scenarios and indices, indicating that coastal and transitional waters were easily mapped by all cases. Unlike Lough Derg, NDVI and NDWI can map most of the river network (compared with other methods). However, there was significant noise from these methods where pixels were incorrectly identified as water when they corresponded to the built environment. This noise was minimal for MNDWI and AWEI, which was expected, as these methods were designed to reduce this type of noise and misclassification. For Sentinel-2 and the Dublin Bay region, $AWEI_{nsh}$ was the best for both scenarios, 0outperforming Landsat water masks.

3.2. Comparison with In Situ Monitoring Points

Figure 8 shows the mapped water monitoring stations under different indices and different satellites, with Table 2 stating the percentage of water monitoring stations visible via the different platforms. Figure 8b shows the full extent of the surface water monitoring stations operated under Cycle 3 of the WFD across the Republic of Ireland, with over half (9736 out of 13,706) of the stations located on rivers; stations located on lakes account for over 20% (2883); while coastal and transitional water bodies account for only 10% of the total number of stations. Figure 8a,c shows the mapped stations calculated from MNDWI1 via Landsat-8 and $AWEI_{nsh}$ via Sentinel-2, respectively. These were the best performing indices for their corresponding platform. Figure 8 shows that a far greater percentage of costal and transitional stations were mapped with remote sensing than other water bodies, with river stations being the water bodies which were harder to see via remote sensing. This visual finding is supported by Table 2.

Table 2 shows the percentage of monitoring stations mapped by Landsat-8 and Sentinel-2 for all six indices and for Scenario 1 and Scenario 2. Overall, Sentinel-2 and $AWEI_{nsh}$ performed best at capturing the most monitoring stations, with 24.78% of station mapped under the Scenario 1, and this dropped to 17.8% under Scenario 2. For Landsat-8, MNDWI1 was the best performing index, mapping 20.59% under the Scenario 1, and this dropped to 12.81% for Scenario 2.

As expected, coastal stations and their monitoring stations were the easiest to map via remote sensing, with an average rate of 80% across all platforms and indices. This dropped to ~57% of transitional water bodies and dropped again to ~43% for lakes. River water bodies were the hardest to map, with an overall average of 1.63%. However, for Scenario 1, Sentinel-2 and $AWEI_{nsh}$ were able to map 6.19%, while Landsat-8 and MNDWI1 were only able to map 2.43% of monitoring stations on river water bodies.

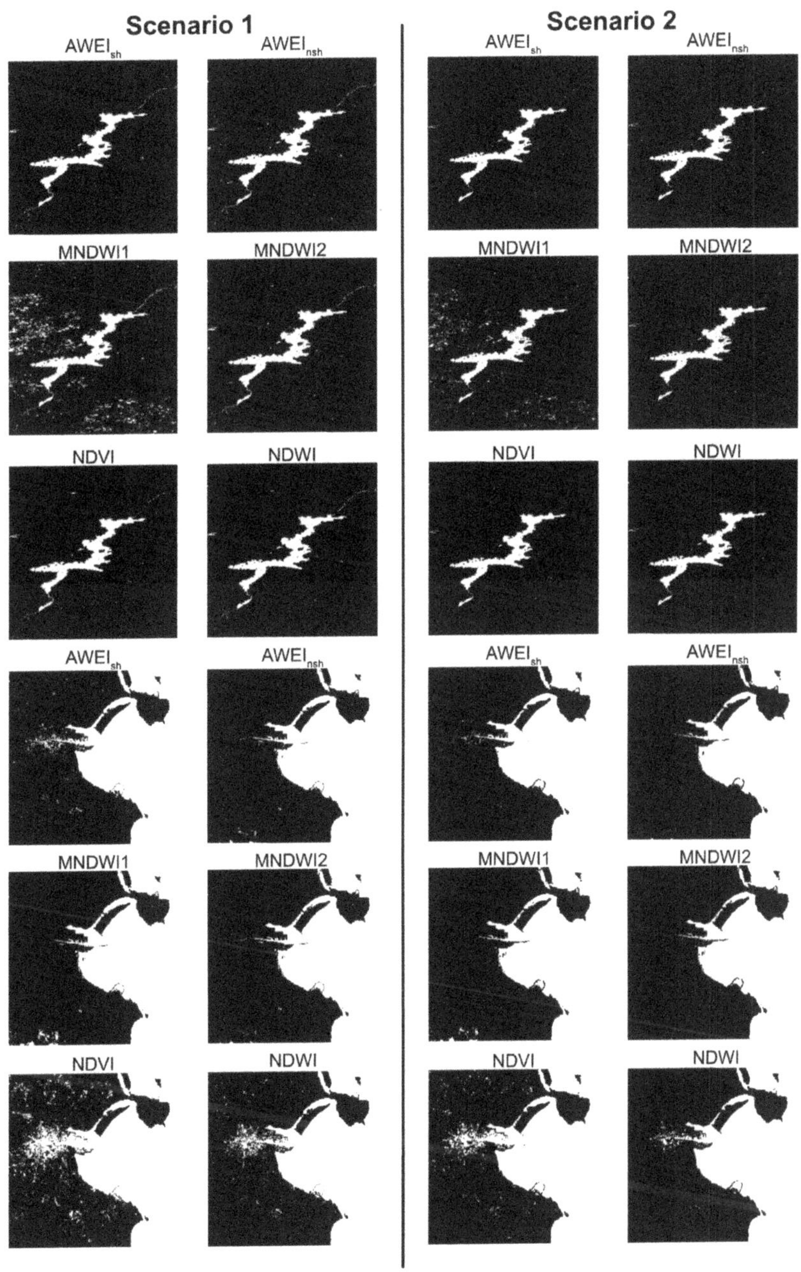

Figure 6. Landsat-8-derived water masks of Lough Derg (**top 12 panels**) and Dublin Bay (**bottom 12 panels**) for Scenario 1 and Scenario 2.

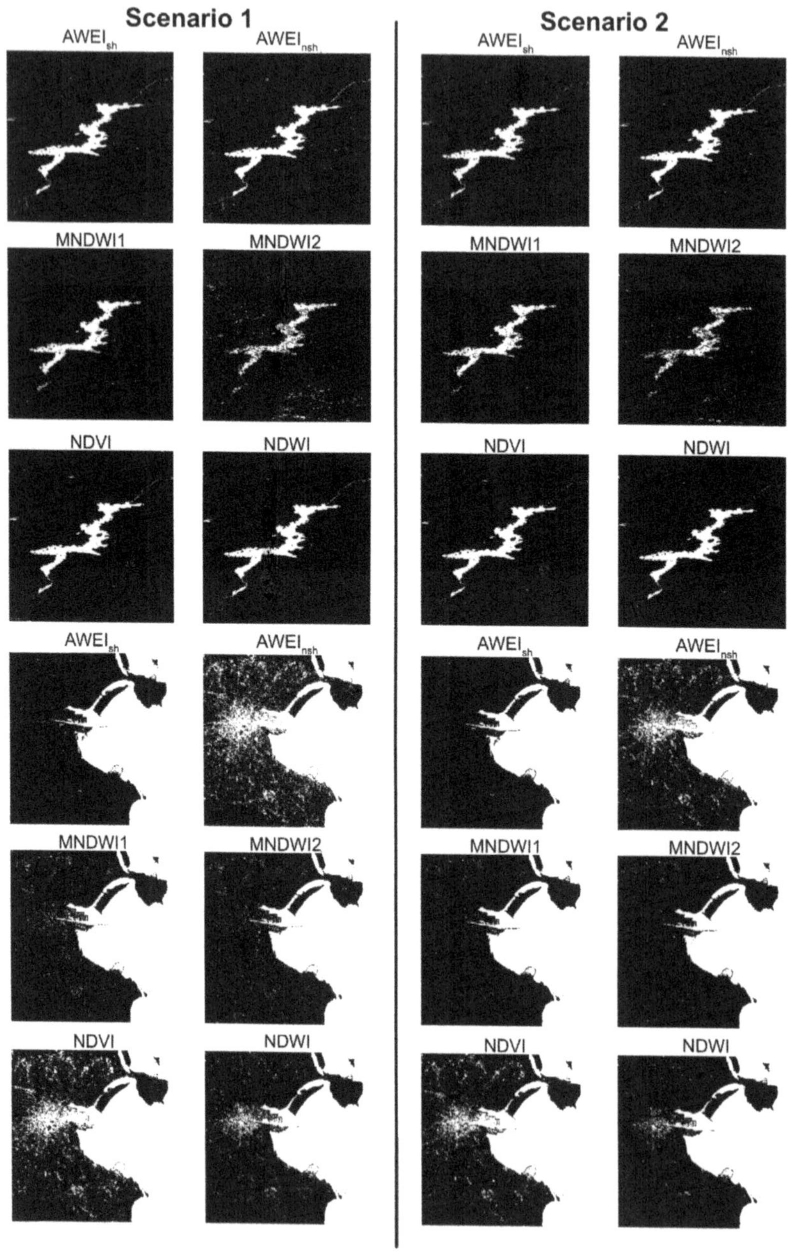

Figure 7. Sentinel-2-derived water masks of Lough Derg (**top 12 panels**) and Dublin Bay (**bottom 12 panels**) for Scenario 1 and Scenario 2.

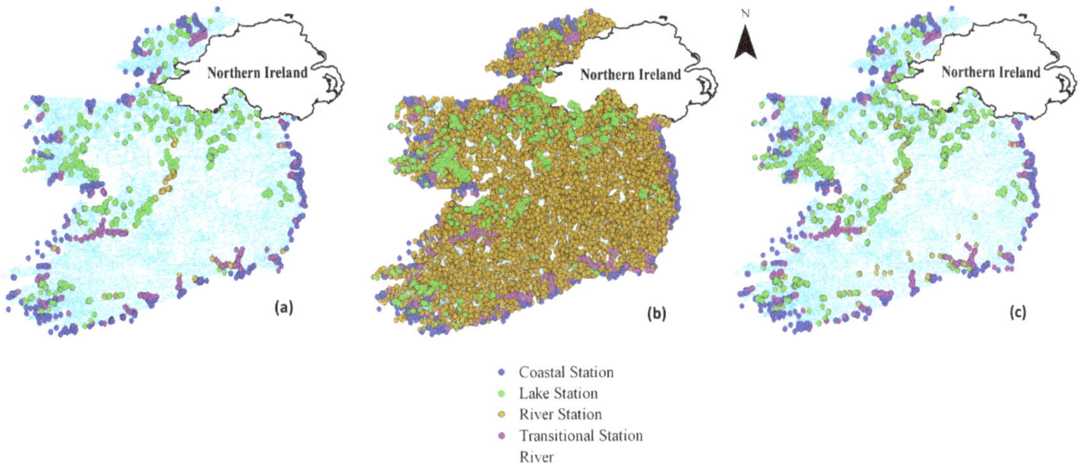

Figure 8. Monitoring stations mapped using the best indices for each platform: (**a**) Landsat-8 and MNDWI1; (**b**) all monitoring stations; (**c**) Sentinel-2 and $AWEI_{sh}$.

Table 2. Percentage (%) of mapped water monitoring stations from Landsat-8 and Sentinel-2 for both Scenario 1 and Scenario 2. Scenario 2 percentages are within brackets.

Platform	Indices	Total	Coastal	Lake	River	Transition
Landsat-8	NDWI	17.7 (12.02)	83.68 (73.76)	50.99 (32.64)	1.83 (0.7)	61.86 (46.6)
	NDVI	17.68 (12.22)	84.71 (75.62)	49.29 (32.74)	2.07 (0.68)	64.68 (49.59)
	MNDWI1	20.59 (12.81)	85.12 (75.21)	61.95 (35.62)	2.43 (0.73)	64.18 (48.76)
	MNDWI2	18.1 (12.12)	83.47 (73.97)	53.9 (33.54)	1.64 (0.62)	60.2 (45.77)
	$AWEI_{sh}$	19.26 (12.54)	86.57 (75.83)	56.64 (33.96)	2.06 (0.76)	64.18 (49.59)
	$AWEI_{nsh}$	19.36 (12.46)	85.95 (75.41)	58.06 (34.41)	1.93 (0.67)	62.35 (47.43)
Sentinel-2	NDWI	17.1 (13.73)	77.69 (74.59)	49.12 (37.25)	2.11 (1.45)	57.55 (50.75)
	NDVI	18.14 (14.29)	80.99 (77.48)	50.47 (37.32)	2.61 (1.67)	63.85 (57.21)
	MNDWI1	12.88 (10.45)	83.26 (77.27)	32.74 (24.63)	0.65 (0.26)	58.87 (53.57)
	MNDWI2	14.91 (10.9)	86.16 (78.72)	37.98 (25.42)	1.46 (0.44)	64.51 (55.89)
	$AWEI_{sh}$	17.66 (14.13)	82.44 (76.86)	50.78 (38.61)	1.99 (1.28)	60.36 (54.06)
	$AWEI_{nsh}$	24.78 (17.8)	85.33 (80.17)	67.74 (48.21)	6.19 (2.92)	70.98 (62.52)

3.3. River Network

In the shapefile of river net-routes available from the Irish EPA, a total of 108,180 segments are fully contained inside the Republic of Ireland. The stream order for each segment is available, and they were used in this study to explore how the size of a river (stream order) impacts the visibility of that river via remote sensing. The majority of river segments corresponded to smaller stream orders, with ~51% corresponding to order 1 stream and ~25% corresponding to order 2 stream. These percentages continued to reduce by approximately half for each increase in stream order, with only ~1.5% of segments corresponding to stream order 6 and 7.

Table 3 illustrates how many river segments were mapped under the different scenarios and by different approaches with respect to their river order. Except for the indices MNDWI1 and MNDWI2, the overall performance of Sentinel-2 was better than that of Landsat-8 for river segments under both scenarios.

Table 3. Percentage (%) of river net-routes which were mapped from Landsat-8 and Sentinel-2 with respect to the river order for both scenarios. Scenario 2 percentages are within brackets.

Platform	River Order	NDVI	NDWI	MNDWI1	MNDWI2	$AWEI_{sh}$	$AWEI_{nsh}$
Landsat-8	1	2.17 (1.5)	2.05 (1.43)	4.51 (2.42)	2.29 (1.55)	2.37 (1.64)	2.56 (1.69)
	2	3.69 (2.78)	3.55 (2.7)	5.26 (3.38)	3.73 (2.81)	3.87 (2.95)	3.96 (2.97)
	3	5.17 (4.06)	5.01 (3.95)	6.42 (4.5)	5.11 (4.0)	5.35 (4.23)	5.38 (4.21)
	4	8.42 (7.17)	8.17 (6.92)	9.17 (7.42)	8.23 (7.01)	8.53 (7.32)	8.5 (7.26)
	5	17.67 (14.68)	17.63 (14.48)	20.42 (15.08)	18.12 (14.57)	18.65 (15.07)	18.93 (15)
	6	37.36 (28.13)	37.86 (28.02)	43.25 (30.04)	38.82 (27.98)	39.58 (29.17)	40.49 (29.05)
	7	82.01 (71.52)	82.76 (72.12)	87.37 (76.08)	83.81 (73.34)	84.6 (74.05)	85.32 (74.48)
Sentinel-2	1	2.2 (1.87)	2.02 (1.75)	1.42 (1.17)	2.46 (1.51)	2.09 (1.77)	2.65 (2.23)
	2	3.7 (3.32)	3.42 (3.1)	2.57 (2.21)	3.35 (2.41)	3.53 (3.14)	4.23 (3.74)
	3	5.22 (4.72)	4.88 (4.44)	3.74 (3.21)	4.46 (3.41)	5.02 (4.51)	5.9 (5.29)
	4	8.56 (7.91)	7.97 (7.37)	6.83 (6.14)	7.36 (6.12)	8.11 (7.58)	9.7 (8.57)
	5	18.93 (16.94)	18.46 (16.24)	14.6 (13.2)	15.6 (13.09)	18.77 (16.45)	25.07 (19.21)
	6	38.28 (34.63)	37.83 (33.28)	28.82 (25.49)	31.62 (26.39)	42.01 (35.66)	56.66 (42.06)
	7	82.14 (79.06)	84.7 (80.39)	67.75 (65.38)	65.68 (61.47)	85.77 (81.55)	89.75 (85.31)

As river orders increased, the percentage of segments that were mapped also increased. This was because as the stream order became higher, the width of the river also increased, making it easier for satellites to detect water. Normally, 80% of the areas of order 7 rivers were mapped by most indices via both platforms. MNDWI1 was the higher performing index for Landsat-8, detecteding over 85% of order 7 river segments, while $AWEI_{nsh}$ performed best with Sentinel-2, with 88.87% of the order 7 area mapped. One surprising result was that MNDWI1 and Landsat-8 outperformed $AWEI_{nsh}$ and Sentinel-2 for stream orders 3 or lower for Scenario 1, and the reverse was true for Scenario 2.

3.4. Lake Segments

There are 810 lakes identified under Cycle 3 WFD in the Republic of Ireland, with the majority of these being smaller than 1 km^2 in area (see Figures 9 and 10). To account for this uneven distribution in lake areas and to make visualization easier, each lake was ranked based on its surface area, with 1 denoting the smallest and 810 denoting the largest lake. From Figures 9 and 10, there was a logarithmic trend for most indices and platforms, with larger lakes being more likely to be mapped than the smaller ones; over 70% of lakes greater than 1 km^2 were mapped, irrespective of platform or indices used. Similar to the analysis for the river network, MNDWI1 and $AWEI_{nsh}$ were the best preforming indices for Landsat-8 and Sentinel-2, respectively, and again, MNDWI and Sentinel-2 produced the worst performing combination.

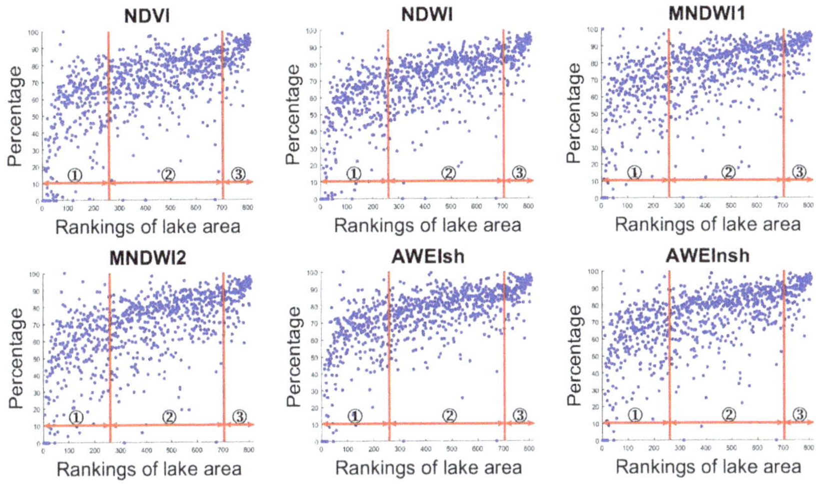

Figure 9. The percentage of mapped areas for each lake segment with respect to their area for Landsat-8 for Scenario 1. Three different ranges of lake area are highlighted: ① Lake area < 0.1 km^2; ② 0.1 km^2 < Lake area < 1 km^2; and ③ Lake area > 1 km^2.

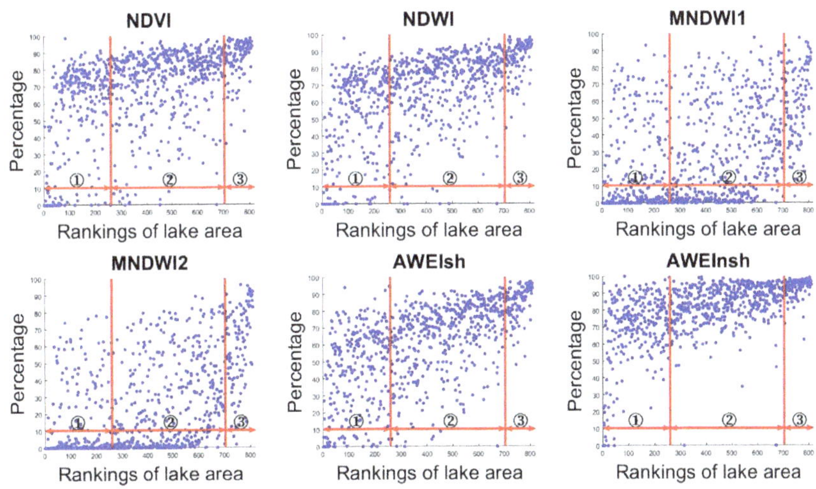

Figure 10. The percentage of mapped areas for each lake segment with regards to their area for Sentinel-2 for Scenario 1. Three different ranges of lake area are highlighted: ① Lake area < 0.1 km^2; ② 0.1 km^2 < Lake area < 1 km^2; and ③ Lake area > 1 km^2.

Figure 11 shows that MNDWI1 and $AWEI_{nsh}$ were the best performing indices for Landsat-8 and Sentinel-2, respectively. No matter which indices were used, on average, over 70% of lake areas were mapped via Landsat-8 under Scenario 1, and this percentage dropped to 40% when potential land contamination was considered (Scenario 2). However, all indices produced similar results. For Sentinel-2, there was more variation across indices, with the MNDWI index only capturing less than 25% on average, while $AWEI_{nsh}$ captured over 80% of the lake areas for Scenario 1. For Scenario 2, the variation was larger, with MNDWI1 and MNDWI2 capturing, on average, less than 12% of lake areas. This contrasts with the other indices, which mapped over 55% for lake areas.

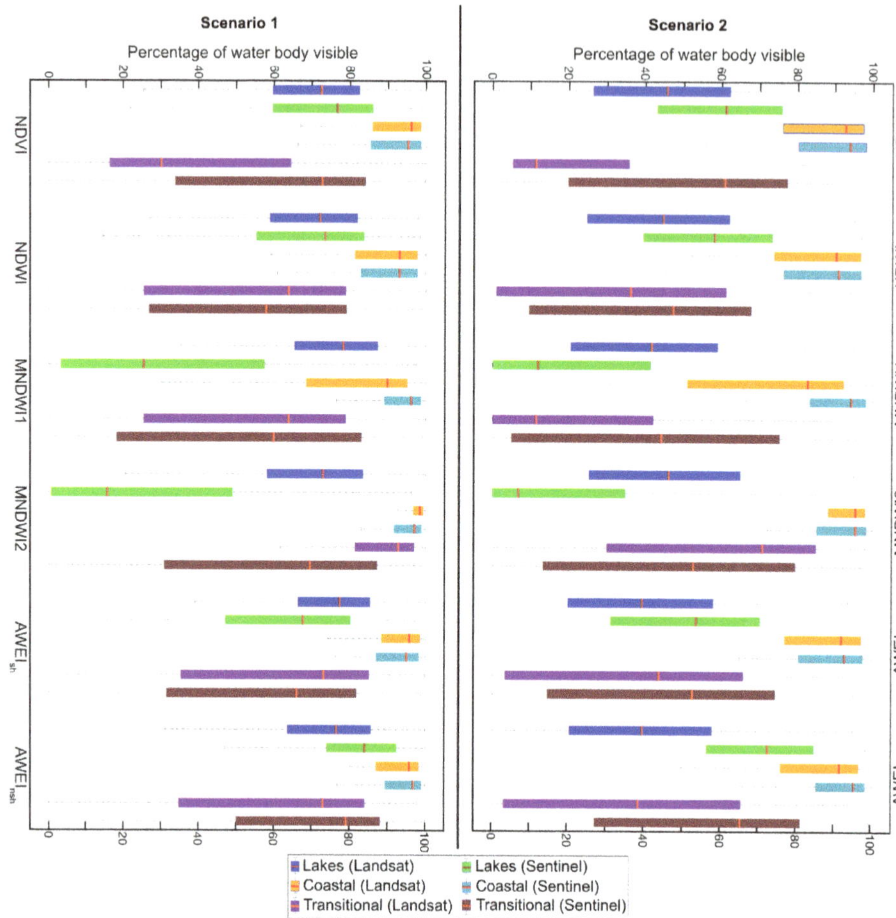

Figure 11. Boxplots showing percentage area of water body mapped by Landsat-8 and Sentinel-2 under Scenario 1 and Scenario 2.

3.5. Coastal and Transitional Water

3.5.1. Coastal Water

Figure 11 depicts the percentage of coastal areas mapped with the different platforms and indices. Coastal areas had a higher percentage area mapped than the other water types. For all scenarios, on average, over 80% of all coastal water bodies were mapped for both Landsat and Sentinel. MNDWI2 with Landsat-8 was the best performing combination of all platforms and indices, with the smallest variations in area mapped and the highest percentage mapped. One surprising result was the performance of the MNDWI index combined with Sentinel-2: for all other water bodies, it was among the poorest performers; but for Coastal bodies, it was among the best performers.

3.5.2. Transitional Water

There are 194 transitional waters monitored under the third WFD cycle. Figure 11 shows that Sentinel-2 outperformed Landsat-8, with the exception for MNDWI2 and Landsat-8, which outperformed all other combinations. On average, over 90% of transitional waters were mapped to MNDWI2 and Landsat-8 for Scenario 1, and this dropped to over 70% for the Scenario 2. Similar to previous water bodies, $AWEI_{nsh}$ was the best index for Sentinel-2 images, with just below 80% of the transitional water body areas mapped in Scenario 1 and just over 65% for Scenario 2.

4. Discussion

The results showed how well the different indices and satellite platforms map the existing in situ monitoring stations for WFD water quality monitoring and the shapefiles of the four main water body types (river, lakes, transitional and coastal) assessed in this study. The results showed that Sentinel-2 outperforms Landsat-8 in mapping water bodies. Comparing just the water masks (Figures 6 and 7), the benefit of the higher resolution of Sentinel-2, with its 10 m pixel size compared with Landsat-8's 30 m resolution, is shown. In these two figures, overall, Sentinel-2 can map more details than Landsat-8. This is also supported by the other results, and the benefit of using the highest spatial resolution possible are shown in Table 2. This table shows the percentage of in situ water quality monitoring stations that can be mapped by the different platforms, and the benefit of using Sentinel-2 are shown. While there is very little difference between the best indices for Landsat-8 and Sentinel-2 for Coastal water bodies, Sentinel-2 outperforms Landsat-8 for transition and lake water bodies and especially for river. For river water bodies alone, Sentinel-2 mapped nearly 4% more of the river monitoring stations, or a ~155% increase over Landsat-8. This increase for river monitoring stations can be explained by the higher resolution of Sentinel-2; it is able to map more river bodies, especially the smaller rivers, as shown in Table 3.

The choice of indices used had a major impact on the results. Six different indices were investigated in this study: NDVI, NDVI, two forms of MNDWI and AWEI. Our results showed that Sentinel-2 and $AWEI_{nsh}$ was the best combination across all measures explored in this study, and MNDWI was the best index for Landsat-8. This result supports the findings of Liu et al. [46] and Du et al. [47], but conflicts with the findings of Acharya, Subedi and Lee [43], who found that NDVI and NDWI outperformed MNDWI and AWEI. However, Acharya, Subedi and Lee [43] only looked at Landsat-8, while we investigated both Landsat-8 and Sentinel-2. Our results do show the both versions of MNDWI performed significantly worse for Sentinel-2 compared to Landsat-8. For all water bodies, the combination of MNDWI and Sentinel-2 was outperformed by all other combinations of index and platform. MNDWI was designed for use with Landsat [75]; however, so was AWEI, and that was shown to the best index to use with Sentinel-2.

Two different scenarios were also tested to explore the potential impact of a conservative approach for dealing with the potential of land contamination. The first scenario (Scenario 1) classified a pixel as water if, after applying the Bottom Valley threshold approach, it was deemed water, while the second scenario (Scenario 2) required that a pixel was classified as water and for the surrounding nine pixels to also be classified as water. The results showed that there is a measurable difference in performance between the two scenarios, for Landsat-8 there was, on average across all indices, a 6.4% drop in the number of in situ monitoring stations mapped, while there was only a 4% drop for Sentinel-2. This again highlights the benefit of using the highest possible spatial resolution.

5. Conclusions

The overall objective of this study was to assess the value of remote sensing optical imagery for water quality monitoring across a range of water body types in the Republic of Ireland. To assess the value, it was first necessary to know how well the various satellite platforms map water bodies and monitoring points across the Republic of Ireland. To quantify this, this study aimed to address the following objectives:

1. What percentages of water bodies are mapped from the different remote sensing platforms? What is the difference between using Landsat-8 or Sentinel-2?
2. Which is the best water index for detecting water pixels across Ireland? Does this vary by water body type?

3. How well does remote sensing map the existing in situ monitoring points? This is critical for the calibration of water quality estimates from remote sensing.

Water masks were created using six different indices for both Sentinel-2 and Landsat-8 and for two different scenarios. The water masks were then compared with the existing monitoring station for water quality and shapefiles describing the river network, lakes and transitional and coastal water bodies.

The results showed that optical remote sensing has potential for water quality monitoring across a range of water body types in the Republic of Ireland. As expected, rivers were the most difficult for both Sentinel-2 and Landsat-8 to map, followed by transitional water bodies, and then lakes with coastal water bodies.

The results also showed that Sentinel-2 was better able to map the different water bodies than Landsat-8, except for coastal water bodies. This is mainly due to the higher spatial resolution of Sentinel-2 compared with Landsat-8. Comparing the different water masks to in situ monitoring stations, Sentinel-2 was able to map ~25% of all stations compared with ~21% for Landsat-8. These percentages were heavily impacted by the large number of river monitoring stations that neither Sentinel-2 nor Landsat-8 could map, with Sentinel-2 mapping ~6.2% compared with ~2.4% for Landsat-8. This highlights the importance of spatial resolution in monitoring water bodies in Ireland, and that Sentinel-2 with its 10 m resolution, outperformed Landsat-8 and its 30 m-resolution imagery.

The choice of indices in identifying water and land pixel was also a critical aspect to consider. We investigated six commonly used indices. Our results showed that the best performing index for Landsat-8 was MNDWI2—excluding for rivers where, MNDWI1 produced slightly better results. For Sentinel-2, $AWEI_{nsh}$ was the best index across all water bodies.

Overall, Sentinel-2 and $AWEI_{nsh}$ were the best-performing combination across all water bodies and, on average, were able to map ~80% of the total area of transitional water bodies; >95% of coastal waters; and >85% of total lake areas. All combinations of platform and index found it difficult to see Irish rivers due to their small size, but $AWEI_{nsh}$ and Sentinel-2 were able to map more than 25% of rivers with a stream order of 5, which increased to >56% for stream orders equal to 6 and ~90% for larger stream orders. This study showed that optical remote sensing has potential in water quality monitoring in the Republic of Ireland, especially for larger rivers, lakes and transitional and coastal water bodies, which are easier to map.

Author Contributions: Conceptualization, methodology, software, formal analysis, writing, review and editing: M.Z. and F.O.; Supervision: F.O. All authors have read and agreed to the published version of the manuscript.

Funding: This research was funded by the joint scholarship between University College Dublin and the Chinese Scholarship Council. The grant number is '201908300017'.

Data Availability Statement: The water maps (water monitoring station map, river, lake, coastal and transitional shapefile) which we used for comparison are provided by Irish EPA (https://gis.epa.ie/EPAMaps/).

Acknowledgments: We would like to thank our anonymous reviewers for their time and effort in reviewing our manuscript.

Conflicts of Interest: The authors declare no conflict of interest.

References

1. Ali, I. New Generation Adsorbents for Water Treatment. *Chem. Rev.* **2012**, *112*, 5073–5091. [CrossRef] [PubMed]
2. Weerasekara, P. The United Nations World Water Development Report 2017-Wastewater: The Untapped Resource. *Future Food J. Food Agric. Soc.* **2017**, *5*, 80–81.
3. Tebbutt, T.H.Y. *Principles of Water Quality Control*; Elsevier: Amsterdam, The Netherlands, 1997.
4. Todo, K.; Sato, K. Directive 2000/60/EC of the European Parliament and of the Council of 23 October 2000 Establishing a Framework for Community Action in the Field of Water Policy. *Environ. Res. Q.* **2000**, *327*, 1–73.

5. Antwi, S.H.; Rolston, A.; Linnane, S.; Getty, D. Communicating Water Availability to Improve Awareness and Implementation of Water Conservation: A Study of the 2018 and 2020 Drought Events in the Republic of Ireland. *Sci. Total Environ.* **2022**, *807*, 150865. [CrossRef]
6. O'Boyle, S.; Trodd, W.; Bradley, C.; Tierney, D.; Wilkes, R.; Longphuirt, S.N.; Smith, J.; Stephens, A.; Barry, J.; Maher, P. *Water Quality in Ireland 2013–2018*; Environmental Protection Agency: Wexford, Ireland, 2019.
7. Yang, H.; Kong, J.; Hu, H.; Du, Y.; Gao, M.; Chen, F. A Review of Remote Sensing for Water Quality Retrieval: Progress and Challenges. *Remote Sens.* **2022**, *14*, 1770. [CrossRef]
8. Gholizadeh, M.H.; Melesse, A.M.; Reddi, L. A Comprehensive Review on Water Quality Parameters Estimation Using Remote Sensing Techniques. *Sensors* **2016**, *16*, 1298. [CrossRef]
9. Ritchie, J.C.; Zimba, P.V.; Everitt, J.H. Remote Sensing Techniques to Assess Water Quality. *Photogramm. Eng. Remote Sens.* **2003**, *69*, 695–704. [CrossRef]
10. Ritchie, J.C.; McHenry, J.R.; Schiebe, F.R.; Wilson, R.B. Relationship of Reflected Solar Radiation and the Concentration of Sediment in the Surface Water of Reservoirs. *Remote Sens. Earth Resour.* **1975**, *3*, 57–71.
11. Schultz, G.A.; Engman, E.T. *Remote Sensing in Hydrology and Water Management*; Springer Science & Business Media: Berlin/Heidelberg, Germany, 2012; ISBN 978-3-642-59583-7.
12. Agarwal, A.; Taveneau, A.; Olbert, I.A. Remote Sensing of Surface Waters in Ireland. In Proceedings of the 2018 CERI, University College Dublin, Dublin, Ireland, 29–30 August 2018; p. 7.
13. Karki, S.; Bermejo, R.; Wilkes, R.; Monagail, M.M.; Daly, E.; Healy, M.; Hanafin, J.; McKinstry, A.; Mellander, P.-E.; Fenton, O.; et al. Mapping Spatial Distribution and Biomass of Intertidal Ulva Blooms Using Machine Learning and Earth Observation. *Front. Mar. Sci.* **2021**, *8*, 1–20. [CrossRef]
14. McCullough, I.M.; Loftin, C.S.; Sader, S.A. High-Frequency Remote Monitoring of Large Lakes with MODIS 500 m Imagery. *Remote Sens. Environ.* **2012**, *124*, 234–241. [CrossRef]
15. Klein, I.; Mayr, S.; Gessner, U.; Hirner, A.; Kuenzer, C. Water and Hydropower Reservoirs: High Temporal Resolution Time Series Derived from MODIS Data to Characterize Seasonality and Variability. *Remote Sens. Environ.* **2021**, *253*, 112207. [CrossRef]
16. Han, W.; Huang, C.; Gu, J.; Hou, J.; Zhang, Y. Spatial-Temporal Distribution of the Freeze–Thaw Cycle of the Largest Lake (Qinghai Lake) in China Based on Machine Learning and MODIS from 2000 to 2020. *Remote Sens.* **2021**, *13*, 1695. [CrossRef]
17. Rao, P.; Jiang, W.; Hou, Y.; Chen, Z.; Jia, K. Dynamic Change Analysis of Surface Water in the Yangtze River Basin Based on MODIS Products. *Remote Sens.* **2018**, *10*, 1025. [CrossRef]
18. Sidle, R.C.; Ziegler, A.D.; Vogler, J.B. Contemporary Changes in Open Water Surface Area of Lake Inle, Myanmar. *Sustain. Sci.* **2007**, *2*, 55–65. [CrossRef]
19. Sawaya, K.E.; Olmanson, L.G.; Heinert, N.J.; Brezonik, P.L.; Bauer, M.E. Extending Satellite Remote Sensing to Local Scales: Land and Water Resource Monitoring Using High-Resolution Imagery. *Remote Sens. Environ.* **2003**, *88*, 144–156. [CrossRef]
20. Cannistra, A.F.; Shean, D.E.; Cristea, N.C. High-Resolution CubeSat Imagery and Machine Learning for Detailed Snow-Covered Area. *Remote Sens. Environ.* **2021**, *258*, 112399. [CrossRef]
21. Cooley, S.W.; Smith, L.C.; Ryan, J.C.; Pitcher, L.H.; Pavelsky, T.M. Arctic-Boreal Lake Dynamics Revealed Using CubeSat Imagery. *Geophys. Res. Lett.* **2019**, *46*, 2111–2120. [CrossRef]
22. Cooley, S.W.; Smith, L.C.; Stepan, L.; Mascaro, J. Tracking Dynamic Northern Surface Water Changes with High-Frequency Planet CubeSat Imagery. *Remote Sens.* **2017**, *9*, 1306. [CrossRef]
23. Junqueira, A.M.; Mao, F.; Mendes, T.S.G.; Simões, S.J.C.; Balestieri, J.A.P.; Hannah, D.M. Estimation of River Flow Using CubeSats Remote Sensing. *Sci. Total Environ.* **2021**, *788*, 147762. [CrossRef]
24. Dash, J.; Ogutu, B.O. Recent Advances in Space-Borne Optical Remote Sensing Systems for Monitoring Global Terrestrial Ecosystems. *Prog. Phys. Geogr. Earth Environ.* **2016**, *40*, 322–351. [CrossRef]
25. Wulder, M.A.; Loveland, T.R.; Roy, D.P.; Crawford, C.J.; Masek, J.G.; Woodcock, C.E.; Allen, R.G.; Anderson, M.C.; Belward, A.S.; Cohen, W.B.; et al. Current Status of Landsat Program, Science, and Applications. *Remote Sens. Environ.* **2019**, *225*, 127–147. [CrossRef]
26. Pahlevan, N.; Sarkar, S.; Franz, B.A.; Balasubramanian, S.V.; He, J. Sentinel-2 MultiSpectral Instrument (MSI) Data Processing for Aquatic Science Applications: Demonstrations and Validations. *Remote Sens. Environ.* **2017**, *201*, 47–56. [CrossRef]
27. Allen, G.H.; Pavelsky, T.M. Global Extent of Rivers and Streams. *Science* **2018**, *361*, 585–588. [CrossRef] [PubMed]
28. Yamazaki, D.; Trigg, M.A.; Ikeshima, D. Development of a Global ~90 m Water Body Map Using Multi-Temporal Landsat Images. *Remote Sens. Environ.* **2015**, *171*, 337–351. [CrossRef]
29. O'Loughlin, F.; Trigg, M.A.; Schumann, G.J.-P.; Bates, P.D. Hydraulic Characterization of the Middle Reach of the Congo River. *Water Resour. Res.* **2013**, *49*, 5059–5070. [CrossRef]
30. Allen, G.H.; Pavelsky, T.M. Patterns of River Width and Surface Area Revealed by the Satellite-Derived North American River Width Data Set. *Geophys. Res. Lett.* **2015**, *42*, 395–402. [CrossRef]
31. Bergsma, E.W.J.; Almar, R. Coastal Coverage of ESA' Sentinel 2 Mission. *Adv. Space Res.* **2020**, *65*, 2636–2644. [CrossRef]
32. Yang, X.; Qin, Q.; Yésou, H.; Ledauphin, T.; Koehl, M.; Grussenmeyer, P.; Zhu, Z. Monthly Estimation of the Surface Water Extent in France at a 10-m Resolution Using Sentinel-2 Data. *Remote Sens. Environ.* **2020**, *244*, 111803. [CrossRef]
33. Chang, M.; Li, P.; Li, Z.; Wang, H. Mapping Tidal Flats of the Bohai and Yellow Seas Using Time Series Sentinel-2 Images and Google Earth Engine. *Remote Sens.* **2022**, *14*, 1789. [CrossRef]

34. Pardo-Pascual, J.E.; Sánchez-García, E.; Almonacid-Caballer, J.; Palomar-Vázquez, J.M.; Priego de los Santos, E.; Fernández-Sarría, A.; Balaguer-Beser, Á. Assessing the Accuracy of Automatically Extracted Shorelines on Microtidal Beaches from Landsat 7, Landsat 8 and Sentinel-2 Imagery. *Remote Sens.* **2018**, *10*, 326. [CrossRef]
35. Wang, Y.; Li, Z.; Zeng, C.; Xia, G.-S.; Shen, H. An Urban Water Extraction Method Combining Deep Learning and Google Earth Engine. *IEEE J. Sel. Top. Appl. Earth Obs. Remote Sens.* **2020**, *13*, 769–782. [CrossRef]
36. Huang, C.; Chen, Y.; Zhang, S.; Wu, J. Detecting, Extracting, and Monitoring Surface Water from Space Using Optical Sensors: A Review. *Rev. Geophys.* **2018**, *56*, 333–360. [CrossRef]
37. Feyisa, G.L.; Meilby, H.; Fensholt, R.; Proud, S.R. Automated Water Extraction Index: A New Technique for Surface Water Mapping Using Landsat Imagery. *Remote Sens. Environ.* **2014**, *140*, 23–35. [CrossRef]
38. Fisher, A.; Flood, N.; Danaher, T. Comparing Landsat Water Index Methods for Automated Water Classification in Eastern Australia. *Remote Sens. Environ.* **2016**, *175*, 167–182. [CrossRef]
39. Wang, S.; Baig, M.H.A.; Zhang, L.; Jiang, H.; Ji, Y.; Zhao, H.; Tian, J. A Simple Enhanced Water Index (EWI) for Percent Surface Water Estimation Using Landsat Data. *IEEE J. Sel. Top. Appl. Earth Obs. Remote Sens.* **2015**, *8*, 90–97. [CrossRef]
40. Shen, L.; Li, C. Water Body Extraction from Landsat ETM+ Imagery Using Adaboost Algorithm. In Proceedings of the 2010 18th International Conference on Geoinformatics, Beijing, China, 18–20 June 2010; pp. 1–4.
41. Yao, F.; Wang, C.; Dong, D.; Luo, J.; Shen, Z.; Yang, K. High-Resolution Mapping of Urban Surface Water Using ZY-3 Multi-Spectral Imagery. *Remote Sens.* **2015**, *7*, 12336–12355. [CrossRef]
42. Li, L.; Su, H.; Du, Q.; Wu, T. A Novel Surface Water Index Using Local Background Information for Long Term and Large-Scale Landsat Images. *ISPRS J. Photogramm. Remote Sens.* **2021**, *172*, 59–78. [CrossRef]
43. Acharya, T.; Subedi, A.; Lee, D. Evaluation of Water Indices for Surface Water Extraction in a Landsat 8 Scene of Nepal. *Sensors* **2018**, *18*, 2580. [CrossRef]
44. Buma, W.G.; Lee, S.-I.; Seo, J.Y. Recent Surface Water Extent of Lake Chad from Multispectral Sensors and GRACE. *Sensors* **2018**, *18*, 2082. [CrossRef]
45. Memon, A.A.; Muhammad, S.; Rahman, S.; Haq, M. Flood Monitoring and Damage Assessment Using Water Indices: A Case Study of Pakistan Flood-2012. *Egypt. J. Remote Sens. Space Sci.* **2015**, *18*, 99–106. [CrossRef]
46. Liu, H.; Hu, H.; Liu, X.; Jiang, H.; Liu, W.; Yin, X. A Comparison of Different Water Indices and Band Downscaling Methods for Water Bodies Mapping from Sentinel-2 Imagery at 10-M Resolution. *Water* **2022**, *14*, 2696. [CrossRef]
47. Du, Y.; Zhang, Y.; Ling, F.; Wang, Q.; Li, W.; Li, X. Water Bodies' Mapping from Sentinel-2 Imagery with Modified Normalized Difference Water Index at 10-m Spatial Resolution Produced by Sharpening the SWIR Band. *Remote Sens.* **2016**, *8*, 354. [CrossRef]
48. Sivanpillai, R.; Jacobs, K.M.; Mattilio, C.M.; Piskorski, E.V. Rapid Flood Inundation Mapping by Differencing Water Indices from Pre- and Post-Flood Landsat Images. *Front. Earth Sci.* **2021**, *15*, 1–11. [CrossRef]
49. Acharya, T.D.; Subedi, A.; Lee, D.H. Evaluation of Machine Learning Algorithms for Surface Water Extraction in a Landsat 8 Scene of Nepal. *Sensors* **2019**, *19*, 2769. [CrossRef]
50. Li, A.; Fan, M.; Qin, G. Comparative Analysis of Machine Learning Algorithms in Water Extraction. *J. Phys. Conf. Ser.* **2021**, *2076*, 012045. [CrossRef]
51. Li, A.; Fan, M.; Qin, G.; Xu, Y.; Wang, H. Comparative Analysis of Machine Learning Algorithms in Automatic Identification and Extraction of Water Boundaries. *Appl. Sci.* **2021**, *11*, 10062. [CrossRef]
52. Peter, K. *Surface Water Quality Monitoring*; European Environment Agency: Copenhagen, Denmark, 1996; pp. 123–126.
53. McGarrigle, M. Assessment of Small Water Bodies in Ireland. *Biol. Environ. Proc. R. Ir. Acad. JSTOR* **2014**, *114*, 119–128. [CrossRef]
54. Feeley, H.; Bruen, M.; Bullock, C.; Christie, M.; Kelly, F.; Kelly-Quinn, M. ESManage Project: Irish Freshwater Resources and Assessment of Ecosystems Services Provision. Available online: https://www.epa.ie/publications/research/water/EPA-RR-207-final-web-2.pdf. (accessed on 22 July 2022).
55. Hartnett, M.; Wilson, J.G.; Nash, S. Irish Estuaries: Water Quality Status and Monitoring Implications under the Water Framework Directive. *Mar. Policy* **2011**, *35*, 810–818. [CrossRef]
56. Connaughton, B. *The Implementation of Environmental Policy in Ireland: Lessons from Translating EU Directives into Action*; Manchester University Press: Manchester, UK, 2019; ISBN 978-1-5261-2756-3.
57. Antwi, S.H.; Linnane, S.; Getty, D.; Rolston, A. River Basin Management Planning in the Republic of Ireland: Past, Present and the Future. *Water* **2021**, *13*, 2074. [CrossRef]
58. O'Brien, D. Public Consultation on the Draft River Basin Management Plan for Ireland 2022–2027. Available online: https://www.gov.ie/en/consultation/2bda0-public-consultation-on-the-draft-river-basin-management-plan-for-ireland-2022-2027/ (accessed on 22 July 2022).
59. Spoto, F.; Sy, O.; Laberinti, P.; Martimort, P.; Fernandez, V.; Colin, O.; Hoersch, B.; Meygret, A. Overview of Sentinel-2. In Proceedings of the 2012 IEEE International Geoscience and Remote Sensing Symposium, Munich, Germany, 22–27 July 2012; pp. 1707–1710.
60. Roy, D.P.; Wulder, M.A.; Loveland, T.R.; Woodcock, C.E.; Allen, R.G.; Anderson, M.C.; Helder, D.; Irons, J.R.; Johnson, D.M.; Kennedy, R.; et al. Landsat-8: Science and Product Vision for Terrestrial Global Change Research. *Remote Sens. Environ.* **2014**, *145*, 154–172. [CrossRef]
61. Mutanga, O.; Kumar, L. Google Earth Engine Applications. *Remote Sens.* **2019**, *11*, 591. [CrossRef]

62. Main-Knorn, M.; Pflug, B.; Louis, J.; Debaecker, V.; Müller-Wilm, U.; Gascon, F. Sen2Cor for Sentinel-2. In Proceedings of the Image and Signal Processing for Remote Sensing XXIII, SPIE, Warsaw, Poland, 4 October 2017; Volume 10427, pp. 37–48.
63. Zhu, Z.; Woodcock, C.E. Object-Based Cloud and Cloud Shadow Detection in Landsat Imagery. *Remote Sens. Environ.* **2012**, *118*, 83–94. [CrossRef]
64. Liu, L.; Xiao, X.; Qin, Y.; Wang, J.; Xu, X.; Hu, Y.; Qiao, Z. Mapping Cropping Intensity in China Using Time Series Landsat and Sentinel-2 Images and Google Earth Engine. *Remote Sens. Environ.* **2020**, *239*, 111624. [CrossRef]
65. Foga, S.; Scaramuzza, P.L.; Guo, S.; Zhu, Z.; Dilley, R.D.; Beckmann, T.; Schmidt, G.L.; Dwyer, J.L.; Joseph Hughes, M.; Laue, B. Cloud Detection Algorithm Comparison and Validation for Operational Landsat Data Products. *Remote Sens. Environ.* **2017**, *194*, 379–390. [CrossRef]
66. Kaplan, G. Sentinel-2 Pan Sharpening—Comparative Analysis. *Proceedings* **2018**, *2*, 345. [CrossRef]
67. Pohl, C.; van Genderen, J. *Remote Sensing Image Fusion: A Practical Guide*; CRC Press: Boca Raton, FL, USA, 2016; ISBN 978-1-4987-3003-7.
68. Wang, Q.; Shi, W.; Li, Z.; Atkinson, P.M. Fusion of Sentinel-2 Images. *Remote Sens. Environ.* **2016**, *187*, 241–252. [CrossRef]
69. Yang, X.; Zhao, S.; Qin, X.; Zhao, N.; Liang, L. Mapping of Urban Surface Water Bodies from Sentinel-2 MSI Imagery at 10 m Resolution via NDWI-Based Image Sharpening. *Remote Sens.* **2017**, *9*, 596. [CrossRef]
70. Schowengerdt, R.A. Reconstruction of Multispatial, Multispectral Image Data Using Spatial Frequency Content. *Photogramm. Eng. Remote Sens.* **1980**, *46*, 1325–1334.
71. Otsu, N. A Threshold Selection Method from Gray-Level Histograms. *IEEE Trans. Syst. Man Cybern.* **1979**, *9*, 62–66. [CrossRef]
72. Yuan, X.; Wu, L.; Peng, Q. An Improved Otsu Method Using the Weighted Object Variance for Defect Detection. *Appl. Surf. Sci.* **2015**, *349*, 472–484. [CrossRef]
73. Hou, X.; Feng, L.; Duan, H.; Chen, X.; Sun, D.; Shi, K. Fifteen-Year Monitoring of the Turbidity Dynamics in Large Lakes and Reservoirs in the Middle and Lower Basin of the Yangtze River, China. *Remote Sens. Environ.* **2017**, *190*, 107–121. [CrossRef]
74. O'Loughlin, F.E.; Neal, J.; Yamazaki, D.; Bates, P.D. ICESat-Derived Inland Water Surface Spot Heights. *Water Resour. Res.* **2016**, *52*, 3276–3284. [CrossRef]
75. Xu, H. Modification of Normalised Difference Water Index (NDWI) to Enhance Open Water Features in Remotely Sensed Imagery. *Int. J. Remote Sens.* **2006**, *27*, 3025–3033. [CrossRef]

remote sensing

MDPI

Article

A Novel Atmospheric Correction for Turbid Water Remote Sensing

Dian Wang [1], Xiangyu Xiang [1], Ronghua Ma [2,3,*], Yongqin Guo [1], Wangyuan Zhu [1,4] and Zhihao Wu [1]

1 Marine Science and Technology College, Zhejiang Ocean University, Zhoushan 316022, China; wangdian@zjou.edu.cn (D.W.)
2 Nanjing Institute of Geography & Limnology Chinese Academy of Sciences, Nanjing 210008, China
3 University of Chinese Academy of Sciences, Beijing 100191, China
4 Zhejiang Haida Marine Survey, Planning and Design Co., Ltd., Zhoushan 316022, China
* Correspondence: rhma@niglas.ac.cn

Abstract: For the remote sensing of turbid waters, the atmospheric correction (AC) is a key issue. The "black pixel" assumption helps to solve the AC for turbid waters. It has proved to be inaccurate to regard all water pixels in the SWIR (Short Wave Infrared) band as black pixels. It is necessary to perform atmospheric correction in the visible bands after removing the radiation contributions of water in the SWIR band. Here, the modified ACZI (m-ACZI) algorithm was developed. The m-ACZI assumes the spatial homogeneity of aerosol types and employs the BPI (Black Pixel Index) and PIFs (Pseudo-Invariant Features) to identify the "black pixel". Then, the radiation contributions of waters in the SWIR band are removed to complete the atmospheric correction for turbid waters. The results showed that the m-ACZI had better performance than the SeaDAS (SeaWiFS Data Analysis System) -SWIR and the EXP (exponential extrapolation) algorithm in the visible band (sMAPE < 30.71%, RMSE < 0.0111 sr^{-1}) and is similar to the DSF (Dark Spectrum Fitting) algorithm in floating algae waters. The m-ACZI algorithm is suitable for turbid inland waters.

Keywords: atmospheric correction; turbid water; ocean color; black pixel

1. Introduction

Atmospheric correction (AC) is a key step of ocean color data processing. The "black pixel" assumption is commonly applied to AC for waters (including open ocean, coastal, and inland waters) [1–7]. The radiance of the "black pixel" is assumed to be dominated by atmospheric radiance, and the contribution from water can be neglected; hence, the AC of satellite images can be completed using the "black pixel" assumption [6,8]. This assumption depends on the fact that the water radiation is dominated by the strong absorption of pure water. Early on, waters in near-infrared (NIR) bands were determined to fit this assumption, and the water-leaving reflectance (ρ_w) (or the remote sensing reflectance (R_{rs})) in the visible bands was derived by removing the aerosol scattering through extrapolation from the NIR bands [6,8,9].

However, follow-up studies showed that the contribution of turbid water in NIR bands could not be neglected [10,11]. The short-wave infrared (SWIR) bands with a longer wavelength and stronger absorption were considered to fit the "black pixel" assumption for turbid waters [9,12–14]. Wang et al. (2019) found that the ρ_w of algae waters was strong in SWIR bands, meaning these bloom waters no longer met the "black pixel" assumption [15]. In Lake Taihu, significant water signals in SWIR bands were detected when algal blooms occurred (Figure 1). Therefore, the contribution of waters in SWIR bands should be removed before the estimation of aerosol scattering in visible bands. Wang et al. (2019) developed the BPI (black pixel index) to obtain the "black pixel" from turbid waters and proposed the ACZI algorithm (atmospheric correction algorithm based on a zero assumption for the short-wave infrared band) for inland turbid waters. The water pixels in the 2201 nm band were considered black pixels by the ACZI algorithm for Landsat 8 images, and aerosol scattering in the visible bands was derived from the assumption of the zero contribution of waters in the 2201 nm band [12]. Figure 1 shows that the Rayleigh-corrected reflectance (ρ_{rc}) in SWIR bands is very different in water areas without algal blooms; it is incorrect that all water pixels in the SWIR band are treated as black pixels. It is necessary to obtain the "black pixel" with a more accurate identification method.

Here, we modified the identification method of the "black pixel" in the ACZI algorithm [12], and the m-ACZI algorithm was developed. In the m-ACZI algorithm, the water contribution in SWIR bands was removed using the combination of BPI [12] and PIFs (pseudo-invariant features) [16,17], and accurate aerosol scattering in SWIR bands was derived. This scheme was evaluated in turbid inland waters, with its performance also compared with other AC algorithms.

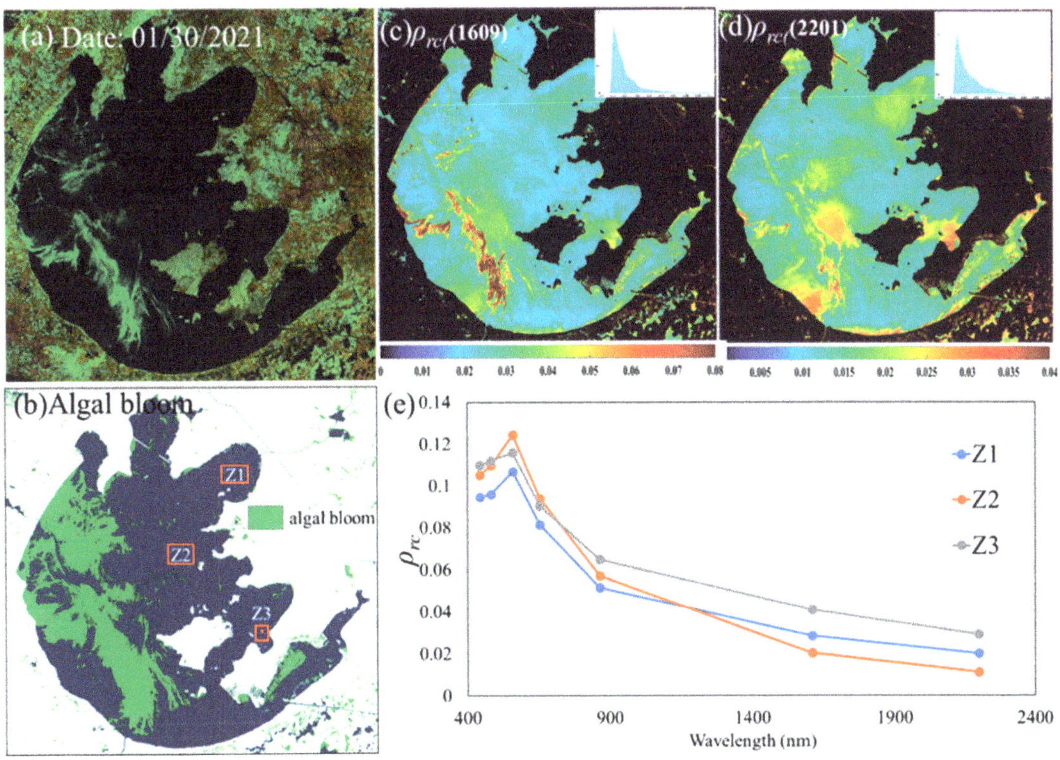

Figure 1. (**a**) False color image (R: SWIR band, G: NIR band, B: Blue band), (**b**) Algal bloom map derived by FAI (floating algae index) [18,19], and the Rayleigh-corrected reflectance (ρ_{rc}) in SWIR bands (**c**,**d**) and its histograms from Lake Taihu with algal bloom on 30 January 2021. (**e**) Average ρ_{rc} from three boxes (Z1, Z2, and Z3) without algal blooms.

2. Dataset and Methods

2.1. Satellite Data

The cloudless Landsat8-OLI (Operational Land Imager) Level-1 data were from the USGS (United State Geological Survey) and were processed to the R_{rs} data via the AC algorithms (details in Section 2.3.2). In this work, three OLI images of Lake Taihu (dates: 11 May 2017, 27 May 2017, and 21 December 2017) were matched with the in situ data to evaluate the performance of the AC algorithms. The time window of data matching was set as ±3 h of the Landsat8 overpass. The AC-driven R_{rs} data were obtained by averaging a 3 × 3 pixel (the coefficient of variation for these valid pixels was < 10%) area surrounding the sample point location [20–23]. A total of 71 sampling points were matched (11 match-up points from 11 May 2017; 22 match-up points from 27 May 2017; and 38 match-up points from 21 December 2017) (Figure 2).

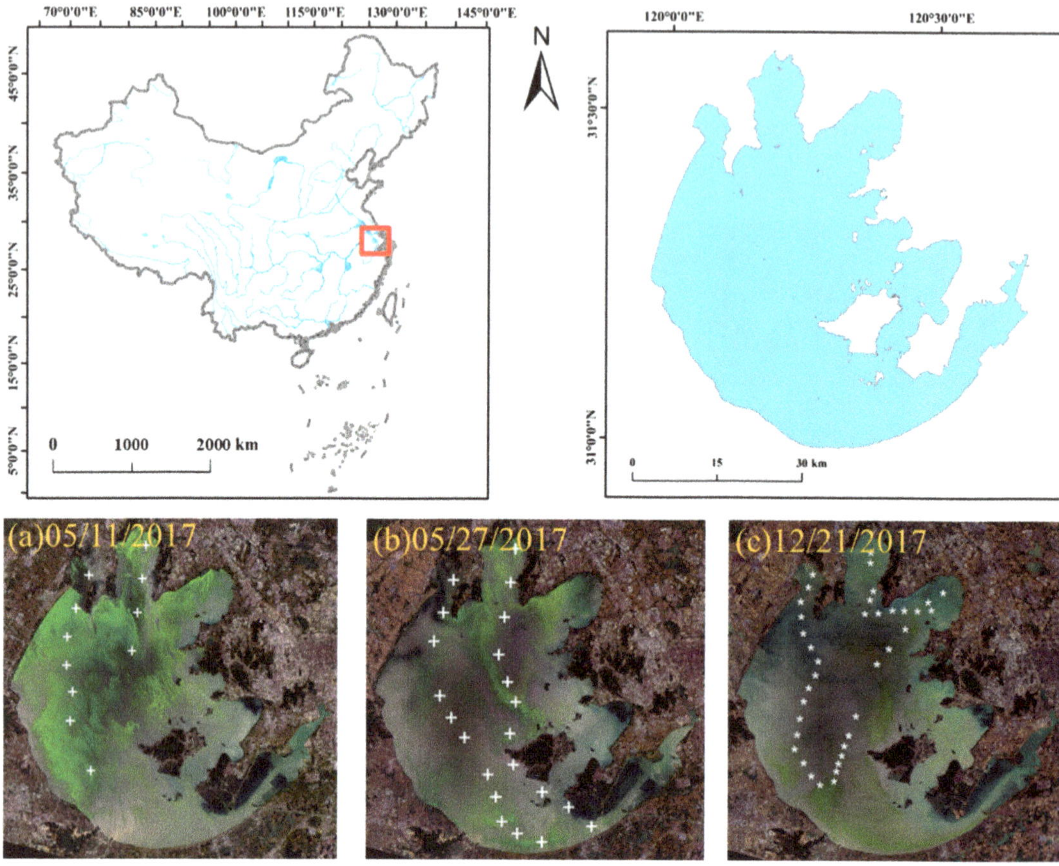

Figure 2. In situ sampling points in Lake Taihu.

2.2. Field Spectral Data

The R_{rs} was measured using the spectrometer (ASD FieldSpec 4) following NASA (National Aeronautics and Space Administration) Ocean optics protocols [24] (Figure 3):

$$R_{rs} = \frac{L_w}{E_d} = \frac{\left(L_t - \sigma \cdot L_{sky}\right) \cdot \rho_p}{\pi \cdot L_p} \tag{1}$$

where L_t is the total water-leaving radiance; ρ_p is the reflectance; L_g is the radiance of the reference panel; L_{sky} is the sky radiance; L_w is the water-leaving radiance; and E_d is the downwelling plane irradiance. Additionally, the water surface reflectance factor σ was assumed to be 0.028 [25].

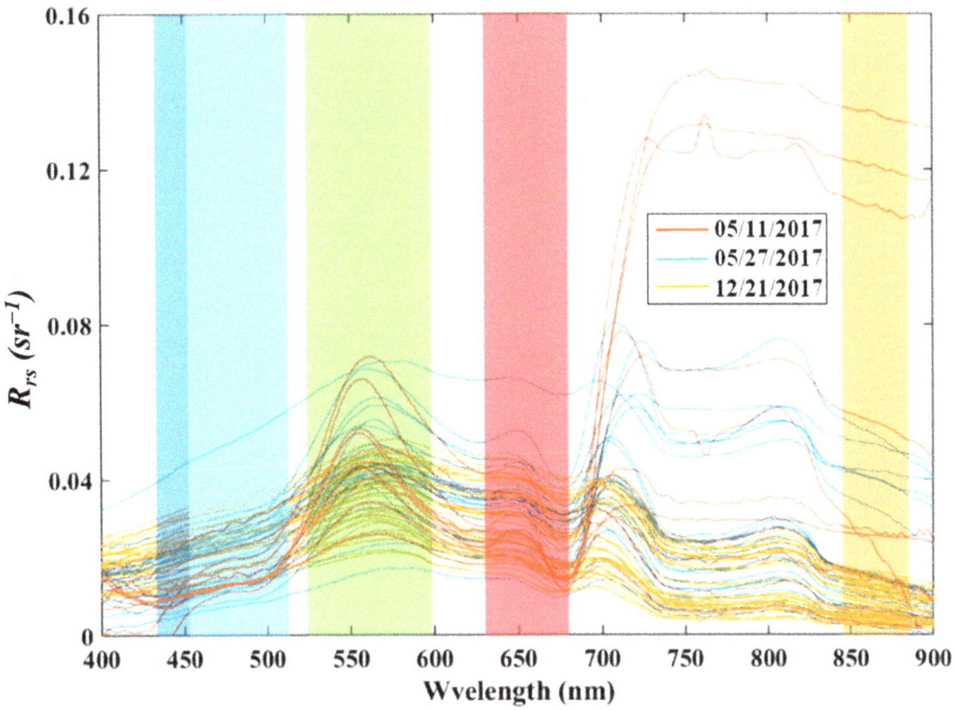

Figure 3. Measured remote sensing reflectance (R_{rs}) from Lake Taihu (dates: 11 May 2017, 27 May 2017, and 21 December 2017). The line color represents different sampling dates. The color column represents the wavelength range of each band of OLI (Deep blue: 443 nm band, Light blue: 483 nm band, Green: 561 nm band, Red: 655 nm band, Yellow: 865 nm band).

2.3. Methods

2.3.1. The m-ACZI Algorithm

The black pixel of the ACZI algorithm was identified using the combination of BPI (Equation (2)) [12] and FAI (the floating algae index) [18,19], and this algorithm assumed that all water pixels in the 2201 nm band accord with the "black pixel" hypothesis. However, the ρ_w in SWIR bands was significant and non-negligible when algal blooms occurred (Figure 1). In this study, the m-ACZI algorithm changed ACZI's black pixel identification method of the black pixels of ACZI to the BPI-PIF combination, and these black pixels are applied to remove the water contribution in SWIR bands. After this stage, the following process is the same as the ACZI algorithm [12]. The BPI was developed by Wang et al. (2019a) [12] to identify the black pixel from turbid waters (Equation (2)). Schott et al. (1988) [17] considered that urban construction land has pseudo-invariant features (PIFs) and proposed the PIF mask method. The BPI and PIF mask followed the methods described by Wang et al. (2019a) [12] and Schott et al. (1988) [17], respectively.

$$BPI = \frac{|\rho_{rc}(655) - \rho_{rc}(561)|}{\rho_{rc}(655) - \rho_{rc}(865)} \tag{2}$$

The PIFs are with features that do not change their reflectivity properties drastically over time in a scene [17]. The surface reflectance of PIFs is similar for the two dates. Hence, the difference between the Rayleigh-corrected reflectance (ρ_{rc}) of PIFs from two dates can be considered as the difference between two values of aerosol reflectance (ρ_a) of PIF. The black pixels are pixels consisting only of atmospheric radiance (Rayleigh scattering and aerosol scattering). If a pixel is a true "black pixel", the difference between its ρ_{rc} for two dates is the same as the difference between its ρ_a for the same two dates (Equations (5)–(7)). Furthermore, theoretically, if the difference in the ρ_{rc} of a water pixel between two dates in the SWIR band is equal or approximate to the difference in the ρ_{rc} of PIFs between two dates, this water pixel can be taken as a "black pixel". Therefore, the characteristics of PIFs can be used to more accurately obtain the BPIs for atmospheric correction. The flowchart of the m-ACZI algorithm is shown in Figure 4.

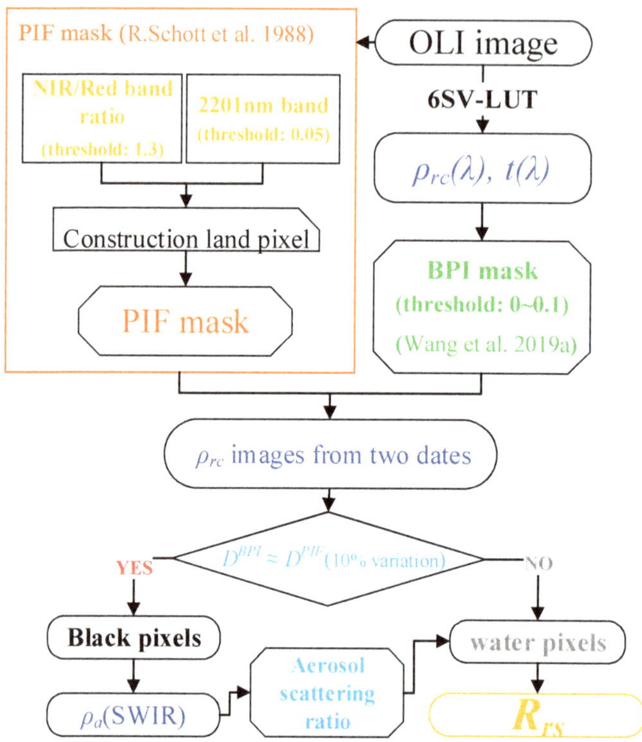

Figure 4. Processing procedure of the m-ACZI algorithm. Note: These thresholds are for Lake Taihu and may need to be adjusted for other waters [12,17].

(1) The ρ_{rc} and transmittance (t) are obtained from a LUT (lookup table) generated for all OLI bands using 6SV (the vector version of the Second Simulation of the Satellite Signal in the Solar Spectrum) according to sun and sensor geometry [26,27]. The ρ_{rc} can be derived from the reflectance of the atmosphere (ρ_t). The sea surface reflectance from whitecaps and sunglint were ignored [1,2,28,29].

$$\rho_{rc} = \rho_t - \rho_r \tag{3}$$

For land pixels:

$$\rho_{rc} = \rho_a + t\rho_s \tag{4}$$

where ρ_s is the surface reflectance.

For a water pixel:

$$\rho_{rc} = \rho_a + t\rho_w \tag{5}$$

(2) Two masks are made by the BPI for two date images (*T1* and *T2*) [4,12], and the overlapped pixels in two masks (BPI mask) are retained. The difference in ρ_{rc} (SWIR) for two date images (D^{BPI}) is calculated and masked by the BPI mask.

$$\begin{aligned} D^{BPI} &= \rho_{rc_{T1}}^{BPI}(SWIR) - \rho_{rc_{T2}}^{BPI}(SWIR) \\ &= \left(\rho_{a_{T1}}^{BPI}(SWIR) + t_{T1}\rho_{w_{T1}}^{BPI}(SWIR)\right) - \left(\rho_{a_{T2}}^{BPI}(SWIR) + t_{T2}\rho_{w_{T2}}^{BPI}(SWIR)\right) \end{aligned} \quad (6)$$

(3) The PIF mask is extracted according to Schott et al. (2016) [16]. The construction land features are considered as PIFs, and this extraction is applied over the construction area within the image (Figure 5). For PIFs, the ρ_s is invariable, and the difference in ρ_{rc} (SWIR) for two date images (D^{PIF}) is masked by the PIF mask.

$$\begin{aligned} D^{PIF} &= \rho_{rc_{T1}}^{PIF}(SWIR) - \rho_{rc_{T2}}^{PIF}(SWIR) \\ &= (\rho_{a_{T1}}^{PIF}(SWIR) + t_{T1}\rho_{s_{T1}}^{PIF}(SWIR)) - (\rho_{a_{T2}}^{PIF}(SWIR) + t_{T2}\rho_{s_{T2}}^{PIF}(SWIR)) \\ &= \rho_{a_{T1}}^{PIF}(SWIR) - \rho_{a_{T2}}^{PIF}(SWIR) + \Delta t\rho_s^{PIF}(SWIR) \end{aligned} \quad (7)$$

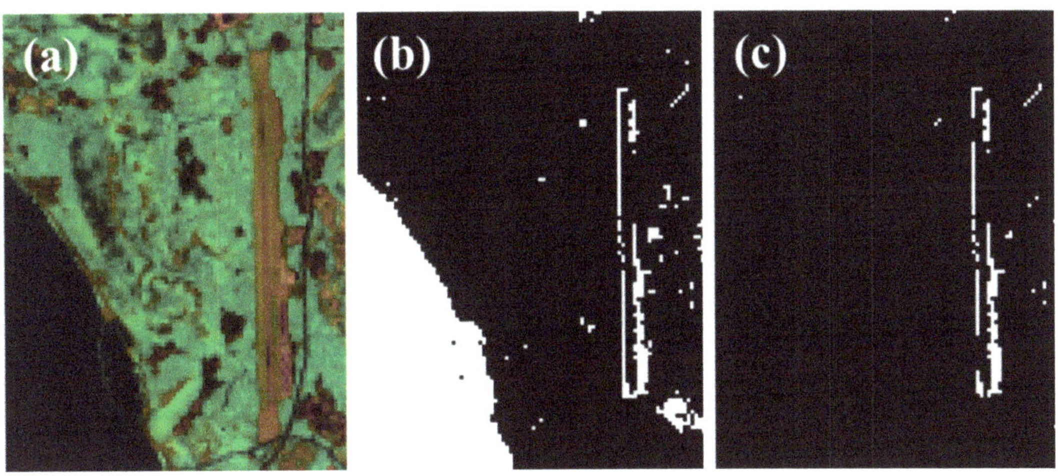

Figure 5. Example of the PIF mask determination. (**a**) False color image (R: SWIR band, G: NIR band, B: Blue band) with vegetation in green; (**b**) Construction land and water; and (**c**) PIF mask over airports near Lake Taihu, China.

Δt will make $\rho_S{}^{PIF}$ residual in D^{PIF} and affect the matching of D^{PIF} and D^{BPI}; for this, we analyze and discuss the impact of Δt on D^{PIF} (1609) (details in Section 4.4). The results showed that when Δt was less than 0.0107, the D^{PIF} change caused by Δt was about 2.70%, which is within the allowable error range.

(4) The $\rho_a{}^{BPI}$ and $\rho_a{}^{PIF}$ are similar in one OLI image based on the assumption of the spatial uniformity distribution of aerosol types. The D^{BPI} matches the mode number of D^{PIF} (with ±10% variation) to obtain the final BPI-PIF mask. The black pixels were identified using the BPI-PIF mask (Example: Figure 6). The aerosol scattering from black pixels is used to remove the water contribution for nonblack pixels.

$$D^{BPI} \cong D^{PIF} \quad (8)$$

(5) The atmospheric correction for all water pixels is completed using the same steps as the ACZI algorithm [12], and R_{rs} (or ρ_w) data are obtained.

Described below is the process of developing the m-ACZI algorithm.

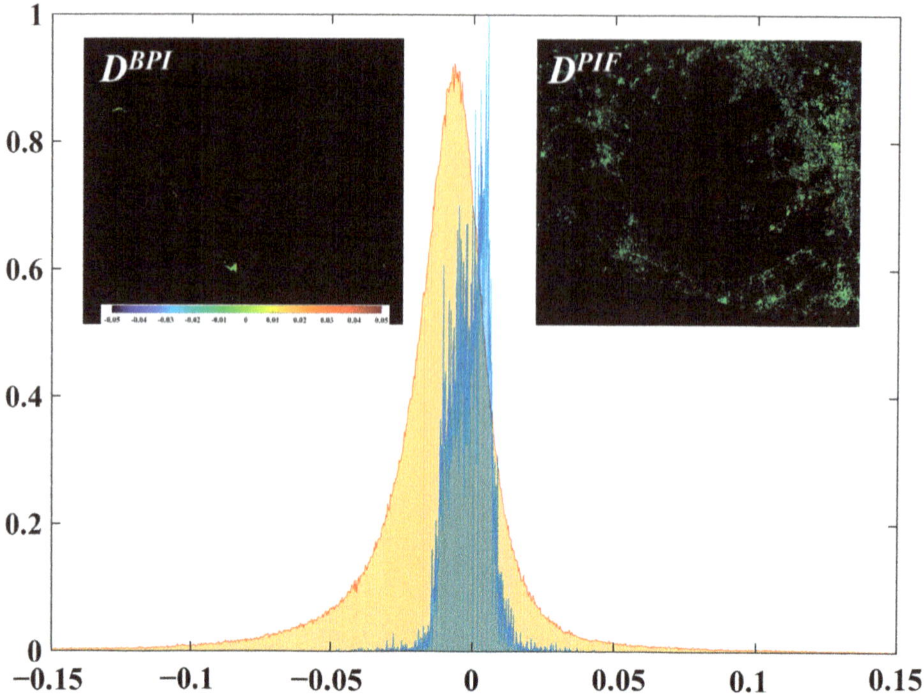

Figure 6. Histogram of D^{BPI} (blue bar) and D^{PIF} (yellow bar) derived by two date images (Dates: 11 May 2017 and 27 May 2017) in Lake Taihu, China.

(1) We extracted the PIF mask according to Schott et al., 1988 [17] and Concha and Schott 2016 [16]. For Lake Taihu, the 2201 nm band with a threshold of 0.05 was used to obtain the water mask, and the threshold of the NIR/RED band ratio was set to 1.3 for obtaining the vegetation mask. The previous two masks were combined using logic. *AND*. gate, creating a "PIF mask" that rejects the vegetation and water and only accepts the construction land features.

(2) In this step, we calculated the ρ_{rc} and t using the 6SV-LUT from ACOLITE (atmospheric correction for OLI lite).

(3) Based on step 2, we adopted the BPI [12] to obtain the BPI mask, and the threshold of the BPI was set to 0~0.1.

(4) Following steps 1, 2, and 3, concerning t values, two images with similar t values were selected to obtain their PIF mask and BPI mask, respectively. Then, these masks were applied to the ρ_{rc} (SWIR) images, respectively, to obtain the PIF and BPI images. We subtracted two of each image type to obtain the D^{PIF} and D^{BPI} images.

(5) We counted the average value of D^{PIF} and used the average D^{PIF} with 10% variation as the judgment value for D^{BPI}, i.e., D^{BPI} pixels greater than the average D^{PIF} with 10% variation were discarded.

(6) Following the ACZI process [12], we counted the median value from the final retained D^{BPI} pixels, and this median value was taken as ρ_a on two SWIR bands. The aerosol scattering ratio (ε), consisting of these two ρ_a(SWIR) values, was applied to all water pixels, and aerosol scattering in the visible and near-infrared bands was obtained using an exponential extrapolation method, thus completing the final atmospheric correction process to obtain R_{rs}.

2.3.2. Other Atmospheric Correction Algorithms

The SeaDAS-SWIR Algorithm

For turbid waters, the "black pixel" assumption in NIR bands is invalid because of the contribution of phytoplankton, detritus, and suspended sediment to water backscatter [2,10]. Research from Hale et al. (1973) indicated that the water in SWIR bands has stronger absorption than that in NIR bands [30], and the turbid water in the former meets the requirement of the "black pixel" assumption [13]. Therefore, Wang and Shi (2005) [10] developed an AC algorithm for turbid water based on extrapolation from SWIR bands and integrated this AC algorithm into SeaDAS (SeaWiFS Data Analysis System, referred to as SeaDAS-SWIR in this paper; more details can be found in Wang and Shi (2005) [10]).

The ACOLITE (atmospheric correction for OLI lite; this study worked with version 20211112) tool was developed for coastal and inland waters by the Remote Sensing and Ecosystem Modelling (REMSEM) team. It performs this process using dark spectrum fitting (DSF) and exponential extrapolation (EXP) [14,26,31–33].

The EXP Algorithm

The idea of the EXP algorithm is consistent with the SeaDAS-SWIR algorithm, both of which are based on exponential extrapolations of the ratio of multiple aerosol scattering in SWIR bands, but the EXP algorithm does not deal with aerosol LUTs (lookup tables) [14]. The EXP algorithm only uses an exponential extrapolation method to estimate aerosol scattering in visible bands and NIR bands [28].

The DSF Algorithm

The DSF algorithm is presented for meter-scale-resolution optical satellite imagery [26,31]. Similar to other AC algorithms, this algorithm was also developed based on the "black pixel" concept [26]. However, the difference is that it defines black pixels as pixels with zero surface reflection in one scene image, including land pixels and water pixels. These black pixels are employed to estimate the atmospheric radiance for other pixels. It can be seen that the DSF depends on two assumptions: (1) the spatial distribution of the atmosphere is uniform in an image; (2) at least one pixel whose surface reflection is about zero exists in one scene image [26,31].

2.3.3. Algorithm Accuracy Analysis

The root-mean-square error (*RMSE*), symmetric mean absolute percentage error (sMAPE), and Bias were used to determine the differences between the in situ data and the AC-driven data:

$$RMSE = \sqrt{\frac{\sum_{i=1}^{n}\left(R^m - R^i\right)^2}{n}} \tag{9}$$

$$sMAPE = \frac{\sum_{i=1}^{n}\frac{\left|R^m - R^i\right|}{\left|R^m + R^i\right|}}{n} \times 2 \times 100\% \tag{10}$$

$$Bias = \frac{\sum_{i=1}^{n}\left[\frac{\left(R^m - R^i\right)}{R^i}\right]}{n} \times 100\% \tag{11}$$

where R^i is the in situ R_{rs} data, R^m is the OLI-R_{rs} data via atmospheric correction, and n is the number of match-up pairs.

3. Results

3.1. Assessment of the m-ACZI Algorithm

As Figures 7 and 8 show, the m-ACZI raised the R_{rs} estimation compared to the ACZI algorithm, so the histograms of R_{rs} generally moved and concentrated to high value areas (Figure 8). Notably, the m-ACZI algorithm did not increase the R_{rs} values for all water pixels, but mainly for algal bloom pixels (green pixels in the RGB true-color images) (Figures 2 and 7). It can also be observed that the m-ACZI algorithm reduced the R_{rs} estimation of the non-algal bloom pixels in the 865 nm band (Figure 7), which were consistent with the characteristics of the in situ R_{rs} (Figure 3).

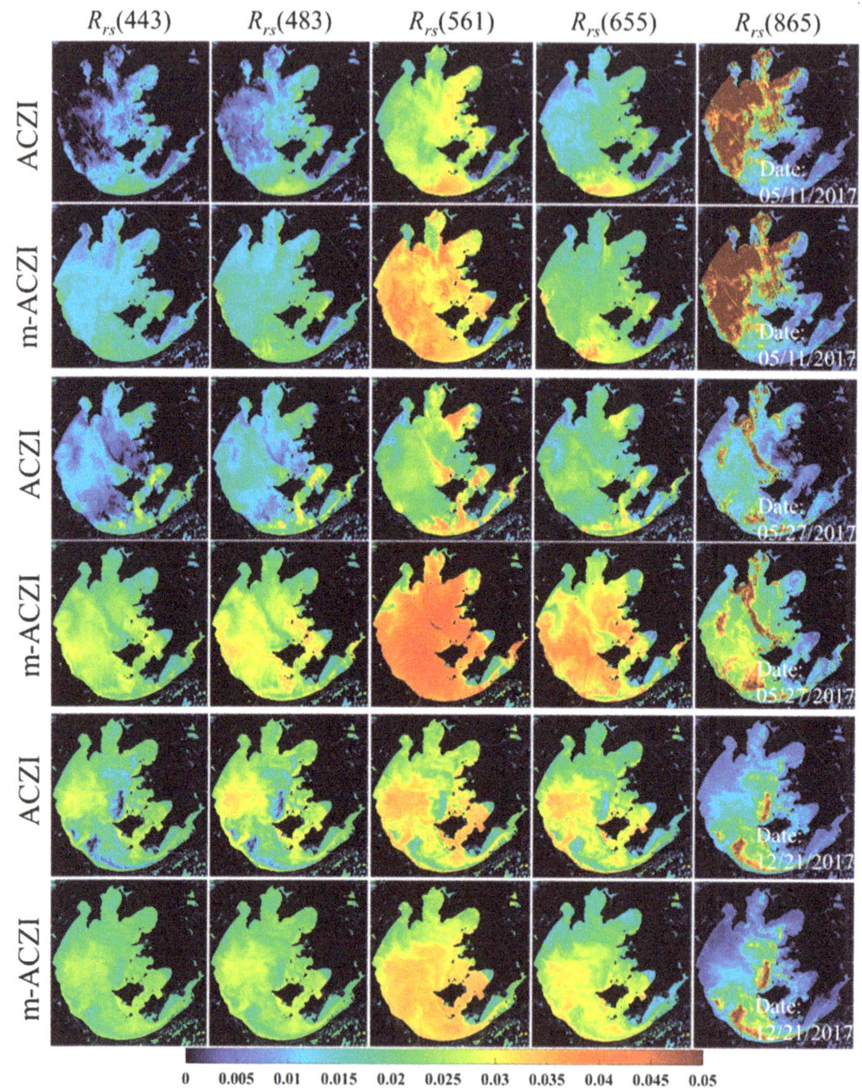

Figure 7. Comparison of the ACZI and the m-ACZI algorithms for OLI R_{rs} over Lake Taihu (dates: 11 May 2017, 27 May 2017, and 21 December 2017).

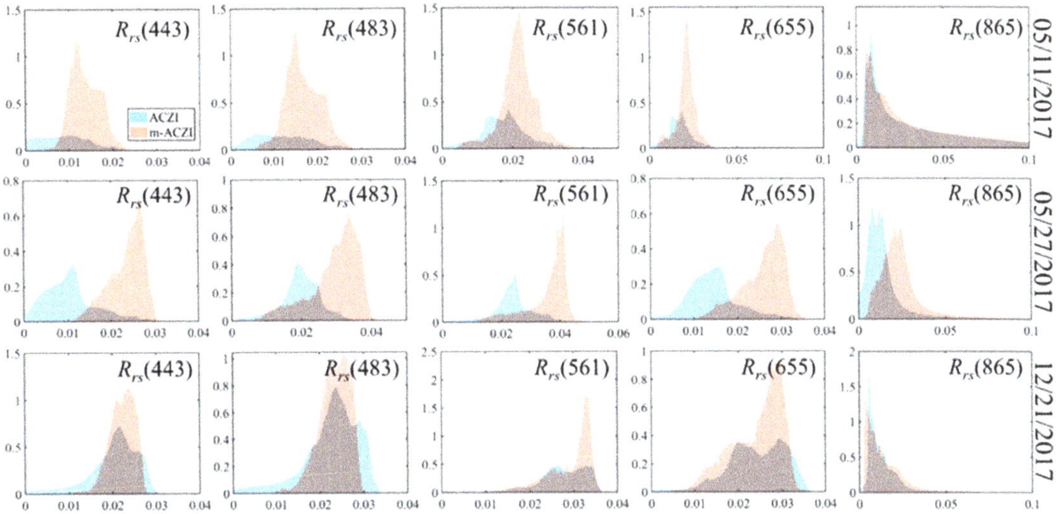

Figure 8. Frequency of R_{rs} retrieved using the ACZI (blue) and the m-ACZI (orange) algorithm in Lake Taihu (Dates: 11 May 2017, 27 May 2017, and 21 December 2017).

Most of the scatterplots of the ACZI were below the 1:1 line, and part of the scatterplots derived in the 865 nm band were above the 1:1 line (Figure 9a). The scatterplots of the m-ACZI were distributed around the 1:1 line, with several significant off-group points in the 865 nm band (Figure 9b). In general, compared with the scatterplots of the ACZI, the scatterplots of the m-ACZI were more concentrated near the 1:1 line. The ACZI-driven R_{rs} was underestimated in all visible bands. The R_{rs} derived by the m-ACZI was also underestimated in the 561 nm and 655 nm bands, but overestimated in two BLUE bands (Figure 9 and Table 1). The minimum sMAPE was obtained by the m-ACZI in the 655 nm band, followed by the m-ACZI in the 561 nm band. Similarly, the minimum RMSE was provided by the m-ACZI in the 483 nm band, followed by the m-ACZI in the 443 nm band. However, the minimum bias was obtained from the data derived by the ACZI in the 443 nm band (Table 1). Overall, from RMSE and sMAPE, the m-ACZI was superior to the ACZI in all visible bands, especially in the 561 nm and 655 nm bands; the performance of the m-ACZI algorithm was greatly improved. However, these 2 algorithms still failed in the NIR band (RMSE > 0.0203 sr^{-1}; sMAPE > 48.31%; bias > 88.83%). From the scatterplots of the 865 nm band in Figure 9, these 2 AC algorithms have overestimations. The assessment results indicate that the m-ACZI algorithm is generally better than the ACZI algorithm.

Table 1. Statistics of the accuracy measures for the ACZI and the m-ACZI algorithms using in situ R_{rs} measurements. The BLUE number represents the minimal statistical value.

Method		443	483	561	655	865
m-ACZI	RMSE(sr^{-1})	0.0063	0.0062	0.0111	0.0082	0.0228
	sMAPE(%)	30.71	23.01	21.41	21.27	48.31
	Bias(%)	52.07	7.59	−16.48	−11.65	88.83
ACZI	RMSE(sr^{-1})	0.0070	0.0076	0.0146	0.0102	0.0203
	sMAPE(%)	35.88	34.07	29.16	29.10	57.48
	Bias(%)	−8.91	−12.32	−22.70	−16.18	94.16

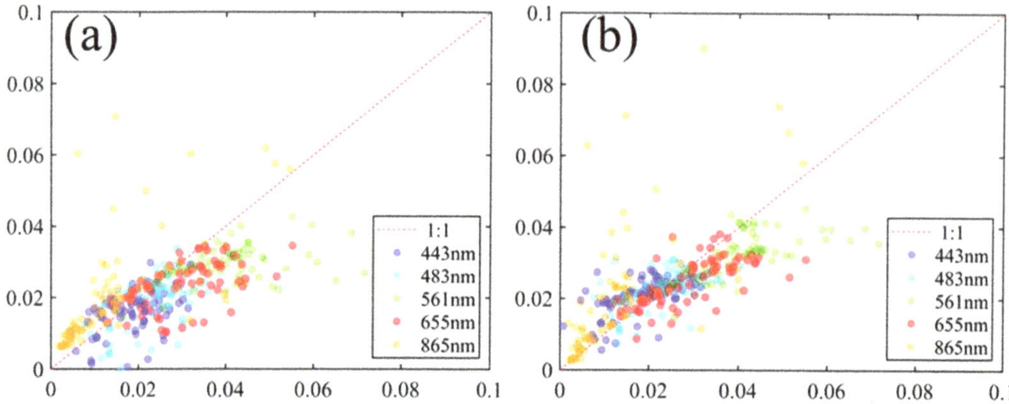

Figure 9. Scatterplots of R_{rs} derived by the ACZI (**a**) and the m-ACZI (**b**) algorithms versus in situ R_{rs}.

3.2. Comparison with other AC Algorithms

Only 54 match-up sample points were used to evaluate the SeaDAS-SWIR algorithm because some water pixels were masked by this algorithm (the "proc_ocean" option of SeaDAS was set to "2-force all pixels to be processed as ocean"). In Figure 10, the scatterplots of the m-ACZI algorithm were more concentrated, while those of other algorithms were discrete, especially in the 443 nm and 483 nm bands. The scatterplot distributions of the m-ACZI and the DSF were very similar in the 561 nm and 655 nm bands and were closer to the 1:1 line than others (Figure 10). The statistical results of the accuracy also supported these findings (Table 2). The assessment results from the m-ACZI and the DSF were very similar: the RMSE difference between them was less than 0.0008 in visible bands, and both were more accurate than the SeaDAS and the EXP algorithms. It should be noted that all the AC algorithms failed in the NIR band (RMSE > 0.0134 sr^{-1}; sMAPE > 49.31%; Bias > 88.83%), and the m-ACZI algorithm showed the best comprehensive performance.

Table 2. Statistics of the accuracy measures for the m-ACZI, SeaDAS, DSF, and EXP algorithms using in situ R_{rs} measurements. The BLUE number represents the minimal statistical value.

Method		443	483	561	655	865
m-ACZI	RMSE(sr^{-1})	0.0063	0.0062	0.0111	0.0082	0.0228
	sMAPE(%)	30.71	23.01	21.41	21.27	48.31
	Bias(%)	52.07	7.59	−16.48	−11.65	88.83
SeaDAS-SWIR	RMSE(sr^{-1})	0.0113	0.0100	0.0108	0.0104	0.0134
(n = 54)	sMAPE(%)	45.11	32.70	20.65	27.33	76.35
	Bias(%)	33.32	13.36	−1.38	4.37	126.68
EXP	RMSE(sr^{-1})	0.0060	0.0070	0.0150	0.0105	0.0217
	sMAPE(%)	34.12	27.76	32.79	28.29	55.56
	Bias(%)	2.50	−10.31	−27.00	−19.72	90.38
DSF	RMSE(sr^{-1})	0.0058	0.0061	0.0098	0.0080	0.0244
	sMAPE(%)	32.27	24.34	17.68	21.80	63.00
	Bias(%)	34.22	4.94	−4.46	0.01	113.55

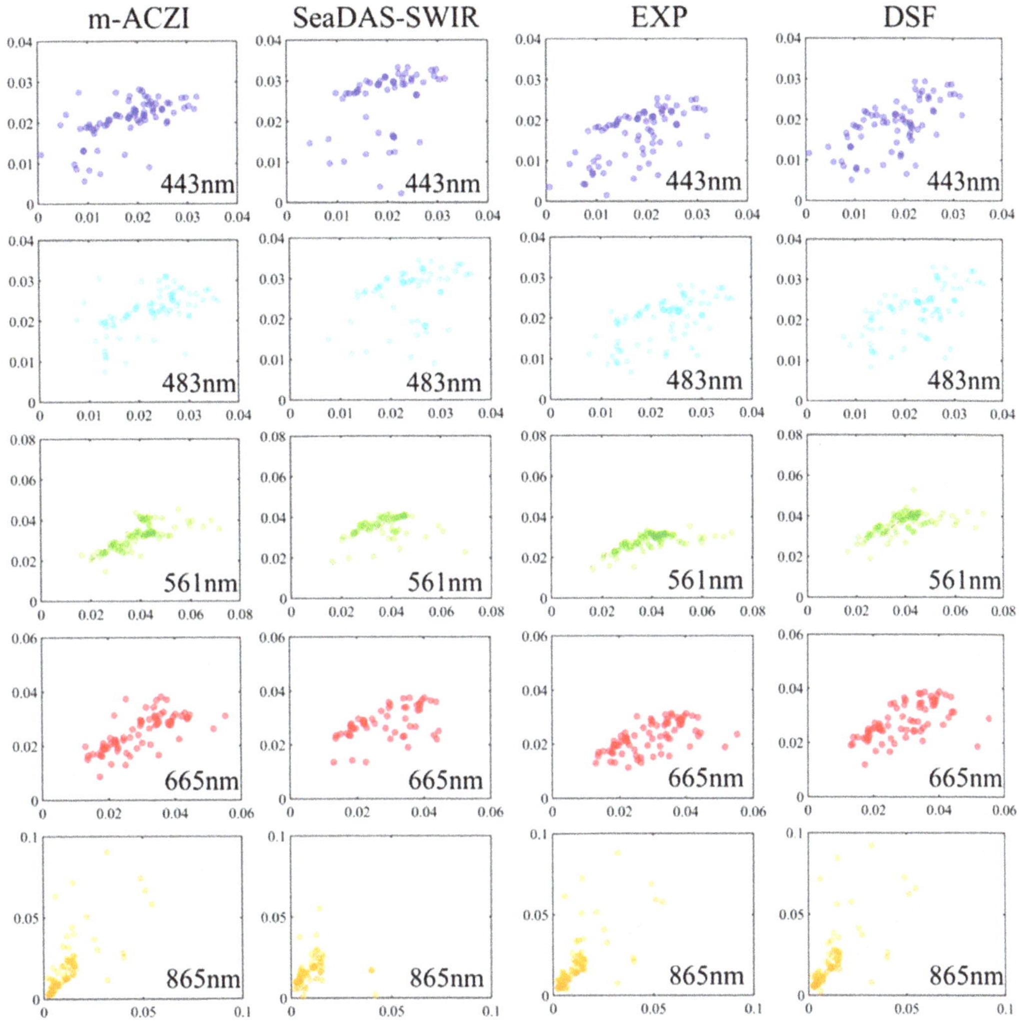

Figure 10. Scatterplots of AC-driven R_{rs} versus in situ R_{rs}.

3.3. Assessment of AC Algorithms for Different Water Types

According to the water classification method of Sun et al. (2012), waters were classified into two types: floating bloom (FB) and un-floating bloom (uFB) [15,34]. Figure 11 shows the scatterplots of R_{rs} between in situ data and the AC algorithm in different water types, with detailed statistics presented in Tables 3 and 4. All AC algorithms had more discrete scatterplots in FB than in uFB, especially in the two blue bands, where the bias of the algorithm increased significantly (Tables 3 and 4). Table 3 shows that the assessment results of the m-ACZI and the DSF algorithms were superior to other algorithms in the uFB. In particular, the m-ACZI and the DSF algorithms showed satisfactory performance in the 561 nm and 655 nm bands (RMSE < 0.0087 sr^{-1}; sMAPE < 19.31%), which was sufficient to support the quantitative remote sensing research for turbid waters [5]. Compared with those in the uFB, the accuracy results of the AC algorithms in FB were reduced (Tables 3 and 4). The DSF algorithm had the best performance in the FB waters, followed by the m-ACZI algorithm. The SeaDAS-SWIR algorithm masked most of the FB pixels, with only 6 in situ points (18 in situ points in total) participating in the evaluation. The performance of these AC algorithms was not satisfactory, especially in the two blue bands, as was that of the m-ACZI algorithm. These AC algorithms (except the ACZI algorithm) overestimated the water-leaving reflection in the 443 nm and 483 nm bands (Table 4). A possible reason was that the m-ACZI algorithm for the FB underestimated the aerosol scattering in the SWIR band, which was further exacerbated by exponential extrapolation. This finally resulted in a serious underestimation of the aerosol scattering and an overestimation of the water-leaving reflection in the 443 nm and 483 nm bands.

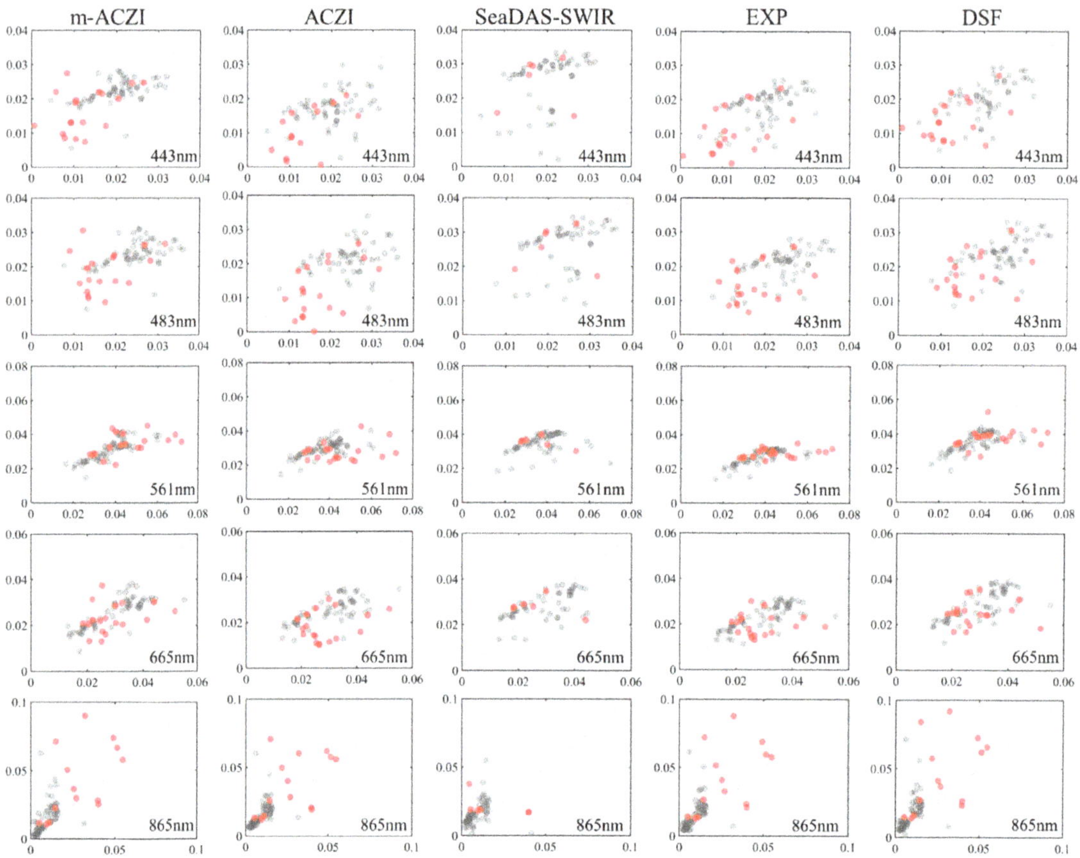

Figure 11. Scatterplots of AC-driven R_{rs} of different water types versus in situ R_{rs}. The red points represent water with floating blooms, and the gray points represent un-floating blooms.

Table 3. Statistics of the accuracy measures for atmospheric correction algorithms for un-floating blooms using in situ R_{rs} measurements. The BLUE number represents the minimal statistical value.

Method		443	483	561	655	865
m-ACZI	RMSE(sr^{-1})	0.0058	0.0058	0.0087	0.0072	0.0127
(n = 53)	sMAPE(%)	25.63	20.49	18.97	17.86	48.62
	Bias(%)	21.32	3.93	−14.76	−10.45	91.30
ACZI	RMSE(sr^{-1})	0.0071	0.0073	0.0109	0.0085	0.0112
(n = 53)	sMAPE(%)	31.77	25.94	23.13	22.59	59.54
	Bias(%)	−6.01	−6.81	−18.20	−10.54	102.18

Table 3. *Cont.*

Method		443	483	561	655	865
SeaDAS-SWIR	RMSE(sr^{-1})	0.0113	0.0101	0.0107	0.0103	0.0123
(n = 48)	sMAPE(%)	44.17	31.74	20.02	26.57	74.39
	Bias(%)	30.91	11.20	−2.26	2.61	119.49
EXP	RMSE(sr^{-1})	0.0058	0.0068	0.0126	0.0093	0.0123
(n = 53)	sMAPE(%)	26.63	24.18	30.28	24.10	55.98
	Bias(%)	−3.30	−10.77	−25.36	−17.42	93.74
DSF	RMSE(sr^{-1})	0.0055	0.0058	0.0075	0.0068	0.0144
(n = 53)	sMAPE(%)	26.92	21.81	15.16	19.31	65.51
	Bias(%)	8.72	3.16	−1.85	1.95	117.79

Table 4. Statistics of the accuracy measures for atmospheric correction algorithms for floating bloom using in situ R_{rs} measurements. The BLUE number represents the minimal statistical value.

Method		443	483	561	655	865
m-ACZI	RMSE(sr^{-1})	0.0078	0.0072	0.0162	0.0107	0.0396
(n = 18)	sMAPE(%)	45.68	30.42	28.61	31.29	47.40
	Bias(%)	142.60	18.39	−21.53	−15.18	81.53
ACZI	RMSE(sr^{-1})	0.0067	0.0086	0.0222	0.0140	0.0355
(n = 18)	sMAPE(%)	51.47	58.00	46.92	48.25	51.41
	Bias(%)	−19.90	−28.52	−35.93	−32.81	70.54
SeaDAS-SWIR	RMSE(sr^{-1})	0.0111	0.0097	0.0115	0.0110	0.0204
(n = 6)	sMAPE(%)	52.71	40.43	25.69	33.33	92.05
	Bias(%)	52.55	30.56	5.67	18.46	184.26
EXP	RMSE(sr^{-1})	0.0067	0.0075	0.0205	0.0134	0.0376
(n = 18)	sMAPE(%)	56.19	38.30	40.20	40.63	54.32
	Bias(%)	19.58	−8.93	−31.81	−26.47	80.47
DSF	RMSE(sr^{-1})	0.0067	0.0070	0.0147	0.0107	0.0416
(n = 18)	sMAPE(%)	48.00	31.78	25.10	29.13	55.60
	Bias(%)	109.31	10.17	−12.16	−5.71	101.07

4. Discussion

The m-ACZI algorithm depends on two prerequisite assumptions: (1) the spatial distribution of aerosol types is uniform; (2) the black pixels from two date images overlap. These are the determining conditions for the success of the m-ACZI algorithm.

4.1. The Assumption of the Spatially Uniformity Distribution of Aerosol Types

Solving the aerosol scattering of nonblack pixels relies on the assumption of the spatially uniform distribution of aerosol types for the m-ACZI algorithm. This assumption has been recognized by a great amount of research [1,4,14,26,28]. For example, the DSF and EXP adopted this assumption for the atmospheric correction process. Taking the Lake Taihu image without algal blooms (obtained on 2 February 2016) as an example (remove cloud pixels) (Figure 12), the spatial distributions of ρ_{rc} (1609) and ρ_{rc} (2201) were uniform, and the standard deviations of ρ_{rc} (1609) and ρ_{rc} (2201) were 0.0026 (average: 0.0218) and 0.0016 (average: 0.0137), respectively. The standard deviation of ρ_{rc} (1609)/ρ_{rc} (2201) was 0.0546 (average: 1.5944) (Figure 12). This suggests that Lake Taihu is consistent with the assumption of the spatially homogeneous distribution of aerosol types.

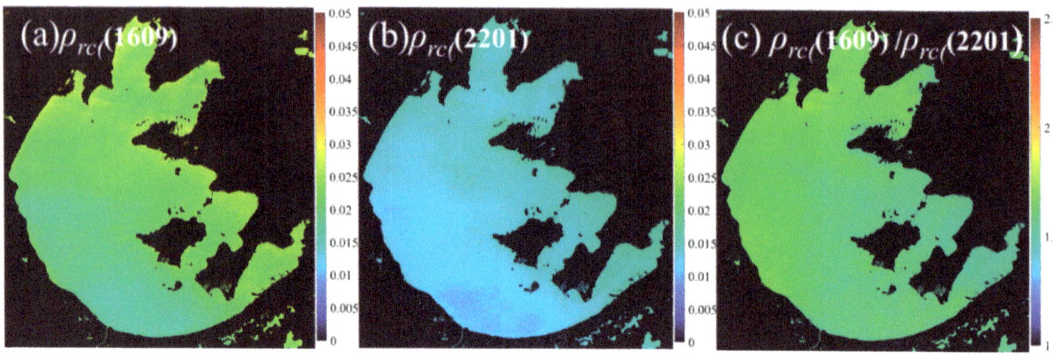

Figure 12. Rayleigh-corrected reflectance (ρ_{rc}) in SWIR bands and its ratio (ρ_{rc} (1609)/ρ_{rc} (2201)) from Lake Taihu on 2 February 2016.

4.2. *The Black Pixels from Two Date Images Overlap*

The m-ACZI algorithm searches and obtains the "black pixel" based on the combination of BPI and PIF for two date images, which requires the BPI and PIF pixels of the two periods to overlap. The PIF pixels are from urban construction land, which is invariable in a year. Therefore, the PIF pixels of the two periods should overlap. The occurrence probability of BPI pixels is calculated in Lake Taihu based on cloudless images from 2013 to 2018 (a total of 36 images) (Figure 13), showing more than 30% in a large area of central Lake Taihu. This means that there is a very high probability of obtaining overlapping BPI pixels from three images, which is sufficient to support the m-ACZI algorithm. Landsat 8-OLI has accumulated a large number of images to complete the m-ACZI algorithm.

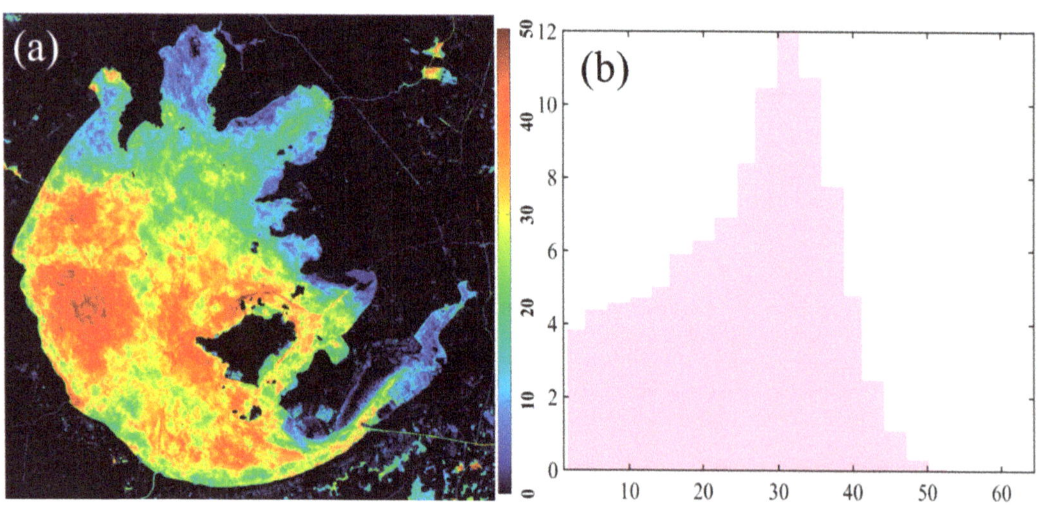

Figure 13. Occurrence probability of BPI pixels (**a**) and its histogram (**b**) in Lake Taihu (2013–2018).

4.3. *The m-ACZI Algorithm Depends on Pure Pixels*

The m-ACZI algorithm needs PIF pixels to identify the "black pixel" because they do not drastically change their reflectivity properties over time. The reflectivity properties of PIFs are derived from construction land, and mixing from other variable land types (such as water, forest land, and agricultural land) weakens the invariant reflectivity properties of PIFs. The theoretical basis of the m-ACZI algorithm requires minimizing the uncertainty caused by a mixture of pixels, and the high spatial resolution can reduce the uncertainty of a mixture of pixels [35]. From this point of view, the m-ACZI algorithm depends on pure pixels.

4.4. The Impact of Δt on D^{PIF}

To analyze the influence of the Δt on D^{PIF}, the t was calculated using 6SV-LUT (from ACOLITE software (version 20211112)) based on cloudless images from 2013 to 2018 (a total of 36 images). In the 1609 nm band, the variation range of t was 0.9981 to 0.9543, the average t was 0.9632, the standard deviation (SD) of t was 0.0111 (Figure 14), the range of Δt was 0.04375 to 0.00002, the average Δt was 0.00971, and the SD of Δt was 0.01255. In the 2201 nm band, the variation range of t was 0.9663 to 0.8982, the average t was 0.9149, the SD of t was 0.0165 (Figure 14), the range of Δt was 0.06810 to 0.00003, the average Δt was 0.01656, and the SD Δt was 0.01699. The variation in t and Δt in the 1609 nm band was less than in 2201 nm; this band was selected as the judgment band for $D^{PIF} \approx D^{BPI}$. Suzhou City (the red box in Figure 15a) near Lake Taihu was taken as an example to illustrate the influence of the Δt on D^{PIF} (1609). Furthermore, to ensure the independence of verification, ρ_s was driven by the ACOLITE. To reduce the error caused by the ACOLITE-AC algorithm, the average ${\rho_s}^{PIF}$ was considered as the real ${\rho_s}^{PIF}$ (Figure 15c). This ${\rho_s}^{PIF}$ deviated from the true value, but it can still be considered as a reference for analyzing the influence of Δt on D^{PIF}. Since there were no images that completely corresponded to the average Δt value, we selected the corresponding two images of the approximate average Δt for analysis. The maximum Δt ($\Delta t = 0.04375$) in the 1609 nm band was from 2 images from 12/29/2014 and 05/14/2018, and the $\Delta t = 0.01065$ (approximate average Δt) in the 1609 nm band was from 2 images from 12/29/2014 and 06/15/2018. As shown in Figure 15, when $\Delta t = 0.04375$, the average $\Delta t{\rho_s}^{PIF}$ was 0.0058, the median $\Delta t{\rho_s}^{PIF}$ was 0.0055, the average $\Delta t{\rho_s}^{PIF}/D^{PIF}$ was 14.85%, and the median $\Delta t{\rho_s}^{PIF}/D^{PIF}$ was 10.84%. When $\Delta t = 0.01065$, the average $\Delta t{\rho_s}^{PIF}$ was 0.0014, the median $\Delta t{\rho_s}^{PIF}$ was 0.0013, the average $\Delta t{\rho_s}^{PIF}/D^{PIF}$ was 4.62%, and the median $\Delta t{\rho_s}^{PIF}/D^{PIF}$ was 2.68%. It can be seen that Δt can introduce an error into the algorithm, but when Δt is less than 0.01065 (this was a common occurrence), the error it introduced was small and within the allowable error range. We propose selecting two images with similar t values for the m-ACZI algorithm, and this t value can be pre-estimated by the LUT.

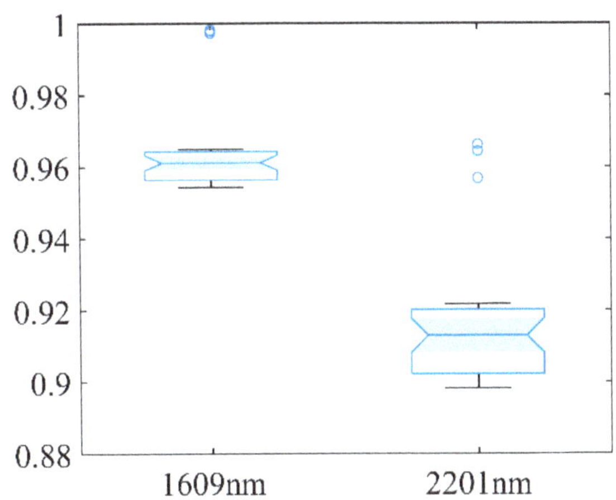

Figure 14. Variations in the transmittance (t) in SWIR bands (based on cloudless images from 2013 to 2018, a total of 36 images).

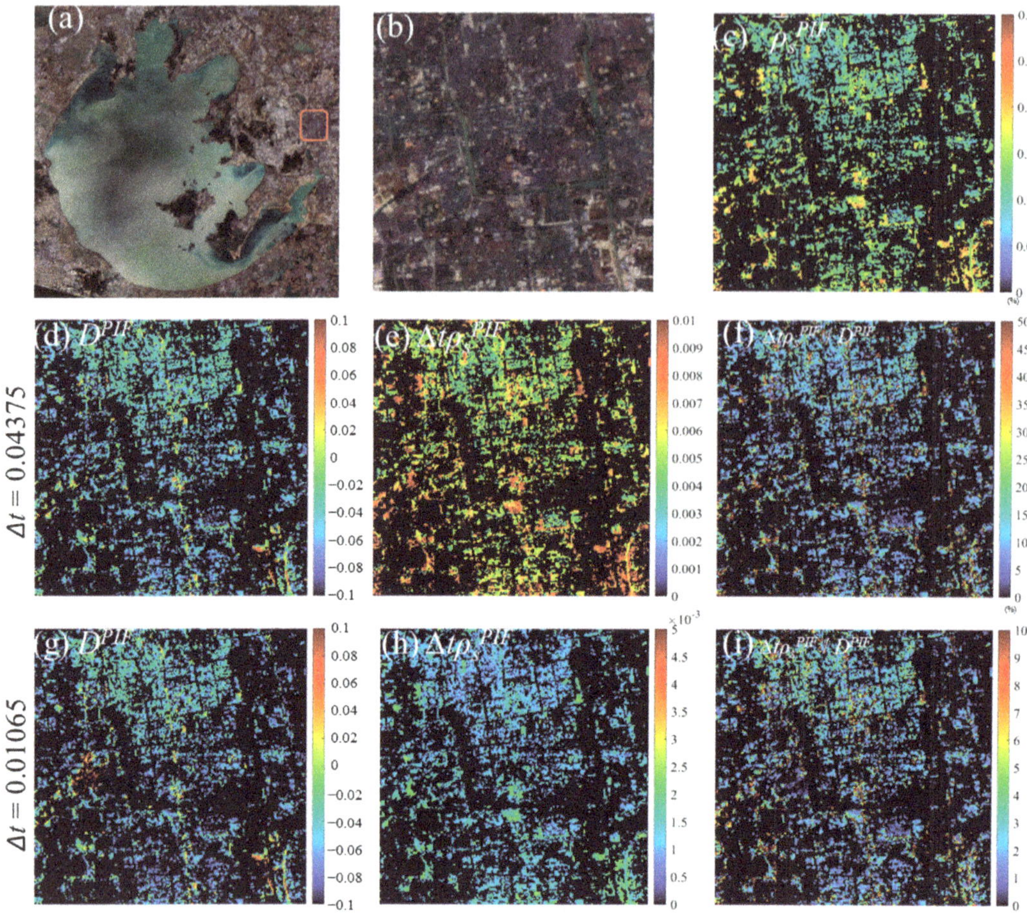

Figure 15. Impact of Δt (means different between t from two date images) on D^{PIF} (1609) (taking Suzhou City (red box) near Lake Taihu as an example). (**b**) RGB image of Suzhou City (red box in (**a**)). The average ${\rho_s}^{PIF}$ (1609) (**c**) was driven by ACOLITE based on cloudless images from 2013 to 2018 (total of 36 images). The D^{PIF} is the difference in ρ_{rc} (SWIR) for two dates images; the $\Delta t{\rho_s}^{PIF}$ represents the residual surface reflectance caused by Δt in D^{PIF}, and $\Delta t{\rho_s}^{PIF}/D^{PIF}$ is the ratio of $\Delta t{\rho_s}^{PIF}$ to D^{PIF}, which represents the error of D^{PIF} caused by Δt. These two Δt values were the maximum Δt (0.0438) and Δt = 0.0107 (approximate average Δt) from 36 images (2013 to 2018), respectively.

5. Conclusions

The "black pixel" assumption plays a key role in providing atmospheric (especially aerosol) scattering information to solve the atmospheric correction process for water. We modified the identification method for the "black pixel" in the ACZI algorithm as a combination of BPI and PIF and provided the m-ACZI algorithm. The m-ACZI algorithm assumes the spatial homogeneity of aerosol types and employs BPI and PIF to identify the "black pixel", which together make this algorithm suitable for inland water (such as lakes, rivers, and reservoirs). Compared with the ACZI algorithm, m-ACZI had better performance in the visible bands, especially in the 561 nm and 655 nm ones. The evaluation results showed that the performance of the m-ACZI algorithm is superior to the SeaDAS-SWIR, ACZI, and EXP algorithms in waters with algal blooms. This algorithm is a useful attempt to improve the accuracy of AC for turbid water and provides a new scheme for obtaining black pixels from turbid inland waters.

Author Contributions: Conceptualization, D.W. and R.M.; methodology, D.W.; validation, D.W., X.X. and W.Z.; formal analysis, D.W.; investigation, D.W.; resources, R.M.; writing—original draft preparation, D.W.; writing—review and editing, D.W., R.M., X.X. and Z.W.; funding acquisition, D.W., Y.G. and R.M. All authors have read and agreed to the published version of the manuscript.

Funding: This research was funded by the Project of Education of Zhejiang province (Grant No. Y202250625), the National Natural Science Foundation of China (Grant No. 42106019), and the Open Foundation from Marine Sciences in the First-Class Subjects of Zhejiang (Grant No. OFMS002). The work was partly supported by the National Earth System Science Data Center, the 14th Five-year Network Security and Informatization Plan of Chinese Academy of Sciences (Grant No. CAS-WX2021SF-0306), and the Scientific Data Center of Nanjing Institute of Geography & Limnology Chinese Academy of Sciences (CAS-WX2022SDC-SJ05).

Data Availability Statement: Not Applicable.

Acknowledgments: We thank the study participants from NIGLAS (Zhigang Cao, Ming Shen, Tianci Qi and Junfeng Xiong, and Xu Fang). Thanks to USGS for the OLI data and REMSEM for the ACOLITE software. We also acknowledge the data support from Lake-Watershed Science Data Center, National Earth System Science Data Sharing Infrastructure, National Science & Technology Infrastructure of China (http://lake.geodata.cn) (accessed in April 2019).

Conflicts of Interest: The authors declare no conflict of interest.

References

1. Hu, C.; Carder, K.L.; Muller-Karger, F.E. Atmospheric Correction of SeaWiFS Imagery over Turbid Coastal Waters. *Remote Sens. Environ.* **2000**, *74*, 195–206. [CrossRef]
2. Ruddick, K.G.; Ovidio, F.; Rijkeboer, M. Atmospheric correction of SeaWiFS imagery for turbid coastal and inland waters. *Appl. Opt.* **2000**, *39*, 897–912. [CrossRef] [PubMed]
3. He, Q.; Chen, C. A new approach for atmospheric correction of MODIS imagery in turbid coastal waters: A case study for the Pearl River Estuary. *Remote Sens. Lett.* **2014**, *5*, 249–257. [CrossRef]
4. Wang, J.; Lee, Z.; Wang, D.; Shang, S.; Wei, J.; Gilerson, A. Atmospheric correction over coastal waters with aerosol properties constrained by multi-pixel observations. *Remote Sens. Environ.* **2021**, *265*, 112633. [CrossRef]
5. IOCCG. *Atmospheric Correction for Remotely-Sensed OceanColour Products*; International Ocean Colour Coordinating Group: Dartmouth, NS, Canada, 2010.
6. Wang, M.; Gordon, H.R. A Simple, Moderately Accurate, Atmospheric correction algorithn for SeaWiFS. *Remote Sens. Environ.* **1994**, *50*, 231–239. [CrossRef]
7. Gordon, H.R.; Menghua, W. Influence of oceanic whitecaps on atmospheric correction of ocean-color sensors. *Appl. Opt.* **1994**, *33*, 7754–7763. [CrossRef]
8. Gordon, H.R.; Clark, D.K. Clear Water Radiance for Atmospheric Correction of Coastal Zone Color Scanner Imegery. *Appl. Opt.* **1981**, *20*, 4175–4226. [CrossRef]
9. Wang, M.; Shi, W. The NIR-SWIR combined atmospheric correction approach for MODIS ocean color data processing. *Opt. Express* **2007**, *24*, 15722–15733. [CrossRef]
10. Wang, M.; Shi, W. Estimation of ocean contribution at the MODIS near-infrared wavelengths along the east coast of the U.S.: Two case studies. *Geophys. Res. Lett.* **2005**, *32*, L13606. [CrossRef]
11. Wang, M. Remote sensing of the ocean contributions from ultraviolet to near-infrared using the shortwave infrared bands: Simulations. *Appl. Opt.* **2007**, *46*, 1535–1547. [CrossRef]
12. Wang, D.; Ma, R.; Xue, K.; Li, J. Improved atmospheric correction algorithm for Landsat 8–OLI data in turbid waters: A case study for the Lake Taihu, China. *Opt. Express* **2019**, *27*, A1400–A1418. [CrossRef]
13. Shi, W.; Wang, M. An assessment of the black ocean pixel assumption for MODIS SWIR bands. *Remote Sens. Environ.* **2009**, *113*, 1587–1597. [CrossRef]
14. Vanhellemont, Q.; Ruddick, K. Turbid wakes associated with offshore wind turbines observed with Landsat 8. *Remote Sens. Environ.* **2014**, *145*, 105–115. [CrossRef]
15. Wang, D.; Ma, R.; Xue, K.; Loiselle, S. The Assessment of Landsat-8 OLI Atmospheric Correction Algorithms for Inland Waters. *Remote Sens.* **2019**, *11*, 169. [CrossRef]
16. Concha, J.A.; Schott, J.R. Retrieval of color producing agents in Case 2 waters using Landsat 8. *Remote Sens. Environ.* **2016**, *185*, 95–107. [CrossRef]

17. Schott, J.R.; Salvaggio, C.; Volchok, W.J. Radiometric scene normalization using pseudoinvariant features. *Remote Sens. Environ.* **1988**, *26*, 1–16. [CrossRef]
18. Hu, C.; Lee, Z.; Ma, R.; Yu, K.; Li, D.; Shang, S. Moderate Resolution Imaging Spectroradiometer (MODIS) observations of cyanobacteria blooms in Taihu Lake, China. *J. Geophys. Res.* **2010**, *115*, C04002. [CrossRef]
19. Hu, C. A novel ocean color index to detect floating algae in the global oceans. *Remote Sens. Environ.* **2009**, *113*, 2118–2129. [CrossRef]
20. Shen, M.; Duan, H.; Cao, Z.; Xue, K.; Qi, T.; Ma, J.; Liu, D.; Song, K.; Huang, C.; Song, X. Sentinel-3 OLCI observations of water clarity in large lakes in eastern China: Implications for SDG 6.3.2 evaluation. *Remote Sens. Environ.* **2020**, *247*, 111950. [CrossRef]
21. Xue, K.; Ma, R.; Shen, M.; Li, Y.; Duan, H.; Cao, Z.; Wang, D.; Xiong, J. Variations of suspended particulate concentration and composition in Chinese lakes observed from Sentinel-3A OLCI images. *Sci. Total Environ.* **2020**, *721*, 137774. [CrossRef]
22. Cao, Z.; Duan, H.; Feng, L.; Ma, R.; Xue, K. Climate- and human-induced changes in suspended particulate matter over Lake Hongze on short and long timescales. *Remote Sens. Environ.* **2017**, *192*, 98–113. [CrossRef]
23. Feng, L.; Hu, C.; Chen, X.; Tian, L.; Chen, L. Human induced turbidity changes in Poyang Lake between 2000 and 2010: Observations from MODIS. *J. Geophys. Res. Ocean.* **2012**, *117*, C07006. [CrossRef]
24. Mueller, J.L.; Fargion, G.S.; Charles, R.; McClain, C.R. Biogeochemical and bio-optical measurements and data analysis protocols. In *Ocean Optics Protocols for Satellite Ocean Color Sensor Validation, Revision 5*; Tech, N., Ed.; NASA Tech: Washington, DC, USA, 2003; Volume V.
25. Mobley, C.D. Estimation of the remote-sensing reflectance from above-surface measurements. *Appl. Opt.* **1999**, *38*, 7442–7455.
26. Vanhellemont, Q.; Ruddick, K. Atmospheric correction of metre-scale optical satellite data for inland and coastal water applications. *Remote Sens. Environ.* **2018**, *216*, 586–597. [CrossRef]
27. Vermote, E.F.; Member, I.; Tan, D.; Deuze, J.L.; Herman, M. Second Simulation of the Satellite Signal in the Solar Spectrum, 6s: An Overview. *IEEE Trans. Geosci. Remote Sens.* **1997**, *35*, 675–687. [CrossRef]
28. Vanhellemont, Q.; Ruddick, K. Advantages of high quality SWIR bands for ocean colour processing: Examples from Landsat-8. *Remote Sens. Environ.* **2015**, *161*, 89–106. [CrossRef]
29. He, X.; Bai, Y.; Pan, D.; Tang, J.; Wang, D. Atmospheric correction of satellite ocean color imagery using the ultraviolet wavelength for highly turbid waters. *Opt. Express* **2012**, *20*, 20754–20770. [CrossRef]
30. Hale, G.M.; Querry, M.R. Optical constants of water in the 200 nm to 200 μm wavelength region. *Appl. Opt.* **1973**, *12*, 555–563. [CrossRef]
31. Vanhellemont, Q. Adaptation of the dark spectrum fitting atmospheric correction for aquatic applications of the Landsat and Sentinel-2 archives. *Remote Sens. Environ.* **2019**, *225*, 175–192. [CrossRef]
32. Vanhellemont, Q.; Ruddick, K. Atmospheric correction of Sentinel-3/OLCI data for mapping of suspended particulate matter and chlorophyll-a concentration in Belgian turbid coastal waters. *Remote Sens. Environ.* **2021**, *256*, 112284. [CrossRef]
33. Vanhellemont, Q. ACOLITE processing for Sentinel-2 and Landsat-8 atmospheric correction and aquatic applications. In Proceedings of the 2016 Ocean Optics Conference, Victoria, BC, Canada, 23–28 October 2016.
34. Sun, D.; Li, Y.; Wang, Q.; Le, C.; Lv, H.; Huang, C.; Gong, S. Specific inherent optical quantities of complex turbid inland waters, from the perspective of water classification. *Photochem. Photobiol Sci.* **2012**, *11*, 1299–1312. [CrossRef] [PubMed]
35. IOCCG. *Uncertainties in Ocean Colour Remote Sensing;* International Ocean Colour Coordinating Group: Dartmouth, NS, Canada, 2019.

remote sensing

MDPI

Article

Estimating Effects of Natural and Anthropogenic Activities on Trophic Level of Inland Water: Analysis of Poyang Lake Basin, China, with Landsat-8 Observations

Jianzhong Li [1,2], Zhubin Zheng [1,*], Ge Liu [3], Na Chen [4], Shaohua Lei [5], Chao Du [1], Jie Xu [6], Yuan Li [7], Runfei Zhang [1,2] and Chao Huang [1]

1 School of Geography and Environmental Engineering, Gannan Normal University, Jiangxi Provincial Key Laboratory of Low-Carbon Solid Waste Recycling, Ganzhou 341000, China
2 School of Geography, Nanjing Normal University, Key Laboratory of Virtual Geographic Environment of Education Ministry, Nanjing 210023, China
3 Northeast Institute of Geography and Agroecology, CAS, Changchun 130102, China
4 Laboratory of Geo-Information Science and Remote Sensing, Wageningen University & Research, Droevendaalsesteeg 3, 6708 PB Wageningen, The Netherlands
5 State Key Laboratory of Hydrology-Water Resources and Hydraulic Engineering, Nanjing Hydraulic Research Institute, Nanjing 210029, China
6 Yangtze River Basin Ecological Environment Monitoring and Scientific Research Center, Yangtze River Basin Ecological Environment Supervision and Administration Bureau, Ministry of Ecological Environment, Wuhan 430010, China
7 School of Tourism and Urban & Rural Planning, Zhejiang Gongshang University, Hangzhou 310018, China
* Correspondence: zhengzhubin@gnnu.edu.cn

Abstract: The intensification of anthropogenic activities has led to the infiltration of enormous quantities of pollutants into rivers and lakes, resulting in significant deterioration in water quality and a more prominent occurrence of eutrophication. Poyang Lake, the largest freshwater lake in China, is facing a severe challenge related to eutrophication, which seriously threatens the delivery of the ecosystem service and the safety of drinking water. To address this challenge, Landsat-8 Operational Land Imager (OLI) data for the Poyang Lake Basin (PLB) from May 2013 to December 2020 were used. Since inland water bodies with complex optical characteristics, we developed a semi-analytical algorithm to assess the trophic state of the water based on two cruise field measurements in 2016 and 2019. Combining the semi-analytical trophic level index (TLI) with an atmospheric correction model is the most suitable model for OLI images of the PLB, this model was then applied to Landsat-8 time series observations. The trends of the trophic state of water bodies in PLB were revealed, and the annual, quarterly and monthly percentages of eutrophic water bodies were calculated. Natural and anthropogenic factors were then used to explain the changes in the trophic state of the PLB waters. The main findings are as follows: (1) From the 8-year observation results, it can be seen that the variation of trophic level of water in PLB showed obviously spatial and temporal variations, characterized by higher in the north than in the south and higher in winter than in summer. (2) Temperature promoted the growth of harmful algae and plays an essential role in affecting changes in the trophic level of the water. (3) Changes in the trophic level of water bodies in PLB were mainly affected by human activities. The results of spatial and temporal variation of the trophic level of water and the driving factors in PLB can extend our knowledge of water quality degradation and provide essential references for relevant policy-making institutions.

Keywords: eutrophication; Poyang Lake Basin (PLB); Landsat-8 OLI; spatio-temporal patterns

1. Introduction

Climate change and human activities have led to significant challenges for global lakes. As the global climate continues to warm, lakes are experiencing a range of impacts that threaten their ecological health and functioning [1,2]. Global lake productivity is increasing, and eutrophication is a growing problem [3,4]. Lake eutrophication is a phenomenon caused by a combination of natural and human factors, which leads to the excessive growth of algae, a sharp decrease in water transparency and dissolved oxygen levels, resulting in the degradation of water quality, aquatic ecosystem, and the overall functioning of the lake [5,6]. Although eutrophication of water bodies naturally occurs [7], human activities and global warming have accelerated the process [4,8], causing a tremendous challenge to human drinking water safety, aquatic biodiversity, and aquatic ecosystems.

Traditionally, monitoring and assessment of eutrophication were based on periodic sampling and analysis of water quality parameters determining trophic status [9], namely, water temperature, Secchi disk depth (SDD), chlorophyll a (Chla) concentration, and other non-optically active water quality parameters [9,10]. Carlson [11] developed a trophic state index (TSI) to evaluate the trophic state of inland water based on measured water quality parameter data. Aizaki et al. [12] improved the TSI model using in situ measurement data from lakes in Japan and developed a modified trophic state index (TSI_M). Jin and Tu [13] developed a trophic level index (TLI) to evaluate the trophic state of Chinese lakes based on the TSI model using survey data from 35 major lakes in China. The traditional method is based on measured data and laboratory analysis, making it difficult to accurately monitor the trophic levels of the lake as a whole and lacking the capability of consistent observation [14].

Remote sensing is widely used in inland aquatic remote sensing studies owing to its wide observation range, periodic observation, and data accessibility [4,7,8]. Remote sensing data have been used to monitor the changing trophic levels of shallow lakes in eastern China [8,14–16]. Although satellite remote sensing has made significant progress in the study of eutrophication in large lakes in recent years, most of the data used in lake eutrophication monitoring and evaluation studies are the Moderate Resolution Imaging Spectrometer (MODIS), Medium-Resolution Imaging Spectroradiometer (MERIS), Sea-Viewing Wide Field-of-View Sensor (SeaWiFS), and Ocean and Land Colour Instrument (OLCI) [8,17–21]. However, the coarse spatial resolution (>100 m) makes it difficult to monitor small lakes and rivers with these remote sensing data. Furthermore, it is impossible to take into account numerous parameters [22]. For instance, MODIS data have a high temporal resolution; however, the spatial resolution is hard to match the demands of inland water environment monitoring, limiting its usefulness in inland water eutrophication monitoring.

Landsat-8 is widely used in inland lake water quality monitoring and eutrophication assessment due to its excellent performance and spatial resolution [7,16,22–24]. Hu et al. [7] assessed the trophic state of lakes in eastern China using Landsat data and a water optical classification model and analyzed their spatiotemporal variations and driving factors. Zheng et al. [22] developed a band ratio model for estimating the diffuse attenuation coefficient (K_d (490)) of inland turbid waters and successfully obtained the spatial and temporal distribution of K_d (490) in Dongting Lake. Li et al. [24] developed an empirical model for estimating water transparency using Landsat data and successfully obtained the spatiotemporal variations of water transparency in Qiandao Lake. The low spectral resolution (>50 nm) of the Landsat-8 satellite restricts its capacity to represent spectral properties of the features of the ground and limits its efficacy in estimating the material composition and concentration of inland waters. However, several studies have demonstrated its potential to assess the trophic levels of water in specific lakes and the precision of the results [7,22,25,26].

In this study, we analyzed the drivers of changes in the trophic level of water in the Poyang Lake Basin (PLB) using Landsat-8 time-series images (2013–2020) and combining natural and anthropogenic data. The following objectives of this study are: (1) to develop a robust semi-analytical algorithm to assess the trophic levels of inland waters in PLB, (2) to characterize the spatial-temporal patterns of changes in the trophic state of waters in the PLB and to quantify the impact of natural and human activities; and (3) to propose the eutrophication control strategies for inland waters in PLB, which will be the theoretical basis for the implementation of environmental protection policies.

2. Materials and Methods

2.1. Overview of the Study Area

Poyang Lake Basin (PLB) is located in the middle and lower reaches of the Yangtze River Basin (24°29′14″–30°04′41″N, 113°34′36″–118°28′58″E, Figure 1). The basin covers 1.62×10^5 km^2. The PLB is a converging water system with Poyang Lake as the center of convergence. The PLB is influenced by the East Asian monsoon climate, with high temperatures and rain in summer and low temperatures and little rain in winter. Poyang Lake, the largest freshwater lake in China, covers an area of 4070 km^2 during the period of abundant water. In addition, two major artificial reservoirs of drinking water sources in Jiangxi Province were selected as study areas. Zhelin Reservoir is an important drinking water source located in the Xiushui River Basin, which is one of the major sub-basins in the Poyang Lake Basin. It plays a crucial role in the regional water supply, flood control, and ecological environment. Yangming Lake provides water for irrigation to more than 40,000 hectares of farmland and supplies drinking water to over 600,000 people in Ganzhou City and the surrounding areas. Tao River is the river most polluted by rare earth mining. However, as human activities such as mining and enclosure aquaculture around the lake have increased, its water bodies have become eutrophic [27], which has threatened aquatic ecosystems, fisheries and recreational purposes, and human health [28].

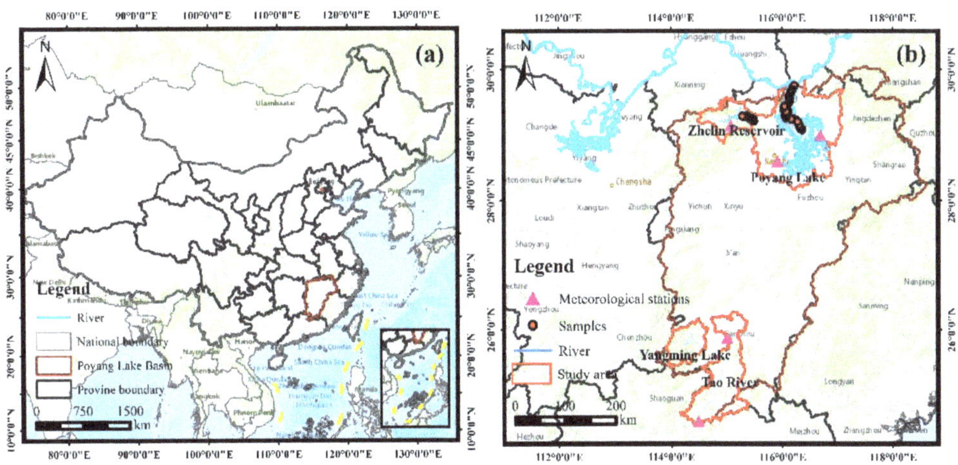

Figure 1. Study area. (**a**) Location of the study area. (**b**) The red points represent the locations of field measurement points at Poyang Lake and Zhelin Reservoir in 2016 and 2019 respectively. Pink triangles represent meteorological stations.

2.2. Data and Processing

2.2.1. In Situ Data Collection

Two cruises were conducted on Poyang Lake from 3–14 October 2016, and 19 April 2019, and one cruise was conducted at Zhelin Reservoir on 19 April 2019 (Figure 1). Remote sensing reflectance ($Rrs(\lambda)$, sr^{-1}) data were obtained from 44 points in Poyang Lake and 10 points in Zhelin Reservoir, respectively. In addition, 54 samples were collected for laboratory analysis to obtain water quality parameters such as chlorophyll a concentration (Chla, mg m^{-3}).

$Rrs(\lambda)$ data were measured using a FieldSpec Spectroradiometer (Analytical Spectral Device, Inc., Boulder, CO, USA) with a spectral collection range of 350–1075 nm and

512 bands [28]. In addition, $Rrs(\lambda)$ were measured following the existing measurement method [29].

Chla concentrations were generally measured using the hot ethanol-spectrophotometer method. The water samples were filtered through pore-size of 0.7 μm Whatman GF/F filters (Whatman, Inc., Maidstone, UK). Then Chla concentrations were determined using Shimadzu UV2550 UV-spectrophotometer (Shimadzu, Inc., Kyoto, Japan) [22]. The absorption coefficients at 750 nm and 665 nm were measured to determine the chlorophyll a concentration.

2.2.2. Landsat OLI Images and Preprocessing

Landsat-8 OLI level-1 data from May 2013 to December 2020 of the PLB were downloaded from the United States Geological Survey (USGS: https://glovis.usgs.gov/, accessed on 10 June 2022) (Figure 2). The data need to be processed into a remote sensing reflectance product as they are pre-processed level-1 data. The pre-processing steps performed on data include radiometric calibration, atmospheric correction (AC), and extraction of water in the study area.

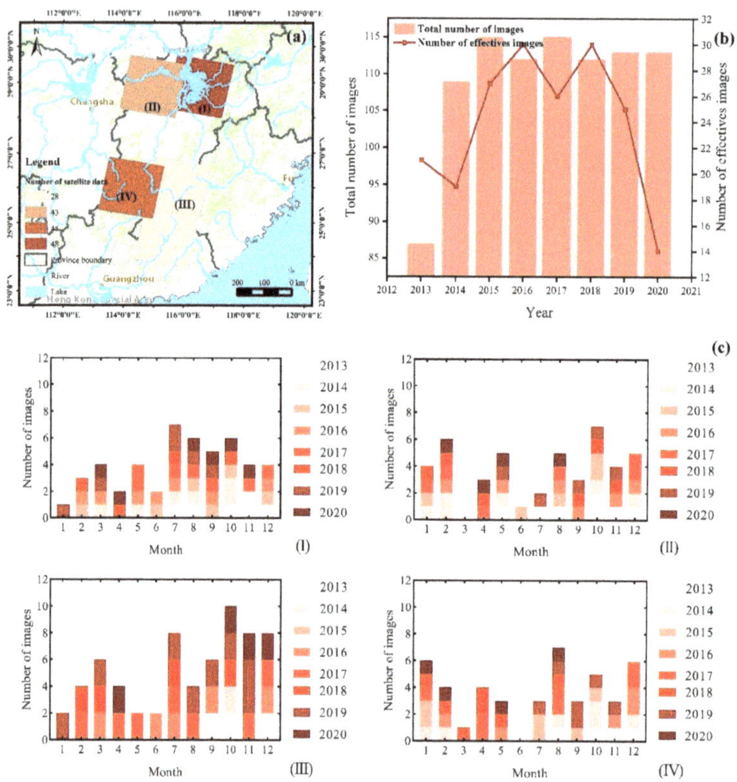

Figure 2. Statistics of Landsat imagery across PLB from 2013 to 2020. (**a**) Spatial distribution of the total Landsat-8 images across the PLB from 2013 to 2020 coverage; (**b**) the annual number of available Landsat images; (**c**) the monthly number of available Landsat images across PLB ((**I**) Poyang Lake region; (**II**) Zhelin Reservoir region; (**III**) Tao River region; (**IV**) Yangming Lake region).

To understand the effect of different atmospheric correction modules on the trophic level assessment model, Fast Line-of-sight Atmospheric Analysis of Spectral Hypercubes (FLAASH), Second Simulation of the Satellite Signal in the Solar Spectrum correction scheme (6S), Dark of subtraction (DOS) and Quick Atmospheric Correction (QUAC) modules were used for OLI data. FLAASH is extensively employed in multispectral and hyperspectral data [30]. 6S takes full account of the topography and sensor altitude to derive atmospheric correction parameters for simulating airborne and satellite observations [31]. DOS assumes the existence of a dark pixel region and the reflectance values of all bands excluding this dark pixel region [32]. QUAC is a method for rapid atmospheric correction of images in the Environment for Visualizing Images (ENVI) [33].

2.2.3. Meteorological, Anthropogenic, and Land Use Data

To analyze the effects of meteorological factors on the variation of trophic levels in waters, temperature, precipitation, and wind speed were used. Meteorological data were obtained from the China Meteorological Administration (http://data.cma.cn/site/index.html/, accessed on 20 June 2022) [34]. The meteorological stations corresponding to Poyang Lake are Poyang and Nanchang, in Zhelin reservoir is Wuning, in Tao River and Yangming Lake are Ganxian and Longnan, respectively.

Human activities have a significant impact on changes in the trophic state of water. To determine the anthropogenic activities on the changes of trophic levels of waters, the total population, gross domestic product (GDP, Chinese Yuan), GDP of primary industry (GDP_1), GDP of secondary industry (GDP_2), GDP of tertiary industry (GDP_3), chemical fertilizer usage (ton), agrochemicals usage (ton), effective irrigation area (km^2), crop planting area (km^2), agriculture, forestry, fisheries and livestock GDP data of the PLB were analyzed from the statistical yearbooks of Nanchang City, Jiujiang City, Shangrao City, and Ganzhou City.

Land use influences the direction and intensity of surface runoff via the alteration of surface topography. Land use and land cover (LULC) data from 2013 to 2020 were interpreted using Landsat-8 data and support vector machine (SVM) classification. The LULC in the PLB was divided into five following categories: bare land, forest and grassland, water, built-up land, and cropland. Training and validation samples were selected using Google Earth high-resolution images.

2.3. *Motivation of Evaluation Factor for Eutrophication*

The comprehensive trophic level index (TLI) is the most extensively utilized among most evaluation processes. Jin and Tu (1990) developed a trophic level index (TLI) to evaluate the trophic state of Chinese lakes based on the TSI model using survey data from 35 major lakes in China.

$$\text{TLI (Chla)} = 25 + 10.86\ln\text{ (Chla)}, \quad (1)$$

$$\text{TLI (TP)} = 94.36 + 16.24\ln\text{ (TP)}, \quad (2)$$

$$\text{TLI (TN)} = 54.53 + 16.94\ln\text{ (TN)}, \quad (3)$$

$$\text{TLI (SD)} = 51.18 - 19.4\ln\text{ (SD)}. \quad (4)$$

Jin et al. [35] assessed the correlations between Chla and TN, TP, and other evaluation parameters. However, due to the large difference in the status of the aquatic environment between lakes and seasons within the same lake, the association between Chla and metrics such as TN, TP, and so on varies to some amount. As a result, if multiple indicators were used to evaluate trophic levels of water, the TLI model should be rectified for different regions and seasons. This challenge limits the application of the trophic level evaluation models, and long-term remote sensing monitoring of the trophic levels of waters is impossible. Chlorophyll a concentration is often described as a scale to evaluate primary productivity and trophic levels in water bodies. In addition, in Case-2 water bodies, the optical properties of Chla are relatively stable and the remote sensing inversion model for Chla concentration is also more mature [36]. There were numerous types of models for the inversion of Chla concentration, such as empirical models, fluorescence line height, semi-analytical models, and bio-optical models [36–44]. Thus, chlorophyll a concentration was chosen as the indicator for the evaluation of the trophic level of water bodies.

2.4. Landsat-Based TLI Calibration

The PLB eutrophication assessment model was developed and effectively implemented to Landsat-8 OLI data, and then long-term remotely sensed monitoring of the trophic level of water was performed. However, since the TLI model was evaluated based on Chla concentration, it is essential to determine the band combinations with the correlation of Chla concentration. Based on previous studies' results, 172 multi-band combination models were constructed for Chla concentration. Table 1 shows the correlation coefficients of the 10 most correlated multi-band combinations with Chla concentration.

Table 1. Correlation coefficients of multi-band combinations with Chla concentration.

Band Combination	Correlation Coefficient	Band Combination	Correlation Coefficient
B1/(B3 + B5)	−0.6239	B2/(B1 + B3)	−0.5914
B2/(B3 + B5)	−0.6034	B3/(B2 + B3)	0.5900
(B3 − B2)/(B1 + B2)	0.6014	(B3 − B2)/(B1 + B3)	0.5897
B3/(B1 + B2)	0.5993	B4/(B1 + B2)	0.5855
(B3 − B1)/(B1 + B2)	0.5966	(B5 − B2)/(B1 − B3)	−0.5850

The central wavelengths of the Band 1, Band 3, and Band 5 of Landsat-8 data were 443 nm, 562 nm, and 865 nm, respectively, which correspond to an absorption canyon (at 420 nm) and two reflection peaks (590 nm and 830 nm) in the Chla spectral curve, respectively. As a result, the ratio of Band 1 (B1) in the absorption canyon and the sum of Band 3 and Band 5 (B3 + B5) in the reflection peak was used as a factor to estimate the level of eutrophication in PLB.

$$\text{TLI}\ (Rrs) = -85.61 \times \text{M} + 75.51,\ \text{M} = \text{B1}/(\text{B3} + \text{B5}). \quad (5)$$

where B1 is the blue band remotely sensed reflectance, B3 is the green band remotely sensed reflectance, and B5 is the near-infrared (NIR) band remotely sensed reflectance of Landsat-8 data.

The remote sensing evaluation model of eutrophication of water bodies in PLB based on Landsat-8 data was constructed (Figure 3). In addition, based on the TLI evaluation criteria, we classified the TLI (*R*rs) into five levels: oligotrophic (TLI (*R*rs) < 30), mesotrophic (30 ≤ TLI (*R*rs) ≤ 50), mildly eutrophic (50 < TLI (*R*rs) ≤ 60), moderately eutrophic (60 < TLI (*R*rs) ≤ 70), severe eutrophic (TLI (*R*rs) > 70) [7,11,18].

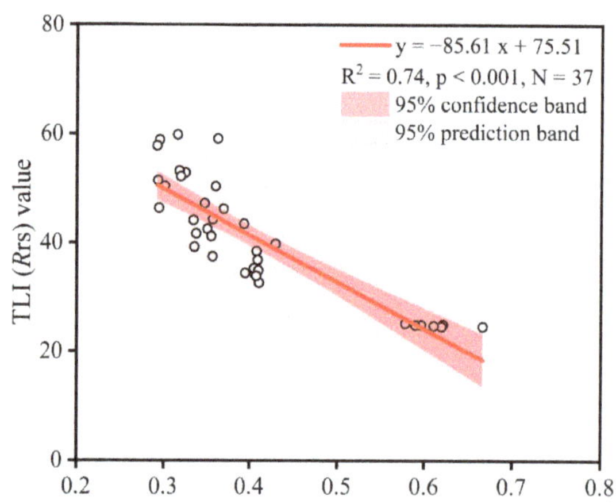

Figure 3. Scatterplots of TLI (*R*rs) calibration between M value and in situ measurement TLI (*R*rs) data.

2.5. Assessments of the TLI Patterns and Driving Variables

The TLI model in Equation (5) was implemented to Landsat-8 OLI surface reflectance data of the PLB to obtain TLI values for lakes, reservoirs, and rivers. The seasonal, annual, and 8-year mean TLI values (i.e., 8-year TLI climatology) from 2013 to 2020 were estimated for each of the study regions in the PLB. To determine the rate of change in water trophic level across the research period, linear regressions were performed on the 8-year annual values.

Meteorological factors have an important influence on changes in the trophic state of water bodies [3,45,46]. Spearman's correlation coefficient values between meteorological factors and the percentage of eutrophic water bodies were calculated using SPSS software. Meteorological factors with significant correlation ($p < 0.05$) were identified for analysis.

The implications of socioeconomic growth on water quality have become a major global issue [47]. However, previous research may have neglected the major influence of socioeconomic variables on water quality changes and intensification of eutrophication [48]. Thus, this study refers to the previous studies [24,49], the processes of socioeconomic variables on water quality were chosen for analysis.

In this study, grey relation analysis (GRA) was used to correlate the proportion of eutrophic water bodies with socioeconomic factors and the percentage of each land use factor, and the factors with the highest reliability were selected for driving force analysis. Deng [50] proposed gray correlation analysis in 1982 as a multi-factor analysis method to reflect the similarity and dissimilarity of development trends among factors. A correlation greater than 0.7, according to the gray correlation analysis method, indicates that the two indicators have a significant correlation [51,52]. The weights of drive factors were analyzed using the entropy weight method (EWM). In water quality evaluation [53,54], ecological and environmental benefit assessment [55], and other studies, the entropy weighting method has been widely used.

2.6. Statistical Analysis and Accuracy Assessment

Mean absolute deviation (*MAD*), root mean square deviation (*RMSD*), and mean absolute percent deviation (*MAPD*) were the main parameters we used to evaluate model accuracy [56]. The *MAD*, *RMSD*, and *MAPD* were calculated as follows:

$$MAD = \frac{1}{N}\sum_{i=1}^{N}\left|TLI_{measured} - TLI_{predicted}\right| \tag{6}$$

$$RMSD = \sqrt{\frac{\sum_{i=1}^{N}\left(TLI_{measured} - TLI_{predicted}\right)^2}{N}} \tag{7}$$

$$MAPD = \frac{1}{N}\sum_{i=1}^{N}\left|\frac{TLI_{measured} - TLI_{predicted}}{TLI_{predicted}}\right| \times 100\% \tag{8}$$

where, $TLI_{measured}$ is the measured value, $TLI_{predicted}$ is the predicted value, and N is the sample size.

3. Results

3.1. Evaluation of the Applicability of Atmospheric Correction Algorithms to the Model of Trophic Level of Inland Waters

What impact may various atmospheric correction models have on inversion results? Four AC models (i.e., FLAASH, 6S, DOS, and QUAC) were used in this study to process Landsat-8 images and assess the effects of different AC methods on inversion results. Due to the presence of persistent clouds in the satellite data of Poyang Lake in October 2016 and April 2019, the image reflectance data of the actual measurement points could not be obtained. Whereas the Landsat-8 data with the nearest sampling time and better image quality were collected within 7 days after the field data. Thus, the impact of different atmospheric correction methods on the model could not be accessed directly. However, previous researchers in the long time series water quality monitoring identified that water quality in a specified duration would be to maintain a steady condition with regularity [16,57–60].

Pearson correlation coefficients were used to calculate the correlation coefficients between remote sensing reflectance and TLI (Chla) for each band of Landsat-8 data on 19 April 2019, via various atmospheric correction algorithms (Figure 4) [61]. The effect of different AC algorithms varies for different bands, with some performing better in red band and others displaying accessibility in the short-wave Infrared (SWIR) band. Overall, the four AC algorithms showed better TLI (Chla) retrieval at the green and the red bands than at the blue and SWIR bands. The results indicated that all AC algorithms in the SWIR regions performed relatively poorly ($r < 0.3$).

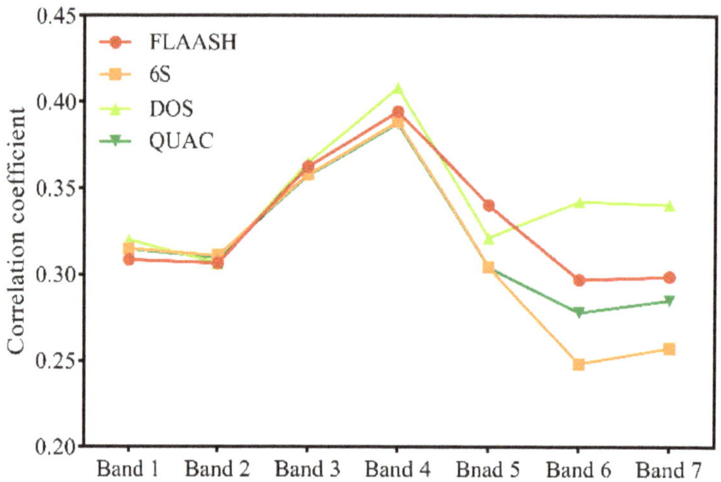

Figure 4. Correlation coefficients between remote sensing reflectance and TLI after different atmospheric correction modules.

Among the four atmospheric correction modules, FLAASH atmospheric correction modules presented the best performance (*MAD* = 4.99, *RMSD* = 5.82, *MAPD* = 19.13%) (Figure 5). The remotely sensed reflectance of each band of OLI retrieved using the DOS atmospheric correction algorithm has the best correlation with TLI (Chla). However, it showed exceptionally worse performance compared to the other AC processors (*MAD* = 24.36, *RMSD* = 24.44, *MAPD* = 93.48%), which might result in poor aerosol models in the NIR band. In summary, the FLAASH atmospheric correction module succeeded the evaluation of the trophic levels of water bodies in the Poyang Lake Basin. As a result, the FLAASH atmospheric correction module was used to process all Landsat-8 data.

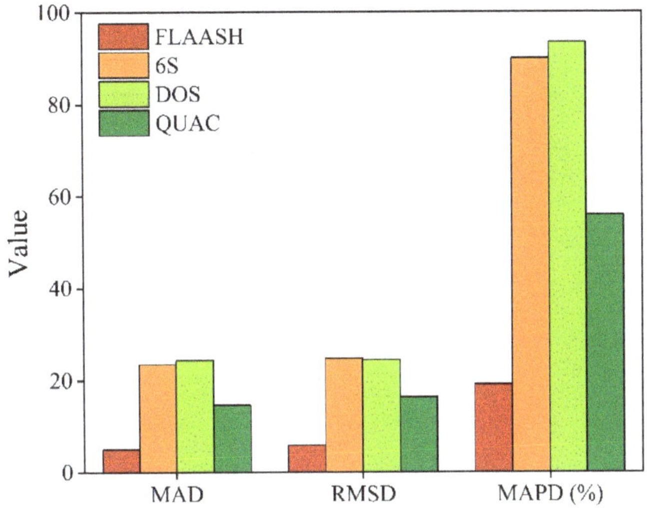

Figure 5. Accuracy of four atmospheric correction algorithms on Landsat-8 OLI.

3.2. Validation of Algorithm in Landsat-Based TLI Calibration

A total of 54 sample points were measured in October 2016 and April 2019, of which 2/3 were used to calibrate the TLI model, and the remaining 1/3 were employed to validate the performance of the model. TLI (*R*rs) calculated from observed chlorophyll a concentrations was used as the measured value, while the TLI (*R*rs) calculated from the remote sensing reflectance (i.e., Equation (5)) was used as the predicted value, and the correlation between the two was evaluated (Figure 6). The results were shown with a *MAD* of 3.88, *RMSD* of 4.43, and *MAPD* of 8.88%.

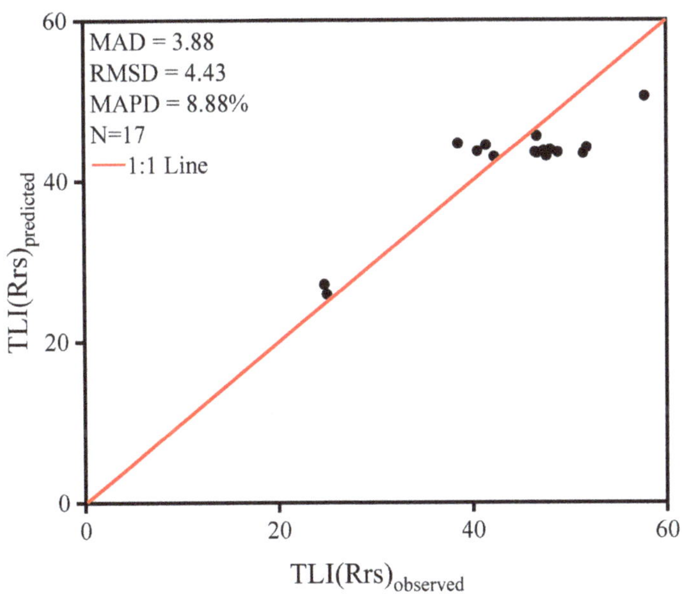

Figure 6. Accuracy validation of TLI model.

3.3. Spatio-Temporal Variations in OLI-Derived TLI

The 8-year TLI climatology values of three study regions (Zhelin Reservoir, Tao River, and Yangming Lake) were >15 and Poyang Lake was >25 (Figure 7a). Poyang Lake was the most eutrophic lake compared to other lake, reservoir, and river, with a long-term average TLI of 27.8, and the mean values for Zhelin Reservoir, Yangming Lake, and Tao River were less than 20. Zhelin Reservoir was comparatively clearer (long-term average TLI = 16.86). In addition, Poyang Lake and Zhelin Reservoir advertised statistical TLI significantly decreasing trends in the past 8 years, with the annual rates of change of $-0.79\ yr^{-1}$ and $-0.15\ yr^{-1}$, respectively. On the contrary, the TLI values of Yangming Lake and Tao River showed a significant increasing trend, with the annual rate of change of $0.90\ yr^{-1}$ and $0.36\ yr^{-1}$, respectively.

Figure 7c–f shows the annual mean values of TLI in four regions of Poyang Lake Basin. For Poyang Lake, the mean TLI value showed a trend from increasing (2014–2018) to decreasing (2018–2020), with the annual average maximum value of TLI appearing in 2018 (TLI = 29.81). Similarly, the annual mean value of TLI in Zhelin Reservoir also showed a trend from increasing (2014–2017) to decreasing (2017–2020), with the annual mean value of TLI occurring in 2017 (TLI = 18.89). In contrast, Tao River and Yangming Lake showed an increasing trend of TLI values in the 8-year period from 2013 to 2020. Compare with the waters in Tao River, the annual mean TLI of Yangming Lake was fluctuating from 2016 to 2020, with the annual mean maximum TLI displayed in 2016 (TLI = 23.36). Overall, the TLI annual value of the Poyang Lake Basin is in a stable state from 2013 to 2020.

The twelve months of the year were divided into four seasons, namely, spring from March to May, summer from June to August, autumn from September to November, and winter from December to next February. Seasonally, TLI values for several studied regions were also determined (Figure 8). We found that the average value of TLI in the Poyang Lake Basin was maximum in the spring (TLI = 23.30), followed by winter (TLI = 20.48), and the lowest in summer and autumn (TLI = 19.02, 17.46, respectively) (Figure 8f). In addition, there were differences in the seasonal variation of mean TLI in each region. For Poyang Lake, the average TLI value remained high in winter (TLI = 32.75) and spring (TLI = 29.64) and displayed an obvious deteriorating trend from summer to winter. The mean value of TLI rose from the lowest value in summer (TLI = 24.22) to the highest value in winter (TLI = 32.75). In contrast, the average the TLI value of Zhelin Reservoir showed a downward trend from spring (TLI = 24.46) to autumn (TLI = 11.53) and increased in winter (TLI = 15.65). Interestingly, the seasonal variation of the Tao River TLI means was opposite to that of the Yangming Lake TLI. The minimum value of the mean TLI value of Tao River appeared in summer (TLI = 12.88), on the contrary, the maximum value of the mean value of TLI of Yangming Lake appeared in summer (TLI = 20.11).

Figure 9a shows the interannual distribution of the percentage of eutrophic water in four typical zones of the PLB. Generally, the proportion of eutrophic water bodies in Yangming Lake was relatively high each year. In particular, in 2018 and 2020, the proportion of eutrophic water bodies exceeded 10%, while in 2018, it accounted for 10.8%. The proportion of eutrophic in Tao River was the lowest; its maximum never exceeded 5% (which appeared in 2016). Figure 9b shows the monthly distribution of the proportion of eutrophic water bodies in four typical regions of PLB. The monthly distribution of the proportion of eutrophic water in Poyang Lake and Yangming Lake was similar, both reaching the maximum (7.56% and 10.77%, respectively) in February–March, then falling to the minimum (1.76% and 2.41%, respectively), and then showing an upward trend. In contrast, the monthly distribution of the proportion of eutrophic water bodies in Zhelin Reservoir and Tao River reached its maximum value and then decreased to the minimum value, and the maximum value appeared May–July.

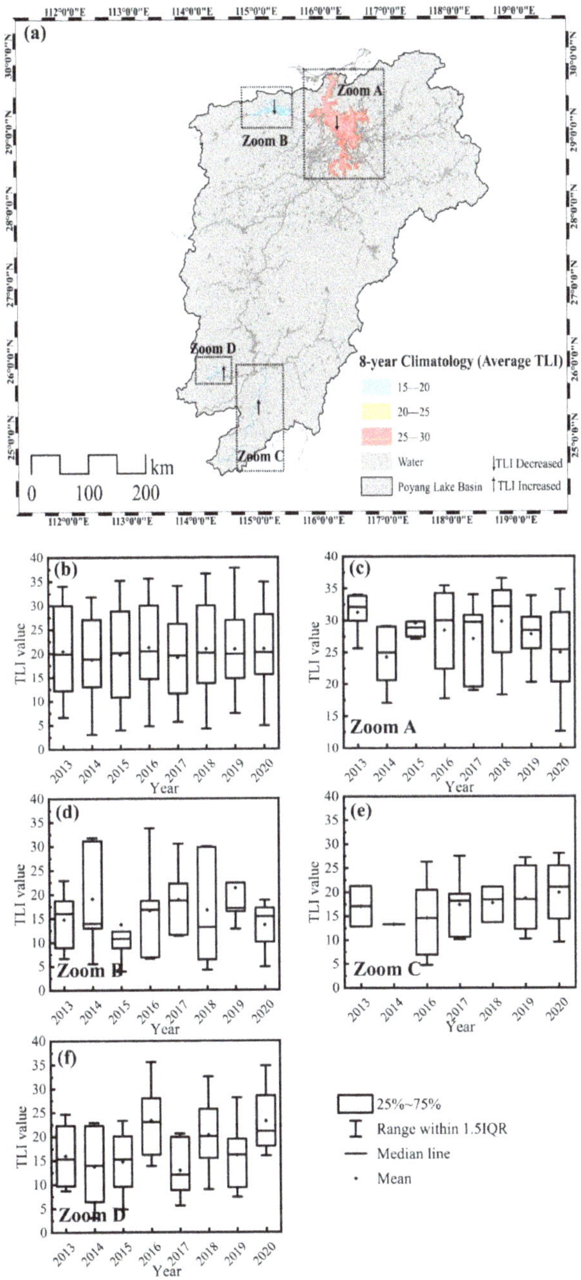

Figure 7. (**a**) Spatial distribution of climatological TLI for the four studied regions on the Poyang Lake Basin (PLB) from the Landsat-8 OLI observations. The annotations "↑" and "↓" represent statistically significant increasing and significant decreasing trends, respectively. Subpanels A to D shows the four studied regions, respectively. (**b**) represent the variations in annual mean TLI in Poyang Lake Basin. (**c**–**f**) represent the variations in annual mean TLI in Poyang Lake, Zhelin Reservoir, Tao River, and Yangming Lake, respectively.

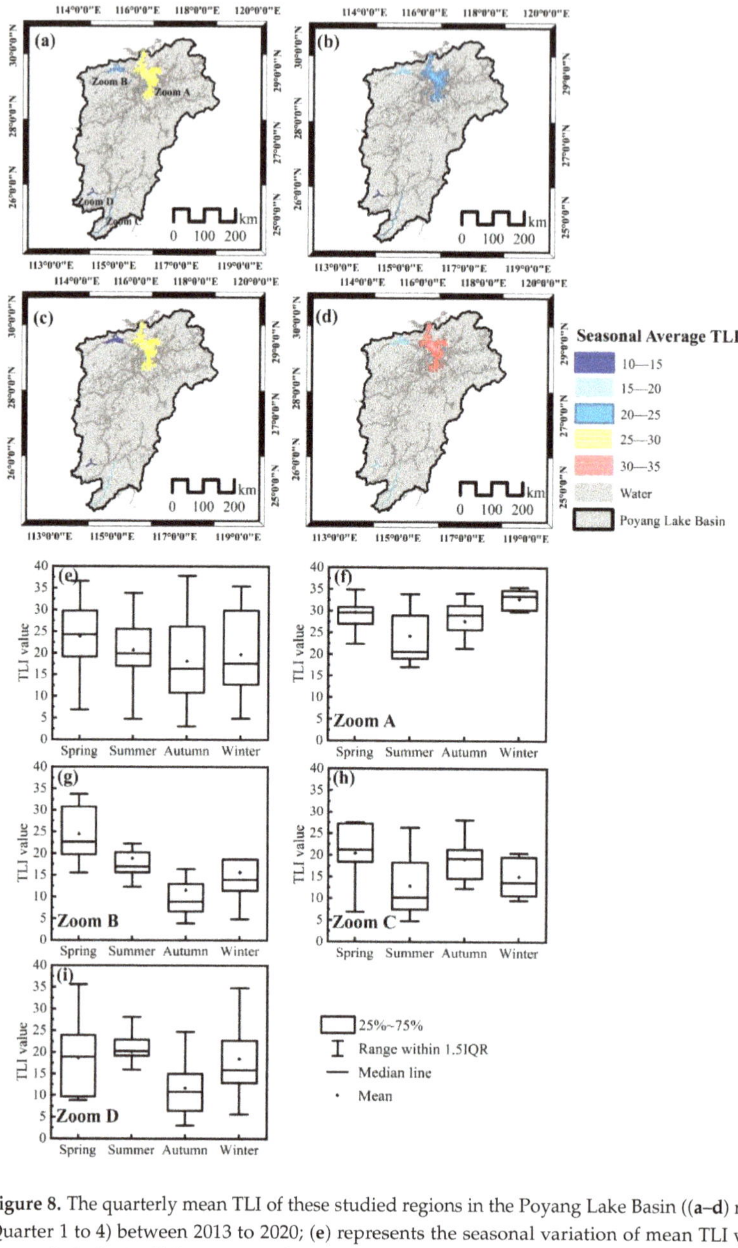

Figure 8. The quarterly mean TLI of these studied regions in the Poyang Lake Basin ((**a**–**d**) represent Quarter 1 to 4) between 2013 to 2020; (**e**) represents the seasonal variation of mean TLI values in Poyang Lake Basin; (**f**–**i**) represents the seasonal variation of mean TLI values in Poyang Lake, Zhelin Reservoir, Tao Lake and Yangming Lake, respectively.

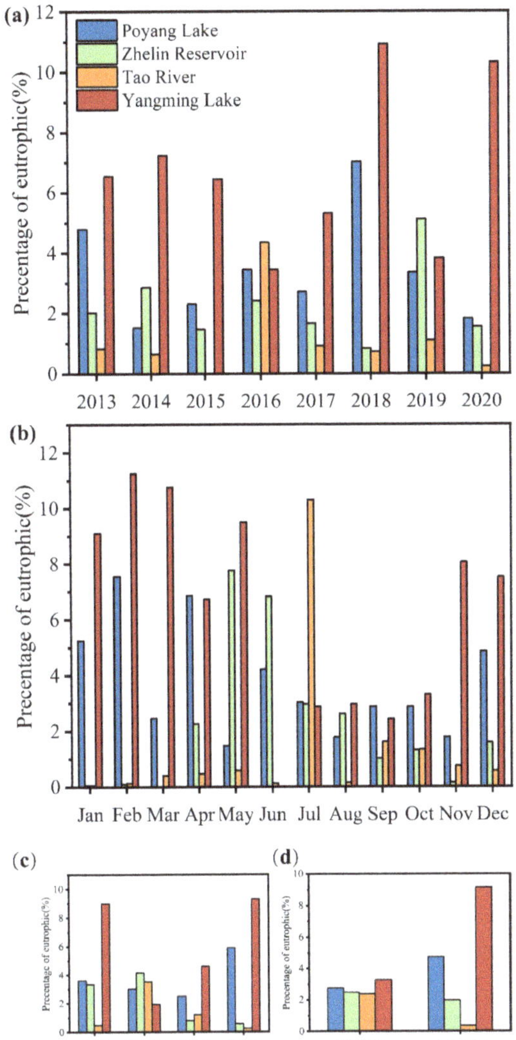

Figure 9. Spatial patterns of eutrophic water body proportion in four typical regions: (**a**) Annual distribution, (**b**) Monthly distribution, (**c**) Quarterly distribution, and (**d**) Pattern distribution between wet and dry seasons.

Figure 9c shows the seasonal distribution of eutrophication proportion in four typical regions of PLB. Poyang Lake and Yangming Lake were higher in winter and spring than in summer and autumn, with the maximum occurring in winter (5.89% and 9.29%, respectively). On the contrary, Zhelin Reservoir and Tao River showed higher performance in spring and summer than in autumn and winter, with the maximum occurring in summer (4.14% and 3.52%, respectively). According to the statistical results of the dry and the wet seasons (Figure 9d), the proportion of eutrophic water in Poyang Lake and Yangming Lake in the dry season (December to next May) was greater than in the wet season (June to November), while Zhelin Reservoir and Tao River showed the opposite trend.

3.4. Relationships between the TLI and Driving Forces

3.4.1. Meteorological Factors

It can be seen from Figure 10 that there was a significant negative correlation between temperature and the proportion of eutrophic water bodies in Poyang Lake and Yangming Lake. The reason for this phenomenon is that the short-term strong precipitation generated during the wet season when the temperature is higher replenishes a large amount of freshwater and accelerated the settling rate of suspended matter [27]. In addition, in Yangming Lake, the negative correlation between the percentage of eutrophic state water and wind speed indicate that lower wind speed is favorable for the accumulation and flotation of cyanobacteria on the water surface, and less disturbance to the water column, higher water clarity, and thus more light entering the water column, allowing higher productivity of algae [62]. Previous studies have shown that the high-level dilution and biodegradation ability of Poyang Lake were the leading reasons for the water bodies to maintain relatively low trophic levels during high water levels and high trophic levels during low water levels, respectively [27,63]. In Zhelin Reservoir and the Tao River, the temperature has a significant impact. Higher temperatures hasten cyanobacterial growth and algae blooms of water.

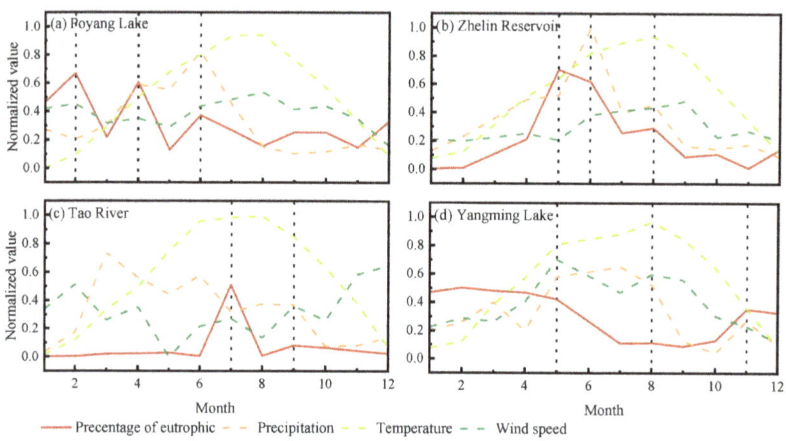

Figure 10. Monthly mean change of the proportion of eutrophic water bodies and three meteorological factors.

3.4.2. Socio-Economic Factors

The correlation degree was estimated employing data from the statistical yearbook, if the correlation degree of the index was low, the index was eliminated. The data of indicators were analyzed and the degree of correlation of each indicator was estimated using gray correlation degree analysis (Table A1). The gray correlation degrees of the total population, chemical fertilizer usage, pesticide use, crop planting area, effective irrigation area, and the proportion of eutrophic state water bodies were greater than 0.7 in the Poyang Lake, and the correlation degree of the remainder indicators does not meet the criteria. The gray correlation between the ten socioeconomic indicators and the proportion of eutrophic state water bodies in the Zhelin Reservoir region would be all greater than 0.7. Except for the gross domestic product of tertiary industry (GDP_3), the gray correlations in the Tao River and Yangming Lake were more than 0.7 between the nine indicators and the percentage of eutrophic state water bodies.

3.4.3. Land Use/Cover Factors

The correlation degree of correlation of the interannual change of the area of each classification of features and the eutrophic state of water body in PLB were calculated separately using the statistical area of each classification in PLB, and the change of area of classifications with lower correlation degrees were not considered. The GRA was used to calculate the degree of correlation of each index in the PLB (Table A2). Except for the built-up land, forest, and grassland, the gray correlation between the other classifications and the eutrophication status of the water body area of Poyang Lake were greater than 0.7. Except for the forest and grassland and bare land, the gray correlation between the classifications and the percentage of eutrophication status to water bodies of Zhelin Reservoir were less than 0.7. In the Tao River, except for bare land, the gray correlation between the classifications and the eutrophication status of the percentage of the water body area was greater than 0.7. Except for built-up land, the gray correlation between the classifications and the eutrophication status of Yangming Lake, a drinking water source, as a proportion of the water body area is greater than 0.7.

4. Discussion

4.1. Driving Forces

The entropy weight method was used to calculate the weights accounted for each factor. Our results showed that natural and anthropogenic factors simultaneously determine the spatiotemporal variation of eutrophication in the PLB (Figure 11). Lakes and reservoirs in the PLB are usually in oligotrophic or mesotrophic state, with average TLI values ranging from 16.86 to 31.26 [7]. Temperature is an important driver that promotes algae growth and plays a crucial role in algal bloom in most lakes of central and eastern China [58,64]. Many studies have confirmed that temperature has an important impact on water eutrophication [9,64]. However, Poyang Lake Basin is controlled by the East Asian monsoon, with hot and rainy summers and cold and dry winters. Precipitation sinks into the lake through surface runoff, resulting in greater depths and an increase in the lake's dilution and purification capacity. It also explains why the proportion of eutrophic water bodies was smaller in months with high temperatures than in months with low temperatures.

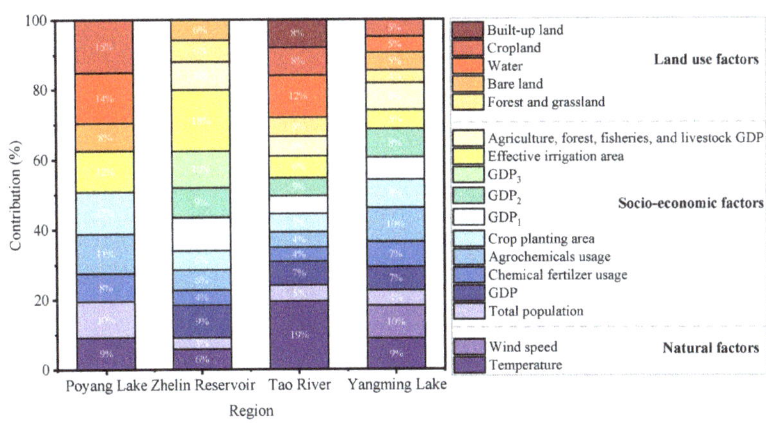

Figure 11. Contribution of drivers on spatial variation of water body eutrophication proportion in typical areas of Poyang Lake Basin using the entropy weight method.

Population growth plays a vital role in the process of water eutrophication. On the one hand, domestic wastewater generated by population growth enters rivers and lakes with surface runoff; On the other hand, population growth leads to urban sprawl, a large amount of forest and grassland were converted into construction land, and the function of soil purification decreases [2,58].

Socioeconomic activities also significantly affect the process of water eutrophication in the Poyang Lake Basin. Agricultural non-point source pollution (i.e., chemical fertilizers and pesticides) seriously affects the water quality of lakes and rivers, causing the lakes to evolve from oligotrophic to eutrophic, with massive growth of algae and severe damage to aquatic ecology [7]. With the rapid economic growth and urbanization process, a large amount of industrial wastewater and domestic sewage were discharged into rivers and lakes, resulting in the deterioration of water quality, which is more obvious in the eastern and southwestern lake regions of China [1,17]. The continuous increase of purse seine farming and mining activities in lakes directly affects water quality and aquatic ecosystems, thus increasing the eutrophication of lakes [27]. However, we need to admit that the impact of urbanization requires more detailed data analysis (e.g., accurate industrial wastewater discharge and domestic sewage discharge).

Cropland only accounts for a small part of the PLB, but it has a significant negative effect on water quality and productivity shifts. Many research results have shown that nitrogen and phosphorus nutrients artificially applied in cultivated soil easily enter rivers and lakes with surface runoff, resulting in an imbalance of N/P ratio in lakes, frequent occurrence of harmful algal blooms, and increased productivity of lakes [24].

Seasonal changes in the proportion of eutrophication in the PLB could be related to the East Asian monsoon climate, which is characterized by hot and rainy summers and cold and dry winters. Previous studies have also illustrated the essential contributions of temperature and precipitation to seasonal changes in water quality [15,22].

4.2. Comparing with the Existing Algorithms

Water bodies in different regions have unique optical properties, and the remote sensing assessment model of water body eutrophication developed for a particular study area might not be suitable to monitor water body eutrophication in other regions. What is the performance of the TLI model of eutrophication in PLB based on Landsat-8 OLI data constructed in this paper? We evaluate published models from Wen, et al. [10], Duan, et al. [65], and Hu, et al. [7].

Table 2 shows the results of comparing the model in this study to other parametrized models. For the TLI model proposed for this study the *MAD* is 3.58, and the *RMSD* is 4.43, both of which was lower than the *MADs* of the existing models, indicating that the model proposed for this study was more applicable to evaluation of water body eutrophication in the PLB. The model proposed in this paper for assessing the trophic level of waters in PLB using remote sensing data is more effective than the currently available models. The optical properties of water bodies in different waters might vary massively, and the dominant factors of water color might differ [64]. However, these models were successfully applied to monitor the eutrophication of inland lakes in China.

Table 2. Comparison of performance between the existing and proposed TLI-evaluated models based on in situ measurement data.

Model	Equation Form	MAD	RMSD	MAPD
Wen, et al. [10]	$TSI_M(Chla) = 48.677\left(\frac{Rrs(561)}{Rrs(655)}\right)^{-0.786}$	4.33	5.31	9.39
	$TSI_M(Chla) = 49.479\left(\frac{\text{OLI3}}{\text{OLI4}}\right)^{-0.603}$	3.74	4.69	7.63
Duan, et al. [65]	$Chla = 5.5973 \times \frac{\text{OLI5}}{\text{OLI4}} + 6.7069$ $TSI_M(Chla) = 10 \times \left(2.46 + \frac{\ln Chla}{\ln 2.5}\right)$	11.25	23.47	22.92
Hu, et al. [7]	$ABI = (\text{OLI4} - \text{OLI2}) \times \frac{\lambda_{\text{Green}} - \lambda_{\text{Blue}}}{\lambda_{\text{Red}} - \lambda_{\text{Blue}}} - (\text{OLI5} - \text{OLI2}) \times \frac{\lambda_{Green} - \lambda_{Blue}}{\lambda_{NIR} - \lambda_{Blue}}$ $TSI = 3.7181 \times ABI + 65.179$	4.49	5.33	9.21
This study	$TLI = -85.612 \times (OLI1/(OLI3 + OLI5)) + 75.511$	3.58	4.43	8.88

Note: OLI1, OLI3, OLI4, and OLI5 are Landsat-8 near-coastal band, blue band, green band, red band, and near-infrared band, respectively. λ_{Green}, λ_{Blue}, λ_{Red}, λ_{NIR} are the corresponding band wavelengths of Landsat-8.

4.3. Limitations of Satellite Monitoring of Water Eutrophication

Satellite monitoring has significantly improved the spatial coverage and time period of eutrophication monitoring in watersheds and can be used as an important complement to field measurements and laboratory analysis, especially for rivers and lakes that lack sampling data. However, this study has some limitations: (i) the TLI model proposed for this study is likely to cause underestimation in high-value areas, especially in areas with TLI greater than 65. Although there are limitations in the generation of long-term large-scale water body eutrophication data sets, TLI estimation models applicable to a wider range of water body types can be developed in the future. (ii) Only uses Landsat-8 data to study water eutrophication in large lakes and reservoirs. In the future, Sentinel-2 MSI data, GaoFen (GF) data, and Huanjing (HJ) data can be considered, and monitoring of small lakes and reservoirs is encouraged. (iii) In this study, only 17 potential explanatory driving factors were used to analyze the temporal and spatial changes of eutrophication in the Poyang Lake Basin. However, more accurate data should be used for analysis (e.g., industrial wastewater and domestic wastewater discharge).

4.4. Implications for the Water Environment Management

We classify the main rivers and lakes in the Poyang Lake Basin into four categories and suggest water quality management plans based on the study results by analyzing the driving forces of eutrophication characteristics of long-term TLI changes. (i) Drinking water source protection regions: All levels of government should effectively supervise and restrict the construction of additional polluting industrial estates in water source protection zones. Enhancing water quality monitoring activities and monitoring water quality in drinking water sources in real-time. (ii) Mineral exploitation regions: To manage the issue of sand mining activity in Poyang Lake, a detailed plan of sand mining in Poyang Lake should be constructed in order to define the mineable area, mineable period, forbidden area, and forbidden period. To solve the problem of the illegal sneak discharge in rare earth mining areas, environmental administrators should strengthen the investigation and supervision of river outfalls, along with punishments. (iii) Agricultural production concentration regions: The structure of the watershed's plantation industry should be modified and improved as soon as feasible to promote the development of ecological green agriculture. (iv) Concentrated human activity regions: Relevant government departments should improve the rate of urban and rural domestic sewage treatment.

5. Conclusions

The spatiotemporal distribution and driving factors of eutrophication in the Poyang Lake Basin were investigated using Landsat-8 OLI images and TLI model. Under the effect of natural and anthropogenic activities factors, the eutrophication of Poyang Lake Basin presented a spatial distribution pattern of high in the north and low in the south. In contrast, the process of water eutrophication showed a trend of declining in the north and rising in the south. The proportion of eutrophication in the Poyang Lake Basin had a seasonal pattern of low in summer and high in winter. The climate of the East Asian monsoon region is characterized by high temperature and rain in summer, which may affect the seasonal variation of water eutrophication.

The eutrophication of the water body in the Poyang Lake Basin was most affected by human activities, with an average contribution of more than 60%. However, the proportion of natural factors and human factors in different regions was not balanced. Human activities contributed the most to the eutrophication of Zhelin Reservoir, reaching 82.1%, and natural factors contributed more to the eutrophication of Tao River than other regions.

To improve the water quality of lakes, we recommend that the government and management departments formulate policies and manage protection measures, strengthen the supervision of activities such as sand mining and mining, and improve the purification capacity of agricultural and urban wastewater.

Author Contributions: Conceptualization, Z.Z. and J.L.; methodology, J.L., Z.Z. and G.L.; software, J.L. and S.L.; validation, J.L., S.L. and R.Z.; formal analysis, J.L., S.L., J.X. and Y.L.; investigation, Z.Z., G.L. and J.L.; resources, Z.Z., G.L. and J.L.; data curation, Z.Z., G.L. and J.L.; writing—original draft preparation, J.L. and Z.Z.; writing—review and editing, Z.Z., N.C., G.L., C.D. and R.Z.; visualization, J.L., N.C. and R.Z.; supervision, G.L., J.L., Z.Z. and C.H.; project administration, Z.Z.; funding acquisition, Z.Z., G.L., C.D., S.L. and Y.L. All authors have read and agreed to the published version of the manuscript.

Funding: This research was funded by the National Natural Science Foundation of China (No. 41701412; 42171385; 41907331; 42101384 & 42071333), the Social Science Foundation of Jiangxi Province (No. 22GL27), the Humanities and Social Sciences Project of Jiangxi Provincial University (No. GL21143), the Science and Technology Project of Jiangxi Educational Committee (No. GJJ2201216), the National Key Research and Development Program of China (No. 2021YFB3901101).

Data Availability Statement: The data presented in this study are available upon request from the corresponding author.

Acknowledgments: We thank the U.S. Geological Survey for providing Landsat-8 satellite data, and China Meteorological Administration for providing meteorological data. The authors would also like to thank four anonymous reviewers for their valuable comments and suggestions for improving the quality of the manuscript.

Conflicts of Interest: The authors declare no conflict of interest.

Appendix A

Table A1. Grey correlation of socio-economic factors.

	The Total Population	GDP	Chemical Fertilizer Usage	Agrochemicals Usage	Crop Planting Area	GDP_1	GDP_2	GDP_3	Effective Irrigation Area	Agriculture, Forestry, Fisheries, and Livestock GDP
Poyang Lake	0.78	0.65	0.86	0.85	0.79	0.69	0.67	0.56	0.82	0.64
Zhelin Reservoir	0.80	0.85	0.81	0.75	0.81	0.84	0.84	0.71	0.81	0.85
Tao River	0.81	0.72	0.81	0.79	0.81	0.79	0.81	0.66	0.81	0.79
Yangming Lake	0.78	0.76	0.76	0.72	0.78	0.78	0.76	0.67	0.78	0.78

Table A2. Grey correlation of land use/cover factors.

	Built-Up Land	Forest and Grassland	Water	Bare Land	Cropland
Poyang Lake	0.59	0.69	0.76	0.89	0.72
Zhelin Reservoir	0.55	0.78	0.62	0.70	0.60
Tao River	0.76	0.83	0.75	0.57	0.78
Yangming Lake	0.66	0.82	0.80	0.72	0.74

References

1. Liu, S.; Hou, J.; Suo, C.; Chen, J.; Liu, X.; Fu, R.; Wu, F. Molecular-level composition of dissolved organic matter in distinct trophic states in Chinese lakes: Implications for eutrophic lake management and the global carbon cycle. *Water Res.* **2022**, *217*, 118438. [CrossRef] [PubMed]
2. Li, S.; Xu, S.; Song, K.; Kutser, T.; Wen, Z.; Liu, G.; Shang, Y.; Lyu, L.; Tao, H.; Wang, X.; et al. Remote Quantification of the Trophic Status of Chinese Lakes. *Hydrol. Earth Syst. Sci. Discuss.* **2022**, *2022*. [CrossRef]
3. Paerl, H.W.; Huisman, J. Blooms like it hot. *Science* **2008**, *320*, 57–58. [CrossRef] [PubMed]
4. Gilarranz, L.J.; Narwani, A.; Odermatt, D.; Siber, R.; Dakos, V. Regime shifts, trends, and variability of lake productivity at a global scale. *Proc. Natl. Acad. Sci. USA* **2022**, *119*, e2116413119. [CrossRef]
5. Conley, D.J.; Paerl, H.W.; Howarth, R.W.; Boesch, D.F.; Seitzinger, S.P.; Havens, K.E.; Lancelot, C.; Likens, G.E. Controlling eutrophication: Nitrogen and phosphorus. *Science* **2009**, *323*, 1014–1015. [CrossRef]
6. Schindler, D.W.; Hecky, R.E.; Findlay, D.L.; Stainton, M.P.; Parker, B.R.; Paterson, M.J.; Beaty, K.G.; Lyng, M.; Kasian, S.E.M. Eutrophication of lakes cannot be controlled by reducing nitrogen input: Results of a 37-year whole-ecosystem experiment. *Proc. Natl. Acad. Sci. USA* **2008**, *105*, 11254–11258. [CrossRef]

7. Hu, M.; Ma, R.; Xiong, J.; Wang, M.; Cao, Z.; Xue, K. Eutrophication state in the Eastern China based on Landsat 35-year observations. *Remote Sens. Environ.* **2022**, *277*, 113057. [CrossRef]
8. Guan, Q.; Feng, L.; Hou, X.; Schurgers, G.; Zheng, Y.; Tang, J. Eutrophication changes in fifty large lakes on the Yangtze Plain of China derived from MERIS and OLCI observations. *Remote Sens. Environ.* **2020**, *246*, 111890. [CrossRef]
9. Shi, K.; Zhang, Y.; Song, K.; Liu, M.; Zhou, Y.; Zhang, Y.; Li, Y.; Zhu, G.; Qin, B. A semi-analytical approach for remote sensing of trophic state in inland waters: Bio-optical mechanism and application. *Remote Sens. Environ.* **2019**, *232*, 111349. [CrossRef]
10. Wen, Z.; Song, K.; Liu, G.; Shang, Y.; Fang, C.; Du, J.; Lyu, L. Quantifying the trophic status of lakes using total light absorption of optically active components. *Environ. Pollut.* **2019**, *245*, 684–693. [CrossRef]
11. Carlson, R.E. A trophic state index for lakes 1. *Limnol. Oceanogr.* **1977**, *22*, 361–369. [CrossRef]
12. Aizaki, M.; Otsuki, A.; Fukushima, T.; Hosomi, M.; Muraoka, K. Application of Carlson's trophic state index to Japanese lakes and relationships between the index and other parameters. *SIL Proc. 1922–2010* **1981**, *21*, 675–681. [CrossRef]
13. Jin, X.; Tu, Q. *Specification of Lake Eutrophication Survey (in Chinese)*; Environmental Science Press: Beijing, China, 1990.
14. Lin, S.-S.; Shen, S.-L.; Zhou, A.; Lyu, H.-M. Assessment and management of lake eutrophication: A case study in Lake Erhai, China. *Sci. Total Environ.* **2021**, *751*, 141618. [CrossRef]
15. Ren, J.; Zheng, Z.; Li, Y.; Lv, G.; Wang, Q.; Lyu, H.; Huang, C.; Liu, G.; Du, C.; Mu, M.; et al. Remote observation of water clarity patterns in Three Gorges Reservoir and Dongting Lake of China and their probable linkage to the Three Gorges Dam based on Landsat 8 imagery. *Sci. Total Environ.* **2018**, *625*, 1554–1566. [CrossRef]
16. Zheng, Z.; Li, Y.; Guo, Y.; Xu, Y.; Liu, G.; Du, C. Landsat-based long-term monitoring of total suspended matter concentration pattern change in the wet season for Dongting Lake, China. *Remote Sens.* **2015**, *7*, 13975–13999. [CrossRef]
17. Shen, M.; Duan, H.; Cao, Z.; Xue, K.; Qi, T.; Ma, J.; Liu, D.; Song, K.; Huang, C.; Song, X. Sentinel-3 OLCI observations of water clarity in large lakes in eastern China: Implications for SDG 6.3.2 evaluation. *Remote Sens. Environ.* **2020**, *247*, 111950. [CrossRef]
18. Liu, H.; He, B.; Zhou, Y.; Kutser, T.; Toming, K.; Feng, Q.; Yang, X.; Fu, C.; Yang, F.; Li, W.; et al. Trophic state assessment of optically diverse lakes using Sentinel-3-derived trophic level index. *Int. J. Appl. Earth Obs. Geoinf.* **2022**, *114*, 103026. [CrossRef]
19. Zeng, S.; Lei, S.; Li, Y.; Lyu, H.; Dong, X.; Li, J.; Cai, X. Remote monitoring of total dissolved phosphorus in eutrophic Lake Taihu based on a novel algorithm: Implications for contributing factors and lake management. *Environ. Pollut.* **2022**, *296*, 118740. [CrossRef]
20. Wang, S.; Li, J.; Zhang, B.; Spyrakos, E.; Tyler, A.N.; Shen, Q.; Zhang, F.; Kuster, T.; Lehmann, M.K.; Wu, Y.; et al. Trophic state assessment of global inland waters using a MODIS-derived Forel-Ule index. *Remote Sens. Environ.* **2018**, *217*, 444–460. [CrossRef]
21. Pozdnyakov, D.V.; Korosov, A.A.; Petrova, N.A.; Grassl, H. Multi-year satellite observations of Lake Ladoga's biogeochemical dynamics in relation to the lake's trophic status. *J. Great Lakes Res.* **2013**, *39*, 34–45. [CrossRef]
22. Zheng, Z.; Ren, J.; Li, Y.; Huang, C.; Liu, G.; Du, C.; Lyu, H. Remote sensing of diffuse attenuation coefficient patterns from Landsat 8 OLI imagery of turbid inland waters: A case study of Dongting Lake. *Sci. Total Environ.* **2016**, *573*, 39–54. [CrossRef] [PubMed]
23. Cao, Z.; Ma, R.; Liu, M.; Duan, H.; Xiao, Q.; Xue, K.; Shen, M. Harmonized Chlorophyll-a Retrievals in Inland Lakes From Landsat-8/9 and Sentinel 2A/B Virtual Constellation Through Machine Learning. *IEEE Trans. Geosci. Remote Sens.* **2022**, *60*, 1–16. [CrossRef]
24. Li, Y.; Shi, K.; Zhang, Y.; Zhu, G.; Zhang, Y.; Wu, Z.; Liu, M.; Guo, Y.; Li, N. Analysis of water clarity decrease in Xin'anjiang Reservoir, China, from 30-Year Landsat TM, ETM+, and OLI observations. *J. Hydrol.* **2020**, *590*, 125476. [CrossRef]
25. Lee, Z.; Shang, S.; Qi, L.; Yan, J.; Lin, G. A semi-analytical scheme to estimate Secchi-disk depth from Landsat-8 measurements. *Remote Sens. Environ.* **2016**, *177*, 101–106. [CrossRef]
26. Watanabe, F.S.; Alcântara, E.; Rodrigues, T.W.; Imai, N.N.; Barbosa, C.C.; Rotta, L.H. Estimation of Chlorophyll-a Concentration and the Trophic State of the Barra Bonita Hydroelectric Reservoir Using OLI/Landsat-8 Images. *Int. J. Environ. Res. Public Health* **2015**, *12*, 10391–10417. [CrossRef] [PubMed]
27. Li, B.; Yang, G.; Wan, R. Multidecadal water quality deterioration in the largest freshwater lake in China (Poyang Lake): Implications on eutrophication management. *Environ. Pollut.* **2020**, *260*, 114033. [CrossRef]
28. Li, J.; Li, Y.; Bi, S.; Xu, J.; Guo, F.; Lyu, H.; Dong, X.; Cai, X. Utilization of GOCI data to evaluate the diurnal vertical migration of Microcystis aeruginosa and the underlying driving factors. *J. Environ. Manag.* **2022**, *310*, 114734. [CrossRef]
29. Morel, A.; Mueller, J.L. Normalized Water-Leaving Radiance and Remote Sensing Reflectance: Bidirectional Reflectance and. In *Ocean Optics Protocols for Satellite Ocean Color Sensor Validation, Revision 3*; National Aeronautics and Space Administration, Goddard Space Flight Center: Greenbelt, MD, USA, 2002; Volume 210004, p. 183.
30. Xing, Q.; Hu, C. Mapping macroalgal blooms in the Yellow Sea and East China Sea using HJ-1 and Landsat data: Application of a virtual baseline reflectance height technique. *Remote Sens. Environ.* **2016**, *178*, 113–126. [CrossRef]
31. Vermote, E.F.; Tanré, D.; Deuze, J.L.; Herman, M.; Morcette, J.J. Second simulation of the satellite signal in the solar spectrum, 6S: An overview. *IEEE Trans. Geosci. Remote Sens.* **1997**, *35*, 675–686. [CrossRef]
32. Gordon, H.R.; Wang, M. Retrieval of water-leaving radiance and aerosol optical thickness over the oceans with SeaWiFS: A preliminary algorithm. *Appl. Opt.* **1994**, *33*, 443–452. [CrossRef]
33. Bernstein, L.S.; Jin, X.; Gregor, B.; Adler-Golden, S.M. Quick atmospheric correction code: Algorithm description and recent upgrades. *Opt. Eng.* **2012**, *51*, 111719. [CrossRef]

34. National Meteorological Information Center. *Daily Meteorological Dataset of Basic Meteorological Elements of China National Surface Weather Station (V3.0) (1951–2020)*; National Meteorological Information Center: Beijing, China, 2021.
35. Xu, Y.; Li, Y.; Wang, Q.; Lv, H.; Liu, Z.; Xu, X.; Tan, J.; Guo, Y.; Wu, C. Eutrophication evaluation of three lakes and one reservoir using CCD images of HJ-1 satellite (in Chinese). *Acta Sci. Circumstantiae* **2011**, *31*, 81–93. [CrossRef]
36. Gupana, R.S.; Odermatt, D.; Cesana, I.; Giardino, C.; Nedbal, L.; Damm, A. Remote sensing of sun-induced chlorophyll-a fluorescence in inland and coastal waters: Current state and future prospects. *Remote Sens. Environ.* **2021**, *262*, 112482. [CrossRef]
37. Guo, Y.; Huang, C.; Li, Y.; Du, C.; Shi, L.; Li, Y.; Chen, W.; Wei, H.; Cai, E.; Ji, G. Hyperspectral reconstruction method for optically complex inland waters based on bio-optical model and sparse representing. *Remote Sens. Environ.* **2022**, *276*, 113045. [CrossRef]
38. Bi, S.; Li, Y.; Liu, G.; Song, K.; Xu, J.; Dong, X.; Cai, X.; Mu, M.; Miao, S.; Lyu, H. Assessment of algorithms for estimating chlorophyll-a concentration in inland waters: A round-robin scoring method based on the optically fuzzy clustering. *IEEE Trans. Geosci. Remote Sens.* **2021**, *60*, 4200717. [CrossRef]
39. Carpenter, D.J.; Carpenter, S.M. Modeling inland water quality using Landsat data. *Remote Sens. Environ.* **1983**, *13*, 345–352. [CrossRef]
40. Gitelson, A.; Szilagyi, F.; Mittenzwey, K.H. Improving quantitative remote sensing for monitoring of inland water quality. *Water Res.* **1993**, *27*, 1185–1194. [CrossRef]
41. Gons, H.J.; Auer, M.T.; Effler, S.W. MERIS satellite chlorophyll mapping of oligotrophic and eutrophic waters in the Laurentian Great Lakes. *Remote Sens. Environ.* **2008**, *112*, 4098–4106. [CrossRef]
42. Le, C.; Li, Y.; Zha, Y.; Sun, D.; Huang, C.; Lu, H. A four-band semi-analytical model for estimating chlorophyll a in highly turbid lakes: The case of Taihu Lake, China. *Remote Sens. Environ.* **2009**, *113*, 1175–1182. [CrossRef]
43. Warren, M.A.; Simis, S.G.H.; Selmes, N. Complementary water quality observations from high and medium resolution Sentinel sensors by aligning chlorophyll-a and turbidity algorithms. *Remote Sens. Environ.* **2021**, *265*, 112651. [CrossRef]
44. Zhang, R.; Zheng, Z.; Liu, G.; Du, C.; Du, C.; Lei, S.; Xu, Y.; Xu, J.; Mu, M.; Bi, S.; et al. Simulation and assessment of the capabilities of Orbita Hyperspectral (OHS) imagery for remotely monitoring chlorophyll-a in eutrophic plateau lakes. *Remote Sens.* **2021**, *13*, 2821. [CrossRef]
45. Mooij, W.M.; Hülsmann, S.; De Senerpont Domis, L.N.; Nolet, B.A.; Bodelier, P.L.E.; Boers, P.; Pires, L.; Gons, H.J.; Ibelings, B.W.; Noordhuis, R. The impact of climate change on lakes in the Netherlands: A review. *Aquat. Ecol.* **2005**, *39*, 381–400. [CrossRef]
46. Zhang, M.; Duan, H.; Shi, X.; Yu, Y.; Kong, F. Contributions of meteorology to the phenology of cyanobacterial blooms: Implications for future climate change. *Water Res.* **2012**, *46*, 442–452. [CrossRef]
47. Whitehead, P.G.; Barbour, E.; Futter, M.N.; Sarkar, S.; Rodda, H.; Caesar, J.; Butterfield, D.; Jin, L.; Sinha, R.; Nicholls, R.; et al. Impacts of climate change and socio-economic scenarios on flow and water quality of the Ganges, Brahmaputra and Meghna (GBM) river systems: Low flow and flood statistics. *Environ. Sci. Process. Impacts* **2015**, *17*, 1057–1069. [CrossRef] [PubMed]
48. Berkhout, F.; Hertin, J.; Jordan, A. Socio-economic futures in climate change impact assessment: Using scenarios as 'learning machines'. *Glob. Environ. Change* **2002**, *12*, 83–95. [CrossRef]
49. Islam, M.M.M.; Iqbal, M.S.; Leemans, R.; Hofstra, N. Modelling the impact of future socio-economic and climate change scenarios on river microbial water quality. *Int. J. Hyg. Environ. Health* **2018**, *221*, 283–292. [CrossRef]
50. Deng, J. Control problems of grey systems. *Syst. Control Lett.* **1982**, *1*, 288–294. [CrossRef]
51. Huang, Y.; Shen, L.; Liu, H. Grey relational analysis, principal component analysis and forecasting of carbon emissions based on long short-term memory in China. *J. Clean. Prod.* **2019**, *209*, 415–423. [CrossRef]
52. Malek, A.; Ebrahimnejad, S.; Tavakkoli-Moghaddam, R. An improved hybrid grey relational analysis approach for green resilient supply chain network assessment. *Sustainability* **2017**, *9*, 1433. [CrossRef]
53. Mu, M.; Li, Y.; Bi, S.; Lyu, H.; Xu, J.; Lei, S.; Miao, S.; Zeng, S.; Zheng, Z.; Du, C. Prediction of algal bloom occurrence based on the naive Bayesian model considering satellite image pixel differences. *Ecol. Indic.* **2021**, *124*, 107416. [CrossRef]
54. Zhe, W.; Xigang, X.; Feng, Y. An abnormal phenomenon in entropy weight method in the dynamic evaluation of water quality index. *Ecol. Indic.* **2021**, *131*, 108137. [CrossRef]
55. Li, Y.; Zhang, Q.; Wang, L.; Liang, L. Regional environmental efficiency in China: An empirical analysis based on entropy weight method and non-parametric models. *J. Clean. Prod.* **2020**, *276*, 124147. [CrossRef]
56. Neil, C.; Spyrakos, E.; Hunter, P.D.; Tyler, A.N. A global approach for chlorophyll-a retrieval across optically complex inland waters based on optical water types. *Remote Sens. Environ.* **2019**, *229*, 159–178. [CrossRef]
57. Bonansea, M.; Rodriguez, M.C.; Pinotti, L.; Ferrero, S. Using multi-temporal Landsat imagery and linear mixed models for assessing water quality parameters in Río Tercero reservoir (Argentina). *Remote Sens. Environ.* **2015**, *158*, 28–41. [CrossRef]
58. Feng, L.; Hou, X.; Zheng, Y. Monitoring and understanding the water transparency changes of fifty large lakes on the Yangtze Plain based on long-term MODIS observations. *Remote Sens. Environ.* **2019**, *221*, 675–686. [CrossRef]
59. Zhou, Q.; Wang, J.; Tian, L.; Feng, L.; Li, J.; Xing, Q. Remotely sensed water turbidity dynamics and its potential driving factors in Wuhan, an urbanizing city of China. *J. Hydrol.* **2021**, *593*, 125893. [CrossRef]
60. Li, Y.; Shi, K.; Zhang, Y.; Zhu, G.; Qin, B.; Zhang, Y.; Liu, M.; Zhu, M.; Dong, B.; Guo, Y. Remote sensing of column-integrated chlorophyll a in a large deep-water reservoir. *J. Hydrol.* **2022**, *610*, 127918. [CrossRef]
61. Akoglu, H. User's guide to correlation coefficients. *Turk. J. Emerg. Med.* **2018**, *18*, 91–93. [CrossRef]
62. Shi, K.; Zhang, Y.; Liu, X.; Wang, M.; Qin, B. Remote sensing of diffuse attenuation coefficient of photosynthetically active radiation in Lake Taihu using MERIS data. *Remote Sens. Environ.* **2014**, *140*, 365–377. [CrossRef]

63. Li, B.; Yang, G.; Wan, R.; Hörmann, G. Dynamic water quality evaluation based on fuzzy matter—Element model and functional data analysis, a case study in Poyang Lake. *Environ. Sci. Pollut. Res.* **2017**, *24*, 19138–19148. [CrossRef]
64. Shi, K.; Zhang, Y.; Qin, B.; Zhou, B. Remote sensing of cyanobacterial blooms in inland waters: Present knowledge and future challenges. *Sci. Bull.* **2019**, *64*, 1540–1556. [CrossRef] [PubMed]
65. Duan, H.; Zhang, Y.; Zhang, B.; Song, K.; Wang, Z. Assessment of Chlorophyll-a Concentration and Trophic State for Lake Chagan Using Landsat TM and Field Spectral Data. *Environ. Monit. Assess.* **2007**, *129*, 295–308. [CrossRef] [PubMed]

remote sensing

MDPI

Article

Airborne Drones for Water Quality Mapping in Inland, Transitional and Coastal Waters—MapEO Water Data Processing and Validation

Liesbeth De Keukelaere [1,*], Robrecht Moelans [1], Els Knaeps [1], Sindy Sterckx [1], Ils Reusen [1], Dominique De Munck [1], Stefan G.H. Simis [2], Adriana Maria Constantinescu [3], Albert Scrieciu [3], Georgios Katsouras [4], Wim Mertens [5], Peter D. Hunter [6], Evangelos Spyrakos [6] and Andrew Tyler [6]

1 Vlaamse Instelling Voor Technologisch Onderzoek (VITO), Boeretang 200, 2400 Mol, Belgium
2 Plymouth Marine Laboratory (PML), Plymouth PL1 3DH, UK
3 Institutul National De Cercetare—Dezvoltare Pentru Geologie Si Geoecologie Marina (GeoEcoMar), 23–25 Dimitrie Onciul St., RO-024053 Bucharest, Romania
4 Athens Water Suppy and Sewerage Company (EYDAP S.a.), Oropou 156, 11146 Athens, Greece
5 Instituut Natuur-En Bosonderzoek (INBO) Havenlaan 88 Bus 73, 1000 Brussel, Belgium
6 Earth and Planetary Observation Sciences (EPOS), Biological and Environmental Sciences, Faculty of Natural Sciences, University of Stirling, Stirling FK9 4LA, UK
* Correspondence: liesbeth.dekeukelaere@vito.be

Abstract: Using airborne drones to monitor water quality in inland, transitional or coastal surface waters is an emerging research field. Airborne drones can fly under clouds at preferred times, capturing data at cm resolution, filling a significant gap between existing in situ, airborne and satellite remote sensing capabilities. Suitable drones and lightweight cameras are readily available on the market, whereas deriving water quality products from the captured image is not straightforward; vignetting effects, georeferencing, the dynamic nature and high light absorption efficiency of water, sun glint and sky glint effects require careful data processing. This paper presents the data processing workflow behind MapEO water, an end-to-end cloud-based solution that deals with the complexities of observing water surfaces and retrieves water-leaving reflectance and water quality products like turbidity and chlorophyll-a (Chl-a) concentration. MapEO water supports common camera types and performs a geometric and radiometric correction and subsequent conversion to reflectance and water quality products. This study shows validation results of water-leaving reflectance, turbidity and Chl-a maps derived using DJI Phantom 4 pro and MicaSense cameras for several lakes across Europe. Coefficients of determination values of 0.71 and 0.93 are obtained for turbidity and Chl-a, respectively. We conclude that airborne drone data has major potential to be embedded in operational monitoring programmes and can form useful links between satellite and in situ observations.

Keywords: airborne drone; UAV; optical water quality; automated drone image processing; MapEO water; inland and coastal waters; georeferencing; sky glint; iCOR

1. Introduction

Unmanned aerial vehicles (UAVs), more commonly referred to as drones, carrying optical sensors, are already embedded in various land mapping and monitoring applications. Drones are easy-to-use, flexible in deployment and can be flown at low altitudes, even in the presence of clouds. Some airborne drones have been developed to collect in situ water samples which can be further analysed in the lab, e.g., [1–3]. In contrast to the water-sampling drones, drones with camera systems have the advantage of providing near-real-time information at a high spatial resolution, making it possible to observe small-scale variations in optically active water constituents and in nearshore and shoreline zones where satellites suffer from mixed pixels and adjacency effects [4]. Two major types of drone platforms exist: (i) fixed-wing drones typically have long endurance but are relatively expensive and need a runaway for take-off; (ii) multi-copter platforms are relatively cheap and only need a vertical take-off and landing spot. Vertical Take-Off and Landing (VTOL) systems are between fixed-wing drones with vertical take-off and landing capabilities, eliminating the need for a runway. For most studies in mapping surface water resources using drones (77%), the multi-copter system is preferred because of the lower cost and small take-off location [5]. Various cameras can be mounted under drones, ranging from broadband Red–Green–Blue (RGB) images to multi- and hyperspectral sensors. The type of application defines the preferred camera system, with RGB the cheapest but often lacking sufficient spectral information for water-quality applications.

Since 2013, interest in using drone data for optical water quality retrieval is steadily increasing [5], with several publications focussing on turbidity, chlorophyll-a (Chl-a) concentrations and macrophytes, e.g., [6–9], in lakes, ponds, rivers and reservoirs. Most of these studies include generic processing software and tools to convert the collected raw digital number data into reflectance values. Addressing drones to monitor optical water properties is a bit more challenging compared to land-based monitoring. In contrast to land applications, drone imagery collected over water bodies rarely includes non-moving stationary points. The presence of stationary points allows the use of standard (stereo) photogrammetry methods like Structure-from-Motion (SfM) [10]. Through triangulation based on subsequent images with sufficient overlap, the 3D coordinates of points are determined. The accuracy of the position and motion sensor is not crucial for the quality of the photogrammetry output. However, in the absence of stationary points, this technique is not applicable.

Another difference compared to land is that water acts as a mirror. The signal detected by the sensor is thus influenced by surface specular reflection from the sun and/or sky glint. The authors of [11] investigated four different methods to correct for surface reflected light, including Hydrolight simulations [12], the dark pixel assumption (NIR = 0), a baseline function or a "deglinting" approach. More recently, drone-based atmospheric correction tools are becoming available, like DROACOR [13]. This fully automatic physics-based processor uses pre-calculated look-up tables generated with libRadtran radiative transfer code to correct for atmospheric conditions.

While drone observations are very useful in local monitoring studies, they also have the potential to be used in satellite validation efforts. As pointed out by [14], the validation of satellite products requires high-quality in situ measurements, referred to as Fiducial Reference Measurements (FRM). Before a measurement can be labelled as FRM, it should (i) be accompanied by an uncertainty budget for all instruments, derived measurement and validation methods, (ii) adhere to openly available measurement protocols and community-wide management practices, (iii) have documented evidence of International System of Units (SI) traceability and (iv) be independent of the satellite retrieval process [15].

In this study, we present the workflow behind MapEO water: a sensor-independent cloud-based drone solution that provides georeferenced water-leaving reflectance and water quality parameters. MapEO water uses direct georeferencing [16], a technique that works in the absence of stationary points, and iCOR4Drones. This module corrects for glint effects under different cloud conditions using MODTRAN5 [17] based look-up tables (LUT). A validation of MapEO water is performed for DJI Phantom 4pro (DJI PH4) RGB and MicaSense multispectral data in various field campaigns across Europe. Simultaneous with the drone flights, in situ reflectance and water quality parameters were collected for validation. In addition, a comparison between satellite, drone and in situ data is conducted for the North Sea in Belgium.

2. Materials and Methods

2.1. MapEO Water Workflow

The VITO MapEO water (https://remotesensing.vito.be/case/mapeo-water, accessed on 1 January 2023) workflow allows the operational processing of drone data in a cloud environment. Once the drone pilot uploads the data into the cloud environment [18], the processing workflow can be initiated. This workflow comprises five steps: (i) georeferencing, which allows the projection of drone imagery; (ii) radiometric correction to convert raw digital numbers into radiance; (iii) conversion from radiance to water-leaving reflectance; (iv) conversion of water-leaving reflectance into optical water quality products like turbidity, suspended sediments or Chl-a concentrations. Each step is detailed below.

2.1.1. Georeferencing

Georeferencing is performed using a technique called direct georeferencing [16], i.e., making direct use of the information provided by the position and motion sensor mounted on the drone platform for georeferencing. The surface to project on is assumed to be flat, and its elevation only depends on the user-specified height difference between the water level and drone take-off location. The obtained accuracy in pixel position is mainly determined by the quality of the auxiliary sensors. The more accurate they are, the better the projection will be.

2.1.2. Radiometric Correction

The radiometric correction converts the raw digital numbers for each pixel of an image and each spectral band into absolute spectral radiance values with units of W m^{-2} sr^{-1} nm^{-1}. The correction combines sensor black level correction, the sensitivity of the sensor, sensor gain and exposure settings and lens vignette effects. The formula for radiometric correction of drone data, following [19]:

$$L_{camera} = V(x,y) \times \frac{a_1}{g} \times \frac{p - p_{BL}}{t_e + a_2 y - a_3 t_e y} \tag{1}$$

with

L_{camera} the spectral radiance measured by the camera and expressed in W/m^2/sr/nm;
$V(x,y)$ vignette model to correct for the fall-off in light sensitivity that occurs in pixels further from the centre of the image;
a_1, a_2, a_3 are radiometric calibration coefficients;
g the sensor gain setting;
p the normalised raw pixel value, i.e., the raw digital number of the pixel divided by the number of bits in the image (4096 for a 12-bit and 65536 for a 16-bit image);
p_{BL} the normalised black level value;
t_e the image exposure time;
x,y the pixel column and row number.

Some camera systems like MicaSense store all required information in the metadata of the camera-generated image files. Other camera systems, like the DJI PH4pro, lack radiometric calibration coefficients, simplifying Equation (1) to

$$L_{camera} = V_{emp}(x,y) \times \frac{N^2}{g} \times \frac{p}{t_e} \tag{2}$$

with

L_{camera} the pseudo radiance received by the sensor;
$V_{emp}(x,y)$ an empirically defined vignetting model, by normalising all pixels of a homogenous surface and storing the defined ratios which lower towards the edges of the image;
N is the camera f-number, i.e., the ratio of the camera's focal length to the aperture.

2.1.3. Conversion from Radiance to Water-Leaving Reflectance

To derive the water-leaving reflectance (ρ_w), irradiance (E_d) and sky glint (d_1) should be corrected. Water-leaving reflectance is the ratio of water-leaving radiance and water surface downward irradiance and reflects information on the substances in the upper layer of the water body [20]. The sensor captures not only water-leaving radiance but also specular reflection of the sun and sky at the water surface, referred to as sun glint and sky glint, respectively. The influence of sun glint can be reduced through thoughtful flight planning [18], but sky glint remains an important contributor. In accordance with [21] ρ_w can be retrieved as follows:

$$\rho_w(i) = \frac{\pi\, L_{camera}(i)}{E_d(i)} - \frac{\pi\, r(\theta_v) L_{sky}(\theta_v,\, \varphi_v, i)}{E_d(i)} = \rho_{Surf}(i) - d_1(i) \tag{3}$$

with

i the wavelength;
$r(\theta_v)$ the Fresnel reflectance;
θ_v the view zenith angle;
φ_v the view azimuth angle;
L_{sky} the downwelling sky radiance;
E_d the downwelling irradiance above the surface;
ρ_{Surf} the surface reflectance;
d_1 referred to as sky glint.

There are two ways to calculate the surface reflectance: (i) using spectral reference panels in the field or (ii) making use of a downwelling light sensor (DLS).

When using spectral reference panels, i.e., panels that have a known reflectance, the equation can be written as

$$\rho_{Surf}(i) = L_{camera}(i) \times \frac{\rho_{Spec}\ (i)}{L_{camera-spec}\ (i)} \tag{4}$$

where $\rho_{Spec}\ (i)$ stands for the known spectral reflectance of the spectral reference panels at wavelength i, and $L_{camera-spec}$ the radiance of the spectral panels in field measured by the camera.

If a DLS is available, the irradiance coming from the sky is measured directly in units $(\mathrm{W/m^2/nm})$. When the DLS is placed parallel with the ground surface, ρ_{Surf} can be calculated by dividing L_{camera} with the irradiance measured by the DLS (I_{DLS}) and multiplied with π:

$$\rho_{Surf} = \frac{L_{camera}\ \pi}{I_{DLS}} \tag{5}$$

However, the DLS will often not be placed parallel with the ground, and a correction for sensor and illumination angles is required. In addition, a Fresnel correction to correct for the air–sensor interface is needed. As described in [19], the Equation can be rewritten into

$$\rho_{Surf} = \frac{\pi L_{camera} \times r_s \times (perc_{diff} + \cos(\alpha_{ss}))}{I_{DLS} \times (perc_{diff} + \sin(\delta_s))} \tag{6}$$

with

$perc_{diff}$ percentage diffuse irradiance;
r_s Fresnel coefficient for air-sensor interface;
α_{ss} solar-sensor angle, i.e., the angle between the DLS vector and sun vector;
δ_s sun elevation angle.

The remaining unknown term in Equation (3) is the sky glint. This component can be derived using the iCOR4Drones tool, an adapted version of iCOR [22] for drone imagery. The tool relies on the MODTRAN5 [14] look-up tables (LUT) as described in [23]. Required input parameters are:

- Sun zenith angle;
- View zenith angles;

- Relative azimuth angle;
- Cloud type: Open Sky, Cumulus, Stratus, Altostratus;
- Spectral Response curves of the camera.

This approach assumes that the pixel is not affected by direct sun glint (realised by avoiding sun glint as part of flight planning) and that wind speed is sufficiently low to avoid foam or white cap radiance. Remaining effects of sun glint and/or foam can still be masked in output products. Light is considered to be unpolarised, and the sky radiance is distributed in a uniform way. If d_1 and ρ_{Surf} are known, the water-leaving reflectance can be calculated (Equation (3)).

2.1.4. Conversion from Water-Leaving Reflectance to Water Quality Products

Once ρ_w is calculated, other water quality products, like turbidity, Chl-a or suspended sediment concentrations, can be derived. In this work, we use the algorithm as described by [24] to derive turbidity and either OC3 [25] or Gilerson [26] algorithms to calculate Chl-a concentrations. Turbidity is estimated as

$$TUR = \frac{A^{\rho}\, \rho_w}{1 - \frac{\rho_w}{C^{\rho}}} \tag{7}$$

with ρ_w the water-leaving reflectance for a given waveband, and A^{ρ} and C^{ρ} two wavelength dependent calibration coefficients (A^{ρ} = 366.14, C^{ρ} = 0.1956 when using the Red band).

Chl-a for clear water types with low turbidity is derived as follows:

$$CHL_a(OC3) = 10^{(a_0 + a_1 x + a_2 x^2 + a_3 x^3 + a_4 x^4)} \tag{8}$$

with

$x = log_{10}\left(\frac{\rho_{w443}}{\rho_{w560}}\right)$ and a_x calibration coefficients ($a_0 = 0.2412$; $a_1 = -2.0546$; $a_2 = 1.1776$; $a_3 = -0.5538$; $a_4 = -0.457$).

In relatively turbid waters, Chl-a is instead derived following

$$CHL_a(Gilerson) = \left(35.75\,\frac{\rho_{w705}}{\rho_{w668}} - 19.3\right)^{1.124} \tag{9}$$

2.2. Validation in Inland and Coastal Waters

2.2.1. Study Sites

Field campaigns were organised at different locations across Europe:

- Lake Balaton in Hungary is the largest lake in Europe, with an area of 592 km^2. Despite its large surface area, it is very shallow, with a mean depth of approximately 3.2 m. The trophic state of Lake Balaton varies between mesotrophic and eutrophic regimes. In addition, Kiss Balaton varied in trophic status and suspended matter concentration.
- The Belgian North Sea covers an area of about 3454 km^2. It is a sensitive ecosystem under considerable pressure from intensive human activity such as sand and gravel extraction, dumping of dredged material, wind farming, fishing, shipping and tourism. The coastal area is subject to anthropogenic eutrophication from land-based nutrients rinsing in rivers and flowing into the sea [27].
- Razelm-Sinoe Lagoon is a coastal lagoon system attached to the southern part of the Danube Delta, located on the Romanian shoreline of the Black Sea. The lagoon system has been impacted by anthropogenic interventions (e.g., closing the connection with the sea), affecting its natural evolution [28]. The inflow of nutrient-rich Ranuve waters has increased eutrophication in the lagoon, especially in the last 20 years [29].
- Rupelmondse Creek is a ground- and rainwater-fed waterbody connected to the Scheldt river through leaking non-return valves. The Scheldt water differs from the creek water in suspended sediments and Chl-a concentration. Knowledge about the

extent of the inflow plume (including short- and long-term temporal variation) is necessary for ecologically sound water management in the polder.

- Lake Marathon is a water supply reservoir located in Greece. It has a total surface area of about 2.4 km^2 and maximum depth of 54 m. The reservoir is constructed from concrete and operates as a backup source for the water supply system of the greater Attica region and as a primary regulating reservoir. Seasonal algae fluctuations and prevention of pollution caused by trespassing and agricultural activities require careful monitoring of the lake [30].

Figure 1 shows the geographic locations of the field campaigns.

Figure 1. Drone flights performed across Europe for which simultaneous in situ data were collected.

2.2.2. Drone Data

Drone flights were performed with the standard RGB camera under a DJI Phantom 4 Pro (DJI PH4), the multispectral MicaSense RedEdge-M (MSREM) or the MicaSense Dual camera (MSDual), consisting of an MSREM and a MicaSense RedEdge-M Blue camera (MSBLUE). Table 1 summarises the technical characteristics of the used camera systems, while Figure 2 shows the spectral characteristics. The spectral characteristics of the PH4pro and MS-REM were measured in the NERC Field Spectroscopy Facility (FSF) (UK) according to the methodology described by [31]. For the MS-BLUE, a Gaussian spectral response was assumed around the central wavelength and taking into account the full width at half maximum (FWHM). Flight protocols, as described in [18], were followed to reduce the pixels affected by sun glint. The most important considerations were (i) to mount the camera under a pitch angle of $\pm 8^\circ$ from nadir where possible; (ii) during flight, point the camera in the opposite direction to the sun to avoid the majority of sun glint effects; (iii) store data in raw digital number format; (iv) obtain information of the irradiance either using a DLS or place spectrally calibrated Lambertian reference panels within the field of view of the camera, as explained in 2.1.1; (v) estimate or measure the relative height difference between the take-off location of the drone and the water level. This last point is important to the relative height difference between the drone and the water level.

The drone data were processed with MapEO water, as discussed in Section 2.1. For validation of the water-leaving reflectance and water quality products, additional quality checks were performed. Two rules were applied as quality control to exclude non-water pixels like reflectance coming from vessels or shadow:

- $\rho_{w-665} > \rho_{w-865}$;
- $\min(\rho_w) > -0.05$.

Table 1. Summary of three different camera systems used in this study.

Parameter	DJI PH4	MSREM	MSDUAL
Raw data format	DNG	TIFF	TIFF
Image dimensions (width × height)	5472 n × 3648	1280 × 960	1280 × 960
Field of view (horizontal; vertical)	73.7°; 53°	47.2°; 35.4°	47.2°; 35.4°
Image Bit Depth	16-bit	12- or 16-bit	12- or 16-bit
Number of spectral bands	3	5	10
Number of lenses	1	5	10
Number of image files per capture	1	5	10
Radiance calibration parameters in image metadata?	No	Yes	Yes
Compatible with irradiance sensor?	No	Yes	Yes

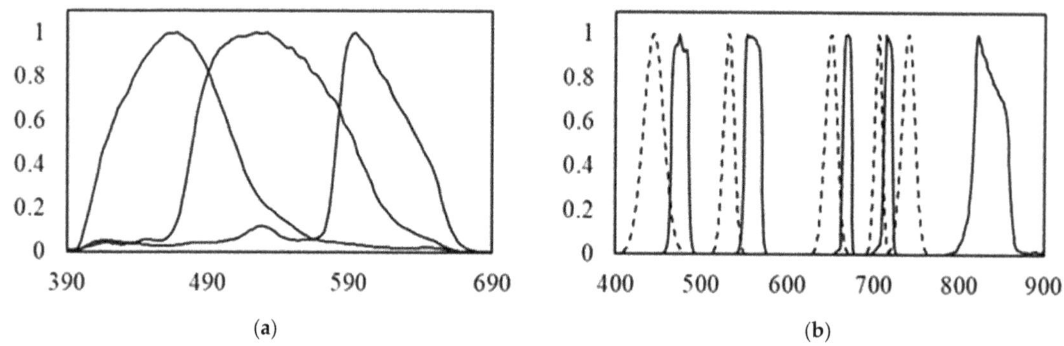

Figure 2. Spectral response curves showing the normalised response as a function of wavelength [27] for (**a**) DJI PH4pro and (**b**) MS-DUAL, consisting of MS-REM (solid lines) and MS-BLUE (striped lines).

In addition, the lowest reflectance in the spectral range from 705 to 865 nm was subtracted from all wavebands. In low to median turbid waters, it can be assumed that the reflectance in the NIR bands is nearly zero, as water strongly absorbs in this spectral range [32]. Non-zero reflectance can be due to residual glint or insufficient correction of irradiance. This subtraction is further referred to as the RedEdge-NIR subtraction. The median value of a 1 × 1 m box around the in situ coordinate was calculated. As one in situ sampling location can fall within different images due to overlap, mean and standard deviations were derived for each in situ measurement. For visualisation, mosaics were generated in a post-processing step, which includes masking non-water pixels to avoid, e.g., pixels affected by vessels, down-sampling to a lower resolution and finally, mosaicking through calculation of mean and standard deviation of all pixels falling within a predefined "super-pixel".

2.2.3. Match Up Data

Reflectance measurements were collected using the TriOS Ramses spectroradiometers, measuring 320–950 nm spectral range [33]. The remote-sensing reflectance, available online, was processed using the Fingerprint method [34], and only measurements passing the FingerPrint Pass Flag were retained in this study. The hyperspectral data were resampled to MS-REM, MS-BLUE and DJI PH4 pro bands using the spectral response functions provided in Figure 2 and multiplied with pi to obtain water-leaving reflectance.

Turbidity and Chl-a in situ values were measured using turbidity meters and HPLC or lab analysis, respectively. Samples and measurements were collected during the drone flights, limiting the maximum time difference to 45 min. All in situ measurements located within one pixel of 2 m of the drone mosaic were grouped, and mean and standard deviation values were calculated for this subset.

A Sentinel-2B MSI image acquired over the Belgian North Sea on 13 April 2021 was used and compared with drone and in situ observations. The Sentinel-2 tiles (31UDT, 31UDS, 31UES and 31UET) were processed using the Terrascope Water processing workflow [35,36]. Terrascope (https://terrascope.be/en, accessed on 24 February 2023) is the Belgian platform for Copernicus satellite data, products, and services. Terrascope uses the semi-empirical algorithm of Nechad [24] and a band-switching approach to allow the derivation of both high turbid and low turbid values. The turbidity products are masked using WorldCover [37] for land masking and IdePix for cloud/cloud shadow masking.

Table 2 provides an overview of the different field campaigns and the corresponding in situ and satellite data used in this study.

Table 2. Overview of drone flights: date, location and sensor are listed, as well as the in situ measurements used for validation. (DJI PH4 = DJI Phantom 4 Pro, MSREM = MicaSense RedEdge-M, MSDUAL = MicaSense Dual camera).

Site	Date	Lat	Lon	Sensor	Rw	Tur	Chl-a	S2	Irradiance
Lake Balaton (HU)	3–5 July 2019	46.83	17.75	MSREM	x	x			Panel
North Sea (BE)	13 April 2021	51.43	3.00	DJI PH4		x		x	Panel
Danube Delta (RO)	12–14 October 2021	44.80	28.99	MSDUAL		x	x		DLS
Rupelmondse Creek (BE)	16 March 2022	51.30	4.31	MSDUAL		x	x		DLS
Lake Marathon (GR)	15 June 2022	38.17	23.90	MSDUAL		x	x		DLS

2.2.4. Validation Metrics

For the validation of the drone-based reflectance and water quality products, a linear least square regression was performed between the set of drone values and the set of in situ measurements. Intercept and slope values are provided, and the coefficient of determination (R^2) and Root Mean Square Error (RMSE) were derived to evaluate the regression model:

$$R^2 = 1 - \frac{\sum(y_i - \hat{y})^2}{\sum(y_i - \overline{y})^2} \quad (10)$$

with $\hat{y}$ the predicted value of y and $\overline{y}$ the mean value of y.

$$RMSE = \sqrt{\frac{1}{N}\sum_{i=1}^{N}(y_i - \hat{y})^2} \quad (11)$$

3. Results

3.1. Validation of Drone-Based Reflectance and Water Quality Products

Figure 3 shows the mean water-leaving reflectance spectra and standard deviation of all matchups collected within one flight at Lake Balaton and one at Kiss Balaton. The upper figures show the spectra without RedEdge-NIR subtraction, while the lower plots show the spectra when the RedEdge-NIR subtraction is applied. While the results for Lake Balaton are very similar in both cases, the spectra of Kiss Balaton improved significantly after performing the RedEdge-NIR subtraction. Although the spectral shape was well retrieved, an offset of nearly 0.03 was observed before applying an additional subtraction. This offset was reduced to <0.01, assuming that the lowest reflectance observed in the RedEdge-NIR spectral range should be zero.

Figure 5 shows the scatter plots of turbidity, and Chl-a derived through MapEO water compared with in situ measurements. While turbidity results are retrieved from DJI PH4, and MSREM sensors, fewer results are available for Chl-a analysis as only flights conducted with MSDUAL cameras are suitable to derive this parameter. The turbidity trendline shows an offset of −2.47, slope of 1.08, R^2 value of 0.71 and RMSE of 10.13. These statistics are based on values derived from the DJI PH4, and MSREM combined. Table 3 shows the turbidity results for each camera sensor separately and when combined. The DJI PH4 follows almost perfectly the 1:1 line with a low intercept. However, the scatter around this line is larger, with an RMSE of 14.35 and R^2 value of 0.58. The MSREM has a slightly steeper slope (1.14) and offset of −4.53; the R^2 and RMSE values are better, with values of 0.85 and 5.61, respectively. Chl-a values could only be retrieved using the MSDUAL camera. A slope of 1.12, an intercept of 2.491, an R^2 of 0.93 and an RMSE of 9.34 was derived.

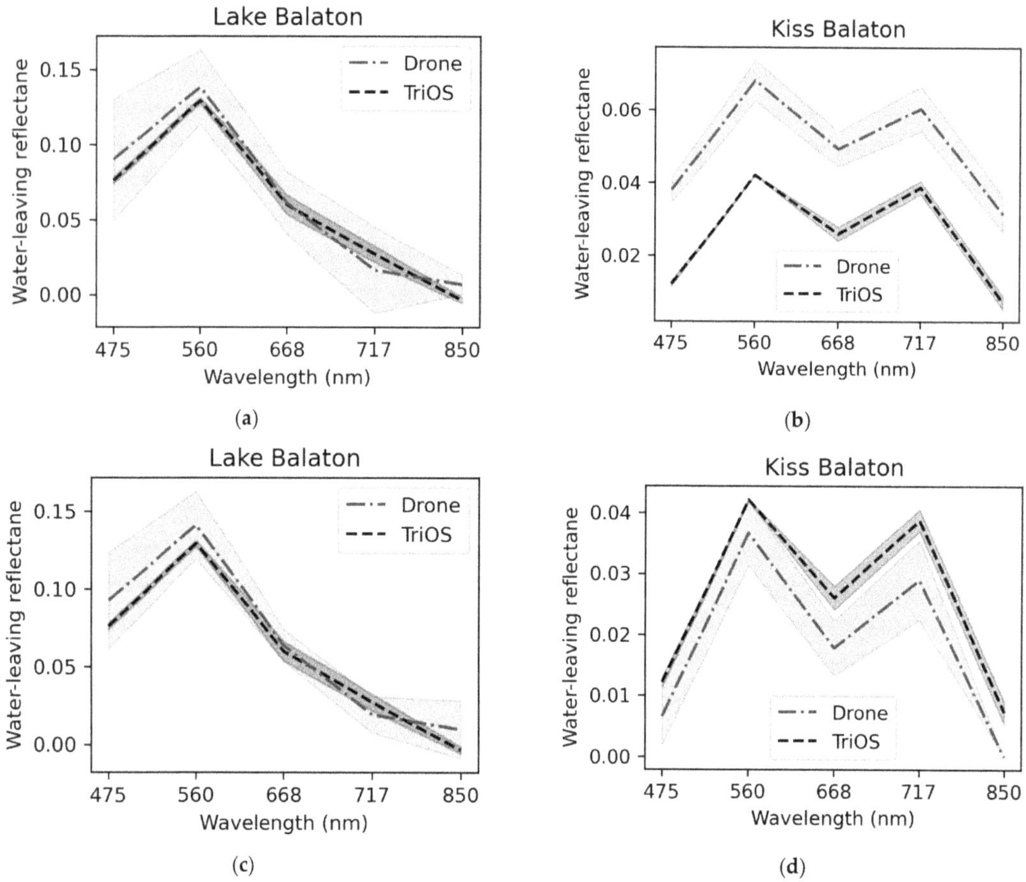

Figure 3. Mean Water-Leaving Reflectance of Drone (grey) and TriOS Ramses (black) data plotted for two different flights at Lake Balaton (Lake Balaton and Kiss Balaton). The upper plots (**a**,**b**) show the reflectance without RedEdge-NIR subtraction. The lower plots (**c**,**d**) show the reflectance with RedEdge-NIR subtraction.

Figure 4 shows the water-leaving reflectance scatterplots (MSREM vs. TriOS) for the four spectral bands of the MSREM camera in Lake Balaton and Kiss Balaton. The NIR band is not included as this band was, in most cases, used for subtraction. All bands have a very low intercept value of < 0.005. The regression statistics of the Blue (475 nm), Green (560 nm) and Red (660 nm) bands yield an $R^2 > 0.93$ and RMSE of 0.01. The slope for the Green band is nearly perfect (1.03), slightly too steep for the Blue band (1.21), and slightly too gentle for the Red band (0.88). The RedEdge band (717 nm) has an R^2 value of 0.85 and a slope of 0.7. In addition, here, an intercept < 0.01 was noted, and an RMSE of 0.01.

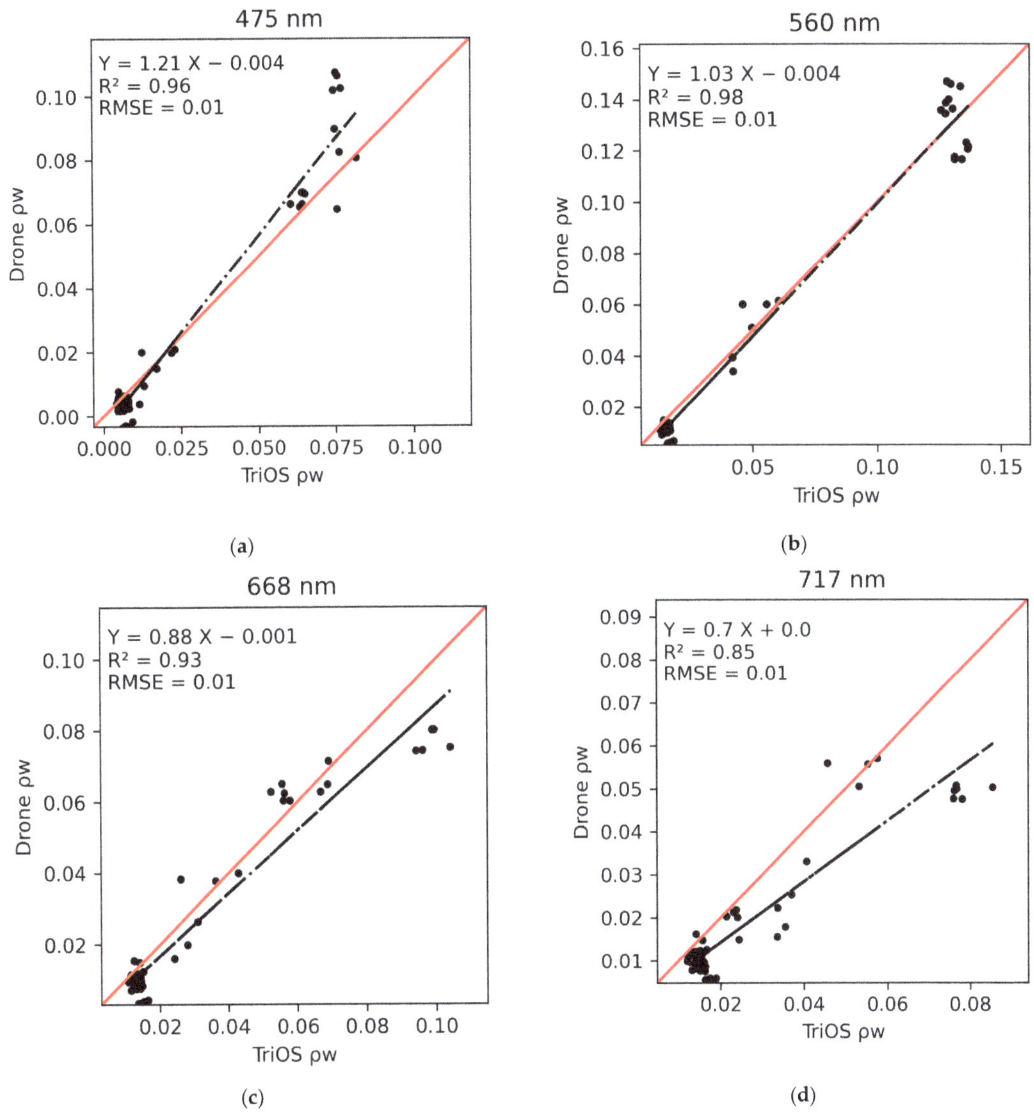

Figure 4. Scatterplots of MS-REM versus in situ (TriOS) measured water-leaving reflectance in Lake Balaton. (**a**) Blue band (475 nm); (**b**) Green band (560 nm); (**c**) Red band (668 nm); and (**d**) Red Edge band (717 nm). Red line = 1:1 line, black line = regression line.

Figures 6 and 7 show examples of derived turbidity and Chl-a maps generated with MapEO water for Rupelmondse Creek, Lake Marathon and the Danube Delta. The Rupelmondse Creek shows low turbidity and high Chl-a values on the northern side, while the opposite is observed in the southern part. These observations are in line with expectations as the southern part is linked with the high turbid Scheldt River and consequently also subjected to tidal influences. The northern part is separated from the Scheldt through sluices. During extreme weather events (spring tides linked with the west of northwest storm wind), these sluices can be opened to avoid floodings at other locations along the Scheldt Estuary (part of SIGMA plan (https://sigmaplan.be/en/, accessed on 1 February

2023). The northern part is consequently hardly affected by tidal effects, and sediments in the water have sufficient time to settle. However, algae blooms can occur, resulting in increased Chl-a concentrations. Lake Marathon serves as a drinking water reservoir, affected from time to time by minor algae blooms. During the drone campaign, very low turbidity and Chl-a concentration were observed on the eastern part of the lake, which was in accordance with real-time field measurements from an unmanned surface vehicle (USV) operated by the Athens Water Supply and Sewage Company [38]. The example of the Danube Delta shows the impact of the Canalul Dunavat (Dunavat Canal) on Lacul Razelm (Lake Razelm): the highly turbid water supplied into the canal by the Danube, through its southern branch, Sfantu Gheorghe, flows into the lake, in a narrow plume. The concentrations of Chl-a in the inflow water are very low compared to the water of the lake. The lake is relatively shallow (0–3.5 m), and the algal blooms increase in intensity from spring to autumn. The plume of the canal and its extent into the lake is mainly controlled by wind, so most of the time, it remains very close to the inlet.

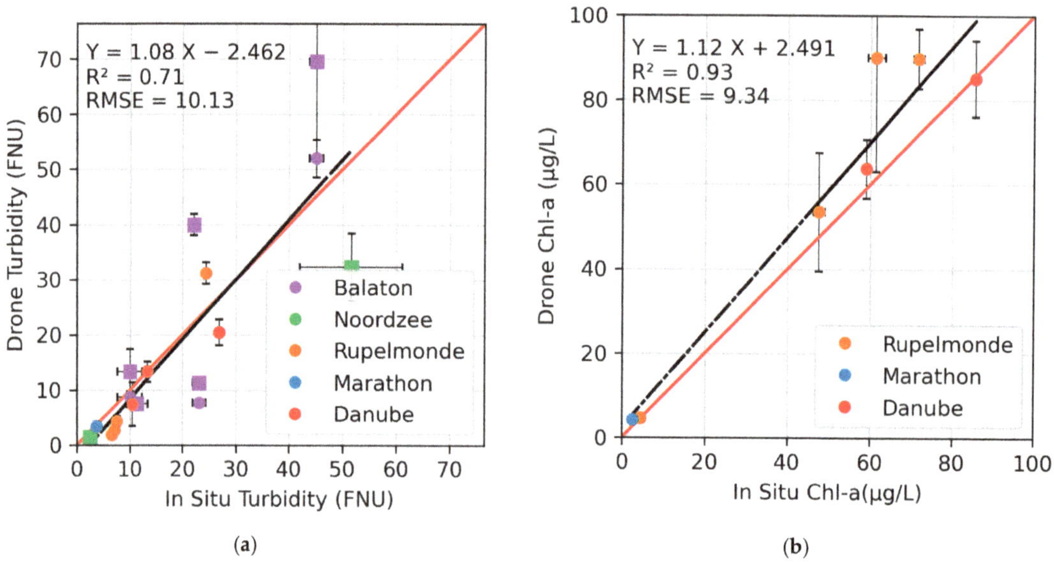

Figure 5. Scatterplot of drone-derived versus in situ measured observations. (**a**) turbidity plot (circles are derived from MSREM, squares are derived from DJI PH4), (**b**) Chl-a plot (derived from MSDUAL). Red line = 1:1 line, black line = regression line.

Table 3. Regression statistics of turbidity products when DJI PH4 and MSREM are combined and for both sensors separate.

Sensor	Slope	Intercept	R^2	RMSE
Combined	1.08	−2.46	0.71	10.13
DJI PH4	0.99	1.5	0.58	14.35
MSREM	1.14	−4.53	0.85	5.61

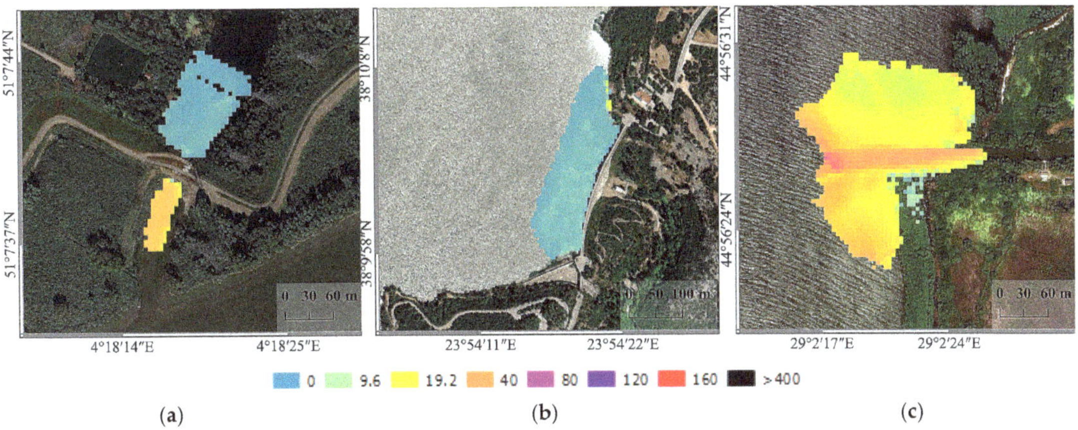

Figure 6. Examples of generated mosaicked turbidity maps. A logarithmic colour scaling is applied with units expressed in FNU. (**a**) Rupelmondse Creek; (**b**) Lake Marathon; (**c**) Danube Delta.

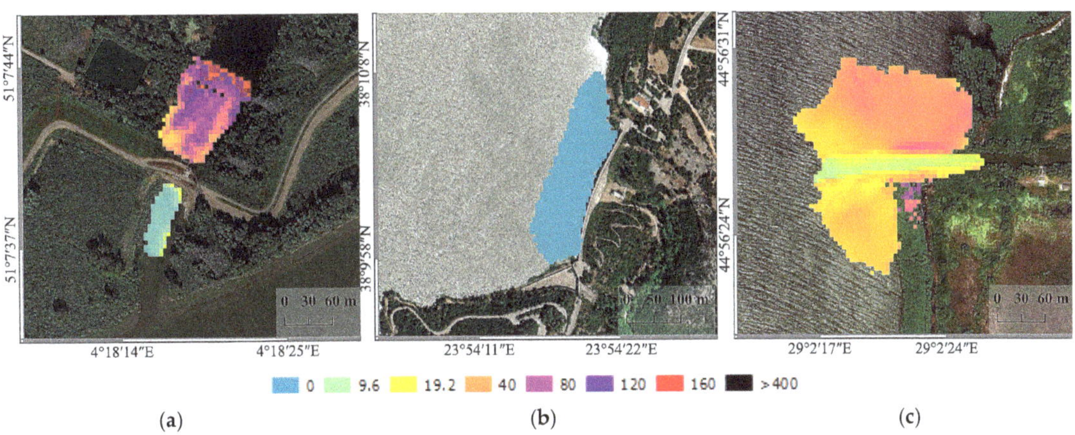

Figure 7. Examples of generated mosaicked Chl-a maps. A logarithmic colour scaling is applied with units expressed in μg/L. (**a**) Rupelmondse Creek; (**b**) Lake Marathon; (**c**) Danube Delta.

3.2. Comparison of Satellite, Drone and In Situ Data

Figure 8 shows the Sentinel-2 derived turbidity product from 13 April 2021 10:59 UTC, obtained from Terrascope (see Section 2.2). The two locations of drone flights are marked in the figure. The drone flights were conducted with a DJI PH4pro drone. Location A was acquired at 10:05 UTC, and location B at 12:32 UTC. The DJI took off and landed on the Simon Stevin Research Vessel. Light fluctuations were corrected using a spectral reference panel onboard, as described in Section 2.1. Figure 9 shows two true colour drone images and the respective drone turbidity map, one for each location. The background colour represents the Sentinel-2 turbidity map. At location A, higher turbidity values were obtained from the drone data compared to the satellite data, while at location B both values are very comparable. Table 4 summarises the turbidity values derived from different sources: Sentinel-2, DJI PH4pro, and the in situ measured turbidity. The turbidity values of the drone data are in line with the in situ observations. For Sentinel-2, the values at location B are similar; for location A, almost half of the measured in situ and drone values were obtained. This can be caused by errors in the atmospheric correction or, as the drone data seems to support, high dynamics in a spatial and temporal context.

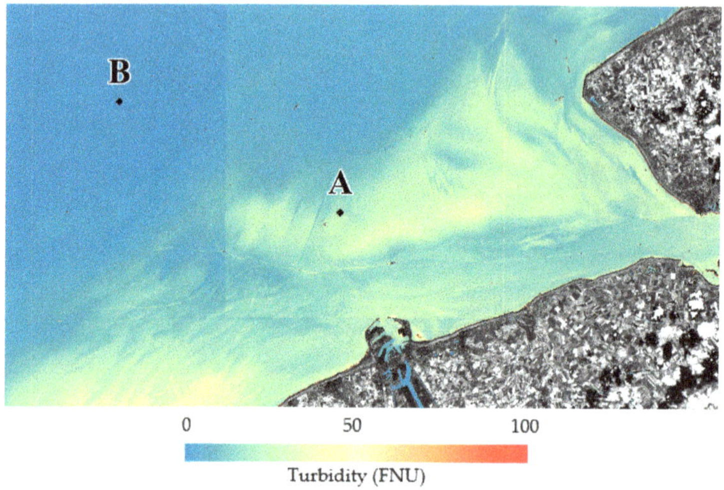

Figure 8. Sentinel-2 derived turbidity product on 13 April 2021 at 10:59 UTC. The two locations of drone flights are marked with A and B.

Table 4. Turbidity values derived from Sentinel-2 MSI, DJI PH4pro and HACH in situ device for the two locations in the Belgian North Sea.

Instrument	Loc A.	Loc B.
S2 MSI	20 FNU	1.5 FNU
DJI PH4pro	Plume: 40–60 FNU Backgr.: 26–38 FNU	Around boat: 0.5–5 FNU Backgr: 1.4 FNU
In Situ	42–62 FNU	2.1–2.9 FNU

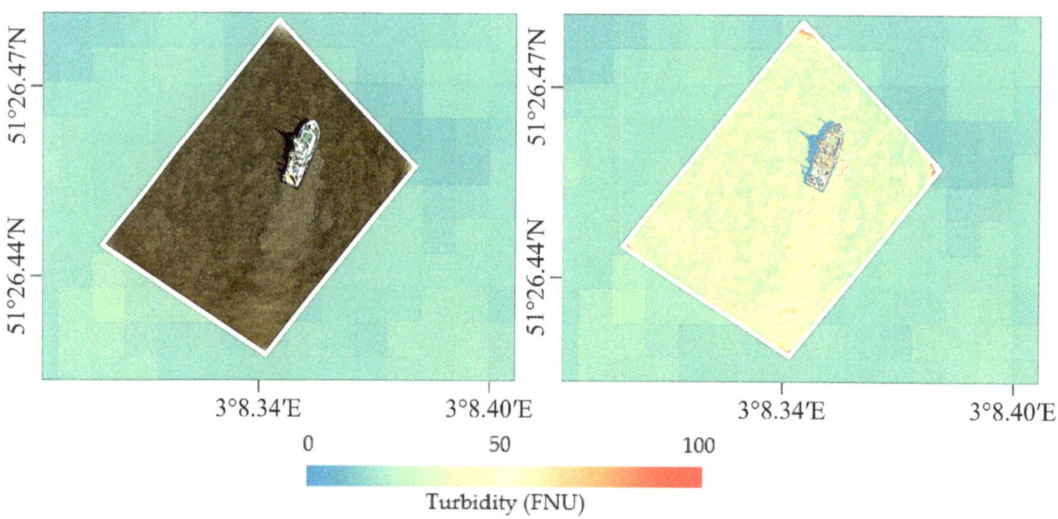

(a) (b)

Figure 9. *Cont.*

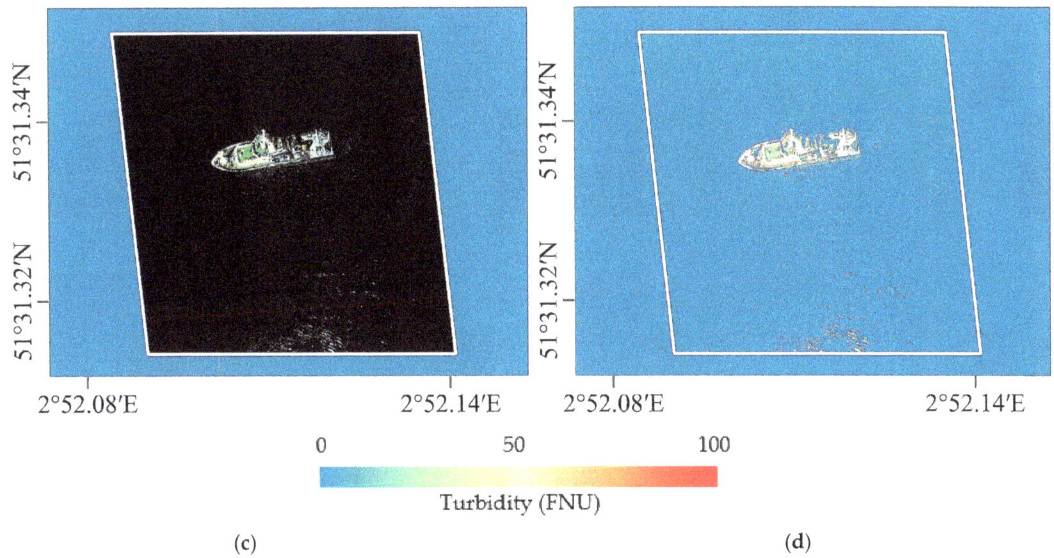

Figure 9. True colour and turbidity products for the two locations. Sentinel-2-derived turbidity products serve as a background. (**a**) True colour drone image at location A, acquired at 10:05 UTC; (**b**) turbidity-derived product at location A; (**c**) True colour drone image at location B, acquired at 12:32 UTC; (**d**) turbidity-derived product at location B.

4. Discussion

4.1. Validation of Drone-Based Reflectance and Water Quality Products

The water-leaving reflectance, the base product of optical remote sensing over water and used to derive water quality parameters, shows a very good fit between in situ radiometric measurements performed from a ship and the observation from the drone platform using the MSREM camera (Figure 4). The water-leaving reflectance in the Blue, Green and Red bands has an R^2 value of 0.93 and RMSE lower than 0.01. Their slopes vary between 0.86 and 1.22; the intercept is <0.005. Compared to these, the RedEdge band performs slightly less but is still reasonably good, with a slope of 0.7 and an R^2 value of 0.85. These results were obtained after an additional correction by subtracting the lowest reflectance value in the 668–850 nm spectral range. This correction is often applied over surface waters, as water strongly absorbs in the NIR spectral range [32]. The offset can be caused by remaining glint effects, adjacency effects, or an insufficient correction for irradiance. As mentioned in Section 2.1, the irradiance is either derived from spectral reference panels placed in the field or from an irradiance sensor mounted vertically upwards on the drone platform. When the spectral reference panels are placed on the ground or a vessel, the signal of the panels, as detected by the camera sensor, can be affected by the surroundings (e.g., vegetation or vessel). In addition, rapidly changing cloud conditions make it challenging to use spectral reference panels, as the values retrieved at certain moments over the panels cannot be considered constant throughout the flight. Therefore, this additional subtraction using the lowest values in the 668–840 nm spectral range cannot be applied at all sites. Furthermore, as pointed out in [39], in extremely turbid waters, the assumption of a zero NIR reflectance is no longer valid. Therefore, caution should be taken on when and where to apply this correction.

The water quality products derived from water-leaving reflectance were validated with in situ measurements (Figure 5). Turbidity could be derived from both camera systems, the DJI PH4 and MSREM (also included in the MSDUAL set-up). When performing a regression analysis, it appeared that the DJI PH4 had good intercept and slope values (0.99

and 1.5, respectively) but low R^2 (0.58) and RMSE values 14.35. For the MSREM, the slope and intercept were 1.14 and −4.53, while the R^2 and RMSE were much better than the DJI PH4, with values of 0.85 and 5.61. These results indicate that it is possible to derive turbidity from a low-cost RGB sensor onboard a drone but with larger errors compared to multispectral data. Chl-a values can only be derived using the MSDUAL camera, limiting the number of match-ups. A slope of 1.12 and an intercept of 2.49 with R2 and RMSE of 0.93 and 9.34, respectively. These results illustrate the usefulness of drones for water quality monitoring. For turbidity, low-cost cameras can be used but at the cost of higher errors. For Chl-a, the sensor should have the correct wavelengths. This was the case for MSDUAL, but the MSBLUE sensor is no longer supported by MicaSense.

4.2. *Comparison of Satellite, Drone and In Situ Data*

In situ point observations are used to define, calibrate and validate algorithms and satellite-derived products. However, often the variability within a pixel is unknown when taking in situ measurements. In the example of the North Sea (Figure 9), a turbidity plume generated by the boat is still visible. However, the boat was located at a fixed position when taking the measurements. Radiometric measurements using an ASD FieldSpec spectroradiometer device were collected at the bow of the research vessel, using optimised viewing geometry with regard to the position of the sun. Water samples were collected at the starboard side of the vessel. Without the drone image, one would believe that both measurements, taken at nearly the same time and location, would represent the same water mass and condition. The drone image, however, shows a spatial heterogeneity in the colour of the water, which is amplified through the disturbance of the research vessel. Caution should thus be used when linking different measurements together. This example shows the importance of FRM for validation studies at multiple levels: drone validation efforts will benefit from in situ FRM data, while drone FRM data can become relevant for satellite validation studies. However, before drone data meet the FRM requirements, a full uncertainty budget should be performed. First, efforts are made for land-based camera applications [40].

4.3. *Overall Considerations*

Expectations about the areal coverage of a drone platform are often unrealistic. The octocopter drones used in this study have a limited endurance of about 30–45 min, resulting in a limited areal coverage of up to 10,000 m^2. Besides flight time, the altitude and visual line of sight (VLOS) define the extent. Long-endurance drone systems are available on the market, which can fly tens of kilometres from the drone pilot at high altitudes. However, these systems are relatively expensive and require additional permissions to operate and local drone flight regulations might still limit operations. As pointed out by [3], in many countries, there are still stringent restrictions regarding where and how drones may be operated.

Working with drone data requires the capability to handle large volumes of data. With an image obtained every 3–5 s, up to 600 images can be easily obtained within one drone flight. In order to handle such large volumes of data, MapEO water is deployed in a cloud environment. Cloud-based processing has multiple advantages, including scalability, flexibility and data loss prevention. Final products can be reduced in size through mosaicking and compression. The MapEO water output product is accompanied by a GeoJSON metadata file [41] containing information on the unique identifier, bounding box, acquisition and processing date, drone platform and sensors, number of spectral bands and cloud conditions. Such a standardised approach is needed to manage large data volumes and catalogue results.

5. Conclusions

This paper presented the workflow behind MapEO water, a cloud-based data processing service that converts airborne drone data into georeferenced water quality products, like water-leaving reflectance, turbidity and Chl-a concentrations. Three different camera systems were tested: the DJI PH4pro containing only broad RGB spectral bands and the MS-REM and MS-DUAL with five and ten spectral bands in the VNIR, respectively. MapEO water contains four processing modules: (i) Georeferencing is conducted through direct georeferencing instead of structure-from-motion techniques common in land applications. These latter techniques are not applicable over water because of their dynamics; (ii) The radiometric correction converts raw digital numbers into radiance. The correction combines sensor black level correction, the sensitivity of the sensor, sensor gain and exposure settings, and lens vignette effects; (iii) conversion from radiance to water-leaving reflectance, taking sky glint effects into account through the iCOR4Drones module; (iv) in the final step, water-leaving reflectance is converted into turbidity and Chl-a water quality products.

Tests were performed at different lakes across Europe. Water-leaving reflectance was validated against TriOS measurements, whereas turbidity and Chl-a products were compared against turbidity meters and in situ water samples analysed in the laboratory. With an R^2 of 0.71 and 0.93 and RMSE of 10.13 and 9.34 for turbidity and Chl-a, respectively, these results show the potential of using drone-based observations for water quality monitoring. In addition, airborne drone data can be used to connect satellites with in situ observations and are very helpful in understanding the spatial conditions when taking in situ measurements for calibration and validation of algorithms or regulatory monitoring.

Author Contributions: Conceptualisation, L.D.K. and R.M.; methodology, R.M. and L.D.K.; software, R.M. and D.D.M.; validation. R.M. and L.D.K.; formal analysis, L.D.K.; investigation, A.M.C., A.S., G.K., W.M., P.D.H., E.S. and A.T.; writing—original draft preparation, L.D.K.; writing—review and editing, R.M., E.K., S.S., I.R. and S.G.H.S.; visualization, L.D.K.; funding acquisition, I.R., E.K., S.S. amd S.G.H.S. All authors have read and agreed to the published version of the manuscript.

Funding: This research was funded by European Union's Horizon 2020 research and innovation programme under grant agreement No 776480 (MONOCLE), grant agreement No 101004186 (Water-ForCE), and was supported by the Belgian Science Policy (Belspo) under Grant SR/00/381 (TIMBERS) and Grant SR/67/311 (DRONESED).

Data Availability Statement: Publicly available datasets were analysed in this study. This data can be found using the Web Coverage Service (WCS) service: https://mapeo.be/geoserver/MONOCLE/wcs, accessed on 1 February 2023. Credentials: monocle/monocle.

Acknowledgments: The authors wish to thank the NERC Field Spectroscopy Facility for making their infrastructure available for data calibration. Both the DJI PH4pro and MS-REM cameras were spectrally calibrated in their lab.

Conflicts of Interest: The authors declare no conflict of interest.

References

1. Koparan, C.; Koc, A.B.; Privette, C.V.; Sawyer, C.B. In Situ Water Quality Measurements Using an Unmanned Aerial Vehicle (UAV) System. *Water* **2018**, *10*, 264. [CrossRef]
2. Rodrigues, P.; Marques, F.; Pinto, E.; Pombeiro, R.; Lourenço, A.; Mendonça, R.; Santana, P.; Barata, J. An open-source watertight unmanned aerial vehicle for water quality monitoring. In Proceedings of the Conference OCEANS 2015—MTS/IEEE, Washington, DC, USA, 19–22 October 2015; pp. 1–6.
3. Ryu, J.H. UAS-based real-time water quality monitoring, sampling, and visualization platform (UASWQP). *HardwareX* **2022**, *11*, e00277. [CrossRef] [PubMed]
4. El Serafy, G.; Schaeffer, B.; Neely, M.-B.; Spinosa, A.; Odermatt, D.; Weathers, K.; Baracchini, T.; Bouffard, D.; Carvalho, L.; Conmy, R.; et al. Integrating Inland and Coastal Water Quality Data for Actionable Knowledge. *Remote. Sens.* **2021**, *13*, 2899. [CrossRef] [PubMed]
5. Sibanda, M.; Mutanga, O.; Chimonyo, V.G.P.; Clulow, A.D.; Shoko, C.; Mazvimavi, D.; Dube, T.; Mabhaudhi, T. Application of Drone Technologies in Surface Water Resources Monitoring and Assessment: A Systematic Review of Progress, Challenges, and Opportunities in the Global South. *Drones* **2021**, *5*, 84. [CrossRef]
6. Flynn, K.F.; Chapra, S. Remote Sensing of Submerged Aquatic Vegetation in a Shallow Non-Turbid River Using an Unmanned Aerial Vehicle. *Remote. Sens.* **2014**, *6*, 12815–12836. [CrossRef]
7. Su, T.-C.; Chou, H.-T. Application of Multispectral Sensors Carried on Unmanned Aerial Vehicle (UAV) to Trophic State Mapping of Small Reservoirs: A Case Study of Tain-Pu Reservoir in Kinmen, Taiwan. *Remote Sens.* **2015**, *7*, 10078–10097. [CrossRef]

8. Larson, M.D.; Milas, A.S.; Vincent, R.K.; Evans, J.E. Multi-depth suspended sediment estimation using high-resolution remote-sensing UAV in Maumee River, Ohio. *Int. J. Remote. Sens.* **2018**, *39*, 5472–5489. [CrossRef]
9. De Keukelaere, L.; Moelans, R.; Strackx, G.; Knaeps, E.; Lemey, E. Mapping water quality with drones: Test case in Texel. *Terra Et Aqua* **2019**, *157*, 6–16.
10. Westoby, M.; Brasington, J.; Glasser, N.F.; Hambrey, M.J.; Reynolds, J.M. 'Structure-from-Motion' photogrammetry: A low-cost, effective tool for geoscience applications. *Geomorphology* **2012**, *179*, 300–314. [CrossRef]
11. Windle, A.E.; Silsbe, G.M. Evaluation of Unoccupied Aircraft System (UAS) Remote Sensing Retrievals for Water Quality Monitoring in Coastal Waters. *Front. Environ. Sci.* **2021**, *9*, 674247. [CrossRef]
12. Mobley, C.D. Polarized reflectance and transmittance properties of windblown sea surfaces. *Appl. Opt.* **2015**, *54*, 4828–4849. [CrossRef]
13. Schläpfer, D.; Popp, C.; Richter, R. Drone Data Atmospheric Correction Concept For Multi—And Hyperspectral Imagery—The Droacor Model. *Int. Arch. Photogramm. Remote Sens. Spat. Inf. Sci.* **2020**, *43*, 473–478. [CrossRef]
14. Ruddick, K.G.; Voss, K.; Boss, E.; Castagna, A.; Frouin, R.; Gilerson, A.; Hieronymi, M.; Johnson, B.C.; Kuusk, J.; Lee, Z.; et al. A Review of Protocols for Fiducial Reference Measurements of Water-Leaving Radiance for Validation of Satellite Remote-Sensing Data over Water. *Remote. Sens.* **2019**, *11*, 2198. [CrossRef]
15. Donlon, C.J.; Wimmer, W.; Robinson, I.; Fisher, G.; Ferlet, M.; Nightingale, T.; Bras, B. A Second-Generation Blackbody System for the Calibration and Verification of Seagoing Infrared Radiometers. *J. Atmos. Ocean. Technol.* **2014**, *31*, 1104–1127. [CrossRef]
16. Mian, O.; Lutes, J.; Lipa, G.; Hutton, J.J.; Gavelle, E.; Borghini, S. Direct georeferencing on small unmanned aerial platforms for improved reliability and accuracy of mapping without the need for ground control points. *ISPRS-Int. Arch. Photogramm. Remote. Sens. Spat. Inf. Sci.* **2015**, 397–402. [CrossRef]
17. Berk, A.; Anderson, G.P.; Acharya, P.K.; Bernstein, L.S.; Muratov, L.; Lee, J.; Fox, M.; Adler-Golden, S.M.; Chetwynd, J.J.H.; Hoke, M.L.; et al. MODTRAN5: 2006 update. In Proceedings of the SPIE 6233, Algorithms and Technologies for Multispectral, Hyperspectral, and Ultraspectral Imagery XII, Orlando, FL, USA, 8 May 2006; Volume 6233, p. 62331F. [CrossRef]
18. De, K.L.; Moelans, R.; De, M.D.; Strackx, G. *RPAS—Remotely Piloted Aircraft Systems—Deployment and Operation Guide*; Zenodo: Geneve, Switzerland, 2022. [CrossRef]
19. MicaSense Inc. "Tutorial 3—DLS Sensor Basic Usage" Micasense/Imageprocessing GitHub. 2017–2019. Available online: https://micasense.github.io/imageprocessing/MicaSenseImageProcessingTutorial1.html (accessed on 20 September 2022).
20. Gao, M.; Li, J.; Zhang, F.; Wang, S.; Xie, Y.; Yin, Z.; Zhang, B. Measurement of Water Leaving Reflectance Using a Digital Camera Based on Multiple Reflectance Reference Cards. *Sensors* **2020**, *20*, 6580. [CrossRef]
21. Sterckx, S.; Knaeps, E.; Ruddick, K. Detection and correction of adjacency effects in hyperspectral airborne data of coastal and inland waters: The use of the near infrared similarity spectrum. *Int. J. Remote. Sens.* **2011**, *32*, 6479–6505. [CrossRef]
22. De Keukelaere, L.; Sterckx, S.; Adriaensen, S.; Knaeps, E.; Reusen, I.; Giardino, C.; Bresciani, M.; Hunter, P.; Neil, C.; Van der Zande, D.; et al. Atmospheric correction of Landsat-8/OLI and Sentinel-2/MSI data using iCOR algorithm: Validation for coastal and inland waters. *Eur. J. Remote. Sens.* **2018**, *51*, 525–542. [CrossRef]
23. Biesemans, J.; Sterckx, S.; Knaeps, E.; Vreys, K.; Adriaensen, S.; Hooyberghs, J.; Nieke, J.; Meuleman, K.; Kempeneers, P.; Deronde, B.; et al. Image processing workflows for airborne remote sensing. In Proceedings of the 5th EARSeL SIG Imaging Spectroscopy Workshop, Bruges, Belgium, 23–25 April 2007.
24. Nechad, B.; Ruddick, K.G.; Neukermans, G. Calibration and validation of a generic multisensor algorithm for mapping of turbidity in coastal waters. In Proceedings of the SPIE—The International Society for Optical Engineering, Berlin, Germany, 9 September 2009; p. 7473.
25. O'Reilly, J.E.; Maritorena, S.; Mitchell, B.G.; Siegel, D.A.; Carder, K.L.; Garver, S.A.; Kahru, M.; McClain, C. Ocean color chlorophyll algorithms for SeaWiFS. *J. Geophys. Res. Earth Surf.* **1998**, *103*, 24937–24953. [CrossRef]
26. Gilerson, A.A.; Gitelson, A.A.; Zhou, J.; Gurlin, D.; Moses, W.; Ioannou, I.; Ahmed, S.A. Algorithms for remote estimation of chlorophyll-a in coastal and inland waters using red and near infrared bands. *Opt. Express* **2010**, *18*, 24109–24125. [CrossRef]
27. Lancelot, C.; Rousseau, V.; Gypens, N. Ecologically based indicators for Phaeocystis disturbance in eutrophied Belgian coastal waters (Southern North Sea) based on field observations and ecological modelling. *J. Sea Res.* **2009**, *61*, 44–49. [CrossRef]
28. Bretcan, P.; Murărescu, M.O.; Samoilă, E.; Popescu, O. The modification of the ecological conditions in the Razim-Sinoie lacustrine complex as an effect of the anthropic intervention. In Proceedings of the XXIVth Conference of the Danubian Countries, Bled, Slovenia, 2–4 June 2008; ISBN 978-961-91090-2-1.
29. Stanica, A.; Scrieciu, A.; Bujini, J.; Teaca, A.; Begun, T.; Ungureanu, C. "Razelm Sinoie Lagoon System—Stat of the Lagoon Report," in Deliverable 2.2. ARCH (Architecture and Roadmap to Manage Multiple Pressures on Lagoons) FP7 Project, EC Grant 282748. 2013; *Unpublished Report*.
30. Katsouras, G.; Chalaris, M.; Tsalas, N.; Dosis, A.; Samios, S.; Lytras, E.; Papadopoulos, K.; Synodinou, A. Integrated ecosystem ecology (Chlorophyll-a) of EYDAP's Reservoirs profiles by using robotic boats. In Proceedings of the 5th International Conference 'Water Resources and Wetlands', Tulcea, Romania, 8–12 September 2021; Available online: http://www.limnology.ro/wrw2020/proceedings/23_Katsouras.pdf (accessed on 9 September 2021).
31. Burggraaff, O.; Schmidt, N.; Zamorano, J.; Pauly, K.; Pascual, S.; Tapia, C.; Spyrakos, E.; Snik, F. Standardized spectral and radiometric calibration of consumer cameras. *Opt. Express* **2019**, *27*, 19075–19101. [CrossRef] [PubMed]

32. Gordon, H.R.; Wang, M. Retrieval of water-leaving radiance and aerosol optical thickness over the oceans with SeaWiFS: A preliminary algorithm. *Appl. Opt.* **1994**, *33*, 443–452. [CrossRef] [PubMed]
33. Hunter, P.; Riddick, C.; Werther, M.; Martinez Vicente, V.; Burggraaff, O.; Blake, M.; Spyrakos, E.; Tóth, V.; Kovács, A.; Tyler, A. *In Situ Bio-Optical Data in Lake Balaton, MONOCLE H2020 Project (Version v1) [Data Set]*; Zenodo: Geneve, Switzerland, 2021. [CrossRef]
34. Simis, S.G.; Olsson, J. Unattended processing of shipborne hyperspectral reflectance measurements. *Remote. Sens. Environ.* **2013**, *135*, 202–212. [CrossRef]
35. De Keukelaere, L.; Landuyt, L.; Knaeps, E. Terrascope Sentinel-2 Algorithm Theoretical Base Document (ATBD) S2–RHOW–V120. 2022. Available online: https://docs.terrascope.be/DataProducts/Sentinel-2/references/VITO_S2_ATBD_S2_RHOW_V120.pdf (accessed on 23 February 2023).
36. De Keukelaere, L.; Landuyt, L.; Knaeps, E. Terrascope Sentinel-2 Algorithm Theoretical Base Document (ATBD) S2–Water Quality–V120. 2022. Available online: https://docs.terrascope.be/DataProducts/Sentinel-2/references/VITO_S2_ATBD_S2_WATER_QUALITY_V120.pdf (accessed on 23 February 2023).
37. Zanaga, D.; Van De Kerchove, R.; De Keersmaecker, W.; Souverijns, N.; Brockmann, C.; Quast, R.; Wevers, J.; Grosu, A.; Paccini, A.; Vergnaud, S.; et al. *ESA WorldCover 10 m 2020 v100*; Zenodo: Geneve, Switzerland, 2021. [CrossRef]
38. Katsouras, G.; Konstantinopoulos, S.; De Keukelaere, L.; Moelans, R.; Christodoulou, C.; Bauer, P.; Tsalas, N.; Hatzikonstantinou, P.; Lioumbas, I.; Katsiapi, M.; et al. Unmanned Vehicles combined with satellite observations as a complement tool for water quality of Lake Marathon. In Proceedings of the EuroGEO Workshop 2022, Athens, Greece, 7–9 December 2022; Available online: http://intcatch.eu/images/Katsouras_poster_EuroGEO2022_v1.pdf (accessed on 23 February 2023).
39. Knaeps, E.; Doxaran, D.; Dogliotti, A.; Nechad, B.; Ruddick, K.; Raymaekers, D.; Sterckx, S. The SeaSWIR dataset. *Earth Syst. Sci. Data* **2018**, *10*, 1439–1449. [CrossRef]
40. Brown, L.A.; Camacho, F.; García-Santos, V.; Origo, N.; Fuster, B.; Morris, H.; Pastor-Guzman, J.; Sánchez-Zapero, J.; Morrone, R.; Ryder, J.; et al. Fiducial Reference Measurements for Vegetation Bio-Geophysical Variables: An End-to-End Uncertainty Evaluation Framework. *Remote. Sens.* **2021**, *13*, 3194. [CrossRef]
41. Butler, H.; Daly, M.; Doyle, A.; Gillies, S.; Hagen, S.; Schaub, T.; The GeoJSON Format. Internet Standards Track document. 2016. Available online: https://www.rfc-editor.org/rfc/rfc7946 (accessed on 23 February 2023).

remote sensing MDPI

Review

A Systematic Review of the Application of the Geostationary Ocean Color Imager to the Water Quality Monitoring of Inland and Coastal Waters

Shidi Shao [1,2], Yu Wang [3], Ge Liu [1,*] and Kaishan Song [1]

1 Northeast Institute of Geography and Agroecology, Chinese Academy of Sciences, Changchun 130102, China; shaoshidi@iga.ac.cn (S.S.); songkaishan@iga.ac.cn (K.S.)
2 University of Chinese Academy of Sciences, Beijing 100049, China
3 Center for Satellite Application on Ecology and Environment, Ministry of Ecology and Environment of the People's Republic of China, Beijing 100029, China; wangyusecmep1220@163.com
* Correspondence: liuge@iga.ac.cn

Abstract: In recent decades, eutrophication in inland and coastal waters (ICWs) has increased due to anthropogenic activities and global warming, thus requiring timely monitoring. Compared with traditional sampling and laboratory analysis methods, satellite remote sensing technology can provide macro-scale, low-cost, and near real-time water quality monitoring services. The Geostationary Ocean Color Imager (GOCI), aboard the Communication Ocean and Meteorological Satellite (COMS) from the Republic of Korea, marked a significant milestone as the world's inaugural geostationary ocean color observation satellite. Its operational tenure spanned from 1 April 2011 to 31 March 2021. Over ten years, the GOCI has observed oceans, coastal waters, and inland waters within its 2500 km × 2500 km target area centered on the Korean Peninsula. The most attractive feature of the GOCI, compared with other commonly used water color sensors, was its high temporal resolution (1 h, eight times daily from 0 UTC to 7 UTC), providing an opportunity to monitor ICWs, where their water quality can undergo significant changes within a day. This study aims to comprehensively review GOCI features and applications in ICWs, analyzing progress in atmospheric correction algorithms and water quality monitoring. Analyzing 123 articles from the Web of Science and China National Knowledge Infrastructure (CNKI) through a bibliometric quantitative approach, we examined the GOCI's strength and performance with different processing methods. These articles reveal that the GOCI played an essential role in monitoring the ecological health of ICWs in its observation coverage (2500 km × 2500 km) in East Asia. The GOCI has led the way to a new era of geostationary ocean satellites, providing new technical means for monitoring water quality in oceans, coastal zones, and inland lakes. We also discuss the challenges encountered by Geostationary Ocean Color Sensors in monitoring water quality and provide suggestions for future Geostationary Ocean Color Sensors to better monitor the ICWs.

Keywords: GOCI; inland and coastal waters; atmospheric correction; algal blooms; water quality parameters

1. Introduction

Water accounts for about 74% of Earth's surface. Water is an important carrier and critical factor in studying the energy exchange of surface materials [1,2]. For research convenience in water color remote sensing, the Earth's water is usually divided into Case I and Case II waters based on the optical properties of water [3–5]. The optical properties of Case I waters (open oceans) are mainly determined by phytoplankton. Case II waters, including inland and coastal waters (denoted as ICWs; all abbreviations or symbols in this study are shown in Table 1 and Table S1), are optical complexes determined by phytoplankton, detritus, and colored dissolved organic matter (CDOM) [6,7]. Most of the

Earth's population lives near ICWs, making them the most economically and socially active regions on Earth. Large quantities of nutrients and pollutants released by human activities are entering various ICWs globally, disrupting their ecological equilibrium. Damaged waterbodies will, in turn, threaten human health and lead to economic losses [8–13]. Remote sensing technology can complement traditional field sampling and laboratory analysis methods by targeting optically active substance concentrations in monitoring water quality, using reflectance signals. With the growing number of water color satellite sensors launched in recent decades, numerous algorithms have been proposed to retrieve water quality parameters from satellite images to monitor ICWs.

Table 1. Abbreviations and symbols used in this study.

	Abbreviations or Symbols		Abbreviations or Symbols
ABs	Algal blooms	OSMI	Ocean Scanning Multispectral Imager
AFAI	Alternative floating algae index	OE	*Optics EXPRESS*
AC	Atmospheric correction	POC	Particulate organic carbon
CNKI	China National Knowledge Infrastructure	PC	Phycocyanin
Chla	Chlorophyll a	RF	Random forest
CDOM	Colored dissolved organic matter	$L_{rc}(\lambda)$	Rayleigh-corrected radiance
CBI	Cyanobacterial bloom intensity	RI	Red tide index
k_d	Diffuse attenuation coefficient	RS	*Remote Sensing*
DOC	Dissolved organic carbon	RSE	*Remote Sensing of Environment*
FAC	Floating algae cover	$R_{rs}(\lambda)$	Remote sensing reflectance
FLH	Fluorescence line height	Sci Total Environ	*Science of the Total Environment*
GABI	Generalized algal bloom index algorithm	SIA	Sea ice area
GLI	Generation Global Imager	SIT	Sea ice thickness
GLIMR	Geostationary Littoral Imaging and Monitoring Radiometer	SSCs	Sea surface currents
HABs	Harmful algal blooms	SSS	Sea surface salinity
IEEE T-GRS	IEEE Transactions on Geoscience and Remote Sensing	SDD	Secchi disk depth
ICWs	Inland and coastal waters	SGLI	Second Generation Global Imager
JAG	*International Journal of Applied Earth Observation and Geoinformation*	SWIR	Shortwave infrared
Int J Remote Sens	*International Journal of Remote Sensing*	SZA	Solar zenith angle
ISPRS	*ISPRS Journal of Photogrammetry and Remote Sensing*	SPM	Suspended particulate matter
LCI	Linear Combination Index	SCI	Synthetic chlorophyll index
MERIS	Medium-Resolution Imaging Spectrometer Instrument	GOCI	The Geostationary Ocean Color Imager
MODIS	Moderate-Resolution Imaging Spectroradiometer	UV	Ultraviolet
NASA	National Aeronautics and Space Administration	VIIRS	Visible Infrared Imaging Radiometer
NIR	Near-infrared	WR	*Water Research*
NPP	Net primary production	YRE	Yalu River estuary
NDVI	Normalized difference vegetation index	YOC	Yellow and East China Sea Ocean Color
NRTI	Normalized red tide index		

Remote sensing of water color from space began in 1978 with the successful launch of the Coastal Zone Color Scanner (CZCS) by the National Aeronautics and Space Administration (NASA) [14]. The CZCS, an experimental radiometer intended to operate for only one year, had six bands ranging from visible to near-infrared (NIR). The CZCS provided valuable experience for subsequent water color satellite development [15–17]. After ten years, several water color sensors were launched, including the Maritime Observation Satellite (MOS) and Ocean Color and Temperature Scanner (OCTS), as well as the Polarization and Directionality of the Earth's Reflectances (POLDER) instrument in 1996 and the Sea-viewing Wide Field-of-view Sensor (SeaWiFS) in 1997. The SeaWiFS was the first water color sensor that could provide continuous data on the bio-optical properties (i.e., those properties of organisms that can be visually recorded) of the oceans [18–20]. In 1999, the United States successfully launched the Moderate Resolution Imaging Spectroradiometer (MODIS) on the Terra satellite, the first advanced polar-orbiting environmental remote-sensing satellite with nine specific water color bands from 412 to 865 nm. MODIS also had three shortwave infrared (SWIR) bands (1240, 1640, and 2130 nm), which had very low reflectance due to the high absorption of pure water [21,22]. The SWIR-based atmospheric correction (AC) method is critical for obtaining reliable and high-accuracy remote sensing reflectance ($R_{rs}(\lambda)$) products of highly turbid ICWs. In 2002, the MODIS sensor on the Aqua satellite was launched and began to observe the Earth along with MODIS from Terra (Terra in the morning, Aqua in the afternoon), providing continuous water color observations. MODIS has been the most critical water color satellite of the past two decades [23–28]. The Visible Infrared Imaging Radiometer (VIIRS) extends and enhances the AVHRR and MODIS observation series with six water color wavelengths. The three VIIRS sensors (Suomi-NPP, NOAA-20, and NOAA-21) were launched in 2011, 2017, and 2022, respectively. Since VIIRS has spectral bands similar to MODIS, the water quality parameters that MODIS can extract or derive can also be directly monitored by VIIRS. VIIRS can provide water color products for highly turbid ICWs, improving the observations capabilities of MODIS, whose water color bands tend to saturate [29–34]. In December 2018, the United States launched SeaHawk-1, a CubeSat (small, cube-shaped satellites) fitted with a low-cost, miniature ocean color sensor called HawkEye that enables fine spatial resolution observations of the ocean [35].

The Medium-Resolution Imaging Spectrometer Instrument (MERIS) is a sensor deployed in March of 2002 onboard the polar-orbiting Envisat-1 environmental research satellite by the European Space Agency (ESA). The most significant advantage of MERIS is its various water color band selections (fifteen bands from 412 to 900 nm). For example, the 620 nm band can be used to estimate cyanobacteria biomass in ICWs by targeting 86 characteristic pigments (phycocyanin (PC)) [36–38]. The Sentinel-3 Ocean and Land Colour Instrument (OLCI) is based on ENVISAT MERIS's opto-mechanical and imaging design and was launched in 2016 (S3A) and 2018 (S3B). The OLCI presents many improvements compared with MERIS, such as an increase in the number of spectral bands (from 15 to 21), mitigation of sun-glint contamination by tilting cameras in the westerly direction, and increased coverage of global ocean (<4 days) and land (<3 days with one satellite, ignoring the effect of clouds), compared to MERIS, which is approximately 15 days [39–42].

In addition to the United States and the European Union, other countries, such as Japan, China, India, and South Korea, have also released satellites with water color bands. Japan launched the ocean color satellite Advanced Earth Observation Satellite (ADEOS) in 1996. ADEOS had twelve bands in the wavelength range of 400–900 nm with a spatial resolution of 700 m [43–45]. Japan launched the ADEOS-II satellite with the Generation Global Imager (GLI) sensor, with thirty-three bands in the 375–12500 nm wavelength range in 2002. Among these thirty-three bands, six had a spatial resolution of 250 m. The Second-Generation Global Imager (SGLI) satellite is an Earth observation satellite developed by the Japan Aerospace Exploration Agency (JAXA) and launched in 2017. The SGLI can observe 19 bands of radiation from ultraviolet (UV) to the thermal infrared region (380 nm–12 μm). The SGLI has two polarization channels (673.5 and 868.5 nm) [46–48]. Since 2002, China has launched a series of ocean water color satellites, including HY-1A (2002), HY-1B (2007), HY-1C (2018), HY-1D (2022), and HY-1E (2023). The water color bands of the early HY satellite series (e.g., HY-1A, HY-1B) are very close to the MODIS water color bands. Starting from HY-1C, the HY series satellites have specific water color bands (e.g., UV bands). The recently launched HY-1E has two sensors, Chinese Ocean Color and Temperature Scanner 2 (COCTS2) (with eighteen bands and a one-day revisit cycle) and Playful Magnetic Resonance Imaging Simulator (PMRIS) (with nineteen bands and a three-day revisit cycle) [49,50]. India has launched several water color satellites, including Oceansat-1 (1996), Oceansat-2 (2009), and Oceansat-3 (2022). The bands have increased from eight (Oceansat-1) to thirteen bands with a spatial resolution of 360 m (Oceansat-3) [51–53]. The Ocean Scanning Multispectral Imager (OSMI), mounted on the polar-orbiting satellite Korea Multi-Purpose Satellite-1 (KOMPSAT-1), was launched by South Korea in 1999. OSMI was launched in 2010 and had six bands in the 400–900 nm wavelength range and a spatial resolution of 1 km [54,55].

The Geostationary Ocean Color Imager (GOCI) was the world's first geostationary water color satellite. It was positioned in a geostationary orbit at approximately 128.2 degrees east longitude, providing continuous monitoring of the oceanic and ICW conditions over the East Asian region. Subsequently, South Korea launched the GOCI-II in 2020. The GOCI-II has many improvements over the GOCI, with more water color bands and higher spatial resolution [56–58].

Compared with MODIS, VIIRS, or other commonly used water color sensors, the most attractive characteristic of the GOCI was its high temporal resolution (1 h, eight times daily from 0 UTC to 7 UTC), which is important for the accurate monitoring of water quality, which can change on an hourly or even sub-hourly basis in ICWs [19,59–63]. The GOCI was the first sensor with the ability to continuously monitor water quality variations in the ICWs with an hourly temporal resolution. The GOCI provided images to monitor the water quality of oceans and ICWs in its target area (about 2500 km × 2500 km centered on the Korean Peninsula) for nearly ten years, from the first image in 2011 to 2021, when it stopped observations.

Systematic reviews of the research progress and challenges in applying the GOCI to inland coastal waters (ICWs) have been scarcely documented until now [64–69]. Many countries have launched or plan to launch geostationary water color satellites such as the GOCI-II and the Geostationary Littoral Imaging and Monitoring Radiometer (GLIMR) [62]. The insights garnered by reviewing the GOCI's functionality have helped design sensors and construct water quality algorithms. This study aims to comprehensively review the GOCI, providing information on the configuration parameters of the GOCI sensor, an overview of the GOCI, an AC algorithm, AB monitoring, estimation of water quality parameters (chlorophyll a (Chla), suspended particulate matter (SPM), water clarity. In addition, we also reviewed other parameters such as CDOM, particulate organic carbon (POC), dissolved organic carbon (DOC), net primary production (NPP), sea surface salinity (SSS), sea surface currents (SSCs), sea ice, sea fog, lake ice, and the diffuse attenuation coefficient (k_d), etc.). We also discuss advantages and disadvantages as well as future trends (Figure 1). We compare its spatial and temporal resolutions with those of MODIS, Landsat, Sentinel, and other water color satellites, highlighting the high observation frequency of the GOCI in Table 2.

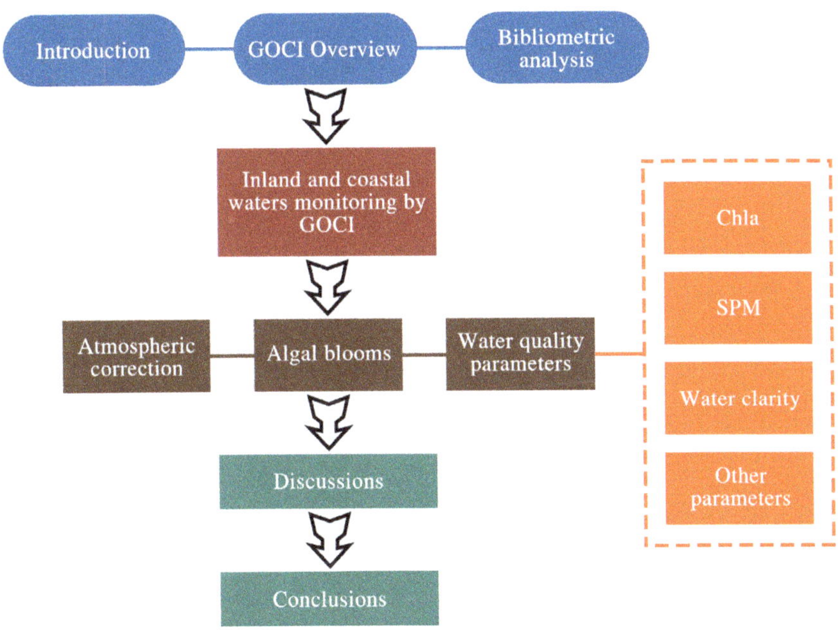

Figure 1. Structure of this paper.

Table 2. Comparison of spatial and temporal resolutions of typical water color remote sensing satellites and their launch dates.

Number	Data	Spatial Resolution (m)	Temporal Resolution	Launched Time
1	CZCS	1000	One day	1978
2	SeaWiFS	1100, 4500	One day	1997
3	MODIS_TERRA	250, 500, 1000	One day	1999
4	MODIS_AQUA	250, 500, 1000	One day	2002
5	VIIRS Suomi NPP	375, 750	One day	2011
6	VIIRS NOAA-20	375, 750	One day	2017
7	VIIRS NOAA-21	375, 750	One day	2021
8	MERIS	300, 1200	Three days	2002
9	Sentinel-3A OLCI	300	<Two days	2016
10	Sentinel-3B OLCI	300	<Two days	2018
11	ADEOS	700	Ten days	1996
12	ADEOS-II	250, 1000	Four days	2002
13	SGLI	250	One day	2017
14	HY-1A	250	Three days/Seven days	2002
15	HY-1B	250	One day/Seven days	2007
16	HY-1C	250	One day/Three days	2018
17	HY-1D	250	One day/Three days	2022
18	HY-1E	100	One day/Three days	2023
19	Oceansat-1	360	Two days	1996
20	Oceansat-2	360	Two days	2009
21	Oceansat-3	360	One day/Three days	2022
22	OSMI	1000	Three days	1999
23	GOCI	500	One hour	2010
24	GOCI-II	250	One hour	2020

2. GOCI Overview

The Communication Ocean and Meteorological Satellite (COMS), launched in 2010, is South Korea's first geostationary orbit meteorological satellite and was designed for meteorological services and ocean monitoring research. It was equipped with a new generation of GOCI sensors, which acquired data in eight bands (including six visible light bands and two NIR bands), as shown in Table 3. Its main advantage was to provide satellite observation of the target area eight times daily, from 0 UTC to 7 UTC. Real-time monitoring was centered on South Korea (36°N 130°E) and also encompassed North Korea, Japan, Russia, and China, including the Bohai Sea, the Yellow Sea, and the East China Sea (Figure 2). The total coverage area was about 2500 km × 2500 km [70].

Table 3. GOCI bands and configuration parameter information.

Bands	Center Wavelength/nm	Band Width/nm	Spectrum Type	Signal-to-Noise Ratio
B1	412	20	VIS	1077
B2	443	20	VIS	1199
B3	490	20	VIS	1316
B4	555	20	VIS	1223
B5	660	20	VIS	1192
B6	680	10	VIS	1093
B7	745	20	NIR	1107
B8	865	40	NIR	1009

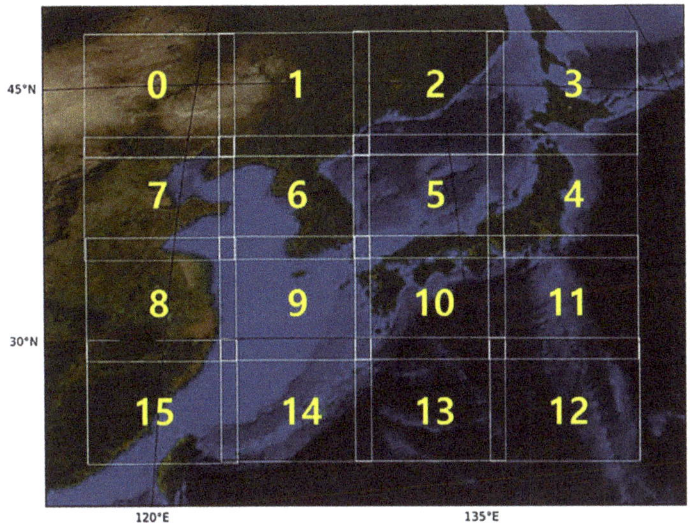

Figure 2. The regional observation area of GOCI (https://kosc.kiost.ac.kr/index.nm?menuCd=43&lang=en, accessed on 25 January 2024).

The GOCI has fulfilled its mission for nearly a decade by playing a significant role in ICW monitoring. The GOCI-II, which succeeded the GOCI, was launched in February 2020 and has up to 13 bands (Table 4). Moreover, the GOCI-II has a spatial resolution of up to 250 m, generating ten views of imagery a day (23 UTC to 8 UTC) [71].

Table 4. Band parameters information for GOCI-II.

Band	Wavelength/nm	Bandwidth/nm	Primary Use
B1	380	20	CDOM
B2	412	20	CDOM, Chla
B3	443	20	Chla absorption maximum
B4	490	20	Chla, other pigments
B5	510	20	Chla, absorbing aerosol in ocean waters
B6	555	20	Turbidity, SPM
B7	620	20	Detect phytoplankton species
B8	660	20	Baseline of fluorescence signal, Chla, SPM
B9	680	10	Fluorescence signal
B10	709	10	Fluorescence base signal, AC, SPM
B11	745	20	AC, vegetation index
B12	865	40	AC, aerosol optical depth
B13	PAN	483	/

3. Bibliometric Analysis

Over the last decade, the GOCI has been increasingly used in remote sensing monitoring research of ICWs due to its advantages and development potential, with a surge in research papers. Our bibliometric analysis, based on Web of Science queries with "GOCI" and "remote sensing" as keywords, retrieved a total of 164 papers, of which 123 met eligibility criteria. The number of publications showed a steady increase, starting from 4 papers published in 2010 to a peak of 13 in 2021 and 2022 (Figure 3 and Table S2). The study areas mentioned in these papers are summarized in Figure 4. Additionally, we find that articles have been published in more than 40 journals, of which 36 were published in *Remote Sensing* (RS); followed by 16 in *Remote Sensing of the Environment* (RSE); 7 in the *Ocean Science Journal* (Ocean Sci J); 6 in *Optics EXPRESS* (OE); 4 in the *Science of the*

Total Environment (Sci Total Environ), *International Journal of Remote Sensing* (Int J Remote Sens), and *Environmental Science and Pollution Research* (Environ Sci Pollut Res); and 3 in *Sensors*, *ISPRS Journal of Photogrammetry and Remote Sensing* (ISPRS), and *IEEE Transaction on Geoscience and Remote Sensing* (IEEE T-GRS) (Figure 5).

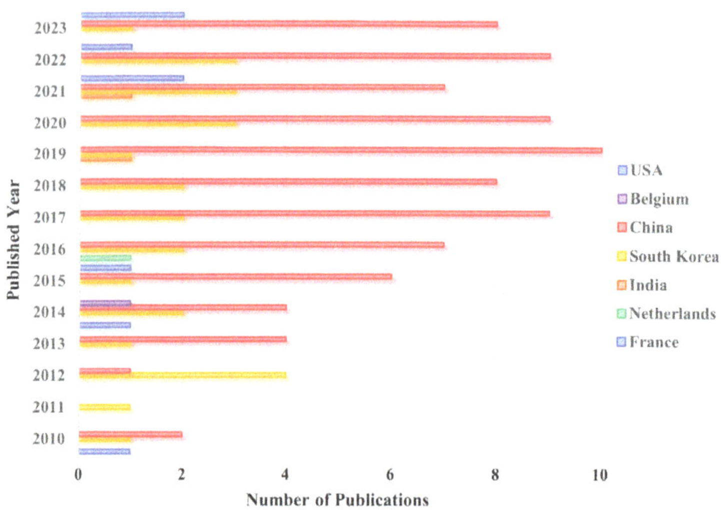

Figure 3. Number and countries of papers published related to GOCI between 2010 and 2023.

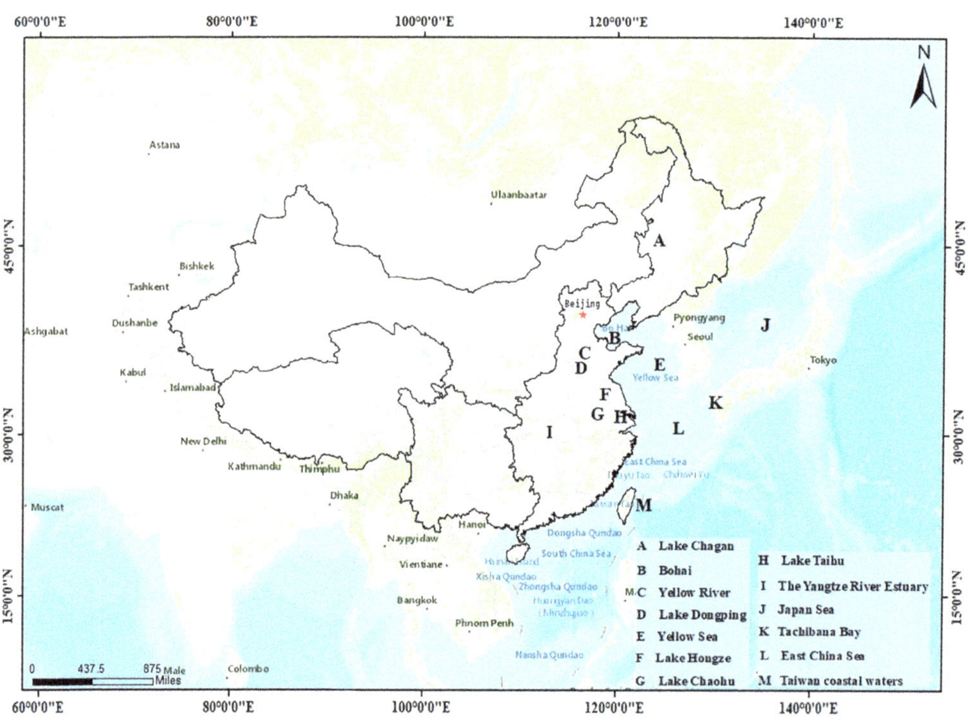

Figure 4. Map of the study area.

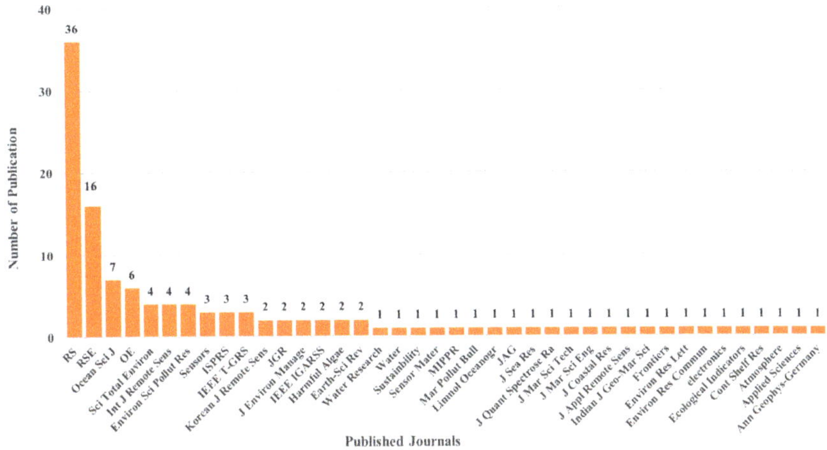

Figure 5. Journals and number of publications.

Using "GOCI", "remote sensing," and "inland and coastal waters" as the subject terms in the "Web of Science" search, we retrieved a total of 1546 eligible authors from the core collection. A VOSviewer statistical analysis of keywords using "author name" was used to form a keyword map (Figure 6). The authors with the highest number of occurrences were Kim, Jhoon, Ryu, and Joo-hyung, totaling 36 occurrences. The highest number of citations was 1090. Recognizing the significant contributions of Chinese scholars, we extended our search to the China National Knowledge Infrastructure (CNKI), merging the search of Chinese and English papers to obtain a clearer picture of the application of the GOCI. A total of 746 documents were retrieved through CNKI with GOCI as the keyword, of which 323 were classified and analyzed according to the "subject". Figure 7 shows that most research was on "SPM" and "Chla", followed by "Estimation studies" and "Spatial and temporal resolution", which can be seen from the size of the box area.

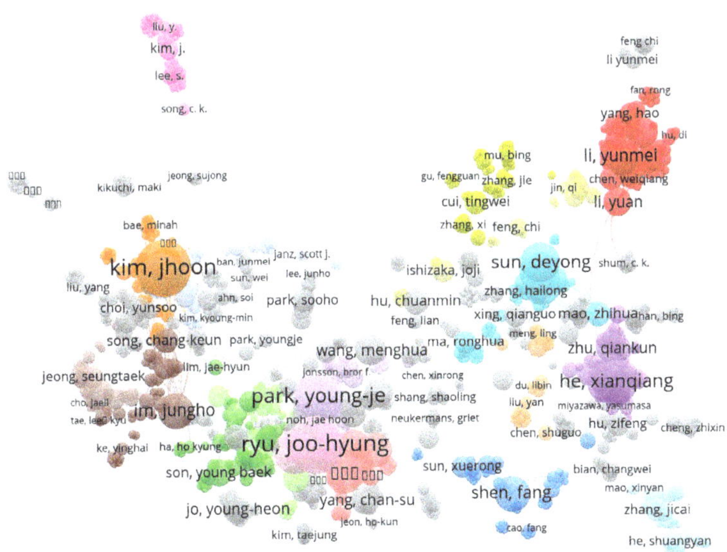

Figure 6. Keyword mapping from the Web of Science, searched by author name.

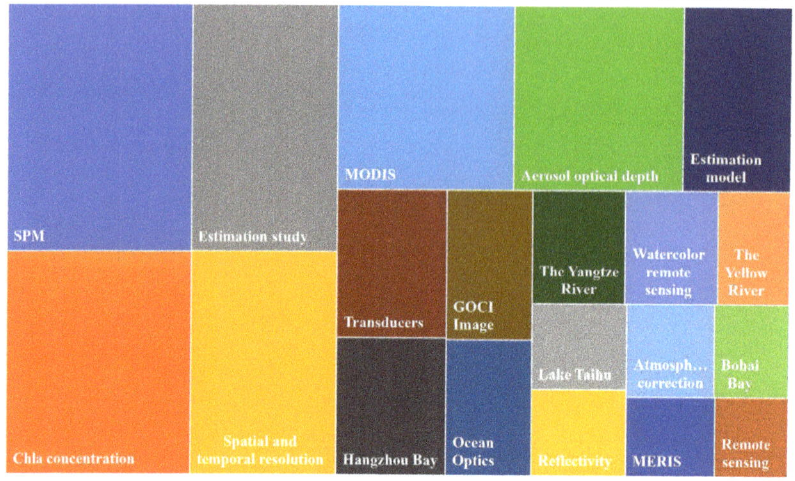

Figure 7. Overview of the application of GOCI retrieved by subject terms from CNKI.

In summary, it can be seen that awareness about GOCI satellites is increasing annually, and their application in various fields is becoming increasingly extensive. In the past decade, the application of the GOCI in ICWs has become more and more comprehensive. We found that the highest number of studies on Chla concentration estimation by scholars was 23 articles, followed by SPM concentration estimation (22) and ABs (17), accounting for 19%, 18%, and 14%, while there were 13 and 10 studies on AC and water clarity, accounting for 11% and 8%, respectively (Figure 8).

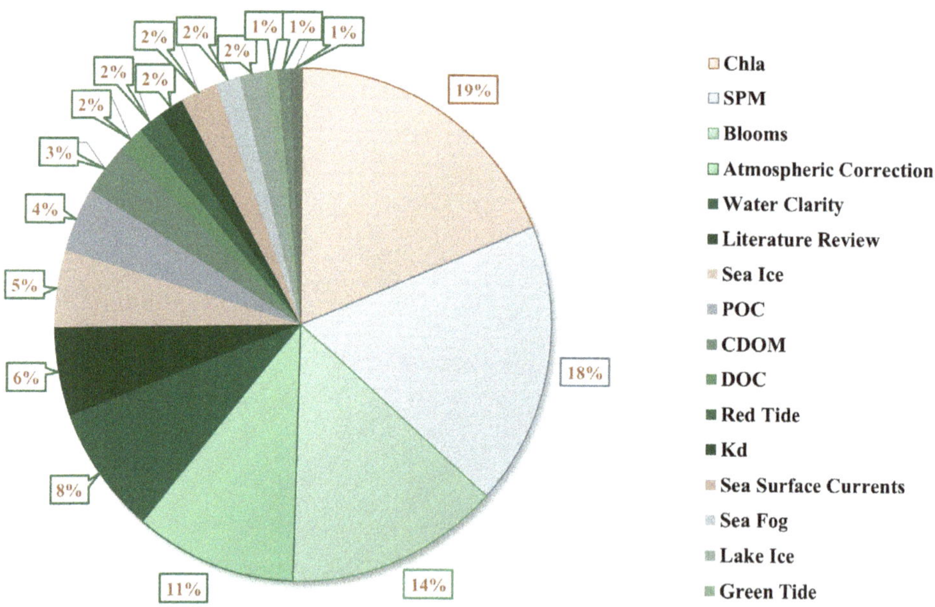

Figure 8. Summary and proportion of the applications of GOCI in inland and coastal waters.

4. Inland and Coastal Waters Monitoring by the GOCI

The GOCI has been entering a rapid development phase in ICW use to offer more services for water quality detection, environmental assessment, and monitoring sudden events. The GOCI will become an important instrument for water color monitoring in the future and will, simultaneously, provide a reference for the design of future water color sensors. In this section, we review the capabilities of the GOCI. We first provide an overview of AC algorithms. We compare the advantages and disadvantages of different methods and their applicability. Secondly, we highlight the research progress of scholars on AB monitoring. Finally, the estimation of water quality parameters is reviewed and divided into four aspects: Chla, SPM, water clarity, and other water quality parameters.

4.1. Atmospheric Correction of GOCI Images

AC is the process of obtaining the reflectance from the top of the atmosphere as observed by satellites. This process is vital because the $R_{rs}(\lambda)$ is the primary input for most water-quality remote sensing algorithms [72–74]. The GOCI Data Processing System (GDPS) is the official data processing software for the GOCI. The standard AC algorithm of the GOCI evolved initially based on the algorithms of the SeaWiFS and MODIS but then expanded and improved according to the characteristics of the GOCI's own instrument design and research fields [70,75–78]. This model followed a regional empirical relationship between the $R_{rs}(660)$ and two NIR bands ($R_{rs}(745)$ and $R_{rs}(865)$) (denoted as SR660). Aerosol properties from neighboring non-turbid waters were also incorporated to finalize the AC process [75]. Ahn [79] improved the SR660 model by introducing two AC models. The first AC model used $R_{rs}(709)$ to replace the $R_{rs}(660)$, taking advantage of the lower contribution of the absorption signal of non-pure water constituents in $R_{rs}(709)$. SR709 has been shown to perform well in waters with only one sediment type. The second AC model used the 620 nm radiation (denoted as SRIOP) to estimate the $R_{rs}(\lambda)$ in the NIR bands. SRIOP showed good performance in waters with a variety of sediment types.

Another software, the SeaWiFS Data Analysis System (SeaDAS), provides an AC module to the GOCI, including the Management Unit of the North Sea Mathematical Models (MUMM) algorithm, which is used to produce $R_{rs}(\lambda)$ products for the turbid Hangzhou Bay and Yellow River Estuary. MUMM assumes that the magnitude of $R_{rs}(\lambda)$ at 745 and 865 nm is dominated by the backscattering of particles. The ratio of $R_{rs}(745)$ and $R_{rs}(865)$ is assumed to be constant. As aerosol types do not usually vary over spatial scales of about 100 km, the ratio of aerosol reflectance in these two bands can also be regarded as a constant. However, this assumption of $R_{rs}(865)/R_{rs}(745)$ is deemed to be invalid for highly turbid waters. Therefore, Ahn [80], Goyens [81], and Jiang [17] used a second-order polynomial relationship to parametrize $R_{rs}(745)$ and $R_{rs}(865)$ for the AC of the GOCI in highly turbid waters.

The Polymer software also offers a GOCI AC module. However, it is seldom used for the GOCI, even though it has been shown to perform well in turbid coastal waters when applied to satellites like MODIS and OLCI. For extremely turbid waters, such as in parts of Changjiang Estuary, Huanghe Estuary, and Hangzhou Bay, where the SPM can be greater than 500 mg L^{-1}, researchers have proposed innovative AC algorithms for the GOCI. Pan [82] proposed an improved GOCI AC algorithm for the Changjiang (Yangtze) estuary and the adjacent coast. This approach, based on a spectral optimization algorithm (ESOA) with a coupled water–atmosphere model, employed the simple semi-empirical radiative transfer (SERT) model to parameterize $R_{rs}(\lambda)$ based on the SPM concentrations and the aerosol models proposed by [56]. SeaDAS was used to calculate the contributions in aerosols, and the genetic algorithm optimized the ESOA AC process. For highly turbid waters, the $R_{rs}(\lambda)$ in the UV band may be much less than $R_{rs}(\lambda)$ in the NIR band, enabling the UV band's use in estimating aerosol scattering. The UV-AC algorithm was applied to the GOCI by using the 412 nm band as the UV band [83]. The retrieved aerosol or apparent optical properties products of other satellite sensors can also be used to help the GOCI obtain the $R_{rs}(\lambda)$ products.

Many researchers have also developed AC methods for the GOCI under non-ideal observation conditions. Men [84] proposed a novel AC algorithm based on deep learning (denoted as DLACC) to improve the quality of $R_{rs}(\lambda)$ data contaminated by cloud edge effects, including stray light, cloud shadows, and cloud adjacent effects. The observation characteristics of GOCI images from morning to evening also determine that many images have a larger solar zenith angle (SZA) (>70°). It has been reported that a high SZA will increase the residual errors in satellite $R_{rs}(\lambda)$ products due to several factors (e.g., high atmospheric multiple scattering effects and underestimation of the Rayleigh scattering contribution). Li [85] proposed a new neural network (NN) AC algorithm for GOCI data with a high SZA that retrieves $R_{rs}(\lambda)$ directly from Rayleigh-corrected radiance ($L_{rc}(\lambda)$). This NN AC algorithm can retrieve $R_{rs}(\lambda)$ in the morning or evening when NIR and KOSC algorithms fail.

4.2. Algal Blooms

ICWs, due to their high population density, are prone to ABs when certain meteorological conditions arise [86–93]. The impact of ABs, such as the reduction in dissolved oxygen and the release of toxins, presents major challenges to aquatic ecosystem management [9,94–96]. Monitoring ABs is essential for improving water quality and the inhabitants' quality of life and reducing unnecessary losses. A variety of strategies have been developed to monitor and understand algal blooms, integrating satellite data, modeling, and machine learning techniques for improved accuracy and efficiency [97].

As presented in Section 4.1, most existing AC algorithms do not apply to the pixels affected by harmful algal blooms (HABs) or red tides [79,98–101]. Kim [102] developed PUK, a red tide quantification algorithm for the GOCI that does not require an independent AC process, incorporating a radiative transfer simulation combined with a machine-learning technique. Other approaches for extracting the red tide area from the GOCI include the linear mixed model (LMM) and the normalized difference vegetation index (NDVI) threshold method [103]. Also, an improved red tide index (RI) was established by Lou [67] using normalized $L_{rc}(\lambda)$ data at 443, 490, and 555 nm to monitor red tide changes dynamically in the East Lake. Additionally, Lee [104] proposed a newly developed normalized red tide index (NRTI) (calculated with Equations (1)–(4)), utilizing a spectroradiometer to observe the spectral characteristics of red tides and seawater in real-time.

$$P555 = R_{rs}555 - \left(R_{rs}660 + \frac{660-555}{660-490}(R_{rs}490 - R_{rs}660)\right) \tag{1}$$

$$P680 = R_{rs}680 - \left(R_{rs}745 + \frac{745-680}{745-660}(R_{rs}660 - R_{rs}745)\right) \tag{2}$$

$$RTI = \frac{P555}{R_{rs}490} \times \frac{P680}{R_{rs}660} \tag{3}$$

$$NRTI = \frac{RTI}{R_{rs}555 - R_{rs}745} \tag{4}$$

Lagrangian particle tracking experiments allow us to understand the paths of floating green algae patches and explain the physical forcing factors that affect the distribution and advection of floating green algae. Using a combination of GOCI data and numerical simulations helped track and monitor extensive areas in the Yellow Sea and East China Sea [105]. Furthermore, Son [106] developed the Index of floating Green Algae for the GOCI (IGAG) (Equation (5)) using multispectral band ratios of three wavelengths (555, 660, and 745 nm) in the Yellow Sea and East China Sea.

$$IGAG = \frac{R(555) + R(660)}{R(745) - R(660)} + \frac{R(745)}{R(660)} \tag{5}$$

A U-net model for the extraction of long-term spatial variations in ABs along the East China Sea was developed using the GOCI and was trained on two different datasets—six-band channels (all visible bands) and RGB-band channels (443, 555, and 680 nm) [107]. Qiu [108] used a Multi-Layer Perceptron (MLP) machine learning algorithm to establish a novel automatic method to continuously monitor ABs in the Yellow Sea. The method consists of two MLP models, which consider both spectral and spatial features of R_{rc} maps. Accuracy assessment and performance comparison showed that the proposed method has the capability to provide highly accurate prediction maps of macroalgal blooms (F1-score approaching 90% or more) with more robustness than the traditional methods. A novel subpixel-level area estimation method for green algae blooms based on spectral unmixing can detect the existence of green algae and determine their proportion in each pixel. A fast endmember extraction method was proposed to calculate the endmember spectral matrix and the abundance map of green algae automatically, which outperformed existing techniques in experiments on GOCI data [109,110]. Qi [111] used an algal index algorithm to quantify equivalent surface algae density (σ, 0–100%) at pixel and synoptic scales from each cloud-free image and offered an analysis of diurnal changing patterns of σ. Ai [112] and Wang [113] used the alternative floating algae index (AFAI) (Equations (6) and (7)), indicated as follows:

$$AFAI = R_{rc}(\lambda_2) - R_{rc}(\lambda_2)' \quad (6)$$

$$R_{rc}(\lambda_2)' = R_{rc}, \lambda_2 + (R_{rc}, \lambda_3 - R_{rc}, \lambda_1) \times \frac{\lambda_{2-}\lambda_1}{\lambda_3 - \lambda_1} \quad (7)$$

where the subscripts refer to bands (λ_1 = 660 nm, λ_2 = 745 nm, and λ_3 = 865 nm). A classification method was then devised to classify the daily change patterns of ABs automatically into four types to monitor the daily dynamics in Lake Taihu. Li [114] used the cyanobacterial bloom intensity (CBI) index (Equation (8)), combined with meteorological factors, water stability, and nutrient factors, to monitor hourly changes in cyanobacterial blooms in Lake Taihu, indicated here:

$$CBI = \sum\nolimits_{i=1}^{n} (AFAI_i + 0.001) \quad (8)$$

Xue [110] developed a new floating algae cover (FAC) model following a logistic curve and applied it to China's two large shallow eutrophic lakes, Lake Taihu and Lake Chaohu.

4.3. Water Quality Parameters

The estimation of water quality parameters involves the use of R_{rc} and the underwater irradiance ratio to estimate Chla concentration, SPM concentration, and yellow substance content by calculating the absorption coefficient of phytoplankton, the absorption coefficient of yellow substances, the absorption coefficient of non-pigmented SPM, and the backscattering coefficient ($b_{bp}(\lambda)$) of SPM [65,115,116]. Most commonly, empirical or semi-empirical and artificial intelligence models have been used [117–121]. This section mainly introduces Chla in Section 4.3.1, SPM in Section 4.3.2, and water clarity in Section 4.3.3, and then it briefly introduces other water quality parameters such as CDOM, POC, DOC, NPP, SSS, SSCs, sea ice, sea fog, lake ice, and k_d, among others, in Section 4.3.4.

4.3.1. Chla

The concentration of Chla is the most prevalent pigment in algae and relates to phytoplankton biomass [97,122–127]. Chla concentration is a key indicator of water quality and aquatic ecological health [128–130].

The GOCI official process software suggests using the empirical ocean Chla algorithm (OC3G) for the GOCI using the 443, 490, and 555 bands (Equation (9)) [131], as follows:

$$Chla = 10^{f_0 + f_1 * R + f_2 * R^2 + f_3 * R^3 + f_4 * R^4} \quad (9)$$

where R is calculated as

$$R = \frac{Max(R_{rs}(443),\ R_{rs}(490))}{R_{rs}(555)} \quad (10)$$

and the values of f_0, f_1, f_2, and f_3 are 0.0831, −1.9941, 0.5629, 0.2944, and −0.5458, respectively.

Various studies have explored alternative models and algorithms for Chla estimation in ICWs. Hong [94] explored the physical–biological interaction between anticyclonic mesoscale eddies and Chla secondary blooms after spring blooms in the East Sea and Japan Sea using the GOCI OC3G Chla products. Zhao [132] found diurnal changes in GOCI-derived Chla, with generally higher values in the afternoon than in the morning, and identified heterogeneities in the temporal and spatial domains. Sakuno [133] applied the Linear Combination Index (LCI) algorithm to the GOCI to map Chla distribution in Tachibana Bay, Japan, and found OC3G to have relatively large uncertainties in turbid ICWs. Park [129] developed MFNN, a neural algorithm that can classify the pixels of GOCI images into three Chla concentration categories (normal, abnormally high, and abnormally low) so that outliers can be identified and provide quality control of GOCI Chla products.

A spectral index method, such as the fluorescence line height (FLH) [134], helped construct a band conversion relationship from 681 to 685 nm to establish a phytoplankton retrieval model suitable for the GOCI. Wang [68] calibrated three empirical spectral index algorithms (OC4, OC5, and Yellow and East China Sea Ocean Color (YOC)) and constructed four machine learning algorithms (BP neural network (BPNN), random forest (RF), AdaBoost, and support vector regression (SVR)) using numerous in situ data collected in the Bohai Sea and Yellow Sea. The comparison illustrated that machine learning methods were more accurate than classic spectral index algorithms and that RF performed best. The waters in Hangzhou Bay were classified into moderately and extremely turbid based on the ratio of $R_{rs}(745)/R_{rs}(490)$ [135]. Then, the OC3G and modified synthetic chlorophyll index (SCI) index were used for these two water types to map Chla distribution in this bay [135]. Fan [97] modified the generalized algal bloom index algorithm (GABI) to map Chla in the turbid Changjiang Estuary. For this water region, the modified GABI algorithm showed better accuracy and stability than the two-band ratio, as well the three-band, and four-band spectral algorithms.

Lake Taihu is a favorite research area for GOCI Chla model construction and application in inland waters. Bao [136] proposed a weighted algorithm to estimate Chla concentrations based on spectral classification and weighted matching using normalized mutual information (NMI). Huang [137] proposed a regional NIR red ratio algorithm ($R_{rs}(745)/R_{rs}(680)$) for Chla retrieval in Lake Taihu. Based on this model, the hourly Chla maps of Lake Taihu were produced from 6 August to 9 August 2013. In this paper, the GOCI-derived Chla products revealed clear evidence of hourly spatial and temporal variations in Chla in the eutrophic inland waters. The $R_{rs}(745)/R_{rs}(680)$ was also used by Du [115] to analyze the spatio-temporal dynamics of Chla and their response to river flow in the estuary of Lake Taihu. Duan [138] proposed a three-band model using GOCI 681, 745, and 865 nm bands. Guo [139] proposed another three-band Chla model ($R_{rs}^{-1}(681) - R_{rs}^{-1}(709) * R_{rs}(753)$) for the GOCI. Simulated and in situ data illustrated that the three-band model outperformed the two-band model under various SPM conditions. Guo [140] proposed a bio-optical hyperspectral reconstruction (BBHR) algorithm to generate simulated hyperspectral $R_{rs}(\lambda)$ ranging in wavelengths from 400 to 800 nm from the GOCI and five widely used water color sensors. Simulated hyperspectral $R_{rs}(\lambda)$ facilitate the construction of more accurate Chla estimation models for optically complex turbid inland waters. Guo [141] compared the performance of different Chla algorithms for the GOCI and the GOCI-II in Lake Taihu, Lake Chaohu, and Lake Hongze, with $L_{rc}(\lambda)$ as input. The results indicated that the RF algorithm outperformed the commonly used spectral index models.

4.3.2. SPM

SPM refers to the substances suspended in a water body and can be used to characterize its cleanliness [9,142–145]. SPM includes inorganic substances (such as minerals, salts, and metals), particles from algae, silt and clay, and other organic matter [146]. SPM directly affects the propagation of light in the water body, which in turn affects the aquatic ecological environment and ultimately determines the primary productivity of the lake [20,147–149].

The initial GOCI SPM standard product was based on the algorithm proposed by Ahn [150] and Moon [151] as Equation (11), indicated here:

$$SPM = 945.07 * (R_{rs}(555))^{1.137} \tag{11}$$

Ruddick [152] applied the SPM model proposed by Nechad [153] to the GOCI as Equation (12) for comparison with the first official SPM algorithm above, as follows:

$$SPM = A_{745} * \frac{\pi * R_{rs}(745)}{1 - \pi R_{rs}(745)/C_{745}} \tag{12}$$

where A_{745} equals 1755, and C_{745} equals 0.198. This paper found that GOCI data can be used to qualitatively study SPM dynamics except in extremely turbid waters, which are masked out of the GOCI official $R_{rs}(\lambda)$ products. The following standard GOCI SPM was retrieved by the YOC algorithm based on Siswanto's SPM retrieval model for the Yellow and East China Sea (as Equation (13)) [154–156]:

$$SPM = 10^{(0.649+25.623*(R_{rs}(555)+R_{rs}(670))-0.646*(\frac{R_{rs}(490)}{R_{rs}(555)}))} \tag{13}$$

He [83] developed a two-band ratio SPM algorithm ($R_{rs}(745)/(R_{rs}(490))$ for Hangzhou Bay and nearby coastal waters. The diurnal variations derived by the GOCI showed a good relationship with the buoy data in the region. Yu [157] proposed an experiential SPM algorithm suitable for the Bohai Sea and Yellow Sea to evaluate the effects of typhoons on SPM by carrying out parameter optimization and genetic programming, as with Equation (14).

$$\lg(SPM) = 4.8581 + 0.8206 * \frac{R_{rs}(745)}{R_{rs}(490)} - 0.9998 * \frac{R_{rs}(555)}{R_{rs}(490)} - 3.6504 * \sqrt{\frac{R_{rs}(490)}{R_{rs}(555) + R_{rs}(660)}} \tag{14}$$

Also, using this model, Cheng [158] studied dynamic variations in SPM in China's macro-tidal Yalu River estuary. Choi [159] and Meng [160] developed a single-red band model ($R_{rs}(660)$) to retrieve SPM on the west coast of the Korean Peninsula and Bohai Sea. Liu [148] developed an empirical SPM retrieval algorithm based on the spectral absorption index (SAI) in highly turbid Hangzhou Bay with Equations (15) and (16).

$$SAI = \frac{d * R_{rs}(490) + \left(1 - \frac{555-490}{745-490}\right) * R_{rs}(745)}{R_{rs}(555)} \tag{15}$$

$$SPM = 10^{2.01814*SAI+0.83774} \tag{16}$$

Lyu [119] developed a two-step SPM retrieval model using in situ simulated GOCI $R_{rs}(\lambda)$ and SPM data collected in Lake Taihu, Lake Chaohu, Lake Dianchi, and the Three Gorges Reservoir. Firstly, a clustering method was applied to classify the turbid inland waters into three types (Type 1, Type 2, and Type 3). Secondly, the best performing of the three SPM algorithms was gauged and determined via comparison of their results (Type 1: $R_{rs}(865)/R_{rs}(555)$, Type 2: $R_{rs}(745)/R_{rs}(555)$, and Type 3: $R_{rs}(745)/R_{rs}(555)$). Similarly, Huang [137] proposed another classification-based SPM inversion algorithm in Lake Taihu.

The waters were classified into two types based on the $R_{rs-depth}(680)$ index calculated with Equation (17), indicated as follows:

$$R_{rs-depth}(680) = \left\{ R_{rs}(660) + [R_{rs}(745) - R_{rs}(660)] * \frac{745 - 680}{745 - 660} \right\} - R_{rs}(680) \quad (17)$$

For the waters with $R_{rs-depth}(680) \leq 0.005$, the two-band ratio (Equation (18)) was used, and for the waters with $R_{rs-depth}(680) > 0.005$, the three-band algorithm (Equation (19)) was used.

$$BR_{SPM} = 10^{2.2135/(1+\exp(-(\frac{R_{rs}(745)}{R_{rs}(555)} - 0.1223/0.1439)))} \quad (18)$$

$$TBA_{SPM} = 10^{2.314/(1+\exp(-(\frac{R_{rs}(745)+R_{rs}(680)}{R_{rs}(555)} - 0.6263)/0.431))} \quad (19)$$

Du [115] proposed a single NIR band algorithm ($R_{rs}(745)$) to retrieve SPM in the estuary of northwest Lake Taihu. Zhang [161] developed a semi-analytical SPM model for estimating SPM using the satellite in situ matchup data collected in Lake Taihu and Hangzhou Bay. He [162] calibrated the SPM model proposed by He [83] to map SPM products using the GOCI in Lake Taihu as Equation (20).

$$SPM = 10^{0.61+1.43*R_{rs}(745)/R_{rs}(490)} \quad (20)$$

The GOCI-derived SPM products are consistent with a two-dimensional model that couples shallow water and SPM transport dynamics. The high-spatial-resolution GOCI SPM product combined with the hydrodynamic model provides the opportunity to monitor the suspended solids transport with high-time frequency monitoring. Xu [145] developed a quadratic polynomial model of one variable ($R_{rs}(745) + R_{rs}(865)$) to map SPM in Lake Taihu (Equation (21)).

$$SPM = 32,517 * (R_{rs}(745) + R_{rs}(865))^2 + 1327.2 * (R_{rs}(745) + R_{rs}(865)) + 0.5 \quad (21)$$

From the hourly GOCI-derived SPM products, they found that the Tiaoxi River greatly influenced the spatial distributions of SPM in Lake Taihu by the plumes after heavy precipitation events. Lei [147] developed a multivariate linear stepwise regression method for the retrieval of the surface horizontal distribution of SPM using the GOCI. Then, the estimation model of the vertical structure of underwater SPM was constructed using layer-by-layer recursion. These efforts retrieved the horizontal and vertical distribution of SPM concentration from GOCI data over Lake Hongze.

4.3.3. Water Clarity

Water clarity is a critical parameter for the optical properties of ICWs [145,163–166], vital for characterizing the turbidity and absorption/scattering of light by water bodies. It directly relates to the inherent optical properties of waterbodies, such as Chla, SPM, and phytoplankton biomass [167–172].

To estimate water clarity using GOCI data, Zhou [172] proposed a regional linear corrected Secchi disk depth (SDD) model based on the mechanical model developed by Lee [173] (denoted as SDD_{Lee}). This model uses the GOCI $R_{rs}(\lambda)$ derived by the MUMM in Jiaozhou Bay, and its formulation is given in Equation (22) as follows:

$$SDD_{Lee} = \frac{1}{2.5Min(K_d^{tr})} In(\frac{|0.14 - R_{rs}^{tr}|}{0.013}) \quad (22)$$

where $Min(K_d^{tr})$ is the minimum K_d in the visible bands (412, 443, 490, 555, 660, and 680 nm), while R_{rs}^{tr} is the $R_{rs}(\lambda)$ corresponding to the waveband with the minimum K_d value. The model considers factors like SZA, tides, and wind force to control diurnal variations. Liu [174] validated the suitability of SDD_{Lee} using in situ $R_{rs}(\lambda)$ products and

SDD in Jiaozhou Bay and Qingdao coastal areas. Liu [175] compared the retrieved SDD by different atmospheric correction products, and the previous conclusions regarding the hourly variations in biochemical parameters using the GOCI were improved. Ref. [168] found that the retrieved SDD by SDD_{Lee} was underestimated for their in situ data collected in the Bohai Sea and Yellow Sea due to the uncertainties in the estimation of $K_d(490)$ in SDD_{Lee} in turbid waters. They replaced the original $K_d(490)$ estimation method using a weighted $K_d(490)$ proposed by Mao [176] combined with two $K_d(490)$ estimation methods (one for clear and one for turbid water) and a weighted coefficient. Ding [177] used the GOCI-derived Chla and SPM products to calculate $a(\lambda)$ and $b_b(\lambda)$ and then input them into the semi-analytical model proposed by He [178] and He [179] (denoted as SDD_{He}) to retrieve SDD in the eastern Chinese seas with Equation (23).

$$SDD_{He} = \frac{1}{4(a(\lambda)+b_b(\lambda))} In(\frac{\rho_p\alpha\beta(a(\lambda)+b_b(\lambda))}{C_e f b_b(\lambda)}) \tag{23}$$

Many studies have shown that SDD_{Lee} is biased in estimating SDD in inland waters [167,180]. Bai [167] used a linear relationship to correct SDD_{Lee} (denoted as SDD_{Bai}) and applied this algorithm to the GOCI in Lake Taihu with as Equation (24).

$$SDD_{Bai} = 2.68 * \frac{1}{2.5Min(K_d^{tr})} In\left(\frac{|0.14 - R_{rs}^{tr}|}{0.013}\right) + 0.0046 \tag{24}$$

Zeng [181] proposed an improved Quasi-Analytical Algorithm (QAA)-estimated SDD algorithm (denoted as SDD_{Zeng}) for the GOCI using in situ data collected in Lake Taihu, Lake Hongze, Lake Dongting, and Lake Erhai. SDD_{Zeng} firstly estimated $b_b(745)$ using QAA with an assumption of the pure water absorption ($a_w(745)$), which is dominated at this band ($a(745) \approx a_w(745)$). $K_d(745)$ was estimated using Equation (25) as follows:

$$K_d(745) = (1 + 0.0124 * \theta_s) * a_w(745) + 3.16 * (1 - 0.52 * \exp(-10.8 * a_w(745))) * b_b(745) \tag{25}$$

where θ_s represents the SZA. Zeng [181] found that the $Min(K_d^{tr})$ is at 555 nm in Lake Taihu and Lake Hongze. Thus, a linear relationship was built to convert $K_d(745)$ to $K_d(555)$.

$$K_d(555) = 0.99 * K_d(745) - 1.96 \tag{26}$$

The retrieved $K_d(555)$ was used to retrieve SDD using Equation (22).

In this section, we summarized the proposed algorithms of the GOCI by researchers in ICWs classified by different water quality parameters. These algorithms can be classified as empirical algorithms, semi-analytical algorithms, analytical algorithms, and machine learning algorithms. For Chla, the most widely used algorithms include OC3G (official algorithm), LCI, FLH, SCI, CABI, YOC, BPNN, RF, SVR, NMI, and BBHR. For SPM, the main used algorithms are three-band semi-analytical algorithms and empirical band combination algorithms (e.g., SAI). For SDD, the QAA and improved QAA algorithms are the most widely used.

4.3.4. Other Parameters

In addition to the main water quality parameters above, scholars have also conducted research on CDOM, POC, DOC, NPP, SSS, SSCs, sea ice, sea fog, lake ice, k_d, and other parameters. CDOM is the main constituent of dissolved organic matter (DOM) and a key indicator of water quality conditions [182–184]. Based on QAA and QAA_CDOM, Wang [185] developed a new algorithm named QAA_cj to estimate CDOM concentration. The algorithm can easily be applied to high-turbidity coastal waters. A simple approach to estimate the CDOM concentration was also proposed based on the datasets derived from in situ measurements during four cruise surveys over the Bohai Sea and Yellow Sea. Eight band combination forms (using Xi as a delegate, where i denotes the numerical order of band combination forms), including single bands, band ratios, and other band combinations by $R_{rs}(\lambda)$ were trained to test the correlations with the CDOM concentrations [186]. POC is an essential form of water carbon, taking part in various biogeochemical processes and influencing both organic and inorganic carbon cycles [187–190]. A remote sensing algorithm that can monitor and accurately estimate the changes in POC fluxes in real-time at the Yangtze Estuary hydrological station was established using GOCI satellite data [191,192]. DOC refers to various dissolved organic molecules ubiquitous in aquatic systems. It is a dynamic reduced carbon pool with important ecological and geochemical functions in inland and coastal waters [193,194]. A regional, red–blue ratio algorithm was proposed to estimate Lake Taihu's surface DOC and analyze its spatiotemporal variation characteristics [195]. Accurate assessments of spatial and temporal changes in NPP in the euphotic zone are critical to understanding the role of the ocean in regulating Earth's climate and assessing future fisheries' production, as well as predicting the impacts of ocean warming, hypoxia, and ocean acidification on marine ecosystems [196–198]. Wu [198] used the Adsorption Based Productivity Model (ABPM) to monitor diurnal and daily surface and euphotic-column integrated NPP. Global changes in SSS can simulate the variations in the exchanges between the atmosphere and the ocean surface, as well as the ocean surface and the deep sea. SSS is also involved in the intensification of the global water cycle. For the estimation of SSS, a multi-layer perceptron neural network (MPNN) was employed to train the nonlinear processes of GOCI-II spectral measurements as inputs and the SSS of the Soil Moisture Active Passive (SMAP) satellite as the target [102]. SSCs are one of the most important physical properties in ocean dynamics and are crucial for understanding ocean physical and biogeochemical processes [69,199,200]. A method based on maximum cross-correlation (MCC) was used to derive changes in SSCs in the highly turbid waters of Hangzhou Bay to capture better tidal phase variations [77]. Sea ice monitoring is vital for marine traffic, fisheries, ports, and weather forecasting. A simple method was proposed that exploited the spatial variability of the sea ice surface and applied spatial convolution filtering, in which the center pixel within the window was replaced by the standard deviation of all pixels within the window, effectively separating thin ice area from open water [201]. Also, accurately detecting sea ice thickness (SIT) is crucial for understanding climate change. Gu [202] proposed an SIT inversion method for Liaodong Bay based on Rayleigh scattering corrected reflectance, which has practical advantages in estimation accuracy and spatiotemporal resolution. This method can also be applied to the global ocean. Also, an object-based feature extraction method and an albedo-based thickness inversion model were used for estimating SIA and SIT, respectively, by Yan [203] to reach a better effect. The emergence of sea fog leads to low visibility and seriously threatens the safety of maritime activities. Therefore, monitoring changes in sea fog is particularly important [204]. Jeon [205] proposed a method to identify sea fog based on the convolutional neural network transfer learning model (CNN-TL). Zhou [204] established a sea fog dataset (SFDD) and a dual-branch sea fog detection network (DB-SFNet). Lake ice is one of the essential climate variables of the cryosphere and is closely related to lake environments, ecological regulation, and the safety of human activities [206–208]. Yang [209] used the enhanced spatial and temporal adaptive reflection fusion model (ESTARFM) to merge Landsat and GOCI images and extracted the length and angles of the linear structure to monitor the hourly changes in the surface morphology of Lake Chagan. k_d is an important optical property of water, which describes the penetration of incident solar radiation in the water column. k_d is a quasi-inherent optical property since it depends on the water constituents and the distribution of the ambient light field [210–213]. An improvement of the parameterization equations in the inverse scheme of the 2Seacolor model was used in the Yangtze Estuary to retrieve k_d [214].

5. Discussions

This section discussed the limitations, uncertainties, and future directions of the GOCI and other geostationary ocean color satellites.

5.1. The Limitations and Uncertainties of Current Studies for the GOCI

Most water color sensors (e.g., MODIS, OLCI, MERIS, and VIIRS) observed the surface water on Earth between the local time of 9:30 a.m. and 2:30 p.m. This is the ideal time window for water color satellite observation. GOCI observation images before or after this time window may face many issues. Firstly, ensuring a sufficient signal-to-noise ratio (SNR) is essential for achieving stable and reliable satellite sensor AC and water quality modeling [215,216]. GOCI images observed in the morning and dusk may have an insufficient SNR, especially in winter [217]. In addition to the SNR, higher observation angles can interfere with the GOCI's AC and water quality modeling processes. For the AC, a high sun zenith angle will increase the residual errors in the satellite $R_{rs}(\lambda)$ products by high atmospheric scattering effects and underestimation of the Rayleigh contribution [24,72]. For water quality modeling, high solar zenith angles (for example, greater than 60°) influence the above-water $R_{rs}(\lambda)$ when extrapolated from the in-water remote sensing ratio. It, in turn, impacts the accuracy of downstream water color retrievals [218]. Until now, we have only found a few papers about obtaining reliable GOCI atmospheric correction and water quality parameter products under unsatisfactory observing conditions [85,131,219].

In addition to high observations, many other issues may reduce the image quality of the GOCI. The sun's glint reflectance of the GOCI is reported to be minor, even in summer [75]. However, GOCI observation images may also be affected by the sun glint through the specular reflection phenomenon [220,221]. The satellite observation angles of the GOCI can simply be assumed to be constant. The sun observation angles changed significantly in one day. Under some specific sun, satellite observation angle combination, and water wave conditions, there may be unignorable sun glint reflectance in some observation regions of the GOCI image. The GOCI L2 data have various flags; one is "Bright_Pxl_Adj" (bright pixel adjacency warn) [222]. We are unsure whether it can determine which water pixels are affected by the adjacency effects. There are still too few studies on the impact of adjacency effects on GOCI nearshore water monitoring [56,84,223]. The improved spatial resolution of the GOCI-II (250 m) provides the possibility for monitoring smaller bays, lakes, reservoirs, and rivers. Studying the adjacency effect removal method of the GOCI-II is significant for its application in the areas above.

5.2. Integrating Geostationary Ocean Color Satellites, Unmanned Aerial Vehicles, and Ground Collaborative Observation

Due to cloud interference, it is difficult for the GOCI to take advantage of its high temporal resolution characteristics, especially in cloudy and humid areas [224]. With the development of satellite remote sensing technology, unmanned aerial vehicles and ground observation technology are also growing rapidly, as shown in Figure 9. The development of unmanned aerial vehicles includes improved sensors and payloads, enhanced flight control systems, miniaturization, lightweight design, and higher battery life [225–227]. Ground observation technology progress includes improving sensors and algorithms, providing more accurate estimates of water quality parameters, and the ability to monitor continuously, in the long term, in real time, and automatically [228,229]. The GOCI can be used to produce water quality products in clear weather. In cloudy weather, unmanned aerial vehicle observation results and point ground observation results for an area of interest can be input into hydrodynamic models to generate a simulated numerical water quality dataset [230,231]. The integration of technology can achieve all-weather water quality monitoring of ICWs.

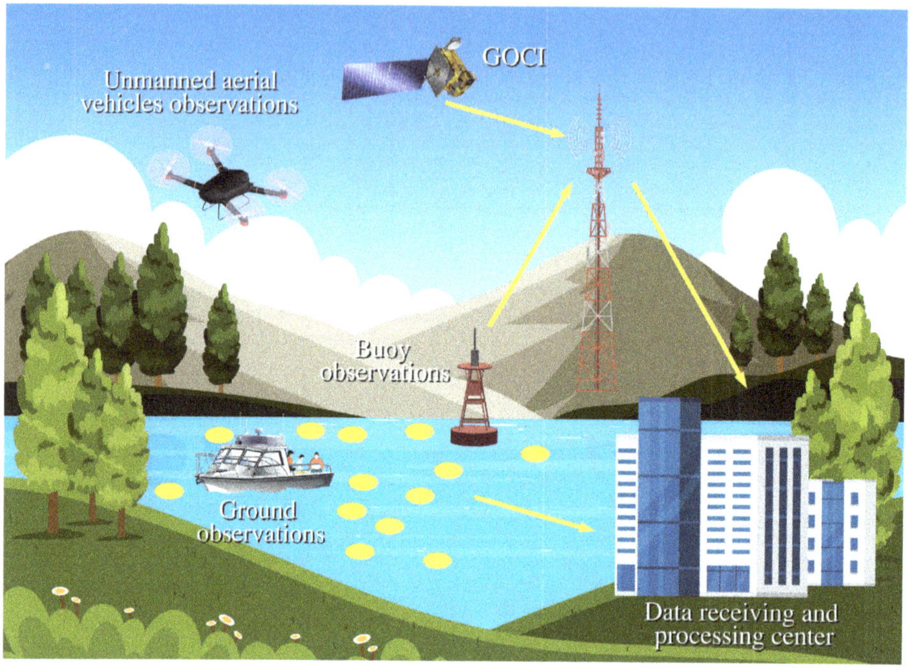

Figure 9. Schematic diagram of integrated ground–air space.

5.3. Fusion of Geostationary Ocean Color Satellites with Other Satellite Products

Image fusion can combine the advantages of different temporal, spatial, and spectral resolution sensors to improve the monitoring of water quality in ICWs [175,232]. Geostationary meteorological satellites (GMSTs) tend to have higher temporal resolution than geostationary ocean color satellites (GOCSs). For example, Himawari 8/9 acquires data every ten minutes, while the GOCI acquires data only once an hour. However, the low signal-to-noise ratio (SNR) and spatial resolution limit the application of GMSTs in water quality monitoring [233]. Taking GOCSs as a reference and fusing them can improve the spatial resolution of GMSTs and enhance spectral information. The fused GMST may produce better water quality products in ICWs than the original images [234]. The schematic diagram of image fusion is shown in Figure 10.

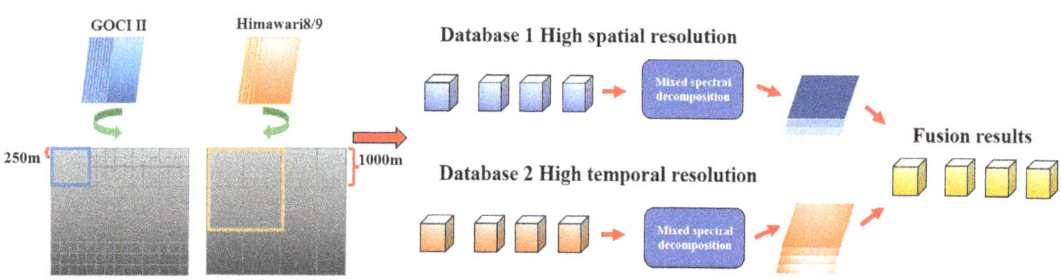

Figure 10. Schematic diagram of image fusion, taking GOCI-II and Himawari8/9 as an example.

Similarly, fusion can also improve GOCS products. The fusion of GOCSs and medium- and high-resolution satellites can improve the spatial resolution of GOCSs, providing more spatial details of ICW quality changes [18]. The fusion of GOCSs and hyperspectral satellites will enhance the spectra of GOCSs, improving the accuracy of the water quality products of a GOCS [235]. In addition to multiple daily observations, a GOCS, like the GOCI, can provide long-time series observation data. Developing technologies to integrate GOCS products with other long-term series satellite products (SeaWiFS, MODIS, and VIIRS) will contribute to long-term series of water quality assessments [236–239].

5.4. Improving Spectral, Spatial, and Temporal Resolution of Geostationary Ocean Color Sensors

5.4.1. Improving Spectral Resolution

The GOCI was equipped with eight bands (six for visible light and two for NIR). The next generation of the GOCI (GOCI-II) increased the number of bands to thirteen (one for UV, eight for visible, three for near-infrared, and one for broad bandwidth). The new 380 nm ultraviolet band can be used to retrieve parameters related to the absorption of CDOM (e.g., DOC, carbon dioxide, or organic pollutants) [117,240]. It also allows for applying ultraviolet AC methods to the GOCI to provide more reliable $R_{rs}(\lambda)$ products [66,241]. The 510 nm band can potentially improve the performance of OC3G or other blue–green Chla algorithms in relatively clear ICWs [242]. PC can characterize cyanobacterial biomass, which has a distinct absorption peak at about 620 nm [243,244]. Many inland lakes on the Korean Peninsula and East China have frequent occurrences of cyanobacterial blooms [245–248]. The newly added 620 nm band allows the GOCI-II to monitor cyanobacteria blooms quantitatively. The last new band of the GOCI-II is 709 nm. Many studies have illustrated this band's usefulness in Chla retrievals [249,250]. The GOCI-II has the potential to provide more accurate Chla products in turbid ICWs.

The newly added five bands provide many advantages to the GOCI-II compared with the GOCI. The GLIMR is a geostationary sensor funded by the NASA Earth Venture Instrument program, which is anticipated to launch within this decade (around 2026–2027) [62]. The GLIMR is planned to provide 340–1025 nm hyperspectral data of aquatic ecosystems. The hyperspectral characteristic of the GLIMR may provide more reliable atmospherically corrected $R_{rs}(\lambda)$ products and more accurate water quality products (e.g., Chla, SPM, or SDD). By targeting absorption or backscatter characteristics at different wavelengths, the GLIMR may provide more new water quality parameters (e.g., algae species, oil spills, microplastics, or greenhouse gases). The GOCI, GOCI-II, and GLIMR (planned) have only visible and NIR bands. The complex optical properties of ICWs require that $R_{rs}(\lambda)$ products are combined with different AC algorithms based on the bands' wavelengths (e.g., UV, visible, NIR, and SWIR bands) [251]. The next-generation GOCSs should have more band choices.

5.4.2. Improving Spatial Resolution

Most of the applications of the GOCI in inland waters have been carried out in Lake Taihu, Lake Hongze, and Lake Chagan. The coarse spatial resolution of 500 m makes it hard to use the GOCI to monitor water quality in small lakes. With an improved spatial resolution of 250 m, the GOCI-II can monitor more inland waters than the GOCI. Inland lakes with small sizes are more easily affected by human activities and global warming than large lakes [252,253]. A higher spatial resolution allows future GOCSs to monitor the water quality of small and medium-sized lakes.

5.4.3. Improving Temporal Resolution

The most significant advantage of a GOCS is its high temporal resolution. The observation times of GLIMR, the GOCI, and the GOCI-II are six (planned), eight, and ten per day. Many water quality changes, such as AB vertical and horizontal movements [9,254], lacustrine greenhouse gas emission fluxes [255], and estuary sediment fluxes [256,257], tend to happen rapidly, at sub-hourly time scales. We partly summarize and provide some possible future directions for the GOCI and other next-generation GOCSs in this study.

5.5. Further Expansion of GOCI-II Products

Through better spatial and temporal resolution, observation area, optical performance, etc., the GOCI-II has 26 types of official products rather than GOCI 13 types (https://kosc.kiost.ac.kr/index.nm?menuCd=44&lang=en accessed on 20 March 2024). The GOCI-II has the potential to provide more products to help protect the environment of ICWs and coastal areas. The new 380 nm ultraviolet band can be used to retrieve parameters related to the absorption of CDOM, DOC, carbon dioxide, organic pollutants, hypoxia, or oil spills [117,240,258–261]. It also allows for applying ultraviolet AC methods to the GOCI to provide more reliable $R_{rs}(\lambda)$ products [66,241]. The 510 nm band can potentially improve the performance of OC3G or other blue–green Chla algorithms in relatively clear ICWs [242]. PC can characterize a cyanobacterial biomass with a distinct absorption peak at about 620 nm [243,244]. Many inland lakes on the Korean Peninsula and East China have frequent occurrences of cyanobacterial blooms [245–248]. The newly added 620 nm band allows the GOCI-II to monitor cyanobacteria blooms quantitatively. The last new band of the GOCI-II is 709 nm. Many studies have illustrated this band's usefulness in Chla retrievals [249,250]. The GOCI-II has the potential to provide more accurate Chla products in turbid ICWs. As we tested, within one hour of observation, the GOCI-II released the image on its official website for users to download. The quasi-real-time data can be used to monitor many pollution events (e.g., blooms, oil spills, and hypoxia) in the ICWs. In addition to being in the water, the GOCI-II can also monitor intertidal zones where the ocean meets the land between high and low tides [262]. They are home to diverse organisms (e.g., algae, mollusks, crustaceans, fish, and birds) specially adapted to survive in this dynamic environment [263]. These zones experience drastic changes in environmental conditions (e.g., fluctuations in temperature, salinity, and exposure to air and sunlight, depending on the tide cycle) [264]. The higher temporal resolution and improved spatial resolution of the GOCI-II can be used to identify and classify intertidal areas.

6. Conclusions

We reviewed the applications of the GOCI for AC and water quality parameter retrieval in ICWs. First, we introduced the GOCI specifications and compared them with those of other satellites, highlighting the advantages of the GOCI in high-temporal water quality monitoring. Then, AC and water quality (such as ABs, Chla, SPM, SDD, and others) algorithms for the GOCI were systematically introduced. The results of the bibliometric analysis showed that many papers were published in the last thirteen years, starting in 2010.

A number of AC or water quality retrieval models for the GOCI have been proposed over the years. These models form a GOCI algorithm library for monitoring water quality in various ICWs. Due to the complex optical properties of ICWs, it is difficult to identify one best performing model of AC or one specific water quality parameter (ABs, Chla, SPM, SDD, or others), as they differ across ICW waterbodies or even across the seasons in the same waterbody. The GOCI provides valuable experience for effectively using a GOCS to monitor the water qualities of ICWs. So far, there is still no official GOCI product for water quality monitoring in ICWs. Classifying the ICWs into different water types and retrieving the water quality by the most suitable model for each type may provide an official, reliable, and high-accuracy water color product for various ICWs.

We partly summarize and introduce future directions for GOCI technology and other next-generation GOCSs to better serve the water quality monitoring of ICWs. Cloudy weather limits the use of high-temporal GOCS satellites. Integrating GOCSs, unmanned aerial vehicles, and ground collaborative observation can achieve all-weather water quality monitoring of ICWs. A GMST tends to have higher temporal resolution than a GOCS and lower spatial and spectral resolutions. The fusion of GOCSs with other GMSTs will improve the spatial and spectral ability of the GMSTs. The fusion of GOCSs with other high spatial or spectral sensors will enhance the image quality of the GOCS itself to monitor ICWs. We suggest that the next generation of GOCSs improve their spatial, spectral, and temporal resolution to better monitor the water quality of ICWs.

Supplementary Materials: The following supporting information can be downloaded at: https://www.mdpi.com/article/10.3390/rs16091623/s1, Table S1: Abbreviation or symbols; Table S2: The number of articles published based on GOCI in the past ten years, the country, the study area, and research direction.

Author Contributions: Conceptualization: G.L., K.S.; Methodology: S.S., Y.W.; Data curation: S.S., G.L.; Formal analysis: S.S., G.L., Y.W.; Investigation: S.S., Y.W.; Writing—original draft: G.L., K.S.; Writing—review and editing: K.S. All authors have read and agreed to the published version of the manuscript.

Funding: The research was jointly supported by the National Key Research and Development Program of China (2021YFB3901101), the National Natural Science Foundation of China (No. 42171385, U23A2008), the Open Fund of State Laboratory of Information Engineering in Surveying, Mapping and Remote Sensing, Wuhan University (Grant No. 22R02) and the Youth Innovation Promotion Association of Chinese Academy of Sciences, China (2022228).

Data Availability Statement: The GOCI and GOCI-II data that support the findings of this study are openly available at https://oceancolor.gsfc.nasa.gov/cgi/browse.pl?per=MO&day=16892&sub=level3&prm=TC&set=80&ndx=0&mon=16161&sen=goci&rad=0&frc=0&cn=90&cs=-90&cw=-180&ce=180&dnm=D and https://www.nosc.go.kr/eng/program/actionTimeSeries.do accessed on 20 March 2024.

Conflicts of Interest: The authors declare that they have no known competing financial interests or personal relationships that could have appeared to influence the work reported in this paper.

References

1. Schofield, O.; Arnone, R.A.; Bissett, W.P.; Dickey, T.D.; Davis, C.O.; Finkel, Z.; Oliver, M.; Moline, M.A. Watercolors in the coastal zone: What can we see? *Oceanography* **2004**, *17*, 30–37. [CrossRef]
2. Shi, J.; Shen, Q.; Yao, Y.; Zhang, F.; Li, J.; Wang, L. Field Radiometric Calibration of a Micro-Spectrometer Based on Remote Sensing of Plateau Inland Water Colors. *Appl. Sci.* **2023**, *13*, 2117. [CrossRef]
3. Morel, A.; Prieur, L. Analysis of variations in ocean color 1. *Limnol. Oceanogr.* **1977**, *22*, 709–722. [CrossRef]
4. Gordon, H.R.; Morel, A.Y. *Remote Assessment of Ocean Color for Interpretation of Satellite Visible Imagery: A Review*; Springer: Berlin/Heidelberg, Germany, 2012.
5. Sathyendranath, S. Reports of the International Ocean-Colour Coordinating Group. *IOCCG Proj. Off. Dartm. Nova Scotia IOCCG Rep.* **2000**, *3*, 140.
6. Schroeder, T.; Schaale, M.; Fischer, J. Retrieval of atmospheric and oceanic properties from MERIS measurements: A new Case-2 water processor for BEAM. *Int. J. Remote Sens.* **2007**, *28*, 5627–5632. [CrossRef]
7. Kyryliuk, D.; Kratzer, S. Evaluation of Sentinel-3A OLCI products derived using the Case-2 Regional CoastColour processor over the Baltic Sea. *Sensors* **2019**, *19*, 3609. [CrossRef] [PubMed]
8. Ding, X.; He, X.; Bai, Y.; Zhu, Q.; Gong, F.; Li, H.; Li, J. High-frequency and tidal period observations of suspended particulate matter in coastal waters by AHI/Himawari-8. *Opt. Express* **2020**, *28*, 27387–27404. [CrossRef] [PubMed]
9. Li, J.; Li, Y.; Bi, S.; Xu, J.; Guo, F.; Lyu, H.; Dong, X.; Cai, X. Utilization of GOCI data to evaluate the diurnal vertical migration of Microcystis aeruginosa and the underlying driving factors. *J. Environ. Manag.* **2022**, *310*, 114734. [CrossRef]
10. Mouw, C.B.; Greb, S.; Aurin, D.; DiGiacomo, P.M.; Lee, Z.; Twardowski, M.; Binding, C.; Hu, C.; Ma, R.; Moore, T.; et al. Aquatic color radiometry remote sensing of coastal and inland waters: Challenges and recommendations for future satellite missions. *Remote Sens. Environ.* **2015**, *160*, 15–30. [CrossRef]
11. Palmer, S.C.J.; Kutser, T.; Hunter, P.D. Remote sensing of inland waters: Challenges, progress and future directions. *Remote Sens. Environ.* **2015**, *157*, 1–8. [CrossRef]
12. Wang, S.; Zhang, X.; Chen, N.; Tian, L.; Zhang, Y.; Nam, W.H. A systematic review and quantitative meta-analysis of the relationships between driving forces and cyanobacterial blooms at global scale. *Environ. Res.* **2023**, *216*, 114670. [CrossRef] [PubMed]
13. Zeng, F.; Song, C.; Cao, Z.; Xue, K.; Lu, S.; Chen, T.; Liu, K. Monitoring inland water via Sentinel satellite constellation: A review and perspective. *ISPRS J. Photogramm. Remote Sens.* **2023**, *204*, 340–361. [CrossRef]
14. Hovis, W.A.; Clark, D.; Anderson, F.; Austin, R.; Wilson, W.; Baker, E.; Ball, D.; Gordon, H.; Mueller, J.; El-Sayed, S. Nimbus-7 Coastal Zone Color Scanner: System description and initial imagery. *Science* **1980**, *210*, 60–63. [CrossRef] [PubMed]
15. Conkright, M.; Gregg, W. Comparison of global chlorophyll climatologies: In situ, CZCS, Blended in situ-CZCS and SeaWiFS. *Int. J. Remote Sens.* **2003**, *24*, 969–991. [CrossRef]
16. Cracknell, A.P. The development of remote sensing in the last 40 years. *Int. J. Remote Sens.* **2018**, *39*, 8387–8427. [CrossRef]
17. Jiang, L.; Wang, M. Improved near-infrared ocean reflectance correction algorithm for satellite ocean color data processing. *Opt. Express* **2014**, *22*, 21657–21678. [CrossRef]

18. Pan, Y.; Shen, F.; Wei, X. Fusion of Landsat-8/OLI and GOCI Data for Hourly Mapping of Suspended Particulate Matter at High Spatial Resolution: A Case Study in the Yangtze (Changjiang) Estuary. *Remote Sens.* **2018**, *10*, 158. [CrossRef]
19. Salisbury, J.; Davis, C.; Erb, A.; Hu, C.; Gatebe, C.; Jordan, C.; Lee, Z.; Mannino, A.; Mouw, C.; Schaaf, C.; et al. Coastal Observations from a New Vantage Point. *Eos* **2016**, *97*. [CrossRef]
20. Wu, J.; Chen, C.; Nukapothula, S. Atmospheric Correction of GOCI Using Quasi-Synchronous VIIRS Data in Highly Turbid Coastal Waters. *Remote Sens.* **2019**, *12*, 89. [CrossRef]
21. Wang, M.; Shi, W. The NIR-SWIR combined atmospheric correction approach for MODIS ocean color data processing. *Opt. Express* **2007**, *15*, 15722–15733. [CrossRef]
22. Shi, W.; Wang, M. Detection of turbid waters and absorbing aerosols for the MODIS ocean color data processing. *Remote Sens. Environ.* **2007**, *110*, 149–161. [CrossRef]
23. Agarwal, N.; Sharma, R.; Thapliyal, P.; Gangwar, R.; Kumar, P.; Kumar, R. Geostationary satellite-based observations for ocean applications. *Curr. Sci.* **2019**, *117*, 506–515. [CrossRef]
24. Bailey, S.W.; Werdell, P.J. A multi-sensor approach for the on-orbit validation of ocean color satellite data products. *Remote Sens. Environ.* **2006**, *102*, 12–23. [CrossRef]
25. Cao, C.; Wang, S.; Li, J.; Zhao, H.; Shen, W.; Xie, Y. MODIS-based monitoring of spatial distribution of trophic status in 144 key lakes and reservoirs of China in summer of 2018. *J. Lake Sci.* **2021**, *33*, 405–413.
26. He, M.; He, S.; Zhang, X.; Zhou, F.; Li, P. Assessment of Normalized Water-Leaving Radiance Derived from GOCI Using AERONET-OC Data. *Remote Sens.* **2021**, *13*, 1640. [CrossRef]
27. Tan, Z.; Cao, Z.; Shen, M.; Chen, J.; Song, Q.; Duan, H. Remote Estimation of Water Clarity and Suspended Particulate Matter in Qinghai Lake from 2001 to 2020 Using MODIS Images. *Remote Sens.* **2022**, *14*, 3094. [CrossRef]
28. Xiong, J.; Lin, C.; Ma, R.; Cao, Z. Remote Sensing Estimation of Lake Total Phosphorus Concentration Based on MODIS: A Case Study of Lake Hongze. *Remote Sens.* **2019**, *11*, 2068. [CrossRef]
29. Baldwin, D.; Tschudi, M.; Pacifici, F.; Liu, Y. Validation of Suomi-NPP VIIRS sea ice concentration with very high-resolution satellite and airborne camera imagery. *ISPRS J. Photogramm. Remote Sens.* **2017**, *130*, 122–138. [CrossRef]
30. Justice, C.O.; Román, M.O.; Csiszar, I.; Vermote, E.F.; Wolfe, R.E.; Hook, S.J.; Friedl, M.; Wang, Z.; Schaaf, C.B.; Miura, T. Land and cryosphere products from Suomi NPP VIIRS: Overview and status. *J. Geophys. Res. Atmos.* **2013**, *118*, 9753–9765. [CrossRef]
31. Lin, L.; Hao, X.; Zhang, B.; Zou, C.-Z.; Cao, C. Assessment of the Reprocessed Suomi NPP VIIRS Enterprise Cloud Mask Product. *Remote Sens.* **2021**, *13*, 2502. [CrossRef]
32. Wang, M.; Jiang, L. VIIRS-derived ocean color product using the imaging bands. *Remote Sens. Environ.* **2018**, *206*, 275–286. [CrossRef]
33. Park, M.-S.; Lee, S.; Ahn, J.-H.; Lee, S.-J.; Choi, J.-K.; Ryu, J.-H. Decadal measurements of the first Geostationary Ocean Color Satellite (GOCI) compared with MODIS and VIIRS data. *Remote Sens.* **2021**, *14*, 72. [CrossRef]
34. Xia, L.; Mao, K.; Ma, Y.; Zhao, F.; Jiang, L.; Shen, X.; Qin, Z. An algorithm for retrieving land surface temperatures using VIIRS data in combination with multi-sensors. *Sensors* **2014**, *14*, 21385–21408. [CrossRef] [PubMed]
35. Morrison, J.M.; Jeffrey, H.; Gorter, H.; Anderson, P.; Clark, C.; Holmes, A.; Feldman, G.C.; Patt, F.S. SeaHawk: An advanced CubeSat mission for sustained ocean colour monitoring. In Proceedings of the Sensors, Systems, and Next-Generation Satellites XX, Edinburgh, UK, 26–29 September 2016; pp. 309–319.
36. Arino, O.; Gross, D.; Ranera, F.; Leroy, M.; Bicheron, P.; Brockman, C.; Defourny, P.; Vancutsem, C.; Achard, F.; Durieux, L. GlobCover: ESA service for global land cover from MERIS. In Proceedings of the 2007 IEEE International Geoscience and Remote Sensing Symposium, Barcelona, Spain, 23–28 July 2007; pp. 2412–2415.
37. Doerffer, R.; Schiller, H. The MERIS Case 2 water algorithm. *Int. J. Remote Sens.* **2007**, *28*, 517–535. [CrossRef]
38. Rast, M.; Bezy, J.; Bruzzi, S. The ESA Medium Resolution Imaging Spectrometer MERIS a review of the instrument and its mission. *Int. J. Remote Sens.* **1999**, *20*, 1681–1702. [CrossRef]
39. Hammond, M.L.; Henson, S.A.; Lamquin, N.; Clerc, S.; Donlon, C. Assessing the Effect of Tandem Phase Sentinel-3 OLCI Sensor Uncertainty on the Estimation of Potential Ocean Chlorophyll-a Trends. *Remote Sens.* **2020**, *12*, 2522. [CrossRef]
40. Moses, W.J.; Gitelson, A.A.; Berdnikov, S.; Povazhnyy, V. Estimation of chlorophyll-aconcentration in case II waters using MODIS and MERIS data—Successes and challenges. *Environ. Res. Lett.* **2009**, *4*, 045005. [CrossRef]
41. Xu, Y.; He, X.; Bai, Y.; Wang, D.; Zhu, Q.; Ding, X. Evaluation of Remote-Sensing Reflectance Products from Multiple Ocean Color Missions in Highly Turbid Water (Hangzhou Bay). *Remote Sens.* **2021**, *13*, 4267. [CrossRef]
42. Tilstone, G.H.; Pardo, S.; Dall'Olmo, G.; Brewin, R.J.W.; Nencioli, F.; Dessailly, D.; Kwiatkowska, E.; Casal, T.; Donlon, C. Performance of Ocean Colour Chlorophyll a algorithms for Sentinel-3 OLCI, MODIS-Aqua and Suomi-VIIRS in open-ocean waters of the Atlantic. *Remote Sens. Environ.* **2021**, *260*, 112444. [CrossRef]
43. Deuzé, J.; Bréon, F.; Devaux, C.; Goloub, P.; Herman, M.; Lafrance, B.; Maignan, F.; Marchand, A.; Nadal, F.; Perry, G. Remote sensing of aerosols over land surfaces from POLDER-ADEOS-1 polarized measurements. *J. Geophys. Res. Atmos.* **2001**, *106*, 4913–4926. [CrossRef]
44. Leroy, M.; Deuzé, J.; Bréon, F.; Hautecoeur, O.; Herman, M.; Buriez, J.; Tanré, D.; Bouffies, S.; Chazette, P.; Roujean, J.-L. Retrieval of atmospheric properties and surface bidirectional reflectances over land from POLDER/ADEOS. *J. Geophys. Res. Atmos.* **1997**, *102*, 17023–17037. [CrossRef]
45. Shimoda, H. ADEOS overview. *IEEE Trans. Geosci. Remote Sens.* **1999**, *37*, 1465–1471. [CrossRef]

46. Kurihara, Y.; Murakami, H.; Ogata, K.; Kachi, M. A quasi-physical sea surface temperature method for the split-window data from the Second-generation Global Imager (SGLI) onboard the Global Change Observation Mission-Climate (GCOM-C) satellite. *Remote Sens. Environ.* **2021**, *257*, 112347. [CrossRef]
47. Matsuoka, A.; Campbell, J.W.; Hooker, S.B.; Steinmetz, F.; Ogata, K.; Hirata, T.; Higa, H.; Kuwahara, V.S.; Isada, T.; Suzuki, K. Performance of JAXA's SGLI standard ocean color products for oceanic to coastal waters: Chlorophyll a concentration and light absorption coefficients of colored dissolved organic matter. *J. Oceanogr.* **2022**, *78*, 187–208. [CrossRef]
48. Tanaka, K.; Okamura, Y.; Mokuno, M.; Amano, T.; Yoshida, J. First year on-orbit calibration activities of SGLI on GCOM-C satellite. In Proceedings of the Earth Observing Missions and Sensors: Development, Implementation, and Characterization V, Honolulu, HI, USA, 25–26 September 2018; pp. 101–110.
49. Qu, L.; Liu, M.; Guan, L. Simulation of Thermal Infrared Brightness Temperatures from an Ocean Color and Temperature Scanner Onboard a New Generation Chinese Ocean Color Observation Satellite. *Remote Sens.* **2023**, *15*, 5059. [CrossRef]
50. Heales, C.J.; Lloyd, E. Play simulation for children in magnetic resonance imaging. *J. Med. Imaging Radiat. Sci.* **2022**, *53*, 10–16. [CrossRef]
51. Chakraborty, A.; Kumar, R.; Stoffelen, A. Validation of ocean surface winds from the OCEANSAT-2 scatterometer using triple collocation. *Remote Sens. Lett.* **2013**, *4*, 84–93. [CrossRef]
52. Parmar, R.; Arora, R.; Rao, M.V.; Thyagarajan, K. OCEANSAT 2: Mission and its applications. In Proceedings of the GEOSS and Next-Generation Sensors and Missions, Goa, India, 13–14 November 2006; pp. 62–73.
53. Singh, R.; Kumar, P.; Pal, P.K. Assimilation of Oceansat-2-scatterometer-derived surface winds in the weather research and forecasting model. *IEEE Trans. Geosci. Remote Sens.* **2011**, *50*, 1015–1021. [CrossRef]
54. Lee, S.-J.; Lee, D.-E.; Choi, S.-Y.; Kwon, O.-S. OSMI-1 enhances TRAIL-induced apoptosis through ER stress and NF-κB signaling in colon cancer cells. *Int. J. Mol. Sci.* **2021**, *22*, 11073. [CrossRef]
55. Hamacher, K.; Buchkremer, R. Measuring online sensory consumer experience: Introducing the Online Sensory Marketing Index (OSMI) as a structural modeling approach. *J. Theor. Appl. Electron. Commer. Res.* **2022**, *17*, 751–772. [CrossRef]
56. Lamquin, N.; Mazeran, C.; Doxaran, D.; Ryu, J.-H.; Park, Y.-J. Assessment of GOCI radiometric products using MERIS, MODIS and field measurements. *Ocean Sci. J.* **2012**, *47*, 287–311. [CrossRef]
57. Shin, J.; Lee, J.-S.; Jang, L.-H.; Lim, J.; Khim, B.-K.; Jo, Y.-H. Sargassum detection using machine learning models: A case study with the first 6 months of GOCI-II imagery. *Remote Sens.* **2021**, *13*, 4844. [CrossRef]
58. Warren, M.; Quartly, G.D.; Shutler, J.; Miller, P.I.; Yoshikawa, Y. Estimation of ocean surface currents from maximum cross correlation applied to GOCI geostationary satellite remote sensing data over the Tsushima (Korea) Straits. *J. Geophys. Res. Ocean.* **2016**, *121*, 6993–7009. [CrossRef]
59. Paduan, J.D.; Washburn, L. High-frequency radar observations of ocean surface currents. *Annu. Rev. Mar. Sci.* **2013**, *5*, 115–136. [CrossRef] [PubMed]
60. Ruddick, K.; Neukermans, G.; Vanhellemont, Q.; Jolivet, D. Challenges and opportunities for geostationary ocean colour remote sensing of regional seas: A review of recent results. *Remote Sens. Environ.* **2014**, *146*, 63–76. [CrossRef]
61. Ryu, J.-H.; Han, H.-J.; Cho, S.; Park, Y.-J.; Ahn, Y.-H. Overview of geostationary ocean color imager (GOCI) and GOCI data processing system (GDPS). *Ocean Sci. J.* **2012**, *47*, 223–233. [CrossRef]
62. Schaeffer, B.A.; Whitman, P.; Vandermeulen, R.; Hu, C.; Mannino, A.; Salisbury, J.; Efremova, B.; Conmy, R.; Coffer, M.; Salls, W.; et al. Assessing potential of the Geostationary Littoral Imaging and Monitoring Radiometer (GLIMR) for water quality monitoring across the coastal United States. *Mar. Pollut. Bull.* **2023**, *196*, 115558. [CrossRef] [PubMed]
63. Son, S.; Kim, Y.H.; Kwon, J.-I.; Kim, H.-C.; Park, K.-S. Characterization of spatial and temporal variation of suspended sediments in the Yellow and East China Seas using satellite ocean color data. *GIScience Remote Sens.* **2014**, *51*, 212–226. [CrossRef]
64. Duan, H.; Ma, R.; Hu, C. Evaluation of remote sensing algorithms for cyanobacterial pigment retrievals during spring bloom formation in several lakes of East China. *Remote Sens. Environ.* **2012**, *126*, 126–135. [CrossRef]
65. Li, J.; Gao, M.; Feng, L.; Zhao, H.; Shen, Q.; Zhang, F.; Wang, S.; Zhang, B. Estimation of chlorophyll-a concentrations in a highly turbid eutrophic lake using a classification-based MODIS land-band algorithm. *IEEE J. Sel. Top. Appl. Earth Obs. Remote Sens.* **2019**, *12*, 3769–3783. [CrossRef]
66. Liu, H.; He, X.; Li, Q.; Kratzer, S.; Wang, J.; Shi, T.; Hu, Z.; Yang, C.; Hu, S.; Zhou, Q. Estimating ultraviolet reflectance from visible bands in ocean colour remote sensing. *Remote Sens. Environ.* **2021**, *258*, 112404. [CrossRef]
67. Lou, X.; Hu, C. Diurnal changes of a harmful algal bloom in the East China Sea: Observations from GOCI. *Remote Sens. Environ.* **2014**, *140*, 562–572. [CrossRef]
68. Wang, J.; Tang, J.; Wang, W.; Wang, Y.; Wang, Z. Quantitative Retrieval of Chlorophyll-a Concentrations in the Bohai–Yellow Sea Using GOCI Surface Reflectance Products. *Remote Sens.* **2023**, *15*, 5285. [CrossRef]
69. Yang, H.; Choi, J.K.; Park, Y.J.; Han, H.J.; Ryu, J.H. Application of the Geostationary Ocean Color Imager (GOCI) to estimates of ocean surface currents. *J. Geophys. Res. Ocean.* **2014**, *119*, 3988–4000. [CrossRef]
70. Li, G.; Wang, L.; Wang, X.; Wang, X.; Sun, G. Geostationary ocean color imager and application progress. *Mar. Environ. Sci.* **2014**, *33*, 966–971.
71. Yin, W.; Huang, D. Applications of geostationary satellite data in the study of ocean and coastal short-term processes: Two cases in the East China Sea. In *Remote Sensing of Ocean and Coastal Environments*; Elsevier: Amsterdam, The Netherlands, 2021; pp. 139–154.

72. Chen, J.; He, X.; Quan, W.; Ma, L.; Jia, M.; Pan, D. A statistical analysis of residual errors in satellite remote sensing reflectance data from oligotrophic open oceans. *IEEE Trans. Geosci. Remote Sens.* **2021**, *60*, 4203912. [CrossRef]
73. Gong, S.; Huang, J.; Li, Y.; Wang, H. Comparison of atmospheric correction algorithms for TM image in inland waters. *Int. J. Remote Sens.* **2008**, *29*, 2199–2210. [CrossRef]
74. Zeng, Q.; Zhao, Y.; Tian, L.-Q.; Chen, X.-L. Evaluation on the atmospheric correction methods for water color remote sensing by using HJ-1A/1B CCD image-taking Poyang Lake in China as a case. *Spectrosc. Spectr. Anal.* **2013**, *33*, 1320–1326.
75. Ahn, J.-H.; Park, Y.-J.; Ryu, J.-H.; Lee, B.; Oh, I.S. Development of atmospheric correction algorithm for Geostationary Ocean Color Imager (GOCI). *Ocean Sci. J.* **2012**, *47*, 247–259. [CrossRef]
76. Concha, J.; Mannino, A.; Franz, B.; Kim, W. Uncertainties in the Geostationary Ocean Color Imager (GOCI) Remote Sensing Reflectance for Assessing Diurnal Variability of Biogeochemical Processes. *Remote Sens.* **2019**, *11*, 295. [CrossRef]
77. Hu, Y.; Dou, T.; Yang, B. A review of research on retrieving the concentration of suspended particulate matter and chlorophyll-a in lake based on GOCI images. *J. Water Resour. Water Eng.* **2017**, *28*, 26–39.
78. Huang, X.; Zhu, J.; Han, B.; Jamet, C.; Tian, Z.; Zhao, Y.; Li, J.; Li, T. Evaluation of Four Atmospheric Correction Algorithms for GOCI Images over the Yellow Sea. *Remote Sens.* **2019**, *11*, 1631. [CrossRef]
79. Ahn, J.-H.; Park, Y.-J. Estimating Water Reflectance at Near-Infrared Wavelengths for Turbid Water Atmospheric Correction: A Preliminary Study for GOCI-II. *Remote Sens.* **2020**, *12*, 3791. [CrossRef]
80. Ahn, J.H.; Park, Y.J.; Kim, W.; Lee, B.; Oh, I.S. Vicarious calibration of the Geostationary Ocean Color Imager. *Opt. Express* **2015**, *23*, 23236–23258. [CrossRef] [PubMed]
81. Goyens, C.; Jamet, C.; Ruddick, K.G. Spectral relationships for atmospheric correction. II. Improving NASA's standard and MUMM near infra-red modeling schemes. *Opt. Express* **2013**, *21*, 21176–21187. [CrossRef] [PubMed]
82. Pan, Y.; Shen, F.; Verhoef, W. An improved spectral optimization algorithm for atmospheric correction over turbid coastal waters: A case study from the Changjiang (Yangtze) estuary and the adjacent coast. *Remote Sens. Environ.* **2017**, *191*, 197–214. [CrossRef]
83. He, X.; Bai, Y.; Pan, D.; Huang, N.; Dong, X.; Chen, J.; Chen, C.-T.A.; Cui, Q. Using geostationary satellite ocean color data to map the diurnal dynamics of suspended particulate matter in coastal waters. *Remote Sens. Environ.* **2013**, *133*, 225–239. [CrossRef]
84. Men, J.; Feng, L.; Chen, X.; Tian, L. Atmospheric correction under cloud edge effects for Geostationary Ocean Color Imager through deep learning. *ISPRS J. Photogramm. Remote Sens.* **2023**, *201*, 38–53. [CrossRef]
85. Li, H.; He, X.; Bai, Y.; Shanmugam, P.; Park, Y.-J.; Liu, J.; Zhu, Q.; Gong, F.; Wang, D.; Huang, H. Atmospheric correction of geostationary satellite ocean color data under high solar zenith angles in open oceans. *Remote Sens. Environ.* **2020**, *249*, 112022. [CrossRef]
86. An, D.; Yu, D.; Zheng, X.; Zhou, Y.; Meng, L.; Xing, Q. Monitoring the Dissipation of the Floating Green Macroalgae Blooms in the Yellow Sea (2007–2020) on the Basis of Satellite Remote Sensing. *Remote Sens.* **2021**, *13*, 3811. [CrossRef]
87. Bing, Z.; Xiaoli, C.; Sen, W.; Xinxin, Y. Analysis of the Causes of Cyanobacteria Bloom: A Review. *J. Resour. Ecol.* **2020**, *11*, 405–413. [CrossRef]
88. Blondeau-Patissier, D.; Gower, J.F.R.; Dekker, A.G.; Phinn, S.R.; Brando, V.E. A review of ocean color remote sensing methods and statistical techniques for the detection, mapping and analysis of phytoplankton blooms in coastal and open oceans. *Prog. Oceanogr.* **2014**, *123*, 123–144. [CrossRef]
89. Cannizzaro, J.P.; Barnes, B.B.; Hu, C.; Corcoran, A.A.; Hubbard, K.A.; Muhlbach, E.; Sharp, W.C.; Brand, L.E.; Kelble, C.R. Remote detection of cyanobacteria blooms in an optically shallow subtropical lagoonal estuary using MODIS data. *Remote Sens. Environ.* **2019**, *231*, 111227. [CrossRef]
90. Huan, Y.; Sun, D.; Wang, S.; Zhang, H.; Qiu, Z.; Bilal, M.; He, Y. Remote sensing estimation of phytoplankton absorption associated with size classes in coastal waters. *Ecol. Indic.* **2021**, *121*, 107198. [CrossRef]
91. Paerl, H.W.; Xu, H.; McCarthy, M.J.; Zhu, G.; Qin, B.; Li, Y.; Gardner, W.S. Controlling harmful cyanobacterial blooms in a hyper-eutrophic lake (Lake Taihu, China): The need for a dual nutrient (N & P) management strategy. *Water Res.* **2011**, *45*, 1973–1983. [PubMed]
92. Sakib, M.H.; Rashid, A.H.A.; Yang, C.S. Comparing performance of inter-sensor NDVI for the detection of floating macroalgal blooms in the Yellow Sea. *Indian J. Geo Mar. Sci.* **2021**, *50*, 613–619.
93. Sun, X.; Wu, M.; Xing, Q.; Song, X.; Zhao, D.; Han, Q.; Zhang, G. Spatio-temporal patterns of Ulva prolifera blooms and the corresponding influence on chlorophyll-a concentration in the Southern Yellow Sea, China. *Sci. Total Environ.* **2018**, *640–641*, 807–820. [CrossRef]
94. Hong, T.T.M.; Park, Y.-G.; Choi, J.M. Divergence Observation in a Mesoscale Eddy during Chla Bloom Revealed in Submesoscale Satellite Currents. *Remote Sens.* **2023**, *15*, 995. [CrossRef]
95. Noh, J.H.; Kim, W.; Son, S.H.; Ahn, J.H.; Park, Y.J. Remote quantification of Cochlodinium polykrikoides blooms occurring in the East Sea using geostationary ocean color imager (GOCI). *Harmful Algae* **2018**, *73*, 129–137. [CrossRef]
96. Sun, D.; Huan, Y.; Qiu, Z.; Hu, C.; Wang, S.; He, Y. Remote-Sensing Estimation of Phytoplankton Size Classes from GOCI Satellite Measurements in Bohai Sea and Yellow Sea. *J. Geophys. Res. Ocean.* **2017**, *122*, 8309–8325. [CrossRef]
97. Fan, C.; Zhang, Y.; Wang, X. Chlorophyll-a concentration inversion and distribution with GOCI images in the Changjiang Estuary. In Proceedings of the First International Conference on Spatial Atmospheric Marine Environmental Optics (SAME 2023), Shanghai, China, 7–9 April 2023; pp. 201–210.

98. Choi, J.-K.; Min, J.-E.; Noh, J.H.; Han, T.-H.; Yoon, S.; Park, Y.J.; Moon, J.-E.; Ahn, J.-H.; Ahn, S.M.; Park, J.-H. Harmful algal bloom (HAB) in the East Sea identified by the Geostationary Ocean Color Imager (GOCI). *Harmful Algae* **2014**, *39*, 295–302. [CrossRef]
99. Shen, F.; Tang, R.; Sun, X.; Liu, D. Simple methods for satellite identification of algal blooms and species using 10-year time series data from the East China Sea. *Remote Sens. Environ.* **2019**, *235*, 111484. [CrossRef]
100. Xu, M.; Gao, Z.; Liu, C. Detecting harmful algal blooms using Geostationary Ocean Color Imager (GOCI) data in Bohai Sea, China. In Proceedings of the Remote Sensing and Modeling of Ecosystems for Sustainability XII, San Diego, CA, USA, 9–13 August 2015; pp. 203–208.
101. Yimin, L.; Zhenyu, T.; Chen, Y.; Feng, H.; Di, M.; Juhua, L.; Hongtao, D. Extraction of Algal Blooms in Dianchi Lake Based on Multi-Source Satellites Using Machine Learning Algorithms. *Adv. Earth Sci.* **2022**, *37*, 1141.
102. Kim, D.-W.; Kim, S.-H.; Baek, J.-Y.; Lee, J.-S.; Jo, Y.-H. GOCI-II based sea surface salinity estimation using machine learning for the first-year summer. *Int. J. Remote Sens.* **2022**, *43*, 6605–6623. [CrossRef]
103. Cao, H.; Han, L. Hourly remote sensing monitoring of harmful algal blooms (HABs) in Taihu Lake based on GOCI images. *Environ. Sci. Pollut. Res. Int.* **2021**, *28*, 35958–35970. [CrossRef]
104. Lee, M.-S.; Park, K.-A.; Micheli, F. Derivation of Red Tide Index and Density Using Geostationary Ocean Color Imager (GOCI) Data. *Remote Sens.* **2021**, *13*, 298. [CrossRef]
105. Son, Y.B.; Choi, B.-J.; Kim, Y.H.; Park, Y.-G. Tracing floating green algae blooms in the Yellow Sea and the East China Sea using GOCI satellite data and Lagrangian transport simulations. *Remote Sens. Environ.* **2015**, *156*, 21–33. [CrossRef]
106. Son, Y.B.; Min, J.-E.; Ryu, J.-H. Detecting massive green algae (Ulva prolifera) blooms in the Yellow Sea and East China Sea using geostationary ocean color imager (GOCI) data. *Ocean Sci. J.* **2012**, *47*, 359–375. [CrossRef]
107. Feng, C.; Wang, S.; Li, Z. Long-term spatial variation of algal blooms extracted using the U-net model from 10 years of GOCI imagery in the East China Sea. *J. Environ. Manag.* **2022**, *321*, 115966. [CrossRef]
108. Qiu, Z.; Li, Z.; Bilal, M.; Wang, S.; Sun, D.; Chen, Y. Automatic method to monitor floating macroalgae blooms based on multilayer perceptron: Case study of Yellow Sea using GOCI images. *Opt. Express* **2018**, *26*, 26810–26829. [CrossRef]
109. Pan, B.; Shi, Z.; An, Z.; Jiang, Z.; Ma, Y. A novel spectral-unmixing-based green algae area estimation method for GOCI data. *IEEE J. Sel. Top. Appl. Earth Obs. Remote Sens.* **2016**, *10*, 437–449. [CrossRef]
110. Xue, K.; Ma, R.; Cao, Z.; Shen, M.; Hu, M.; Xiong, J. Monitoring fractional floating algae cover over eutrophic lakes using multisensor satellite images: MODIS, VIIRS, GOCI, and OLCI. *IEEE Trans. Geosci. Remote Sens.* **2022**, *60*, 4211715. [CrossRef]
111. Qi, L.; Hu, C.; Visser, P.M.; Ma, R. Diurnal changes of cyanobacteria blooms in Taihu Lake as derived from GOCI observations. *Limnol. Oceanogr.* **2018**, *63*, 1711–1726. [CrossRef]
112. Ai, Y.; Lee, S.; Lee, J. Drinking water treatment residuals from cyanobacteria bloom-affected areas: Investigation of potential impact on agricultural land application. *Sci. Total Environ.* **2020**, *706*, 135756. [CrossRef] [PubMed]
113. Wang, S.; Zhang, X.; Chen, N.; Wang, W. Classifying diurnal changes of cyanobacterial blooms in Lake Taihu to identify hot patterns, seasons and hotspots based on hourly GOCI observations. *J. Environ. Manag.* **2022**, *310*, 114782. [CrossRef] [PubMed]
114. Li, J.; Li, Y.; Dong, X.; Wang, H.; Cai, X.; Zhu, Y.; Lyu, H.; Zeng, S.; Bi, S.; Wang, G. Contributions of meteorology and nutrient to the surface cyanobacterial blooms at different timescales in the shallow eutrophic Lake Taihu. *Sci. Total Environ.* **2023**, *894*, 165064. [CrossRef] [PubMed]
115. Du, C.; Li, Y.; Wang, Q.; Liu, G.; Zheng, Z.; Mu, M.; Li, Y. Tempo-spatial dynamics of water quality and its response to river flow in estuary of Taihu Lake based on GOCI imagery. *Environ. Sci. Pollut. Res. Int.* **2017**, *24*, 28079–28101. [CrossRef]
116. Wang, M.; Son, S.; Harding, L.W. Retrieval of diffuse attenuation coefficient in the Chesapeake Bay and turbid ocean regions for satellite ocean color applications. *J. Geophys. Res. Ocean.* **2009**, *114*, C10011. [CrossRef]
117. Li, J.; Yu, Q.; Tian, Y.Q.; Becker, B.L. Remote sensing estimation of colored dissolved organic matter (CDOM) in optically shallow waters. *ISPRS J. Photogramm. Remote Sens.* **2017**, *128*, 98–110. [CrossRef]
118. Lin, S.; Jing, C.; Rui, L. A method of water quality analysis: Chlorophyll a concentration estimation of Dongping Lake based on GOCI image. *Environ. Prot.* **2017**, *45*, 60–63.
119. Lyu, H.; Zhang, J.; Zha, G.; Wang, Q.; Li, Y. Developing a two-step retrieval method for estimating total suspended solid concentration in Chinese turbid inland lakes using Geostationary Ocean Colour Imager (GOCI) imagery. *Int. J. Remote Sens.* **2015**, *36*, 1385–1405. [CrossRef]
120. Sagan, V.; Peterson, K.T.; Maimaitijiang, M.; Sidike, P.; Sloan, J.; Greeling, B.A.; Maalouf, S.; Adams, C. Monitoring inland water quality using remote sensing: Potential and limitations of spectral indices, bio-optical simulations, machine learning, and cloud computing. *Earth-Sci. Rev.* **2020**, *205*, 103187. [CrossRef]
121. Qing, S.; Cui, T.; Lai, Q.; Bao, Y.; Diao, R.; Yue, Y.; Hao, Y. Improving remote sensing retrieval of water clarity in complex coastal and inland waters with modified absorption estimation and optical water classification using Sentinel-2 MSI. *Int. J. Appl. Earth Obs. Geoinf.* **2021**, *102*, 102377. [CrossRef]
122. Bao, Y.; Tian, Q.; Chen, M.; Lü, C. Analysis on Diurnal Variation of Chlorophyll-a Concentration of Taihu Lake Based on Optical Classification with GOCI Data. *Guang Pu Xue Yu Guang Pu Fen Xi = Guang Pu* **2016**, *36*, 2562–2567. [PubMed]
123. Cui, T.W.; Zhang, J.; Wang, K.; Wei, J.W.; Mu, B.; Ma, Y.; Zhu, J.H.; Liu, R.J.; Chen, X.Y. Remote sensing of chlorophyll a concentration in turbid coastal waters based on a global optical water classification system. *ISPRS J. Photogramm. Remote Sens.* **2020**, *163*, 187–201. [CrossRef]

124. Hu, M.; Zhang, Y.; Ma, R.; Xue, K.; Cao, Z.; Chu, Q.; Jing, Y. Optimized remote sensing estimation of the lake algal biomass by considering the vertically heterogeneous chlorophyll distribution: Study case in Lake Chaohu of China. *Sci. Total Environ.* **2021**, *771*, 144811. [CrossRef] [PubMed]
125. Kim, W.; Moon, J.-E.; Park, Y.-J.; Ishizaka, J. Evaluation of chlorophyll retrievals from Geostationary Ocean Color Imager (GOCI) for the North-East Asian region. *Remote Sens. Environ.* **2016**, *184*, 482–495. [CrossRef]
126. Neil, C.; Spyrakos, E.; Hunter, P.D.; Tyler, A.N. A global approach for chlorophyll-a retrieval across optically complex inland waters based on optical water types. *Remote Sens. Environ.* **2019**, *229*, 159–178. [CrossRef]
127. Soomets, T.; Toming, K.; Paavel, B.; Kutser, T. Evaluation of remote sensing and modeled chlorophyll-a products of the Baltic Sea. *J. Appl. Remote Sens.* **2022**, *16*, 046516. [CrossRef]
128. Cao, Z.; Wang, M.; Ma, R.; Zhang, Y.; Duan, H.; Jiang, L.; Xue, K.; Xiong, J.; Hu, M. A decade-long chlorophyll-a data record in lakes across China from VIIRS observations. *Remote Sens. Environ.* **2024**, *301*, 113953. [CrossRef]
129. Park, J.-E.; Park, K.-A. Application of Deep Learning for Speckle Removal in GOCI Chlorophyll-a Concentration Images (2012–2017). *Remote Sens.* **2021**, *13*, 585. [CrossRef]
130. Shi, K.; Li, Y.; Li, L.; Lu, H.; Song, K.; Liu, Z.; Xu, Y.; Li, Z. Remote chlorophyll-a estimates for inland waters based on a cluster-based classification. *Sci. Total Environ.* **2013**, *444*, 1–15. [CrossRef] [PubMed]
131. Li, H.; He, X.; Bai, Y.; Chen, X.; Gong, F.; Zhu, Q.; Hu, Z. Assessment of satellite-based chlorophyll-a retrieval algorithms for high solar zenith angle conditions. *J. Appl. Remote Sens.* **2017**, *11*, 012004. [CrossRef]
132. Zhao, D.; Feng, L. Assessment of the Number of Valid Observations and Diurnal Changes in Chl-a for GOCI: Highlights for Geostationary Ocean Color Missions. *Sensors* **2020**, *20*, 3377. [CrossRef] [PubMed]
133. Sakuno, Y.; Makio, K.; Koike, K. Chlorophyll-a Estimation in Tachibana Bay by Data Fusion of GOCI and MODIS Using Linear Combination Index Algorithm. *Adv. Remote Sens.* **2013**, *2013*, 40465. [CrossRef]
134. Zhao, M.; Bai, Y.; Li, H.; He, X.; Gong, F.; Li, T. Fluorescence Line Height Extraction Algorithm for the Geostationary Ocean Color Imager. *Remote Sens.* **2022**, *14*, 2511. [CrossRef]
135. Yang, Y.; He, S.; Gu, Y.; Zhu, C.; Wang, L.; Ma, X.; Li, P. Retrieval of Chlorophyll a Concentration Using GOCI Data in Sediment-Laden Turbid Waters of Hangzhou Bay and Adjacent Coastal Waters. *J. Mar. Sci. Eng.* **2023**, *11*, 1098. [CrossRef]
136. Bao, Y.; Tian, Q.; Chen, M. A Weighted Algorithm Based on Normalized Mutual Information for Estimating the Chlorophyll-a Concentration in Inland Waters Using Geostationary Ocean Color Imager (GOCI) Data. *Remote Sens.* **2015**, *7*, 11731–11752. [CrossRef]
137. Huang, C.; Shi, K.; Yang, H.; Li, Y.; Zhu, A.-X.; Sun, D.; Xu, L.; Zou, J.; Chen, X. Satellite observation of hourly dynamic characteristics of algae with Geostationary Ocean Color Imager (GOCI) data in Lake Taihu. *Remote Sens. Environ.* **2015**, *159*, 278–287. [CrossRef]
138. Duan, H.; Ma, R.; Zhang, Y.; Loiselle, S.A.; Xu, J.; Zhao, C.; Zhou, L.; Shang, L. A new three-band algorithm for estimating chlorophyll concentrations in turbid inland lakes. *Environ. Res. Lett.* **2010**, *5*, 044009. [CrossRef]
139. Guo, Y.; Huang, C.; Li, Y.; Du, C.; Li, Y.; Chen, W.; Shi, L.; Ji, G. an expanded three band model to monitor inland optically complex water using Geostationary Ocean Color Imager (GOCI). *Front. Remote Sens.* **2022**, *3*, 803884. [CrossRef]
140. Yulong, G.; Changchun, H.; Yunmei, L.; Chenggong, D.; Lingfei, S.; Yuan, L.; Weiqiang, C.; Hejie, W.; Enxiang, C.; Guangxing, J. Hyperspectral reconstruction method for optically complex inland waters based on bio-optical model and sparse representing. *Remote Sens. Environ.* **2022**, *276*, 113045. [CrossRef]
141. Guo, Y.; Wei, X.; Huang, Z.; Li, H.; Ma, R.; Cao, Z.; Shen, M.; Xue, K. Retrievals of Chlorophyll-a from GOCI and GOCI-II Data in Optically Complex Lakes. *Remote Sens.* **2023**, *15*, 4886. [CrossRef]
142. Binding, C.E.; Bowers, D.G.; Mitchelson-Jacob, E.G. Estimating suspended sediment concentrations from ocean colour measurements in moderately turbid waters; the impact of variable particle scattering properties. *Remote Sens. Environ.* **2005**, *94*, 373–383. [CrossRef]
143. Cao, Z.; Ma, R.; Duan, H.; Xue, K.; Shen, M. Effect of Satellite Temporal Resolution on Long-Term Suspended Particulate Matter in Inland Lakes. *Remote Sens.* **2019**, *11*, 2785. [CrossRef]
144. Kang, Y.; Dong, C. Spatio-temporal Analysis of suspended sediment Concentration in the Yongjiang Estuary Based on GOCI. *IOP Conf. Ser. Earth Environ. Sci.* **2018**, *108*, 032017. [CrossRef]
145. Xu, Y.; Qin, B.; Zhu, G.; Zhang, Y.; Shi, K.; Li, Y.; Shi, Y.; Chen, L. High Temporal Resolution Monitoring of Suspended Matter Changes from GOCI Measurements in Lake Taihu. *Remote Sens.* **2019**, *11*, 985. [CrossRef]
146. Jiang, D.; Zhang, H.; Zou, T.; Li, Y.; Tang, C.; Li, R. Suspended particle size retrieval based on geostationary ocean color imager (GOCI) in the Bohai Sea. *J. Coast. Res.* **2016**, *74*, 117–125. [CrossRef]
147. Lei, S.; Xu, J.; Li, Y.; Du, C.; Liu, G.; Zheng, Z.; Xu, Y.; Lyu, H.; Mu, M.; Miao, S.; et al. An approach for retrieval of horizontal and vertical distribution of total suspended matter concentration from GOCI data over Lake Hongze. *Sci. Total Environ.* **2020**, *700*, 134524. [CrossRef]
148. Liu, J.; Liu, J.; He, X.; Chen, T.; Zhu, F.; Wang, Y.; Hao, Z.; Chen, P. Retrieval of total suspended particulate matter in highly turbid Hangzhou Bay waters based on geostationary ocean color imager. In Proceedings of the Remote Sensing of the Ocean, Sea Ice, Coastal Waters, and Large Water Regions 2017, Warsaw, Poland, 11–14 September 2017; pp. 105–111.
149. Padial, A.A.; Thomaz, S.M. Prediction of the light attenuation coefficient through the Secchi disk depth: Empirical modeling in two large Neotropical ecosystems. *Limnology* **2008**, *9*, 143–151. [CrossRef]

150. Ahn, Y.-H.; Moon, J.-E.; Gallegos, S. Development of suspended particulate matter algorithms for ocean color remote sensing. *Korean J. Remote Sens.* **2001**, *17*, 285–295.
151. Moon, J.-E.; Ahn, Y.-H.; Ryu, J.-H.; Shanmugam, P. Development of ocean environmental algorithms for Geostationary Ocean Color Imager (GOCI). *Korean J. Remote Sens.* **2010**, *26*, 189–207.
152. Ruddick, K.; Vanhellemont, Q.; Yan, J.; Neukermans, G.; Wei, G.; Shang, S. Variability of suspended particulate matter in the Bohai Sea from the geostationary Ocean Color Imager (GOCI). *Ocean Sci. J.* **2012**, *47*, 331–345. [CrossRef]
153. Nechad, B.; Ruddick, K.G.; Park, Y. Calibration and validation of a generic multisensor algorithm for mapping of total suspended matter in turbid waters. *Remote Sens. Environ.* **2010**, *114*, 854–866. [CrossRef]
154. Hu, Z.; Pan, D.; He, X.; Bai, Y. Diurnal Variability of Turbidity Fronts Observed by Geostationary Satellite Ocean Color Remote Sensing. *Remote Sens.* **2016**, *8*, 147. [CrossRef]
155. Moon, J.-E.; Park, Y.-J.; Ryu, J.-H.; Choi, J.-K.; Ahn, J.-H.; Min, J.-E.; Son, Y.-B.; Lee, S.-J.; Han, H.-J.; Ahn, Y.-H. Initial validation of GOCI water products against in situ data collected around Korean peninsula for 2010–2011. *Ocean Sci. J.* **2012**, *47*, 261–277. [CrossRef]
156. Siswanto, E.; Tang, J.; Yamaguchi, H.; Ahn, Y.-H.; Ishizaka, J.; Yoo, S.; Kim, S.-W.; Kiyomoto, Y.; Yamada, K.; Chiang, C. Empirical ocean-color algorithms to retrieve chlorophyll-a, total suspended matter, and colored dissolved organic matter absorption coefficient in the Yellow and East China Seas. *J. Oceanogr.* **2011**, *67*, 627–650. [CrossRef]
157. Yu, X. *Retrieval of Suspended Matter Concentration and Reconstruction of Missing Data Based on GOCI in Bohai and Yellow Sea*; Ocean University of China: Qingdao, China, 2013.
158. Cheng, Z.; Wang, X.H.; Paull, D.; Gao, J. Application of the geostationary ocean color imager to mapping the diurnal and seasonal variability of surface suspended matter in a macro-tidal estuary. *Remote Sens.* **2016**, *8*, 244. [CrossRef]
159. Choi, J.-K.; Park, Y.J.; Lee, B.R.; Eom, J.; Moon, J.-E.; Ryu, J.-H. Application of the Geostationary Ocean Color Imager (GOCI) to mapping the temporal dynamics of coastal water turbidity. *Remote Sens. Environ.* **2014**, *146*, 24–35. [CrossRef]
160. Meng, Q.; Xu, J.; Wang, L.; Chen, Y.; Wang, X. Diurnal Changes Monitoring and Analysis of the Total Suspended Matters in Bohai Sea Using Geostationary Ocean Color Imager. *IOP Conf. Ser. Earth Environ. Sci.* **2019**, *234*, 012036. [CrossRef]
161. Zhang, M.; Tang, J.; Dong, Q.; Song, Q.; Ding, J. Retrieval of total suspended matter concentration in the Yellow and East China Seas from MODIS imagery. *Remote Sens. Environ.* **2010**, *114*, 392–403. [CrossRef]
162. He, A.; He, X.; Bai, Y.; Zhu, Q.; Gong, F.; Huang, H.; Pan, D. Simulation of Sedimentation in Lake Taihu with Geostationary Satellite Ocean Color Data. *Remote Sens.* **2019**, *11*, 379. [CrossRef]
163. Amin, R.; Shulman, I. Hourly turbidity monitoring using Geostationary Ocean Color Imager fluorescence bands. *J. Appl. Remote Sens.* **2015**, *9*, 096024. [CrossRef]
164. Kukushkin, A.S. Long-term seasonal variability of water transparency in the surface layer of the deep part of the Black Sea. *Russ. Meteorol. Hydrol.* **2014**, *39*, 178–186. [CrossRef]
165. Wang, S.; Mao, Y.; Zheng, L.; Qiu, Z.; Bilal, M.; Sun, D. Remote sensing of water turbidity in the eastern China seas from geostationary ocean colour imager. *Int. J. Remote Sens.* **2020**, *41*, 4080–4101. [CrossRef]
166. Yan, Z.; Dingfeng, Y.; Xiaoyan, L.; Qian, Y.; Yingying, G. Research on remote sensing retrieval and diurnal variation of Secchi disk, depth of Jiaozhou Bay based on GOCI. *Remote Sens. Nat. Resour.* **2021**, *33*, 108.
167. Bai, S.; Gao, J.; Sun, D.; Tian, M. Monitoring Water Transparency in Shallow and Eutrophic Lake Waters Based on GOCI Observations. *Remote Sens.* **2020**, *12*, 163. [CrossRef]
168. Mao, Y.; Wang, S.; Qiu, Z.; Sun, D.; Bilal, M. Variations of transparency derived from GOCI in the Bohai Sea and the Yellow Sea. *Opt. Express* **2018**, *26*, 12191–12209. [CrossRef] [PubMed]
169. Swift, T.J.; Perez-Losada, J.; Schladow, S.G.; Reuter, J.E.; Jassby, A.D.; Goldman, C.R. Water clarity modeling in Lake Tahoe: Linking suspended matter characteristics to Secchi depth. *Aquat. Sci.* **2006**, *68*, 1–15. [CrossRef]
170. Swan, B.K.; Reifel, K.M.; Tiffany, M.A.; Watts, J.M.; Hurlbert, S.H. Spatial and temporal patterns of transparency and light attenuation in the Salton Sea, California, 1997–1999. *Lake Reserv. Manag.* **2007**, *23*, 653–662. [CrossRef]
171. Testa, J.M.; Lyubchich, V.; Zhang, Q. Patterns and Trends in Secchi Disk Depth over Three Decades in the Chesapeake Bay Estuarine Complex. *Estuaries Coasts* **2019**, *42*, 927–943. [CrossRef]
172. Zhou, Y.; Yu, D.; Cheng, W.; Gai, Y.; Yao, H.; Yang, L.; Pan, S. Monitoring multi-temporal and spatial variations of water transparency in the Jiaozhou Bay using GOCI data. *Mar. Pollut. Bull.* **2022**, *180*, 113815. [CrossRef]
173. Lee, Z.; Shang, S.; Hu, C.; Du, K.; Weidemann, A.; Hou, W.; Lin, J.; Lin, G. Secchi disk depth: A new theory and mechanistic model for underwater visibility. *Remote Sens. Environ.* **2015**, *169*, 139–149. [CrossRef]
174. Liu, X.-Y.; Hu, J.-W.; Tian, L.; Yu, D.-F.; Gao, H.; Yang, L.; An, D.-Y. Comparative study on transparency retrieved from GOCI under four different atmospheric correction algorithms in Jiaozhou Bay and Qingdao Coastal area. *IEEE J. Sel. Top. Appl. Earth Obs. Remote Sens.* **2023**, *17*, 2077–2089. [CrossRef]
175. Liu, X.; Wang, M. Global daily gap-free ocean color products from multi-satellite measurements. *Int. J. Appl. Earth Obs. Geoinf.* **2022**, *108*, 102714. [CrossRef]
176. Mao, Y.; Qiu, Z.; Sun, D.; Wang, S.; Lu, Y.; Wu, C.; Yue, X.; Ye, Z. A Novel Remote Sensing Algorithm for Estimating Diffuse Attenuation Coefficient in the BohaiSea and Yellow Sea. *Guangxi Sci.* **2016**, *23*, 513–519.
177. Ding, X.; Gong, F.; Zhu, Q.; Li, J.; Wang, X.; Bai, R.; Xu, Y. Using geostationary satellite ocean color data and superpixel to map the diurnal dynamics of water transparency in the eastern China seas. *Ecol. Indic.* **2022**, *142*, 109219. [CrossRef]

178. He, X.; Pan, D.; Mao, Z. Water-transparency (Secchi Depth) monitoring in the China Sea with the SeaWiFS satellite sensor. In Proceedings of the Remote Sensing for Agriculture, Ecosystems, and Hydrology VI, Maspalomas, Spain, 13–16 September 2004; pp. 112–122.
179. He, X.; Pan, D.; Bai, Y.; Wang, T.; Chen, C.-T.A.; Zhu, Q.; Hao, Z.; Gong, F. Recent changes of global ocean transparency observed by SeaWiFS. *Cont. Shelf Res.* **2017**, *143*, 159–166. [CrossRef]
180. Feng, L.; Hou, X.; Zheng, Y. Monitoring and understanding the water transparency changes of fifty large lakes on the Yangtze Plain based on long-term MODIS observations. *Remote Sens. Environ.* **2019**, *221*, 675–686. [CrossRef]
181. Zeng, S.; Lei, S.; Li, Y.; Lyu, H.; Xu, J.; Dong, X.; Wang, R.; Yang, Z.; Li, J. Retrieval of Secchi Disk Depth in Turbid Lakes from GOCI Based on a New Semi-Analytical Algorithm. *Remote Sens.* **2020**, *12*, 1516. [CrossRef]
182. Battin, T.J.; Luyssaert, S.; Kaplan, L.A.; Aufdenkampe, A.K.; Richter, A.; Tranvik, L.J. The boundless carbon cycle. *Nat. Geosci.* **2009**, *2*, 598–600. [CrossRef]
183. Berggren, M.; Laudon, H.; Jansson, M. Landscape regulation of bacterial growth efficiency in boreal freshwaters. *Glob. Biogeochem. Cycles* **2007**, *21*, GB4002. [CrossRef]
184. Min, S.-H.; Park, M.-O.; Kim, S.-W.; Han, I.-S.; Kim, W.; Park, Y.-J. Correlation between SST and CDOM during Summer Coastal Upwelling along the Southeast Coast of Korea. *J. Coast. Res.* **2018**, *85*, 1471–1475. [CrossRef]
185. Wang, Y.; Shen, F.; Sokoletsky, L.; Sun, X. Validation and Calibration of QAA Algorithm for CDOM Absorption Retrieval in the Changjiang (Yangtze) Estuarine and Coastal Waters. *Remote Sens.* **2017**, *9*, 1192. [CrossRef]
186. Ling, Z.; Sun, D.; Wang, S.; Qiu, Z.; Huan, Y.; Mao, Z.; He, Y. Remote sensing estimation of colored dissolved organic matter (CDOM) from GOCI measurements in the Bohai Sea and Yellow Sea. *Environ. Sci. Pollut. Res. Int.* **2020**, *27*, 6872–6885. [CrossRef] [PubMed]
187. Bai, Y.; Cai, W.J.; He, X.; Zhai, W.; Pan, D.; Dai, M.; Yu, P. A mechanistic semi-analytical method for remotely sensing sea surface pCO2 in river-dominated coastal oceans: A case study from the East China Sea. *J. Geophys. Res. Ocean.* **2015**, *120*, 2331–2349. [CrossRef]
188. Bauer, J.E.; Cai, W.-J.; Raymond, P.A.; Bianchi, T.S.; Hopkinson, C.S.; Regnier, P.A. The changing carbon cycle of the coastal ocean. *Nature* **2013**, *504*, 61–70. [CrossRef] [PubMed]
189. Liu, D.; Bai, Y.; He, X.; Tao, B.; Pan, D.; Chen, C.-T.A.; Zhang, L.; Xu, Y.; Gong, C. Satellite estimation of particulate organic carbon flux from Changjiang River to the estuary. *Remote Sens. Environ.* **2019**, *223*, 307–319. [CrossRef]
190. Regnier, P.; Friedlingstein, P.; Ciais, P.; Mackenzie, F.T.; Gruber, N.; Janssens, I.A.; Laruelle, G.G.; Lauerwald, R.; Luyssaert, S.; Andersson, A.J. Anthropogenic perturbation of the carbon fluxes from land to ocean. *Nat. Geosci.* **2013**, *6*, 597–607. [CrossRef]
191. Xu, J.; Lei, S.; Bi, S.; Li, Y.; Lyu, H.; Xu, J.; Xu, X.; Mu, M.; Miao, S.; Zeng, S.; et al. Tracking spatio-temporal dynamics of POC sources in eutrophic lakes by remote sensing. *Water Res.* **2020**, *168*, 115162. [CrossRef]
192. Wei, X.; Shen, F.; Pan, Y.; Chen, S.; Sun, X.; Wang, Y. Satellite Observations of the Diurnal Dynamics of Particulate Organic Carbon in Optically Complex Coastal Oceans: The Continental Shelf Seas of China. *J. Geophys. Res. Ocean.* **2019**, *124*, 4710–4726. [CrossRef]
193. Catalán, N.; Obrador, B.; Alomar, C.; Pretus, J.L. Seasonality and landscape factors drive dissolved organic matter properties in Mediterranean ephemeral washes. *Biogeochemistry* **2012**, *112*, 261–274. [CrossRef]
194. Fichot, C.G.; Tzortziou, M.; Mannino, A. Remote sensing of dissolved organic carbon (DOC) stocks, fluxes and transformations along the land-ocean aquatic continuum: Advances, challenges, and opportunities. *Earth-Sci. Rev.* **2023**, *242*, 104446. [CrossRef]
195. Huang, C.; Yunmei, L.; Liu, G.; Guo, Y.; Yang, H.; Zhu, A.X.; Song, T.; Huang, T.; Zhang, M.; Shi, K. Tracing high time-resolution fluctuations in dissolved organic carbon using satellite and buoy observations: Case study in Lake Taihu, China. *Int. J. Appl. Earth Obs. Geoinf.* **2017**, *62*, 174–182. [CrossRef]
196. Gomes, H.d.R.; Xu, Q.; Ishizaka, J.; Carpenter, E.J.; Yager, P.L.; Goes, J.I. The influence of riverine nutrients in niche partitioning of phytoplankton communities—A contrast between the Amazon River Plume and the ChangJiang (Yangtze) River diluted water of the East China Sea. *Front. Mar. Sci.* **2018**, *5*, 343. [CrossRef]
197. Reid, P.C.; Fischer, A.C.; Lewis-Brown, E.; Meredith, M.P.; Sparrow, M.; Andersson, A.J.; Antia, A.; Bates, N.R.; Bathmann, U.; Beaugrand, G. Impacts of the oceans on climate change. *Adv. Mar. Biol.* **2009**, *56*, 1–150. [PubMed]
198. Wu, J.; Goes, J.I.; do Rosario Gomes, H.; Lee, Z.; Noh, J.-H.; Wei, J.; Shang, Z.; Salisbury, J.; Mannino, A.; Kim, W. Estimates of diurnal and daily net primary productivity using the Geostationary Ocean Color Imager (GOCI) data. *Remote Sens. Environ.* **2022**, *280*, 113183. [CrossRef]
199. Cui, H.; Chen, J.; Cao, Z.; Huang, H.; Gong, F. A Novel Multi-Candidate Multi-Correlation Coefficient Algorithm for GOCI-Derived Sea-Surface Current Vector with OSU Tidal Model. *Remote Sens.* **2022**, *14*, 4625. [CrossRef]
200. Sun, D.; Su, X.; Qiu, Z.; Wang, S.; Mao, Z.; He, Y. Remote sensing estimation of sea surface salinity from GOCI measurements in the southern Yellow Sea. *Remote Sens.* **2019**, *11*, 775. [CrossRef]
201. Lee, H.-S.; Lee, K.-S. Capability of geostationary satellite imagery for sea ice monitoring in the Bohai and Yellow seas. *J. Mar. Sci. Technol.* **2016**, *24*, 10.
202. Gu, F.; Zhang, R.; Tian-Kunze, X.; Han, B.; Zhu, L.; Cui, T.; Yang, Q. Sea Ice Thickness Retrieval Based on GOCI Remote Sensing Data: A Case Study. *Remote Sens.* **2021**, *13*, 936. [CrossRef]
203. Yan, Y.; Huang, K.; Shao, D.; Xu, Y.; Gu, W. Monitoring the Characteristics of the Bohai Sea Ice Using High-Resolution Geostationary Ocean Color Imager (GOCI) Data. *Sustainability* **2019**, *11*, 777. [CrossRef]

204. Zhou, Y.; Chen, K.; Li, X. Dual-branch neural network for sea fog detection in geostationary ocean color imager. *IEEE Trans. Geosci. Remote Sens.* **2022**, *60*, 4208617. [CrossRef]
205. Jeon, H.-K.; Kim, S.; Edwin, J.; Yang, C.-S. Sea fog identification from GOCI images using CNN transfer learning models. *Electronics* **2020**, *9*, 311. [CrossRef]
206. Arp, C.D.; Cherry, J.E.; Brown, D.; Bondurant, A.C.; Endres, K.L. Observation-derived ice growth curves show patterns and trends in maximum ice thickness and safe travel duration of Alaskan lakes and rivers. *Cryosphere* **2020**, *14*, 3595–3609. [CrossRef]
207. Pearce, D.A. Antarctic subglacial lake exploration: A new frontier in microbial ecology. *ISME J.* **2009**, *3*, 877–880. [CrossRef] [PubMed]
208. Hampton, S.E.; Galloway, A.W.; Powers, S.M.; Ozersky, T.; Woo, K.H.; Batt, R.D.; Labou, S.G.; O'Reilly, C.M.; Sharma, S.; Lottig, N.R. Ecology under lake ice. *Ecol. Lett.* **2017**, *20*, 98–111. [CrossRef] [PubMed]
209. Yang, H.; Kong, J.; Hu, H.; Du, Y.; Gao, M.; Chen, F. A review of remote sensing for water quality retrieval: Progress and challenges. *Remote Sens.* **2022**, *14*, 1770. [CrossRef]
210. Kirk, J. Dependence of relationship between inherent and apparent optical properties of water on solar altitude. *Limnol. Oceanogr.* **1984**, *29*, 350–356. [CrossRef]
211. Kirk, J.T. Volume scattering function, average cosines, and the underwater light field. *Limnol. Oceanogr.* **1991**, *36*, 455–467. [CrossRef]
212. McKee, D.; Cunningham, A.; Dudek, A. Optical water type discrimination and tuning remote sensing band-ratio algorithms: Application to retrieval of chlorophyll and Kd (490) in the Irish and Celtic Seas. *Estuar. Coast. Shelf Sci.* **2007**, *73*, 827–834. [CrossRef]
213. Zhang, Y.; Xu, Z.; Yang, Y.; Wang, G.; Zhou, W.; Cao, W.; Li, Y.; Zheng, W.; Deng, L.; Zeng, K. Diurnal variation of the diffuse attenuation coefficient for downwelling irradiance at 490 nm in coastal East China Sea. *Remote Sens.* **2021**, *13*, 1676. [CrossRef]
214. Yu, X.; Salama, M.S.; Shen, F.; Verhoef, W. Retrieval of the diffuse attenuation coefficient from GOCI images using the 2SeaColor model: A case study in the Yangtze Estuary. *Remote Sens. Environ.* **2016**, *175*, 109–119. [CrossRef]
215. Tang, R.; Shen, F.; Pan, Y.; Ruddick, K.; Shang, P. Multi-source high-resolution satellite products in Yangtze Estuary: Cross-comparisons and impacts of signal-to-noise ratio and spatial resolution. *Opt. Express* **2019**, *27*, 6426–6441. [CrossRef] [PubMed]
216. Hu, C.; Feng, L.; Lee, Z. Evaluation of GOCI sensitivity for at-sensor radiance and GDPS-retrieved chlorophyll-a products. *Ocean Sci. J.* **2012**, *47*, 279–285. [CrossRef]
217. Qi, L.; Lee, Z.; Hu, C.; Wang, M. Requirement of minimal signal-to-noise ratios of ocean color sensors and uncertainties of ocean color products. *J. Geophys. Res. Ocean.* **2017**, *122*, 2595–2611. [CrossRef]
218. Bi, S.; Röttgers, R.; Hieronymi, M. Transfer model to determine the above-water remote-sensing reflectance from the underwater remote-sensing ratio. *Opt. Express* **2023**, *31*, 10512–10524. [CrossRef]
219. Xu, F.; He, X.; Jin, X.; Cai, W.; Bai, Y.; Wang, D.; Gong, F.; Zhu, Q. Spherical vector radiative transfer model for satellite ocean color remote sensing. *Opt. Express* **2023**, *31*, 11192–11212. [CrossRef] [PubMed]
220. Prakash, W.; Varma, A.; Bhandari, S. An algorithm for the precise location of the solar specular reflection point in the visible band images from geostationary meteorological satellites. *Comput. Geosci.* **1994**, *20*, 1467–1482. [CrossRef]
221. Wu, X.; Lu, Y.; Jiao, J.; Ding, J.; Fu, W.; Qian, W. Using sea wave simulations to interpret the sunglint reflection variation with different spatial resolutions. *IEEE Geosci. Remote Sens. Lett.* **2020**, *19*, 1501304. [CrossRef]
222. Park, Y.; Ahn, Y.; Han, H.; Yang, H.; Moon, J.; Ahn, J.; Lee, B.; Min, J.; Lee, S.; Kim, K. *GOCI Level 2 Ocean Color Products (GDPS 1.3) Brief Algorithm Description*; Korea Ocean Satellite Center (KOSC): Ansan, Republic of Korea, 2014; pp. 24–40.
223. Kim, Y.J.; Kim, W.; Im, J.; Choi, J.; Lee, S. Atmospheric-correction-free red tide quantification algorithm for GOCI based on machine learning combined with a radiative transfer simulation. *ISPRS J. Photogramm. Remote Sens.* **2023**, *199*, 197–213. [CrossRef]
224. Mao, K.; Yuan, Z.; Zuo, Z.; Xu, T.; Shen, X.; Gao, C. Changes in global cloud cover based on remote sensing data from 2003 to 2012. *Chin. Geogr. Sci.* **2019**, *29*, 306–315. [CrossRef]
225. Colomina, I.; Molina, P. Unmanned aerial systems for photogrammetry and remote sensing: A review. *ISPRS J. Photogramm. Remote Sens.* **2014**, *92*, 79–97. [CrossRef]
226. Kwon, D.Y.; Kim, J.; Park, S.; Hong, S. Advancements of remote data acquisition and processing in unmanned vehicle technologies for water quality monitoring: An extensive review. *Chemosphere* **2023**, *343*, 140198. [CrossRef] [PubMed]
227. Yuan, S.; Li, Y.; Bao, F.; Xu, H.; Yang, Y.; Yan, Q.; Zhong, S.; Yin, H.; Xu, J.; Huang, Z. Marine environmental monitoring with unmanned vehicle platforms: Present applications and future prospects. *Sci. Total Environ.* **2023**, *858*, 159741. [CrossRef]
228. Zhao, H.; Zhou, Y.; Wu, H.; Kutser, T.; Han, Y.; Ma, R.; Yao, Z.; Zhao, H.; Xu, P.; Jiang, C. Potential of Mie–fluorescence–Raman lidar to profile chlorophyll a concentration in inland waters. *Environ. Sci. Technol.* **2023**, *57*, 14226–14236. [CrossRef] [PubMed]
229. Wang, W.; Shi, K.; Zhang, Y.; Li, N.; Sun, X.; Zhang, D.; Zhang, Y.; Qin, B.; Zhu, G. A ground-based remote sensing system for high-frequency and real-time monitoring of phytoplankton blooms. *J. Hazard. Mater.* **2022**, *439*, 129623. [CrossRef] [PubMed]
230. Wang, X.; Wang, W.; He, Y.; Zhang, S.; Huang, W.; Woolway, R.I.; Shi, K.; Yang, X. Numerical simulation of thermal stratification in Lake Qiandaohu using an improved WRF-Lake model. *J. Hydrol.* **2023**, *618*, 129184. [CrossRef]
231. Li, Y.; Li, Y.-M.; Wang, Q.; Zhu, L.; Guo, Y.-L. An Observing System Simulation Experiments framework based on the ensemble square root Kalman Filter for evaluating the concentration of chlorophyll a by multi-source data: A case study in Taihu Lake. *Aquat. Ecosyst. Health Manag.* **2014**, *17*, 233–241. [CrossRef]

232. Guo, Y.; Li, Y.; Zhu, L.; Wang, Q.; Lv, H.; Huang, C.; Li, Y. An inversion-based fusion method for inland water remote monitoring. *IEEE J. Sel. Top. Appl. Earth Obs. Remote Sens.* **2016**, *9*, 5599–5611. [CrossRef]
233. Murakami, H. Ocean color estimation by Himawari-8/AHI. In Proceedings of the Remote Sensing of the Oceans and Inland Waters: Techniques, Applications, and Challenges, New Delhi, India, 4–7 April 2016; pp. 177–186.
234. Guo, Y.; Huang, C.; Zhang, Y.; Li, Y.; Chen, W. A novel multitemporal image-fusion algorithm: Method and application to GOCI and himawari images for inland water remote sensing. *IEEE Trans. Geosci. Remote Sens.* **2020**, *58*, 4018–4032. [CrossRef]
235. Chang, N.-B.; Vannah, B.; Yang, Y.J. Comparative sensor fusion between hyperspectral and multispectral satellite sensors for monitoring microcystin distribution in Lake Erie. *IEEE J. Sel. Top. Appl. Earth Obs. Remote Sens.* **2014**, *7*, 2426–2442. [CrossRef]
236. Mélin, F.; Zibordi, G. Optically based technique for producing merged spectra of water-leaving radiances from ocean color remote sensing. *Appl. Opt.* **2007**, *46*, 3856–3869. [CrossRef] [PubMed]
237. Mélin, F.; Zibordi, G.; Djavidnia, S. Merged series of normalized water leaving radiances obtained from multiple satellite missions for the Mediterranean Sea. *Adv. Space Res.* **2009**, *43*, 423–437. [CrossRef]
238. Mélin, F.; Vantrepotte, V.; Chuprin, A.; Grant, M.; Jackson, T.; Sathyendranath, S. Assessing the fitness-for-purpose of satellite multi-mission ocean color climate data records: A protocol applied to OC-CCI chlorophyll-a data. *Remote Sens. Environ.* **2017**, *203*, 139–151. [CrossRef]
239. Cao, Z.; Duan, H.; Shen, M.; Ma, R.; Xue, K.; Liu, D.; Xiao, Q. Using VIIRS/NPP and MODIS/Aqua data to provide a continuous record of suspended particulate matter in a highly turbid inland lake. *Int. J. Appl. Earth Obs. Geoinf.* **2018**, *64*, 256–265. [CrossRef]
240. Cao, F.; Miller, W.L. A new algorithm to retrieve chromophoric dissolved organic matter (CDOM) absorption spectra in the UV from ocean color. *J. Geophys. Res. Ocean.* **2015**, *120*, 496–516. [CrossRef]
241. He, X.; Bai, Y.; Pan, D.; Tang, J.; Wang, D. Atmospheric correction of satellite ocean color imagery using the ultraviolet wavelength for highly turbid waters. *Opt. Express* **2012**, *20*, 20754–20770. [CrossRef] [PubMed]
242. Werdell, P.J.; Bailey, S.W.; Franz, B.A.; Harding, L.W., Jr.; Feldman, G.C.; McClain, C.R. Regional and seasonal variability of chlorophyll-a in Chesapeake Bay as observed by SeaWiFS and MODIS-Aqua. *Remote Sens. Environ.* **2009**, *113*, 1319–1330. [CrossRef]
243. Simis, S.G.; Peters, S.W.; Gons, H.J. Remote sensing of the cyanobacterial pigment phycocyanin in turbid inland water. *Limnol. Oceanogr.* **2005**, *50*, 237–245. [CrossRef]
244. Liu, G.; Simis, S.G.; Li, L.; Wang, Q.; Li, Y.; Song, K.; Lyu, H.; Zheng, Z.; Shi, K. A four-band semi-analytical model for estimating phycocyanin in inland waters from simulated MERIS and OLCI data. *IEEE Trans. Geosci. Remote Sens.* **2017**, *56*, 1374–1385. [CrossRef]
245. Lyu, L.; Song, K.; Wen, Z.; Liu, G.; Fang, C.; Shang, Y.; Li, S.; Tao, H.; Wang, X.; Li, Y. Remote estimation of phycocyanin concentration in inland waters based on optical classification. *Sci. Total Environ.* **2023**, *899*, 166363. [CrossRef] [PubMed]
246. Kwon, Y.S.; Pyo, J.; Kwon, Y.-H.; Duan, H.; Cho, K.H.; Park, Y. Drone-based hyperspectral remote sensing of cyanobacteria using vertical cumulative pigment concentration in a deep reservoir. *Remote Sens. Environ.* **2020**, *236*, 111517. [CrossRef]
247. Pyo, J.; Duan, H.; Baek, S.; Kim, M.S.; Jeon, T.; Kwon, Y.S.; Lee, H.; Cho, K.H. A convolutional neural network regression for quantifying cyanobacteria using hyperspectral imagery. *Remote Sens. Environ.* **2019**, *233*, 111350. [CrossRef]
248. Shen, M.; Cao, Z.; Xie, L.; Zhao, Y.; Qi, T.; Song, K.; Lyu, L.; Wang, D.; Ma, J.; Duan, H. Microcystins risk assessment in lakes from space: Implications for SDG 6.1 evaluation. *Water Res.* **2023**, *245*, 120648. [CrossRef] [PubMed]
249. Liu, G.; Li, L.; Song, K.; Li, Y.; Lyu, H.; Wen, Z.; Fang, C.; Bi, S.; Sun, X.; Wang, Z. An OLCI-based algorithm for semi-empirically partitioning absorption coefficient and estimating chlorophyll a concentration in various turbid case-2 waters. *Remote Sens. Environ.* **2020**, *239*, 111648. [CrossRef]
250. Gitelson, A.A.; Dall'Olmo, G.; Moses, W.; Rundquist, D.C.; Barrow, T.; Fisher, T.R.; Gurlin, D.; Holz, J. A simple semi-analytical model for remote estimation of chlorophyll-a in turbid waters: Validation. *Remote Sens. Environ.* **2008**, *112*, 3582–3593. [CrossRef]
251. Bai, R.; He, X.; Bai, Y.; Gong, F.; Zhu, Q.; Wang, D.; Li, T. Atmospheric correction algorithm based on the interpolation of ultraviolet and shortwave infrared bands. *Opt. Express* **2023**, *31*, 6805–6826. [CrossRef] [PubMed]
252. Pi, X.; Luo, Q.; Feng, L.; Xu, Y.; Tang, J.; Liang, X.; Ma, E.; Cheng, R.; Fensholt, R.; Brandt, M. Mapping global lake dynamics reveals the emerging roles of small lakes. *Nat. Commun.* **2022**, *13*, 5777. [CrossRef]
253. Tao, H.; Song, K.; Liu, G.; Wen, Z.; Wang, Q.; Du, Y.; Lyu, L.; Du, J.; Shang, Y. Songhua River basin's improving water quality since 2005 based on Landsat observation of water clarity. *Environ. Res.* **2021**, *199*, 111299. [CrossRef]
254. Ani, C.J.; Baird, M.; Robson, B. Modelling buoyancy-driven vertical movement of Trichodesmium application in the Great Barrier Reef. *Ecol. Model.* **2024**, *487*, 110567. [CrossRef]
255. Wen, Z.; Shang, Y.; Lyu, L.; Tao, H.; Liu, G.; Fang, C.; Li, S.; Song, K. Re-estimating China's lake CO_2 flux considering spatiotemporal variability. *Environ. Sci. Ecotechnology* **2024**, *19*, 100337. [CrossRef] [PubMed]
256. Sommerfield, C.K.; Wong, K.C. Mechanisms of sediment flux and turbidity maintenance in the Delaware Estuary. *J. Geophys. Res. Ocean.* **2011**, *116*, C01005. [CrossRef]
257. Uncles, R.; Elliott, R.; Weston, S. Observed fluxes of water, salt and suspended sediment in a partly mixed estuary. *Estuar. Coast. Shelf Sci.* **1985**, *20*, 147–167. [CrossRef]
258. Li, Y.; Robinson, S.V.; Nguyen, L.H.; Liu, J. Satellite prediction of coastal hypoxia in the northern Gulf of Mexico. *Remote Sens. Environ.* **2023**, *284*, 113346. [CrossRef]

259. Hu, C.; Lu, Y.; Sun, S.; Liu, Y. Optical remote sensing of oil spills in the ocean: What is really possible? *J. Remote Sens.* **2021**, *2021*, 9141902. [CrossRef]
260. Sun, Z.; Sun, S.; Zhao, J.; Ai, B.; Yang, Q. Detection of Massive Oil Spills in Sun Glint Optical Imagery through Super-Pixel Segmentation. *J. Mar. Sci. Eng.* **2022**, *10*, 1630. [CrossRef]
261. Jiao, J.; Lu, Y.; Liu, Y. Optical quantification of oil emulsions in multi-band coarse-resolution imagery using a lab-derived HSV model. *Mar. Pollut. Bull.* **2022**, *178*, 113640. [CrossRef]
262. Raffaelli, D.G.; Hawkins, S.J. *Intertidal Ecology*; Springer Science & Business Media: Berlin/Heidelberg, Germany, 1996.
263. Jia, M.; Wang, Z.; Mao, D.; Ren, C.; Wang, C.; Wang, Y. Rapid, robust, and automated mapping of tidal flats in China using time series Sentinel-2 images and Google Earth Engine. *Remote Sens. Environ.* **2021**, *255*, 112285. [CrossRef]
264. Kennedy, V.S.; Newell, R.I.; Shumway, S. Natural environmental factors. In *The Eastern Oyster Crassostrea virginica*; Maryland Sea Grant: College Park, MD, USA, 1996; pp. 467–513.

MDPI AG
Grosspeteranlage 5
4052 Basel
Switzerland
Tel.: +41 61 683 77 34

Remote Sensing Editorial Office
E-mail: remotesensing@mdpi.com
www.mdpi.com/journal/remotesensing